100 Chemical Myths

Lajos Kovács • Dezső Csupor
Gábor Lente • Tamás Gunda

100 Chemical Myths

Misconceptions, Misunderstandings, Explanations

 Springer

Lajos Kovács
Department of Medicinal Chemistry
University of Szeged
Szeged
Hungary

Dezső Csupor
Department of Pharmacognosy
University of Szeged
Szeged
Hungary

Gábor Lente
Department of Inorganic and Analytical Chemistry
University of Debrecen
Debrecen
Hungary

Tamás Gunda
Department of Pharmaceutical Chemistry
University of Debrecen
Debrecen
Hungary

ISBN 978-3-319-08418-3   ISBN 978-3-319-08419-0 (eBook)
DOI 10.1007/978-3-319-08419-0
Springer Cham Heidelberg New York Dordrecht London

Library of Congress Control Number: 2014947530

© Springer International Publishing Switzerland 2014
This work is subject to copyright. All rights are reserved by the Publisher, whether the whole or part of the material is concerned, specifically the rights of translation, reprinting, reuse of illustrations, recitation, broadcasting, reproduction on microfilms or in any other physical way, and transmission or information storage and retrieval, electronic adaptation, computer software, or by similar or dissimilar methodology now known or hereafter developed. Exempted from this legal reservation are brief excerpts in connection with reviews or scholarly analysis or material supplied specifically for the purpose of being entered and executed on a computer system, for exclusive use by the purchaser of the work. Duplication of this publication or parts thereof is permitted only under the provisions of the Copyright Law of the Publisher's location, in its current version, and permission for use must always be obtained from Springer. Permissions for use may be obtained through RightsLink at the Copyright Clearance Center. Violations are liable to prosecution under the respective Copyright Law.
The use of general descriptive names, registered names, trademarks, service marks, etc. in this publication does not imply, even in the absence of a specific statement, that such names are exempt from the relevant protective laws and regulations and therefore free for general use.
While the advice and information in this book are believed to be true and accurate at the date of publication, neither the authors nor the editors nor the publisher can accept any legal responsibility for any errors or omissions that may be made. The publisher makes no warranty, express or implied, with respect to the material contained herein.

Cover illustration: kindly provided by Tamás Gunda

Printed on acid-free paper

Springer is part of Springer Science+Business Media (www.springer.com)

Original title: Száz kémiai mítosz. Tévhitek, félreértések, magyarázatok. *Akadémiai Kiadó, Budapest, Hungary, 2011*

Translated from Hungarian by Gábor Lente and Katalin Ösz, University of Debrecen, Hungary

Subject reader: Zoltán Tóth, University of Debrecen, Hungary

Language advisor: Louis Mattia

Nothing in life is to be feared, it is only to be understood

Marie Curie

Foreword

My early years were spent in Hungary so it should come as no surprise that my first venture into the world of chemistry involved paprika. Bread smeared with goose fat was a popular childhood delight, always topped with a sprinkling of paprika. No worries about cholesterol back then! One day, however, I got a spicy surprise as I bit into my snack. The paprika practically set my mouth on fire! My mother, it seems, had bought hot paprika instead of the usual sweet.

I was only about seven years old at the time, but it set me thinking. The hot stuff looked exactly the same as the sweet. What was the difference, I wondered? I didn't realize it at the time, but I was actually dipping my toes into the deep pool of chemistry. Many times since I have thought about how that paprika incident sparked my chemical curiosity which eventually would burst into a flame.

That flame burned brightly through the sixties as I became more and more enthralled by the wonders of chemistry. I remember visiting the DuPont pavilion at the New York World's Fair in 1964 and being thrilled by a Broadway style musical entitled the "Wonderful World of Chemistry." What a great show it was. Everything in the DuPont theatre was made of some newly-invented material. Doors featured alkyd resin paint and polyacetal doorknobs, the ceiling was made of polyvinyl fluoride, floors were carpeted with nylon and seats covered with polyvinyl chloride. The polyester curtain went up to reveal dancers in colorful spandex costumes, tapping their polyurethane shoes on an acrylic-glossed stage.

Next to the DuPont pavilion was a NASA display of rockets and an exhibit that featured the planned trip to the Moon, including samples of the special fabrics and plastics that chemists had designed for the fledgling space program. Chemistry was flying high! Nobody took issue with DuPont's slogan of "Better Things for Better Living Through Chemistry." The word "chemical" was not seen to be synonymous with "poison" or "toxin" and when you were introduced as a student studying chemistry, you were not looked upon as someone who was destined destroy people's health with a plethora of untested toxins and wreak havoc with the environment.

By the time I graduated with my chemistry degree in the seventies, the winds of change were beginning to blow. Chemistry went from being a heroic science that furnished us with new medicines, fibers and plastics, to one associated with napalm, Agent Orange and pollution. Rachel Carson's *Silent Spring* called attention to the misuse of pesticides and before long "chemical" became a dirty word, with some chemists even suggesting that when communicating with the public it be replaced by the term "substance." Somehow "substance" was seen to be more benign than "chemical." By 1982 the image of chemistry had been so tainted that DuPont felt the need to drop the "through chemistry" phrase from its slogan.

While some of the criticism aimed at the chemical industry's insufficient testing of its products and lack of attention to environmental issues was justified, the "chemophobia" generated by various activist groups was not based on facts. But unfortunately emotion usually rushes in to fill the vacuum created by a lack of chemical knowledge and emotion tends to trump science. Often the result is an irrational fear of chemicals that creates confusion and unnecessary stress. A calming element can be introduced into the discussion with an explanation that chemicals are not good or bad, they are just the building blocks of all matter. They don't make any decisions, people do. Chemicals are not to be feared, nor are they to be worshipped. They are to be understood. It is ignorance that breeds fear and knowledge that dispels it.

Therein, though, lies a problem. How do you enlighten the public about chemical issues, which are often very complex, given that many people do not know what a molecule is and look upon a structural diagram as if it were some sort of an ancient hieroglyphic? What is needed is a book that explains controversies in a readily understandable fashion without skimping on the science. That is exactly what the authors of this fascinating compendium of chemical knowledge have managed to put together.

A brief introduction to the nature of molecules is followed by an insightful discussion of a vast array of current and historical chemical topics. The differences between natural and synthetic substances are clearly addressed, as well those between butter and margarine and generic and trade name drugs. Concerns about food additives, sweeteners, phytoestrogens, antioxidants and plastics are explored in an objective fashion. Scams such as alkaline diets, bogus cancer treatments and nonsensical detox regimens are exposed with scientific flair.

If you've ever wondered why ice melts under a skate or why meat is red or where the colours of flowers come from, you will wonder no more. No longer will you be at the mercy of advertisers who make questionable claims about noni juice, glucosamine, bottled water, aromatherapy and homeopathy. Confusion about the placebo effect, brown versus white sugar, blood doping, DDT, dioxins and the ozone layer will be seen to evaporate with the infusion of sound science. You will even discover why you don't need hot water to make a hard-boiled egg and why drugs and grapefruit juice do not mix.

Of course no book by Hungarian authors would be complete without the story of Dr. Albert Szent-Gyorgyi's isolation of vitamin C from paprika. It's here in glorious detail. Finally you will even discover why goose fat may not be the healthiest snack and how hot paprika differs from the sweet version. This amazingly comprehensive and entertaining book is guaranteed to serve up just the right blend of sweet science and spicy anecdote. Bon appétit!

Office for Science and Society Joe Schwarcz
McGill University Director
Montreal

Preface

"For every complex problem there is a simple solution, and it is always wrong," said Dr. Victor Herbert, renowned hematologist and champion of evidence-based medicine. In this book, the authors intend to describe chemical misconceptions from as many different points of views as their talent allows, often times explaining connections between distinctly non-chemical phenomena. A quest for finding the truth is by no means an easy one and has often been compared to peeling an onion. Peer Gynt in Henrik Ibsen's play declares the following when he peels an onion layer by layer:

> There's a most surprising lot of layers!
> Are we never coming to the kernel?
> [Pulls all that is left to pieces.]
> There isn't one! To the innermost bit
> It's nothing but layers, smaller and smaller.
> Nature's a joker!
> (Translation by Edgar Lucas)

This book has been written for open-minded non-specialists, but we also hope to surprise chemists with a few stories never heard before. The text is divided into four main chapters and altogether describes one hundred misconceptions: after a general introduction, myths about food take center stage, followed by a look at misconceptions about medicines, and finally, we investigate all sorts of catastrophes, poisons and other chemical-related stuff. These groups are by no means intended to be exclusive or even unambiguous: there is a lot of overlap between them. The stories were written independently of each other, but we made efforts to make connections between the different stories by inserting the related chapter number into the text as shown here (→**Chapter number**). At the end of the book, the reader will find some literature connected to each story, sometimes the reference provides further information, and sometimes the material was used as sources to write our text.

Many of the concepts developed in this book occur in multiple stories, these are written in **bold letters** at their first appearance and a definition or explanation is given for each one at the end of the book in the *Glossary of terms*.

Fig. 1 Two-dimensional structures of testosterone

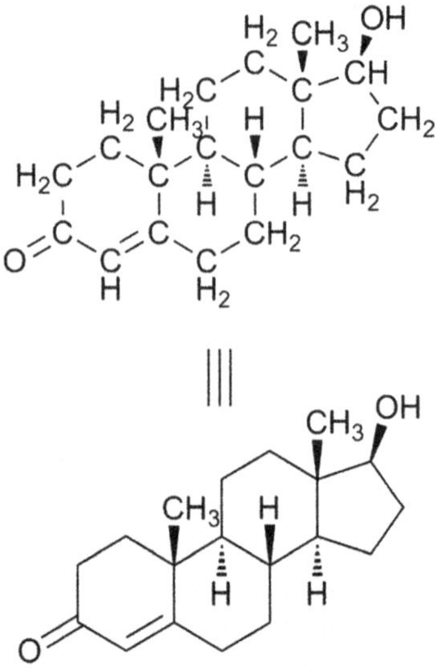

We had to make choices in representing the chemical structures of organic compounds, almost all commonly used methods are represented in one or more stories of the book. In two-dimensional structures, we followed the convention that only occasionally gives the symbols for the carbon or hydrogen atoms, lines were used instead (Fig. 1). In these formulae, lines mean chemical bonds and no symbol at the intersection of lines means a carbon atom with just enough hydrogen atoms to make the valence count four for each carbon. Sometimes, at the end of chains, it is customary to give these symbols to avoid confusions. The wedged bold (━━◤) and wedged hashed (⋯⋯ⅠⅠⅠⅠⅠ) lines used as bonds carry some information about three-dimensional structure. The former means a bond in front of the plane of the rest of the molecule, the latter behind.

In three-dimensional structures, we wish to give some sort of image of the actual shape of the molecule. Individual atoms are often represented by spheres, and bonds may or may not be visible depending on the size of the spheres (Fig. 2).

There are more complicated substances (proteins, for example), where further simplifications were necessary: the main characteristics of the structures are often shown by colored ribbons, helices, or other irregularly shaped lines (Fig. 3).

The authors
Szeged and Debrecen, June 2014

Preface xiii

Fig. 2 Three-dimensional structures of testosterone

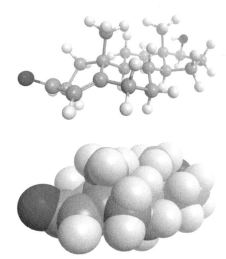

Fig. 3 Simplified three-dimensional structure of botulinum toxin

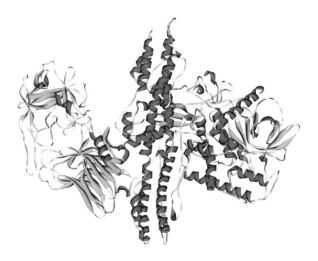

Acknowledgements

The authors would like to thank the following persons and funding bodies who contributed to the successful accomplishment of this project:

Ferenc Bogár, Klára Boros, Boglárka Csupor-Löffler, Györgyi Ferenc, Eriko Hagiya, Katalin Horváthné Almássy, Ovidiu Ivanciuc, Zoltán Kele, Attila Kovács, Hajnalka Kovács, Zoltán Kupihár, Sándor László, Aurora Lopez-Delgado, Louis Mattia, András Máté-Tóth, Ed McLachlan, Sándor Nemes, Tamás Paál, Petra Pádár, Gábor Paragi, Botond Penke, Frank Pine, Ray Robinson, Pál Sipos, Kálmán Szendrei, János Szépvölgyi, Viktória Szucsics, István Tanka, Zoltán Tóth, Béla Vízi, Tom Way.

Gábor Lente's contribution to this book was supported by the EU and co-financed by the European Social Fund under the project ENVIKUT (TÁMOP-4.2.2.A-11/1/KONV-2012-0043).

Contents

1 **Misconceptions in General** 1
   1.1 Fear of the Unknown: Chemicals 1
   1.2 Life as a Risky Business 3
   1.3 Natural Products: A Delusion of Safety 6
   1.4 Man-Made Commodities and Safety Issues 9
   1.5 The Cholera Pandemonium: The Blind Leading the Blind 11
   1.6 Regulating Chemicals: Maybe or Maybe Not 13
   1.7 Biowaste: Biotechnology in Perspective 16
   1.8 Ice Skating and the Science of Sliding 18
   1.9 So What's Happening to the Ozone Hole? 21
   1.10 Predicting the Elements: Success and Failure 23
   1.11 Vedic Wisdom: Lead and Ayurveda 26
   1.12 Manipulating Weather: Ocean Fertilization 30

2 **Food** 35
   2.1 Test Your Cranberry Pie: Vitamin C and Benzoates 35
   2.2 Food Dyes: The Good, the Bad and the Ugly 37
       2.2.1 Azo Dyes 38
       2.2.2 Triphenylmethane Dyes 39
   2.3 The Organic Vegetable Hype 40
       2.3.1 A Question for Eternity: Natural or Artificial Fertilizers? 40
       2.3.2 Nitrates in Vegetables 41
       2.3.3 Are Organic Vegetables More Nutritious than Others? 42
   2.4 "You're the One for Me, Fatty": Concocted Fats 43
   2.5 Fat Matters: Margarine vs. Butter 45
   2.6 Fake Food and Kidney Stones 50
   2.7 The Coming Shortage: Vanilla and Menthol 53
   2.8 Red Alert: Meat Colors 57
   2.9 Moldy Business: Whole-Grain Cereals 60
   2.10 Red Wine and the Vineyard Blues 63
   2.11 Popeye's Spinach and the Search for Iron 69
   2.12 The Perplexing Spice: Curry 71

2.13	Bigger the Better? The Alpha and Omega of Fatty Acids	74
2.14	Sweet as Birch: Xylitol	78
2.15	Bittersweet Effects: Grapefruit Seeds	83
2.16	Sweet Dreams Without Sugar: Artificial Sweeteners	86
	2.16.1 The Saccharin Case in a Nutshell	87
	2.16.2 Aspartame and Methanol	88
2.17	An Ointment for Your Throat: The Secrets of Olive Oil	90
2.18	To Add or Not to Add? Food Additives	93
	2.18.1 Baking Powder	94
	2.18.2 Salicylic Acid	94
	2.18.3 The Chinese Restaurant Syndrome and Monosodium Glutamate	95
2.19	There's Salt, and then There's Salt	96
2.20	Sugar for Your Tea? White or Brown?	98
2.21	The Thickening Stuff: Guar Gum Gumbo	100
2.22	Is Caffeine Free of Risk?	103
2.23	The Biofuel Dilemma	106
2.24	Perfect Timing: Egg Cooking	111
2.25	Food Fraud: Then and Now	114

3 Medicines ... 119

3.1	Are Generic Medicines the Same as the Original?	119
3.2	The Vitamin that Never Was: B_{17}	122
3.3	Joint Efforts: Glucosamine and Chondroitin	127
3.4	A Man's Drug of Choice: Testosterone	130
3.5	Doped Up and Ready to Win. Maybe	132
3.6	Can Placebos Heal?	134
3.7	Synthetic Drugs vs. Natural Herbs	137
3.8	Bitter, Stronger: Herbal Tea Sweeteners	142
3.9	Homeopathy: No Active Ingredient? No Side Effect?	144
3.10	Aromatherapy: Nice but Useless?	147
3.11	Innocent or Guilty? Chamomile	151
3.12	Depressed: Antidepressant Side Effects	153
3.13	Pre-Emptive Action: Vitamin C	156
3.14	Vitamin Megadosing: the Real Truth	160
3.15	Detox Hoax: Poison Removal from Your Body	166
3.16	Cloudy Dreams in Green: Absinthe	169
3.17	Is Green Tea Caffeine-Free?	172
3.18	Red Rice Remedy: Cholesterol Relief?	176
3.19	What Exactly Does Aloë Cure?	178
3.20	Island Pleasure for Businessmen: Noni	182
3.21	Castor Oil: A Painful Story	184
3.22	The Oldest Sugar-Free Candy: Licorice	187

3.23	Cannon Balls for Your Health: Antibiotics	190
3.24	Hearty Soybeans: Phytoestrogens	193
3.25	Willow Fever: Aspirin	198
3.26	Pill Secrets: The Efficiency of Drugs	201
3.27	In Quest of Active Ingredients: Herbs	205
3.28	Getting Rid of Acids: Alkaline Cures	208
3.29	"…upon the Face of the Waters": Mineral and Tap Water	212
3.30	The Illusive Qualities of Plant Preparations	217
3.31	The Antioxidant Story	222
3.32	Stomach Ache: Baking Soda	228
3.33	Does Hot Pepper Cause Ulcer?	229
3.34	Cocaine—Drug, Narcotic, or What?	232

4 Catastrophes, Poisons, Chemicals 235

4.1	Is the Use of Cyanogen Bromide Forbidden in Hospitals?	235
4.2	Poison in Groundwater: Arsenic	237
4.3	Don't Touch the Spilled Mercury!	242
4.4	Was DDT of More Harm Than Use?	245
4.5	Could Dioxin be the Most Toxic Substance?	248
4.6	Was Napoleon Murdered with Arsenic?	252
4.7	The Resin Wars: Formaldehyde	255
4.8	The Great Hungarian Red Mud Deluge	258
4.9	Death in a Single Grain	263
4.10	A Closer Look at Acids	265
4.11	O, the Fresh Smell of Ozone	267
4.12	The Diamond Thermometer	268
4.13	Everlasting Love: Diamonds	270
4.14	Is pH 5.5 Neutral?	272
4.15	Sorrel Soup Scare: Oxalic Acid	274
4.16	The Mysteries of Magnetism	277
4.17	Water: The Extravagant Liquid	279
4.18	Fire in Space	281
4.19	Is Natural Gas Toxic?	283
4.20	Respiratory Affairs: Carbon Dioxide	285
4.21	The True Death Toll of Chernobyl	288
4.22	Were There Nuclear Explosions in Chernobyl or Fukushima?	290
4.23	What Was the Gulf War Syndrome?	292
	4.23.1 Nerve Agent Antidote Pyridostigmine Bromide (PB)	293
	4.23.2 Organophosphate Type Pesticides and Insect Repellents	294
	4.23.3 Nerve Agent Sarin (Fig. 4.25)	294
	4.23.4 Depleted Uranium Shells	294
4.24	The Strange Case of Bisphenol A	295
4.25	Death in the Danube: Cyanide Spills	298
4.26	Sparkling Cyanide: The Ultimate Poison?	300

4.27	Poly... What? Plastics vs. Natural Materials	302
4.28	The Difference between a Rose and a Cornflower	305
4.29	The Erin Brockovich Mystery: Chromium Salts	307

Glossary of Terms 311

Sources and Further Reading 321
General Reading 321
Special Reading Related to the Individual Essays 322

1.1	Fear of the Unknown: Chemicals	322
1.2	Life as a Risky Business	322
1.3	Natural Products: A Delusion of Safety	322
1.4	Man-Made Commodities and Safety Issues	323
1.5	The Cholera Pandemonium: The Blind Leading the Blind	323
1.6	Regulating Chemicals: Maybe or Maybe Not	323
1.7	Biowaste: Biotechnology in Perspective	324
1.8	Ice Skating and the Science of Sliding	324
1.9	So What's Happening to the Ozone Hole?	324
1.10	Predicting the Elements: Success and Failure	324
1.11	Vedic Wisdom: Lead and Ayurveda	324
1.12	Manipulating Weather: Ocean Fertilization	325
2.1	Test Your Cranberry Pie: Vitamin C and Benzoates	325
2.2	Food Dyes: The Good, the Bad and the Ugly	326
2.3	The Organic Vegetable Hype	326
2.4	"You're the One for Me, Fatty": Concocted Fats	326
2.5	Fat Matters: Margarine vs. Butter	326
2.6	Fake Food and Kidney Stones	327
2.7	The Coming Shortage: Vanilla and Menthol	328
2.8	Red Alert: Meat Colors	328
2.9	Moldy Business: Whole-Grain Cereals	329
2.10	Red Wine and the Vineyard Blues	329
2.11	Popeye's Spinach and the Search for Iron	330
2.12	The Perplexing Spice: Curry	331
2.13	Bigger the Better? The Alpha and Omega of Fatty Acids	331
2.14	Sweet as Birch: Xylitol	332
2.15	Bittersweet Effects: Grapefruit Seeds	332
2.16	Sweet Dreams Without Sugar: Artificial Sweeteners	333
2.17	An Ointment for your Throat: The Secrets of Olive Oil	333
2.18	To Add or Not to Add? Food Additives	334
2.19	There's Salt, and then There's Salt	334
2.20	Sugar for Your Tea? White or Brown?	334
2.21	The Thickening Stuff: Guar Gum Gumbo	334
2.22	Is Caffeine Free of Risk?	334
2.23	The Biofuel Dilemma	335
2.24	Perfect Timing: Egg Cooking	335

2.25	Food Fraud: Then and Now	336
3.1	Are Generic Medicines the Same as the Original?	336
3.2	The Vitamin that Never was: B_{17}	336
3.3	Joint Efforts: Glucosamine and Chondroitin	337
3.4	A Man's Drug of Choice: Testosterone	337
3.5	Doped Up and Ready to Win. Maybe	338
3.6	Can Placebos Heal?	338
3.7	Synthetic Drugs vs. Natural Herbs	338
3.8	Bitter, Stronger: Herbal Tea Sweeteners	339
3.9	Homeopathy: No Active Ingredient? No Side Effect?	339
3.10	Aromatherapy: Nice but Useless?	339
3.11	Innocent or Guilty? Chamomile	340
3.12	Depressed: Antidepressant Side Effects	340
3.13	Pre-Emptive Action: Vitamin C	341
3.14	Vitamin Megadosing: the Real Truth	341
3.15	Detox Hoax: Poison Removal from Your Body	341
3.16	Cloudy Dreams in Green: Absinthe	341
3.17	Is Green Tea Caffeine-Free?	342
3.18	Red Rice Remedy: Cholesterol Relief?	343
3.19	What Exactly Does Aloë Cure?	343
3.20	Island Pleasure for Businessmen: Noni	344
3.21	Castor Oil: A Painful Story	344
3.22	The Oldest Sugar-Free Candy: Licorice	344
3.23	Cannon Balls for Your Health: Antibiotics	345
3.24	Hearty Soybeans: Phytoestrogens	345
3.25	Willow Fever: Aspirin	346
3.26	Pill Secrets: the Efficiency of Drugs	346
3.27	In Quest of Active Ingredients: Herbs	347
3.28	Getting Rid of Acids: Alkaline Cures	347
3.29	"...upon the Face of the Waters": Mineral and Tap Water	347
3.30	The Illusive Qualities of Plant Preparations	347
3.31	The Antioxidant Story	348
3.32	Stomach Ache: Baking Soda	349
3.33	Does Hot Pepper Cause Ulcer?	349
3.34	Cocaine—Drug, Narcotic, or What?	350
4.1	Is the Use of Cyanogen Bromide Forbidden in Hospitals?	350
4.2	Poison in Groundwater: Arsenic	351
4.3	Don't Touch the Spilled Mercury!	351
4.4	Was DDT of More Harm Than Use?	352
4.5	Could Dioxin be the Most Toxic Substance?	352
4.6	Was Napoleon Murdered with Arsenic?	353
4.7	The Resin Wars: Formaldehyde	353
4.8	The Great Hungarian Red Mud Deluge	353
4.9	Death in a Single Grain	355
4.10	A Closer Look at Acids	355

4.11	O, the Fresh Smell of Ozone	356
4.12	The Diamond Thermometer	356
4.13	Everlasting Love: Diamonds	356
4.14	Is pH 5.5 Neutral?	356
4.15	Sorrel Soup Scare: Oxalic Acid	356
4.16	The Mysteries of Magnetism	357
4.17	Water: The Extravagant Liquid	357
4.18	Fire in Space	357
4.19	Is Natural Gas Toxic?	357
4.20	Respiratory Affairs: Carbon Dioxide	357
4.21	The True Death Toll of Chernobyl	358
4.22	Were There Nuclear Explosions in Chernobyl or Fukushima?	358
4.23	What Was the Gulf War Syndrome?	358
4.24	The Strange Case of Bisphenol A	358
4.25	Death in the Danube: Cyanide Spills	359
4.26	Sparkling Cyanide: the Ultimate Poison?	359
4.27	Poly… What? Plastics vs. Natural Materials	359
4.28	The Difference Between a Rose and a Cornflower	360
4.29	The Erin Brockovich Mystery: Chromium Salts	360
	Glossary of Terms	361

Index 363

Chapter 1
Misconceptions in General

1.1 Fear of the Unknown: Chemicals

Nathan Zohner, a then 14-year-old high school student from Idaho Falls (USA) wrote a petition in 1997 about the dangers of dihydrogen monoxide (DHMO) calling for a ban on its use (Fig. 1.1). Out of 50 people interviewed by Nathan, only a single one recognized that the substance was actually water. This story received a lot media attention and was resurrected in 2004: city officials in Aliso Viejo, California, proposed a ban on using DHMO. New Zealand politician Jacqui Dean also proved her lack of common sense in 2007 by sending a letter to the Ministry of Health asking if there were any plans to ban dihydrogen monoxide. The story has become an outstanding chapter in the great book of modern human ignorance.

Would you drink a liquid that contains butanol, isoamyl alcohol, hexanol, **tannin**, benzyl alcohol, caffeine, geraniol, quercetin, epicatechin-3-O-gallate, epigallocatechin-3-O-gallate, inorganic salts and water? This question itself may be appalling to the reader, but it only lists the main constituents of tea.

Chemophobia means an irrational fear of chemicals. If the two examples above do not convince the reader about its prevalence in everyday life, the media frenzy about the latest threats to food safety or reporting about accidents in chemical industry will probably do the job.

Chemophobia has rooted itself deeply in society, but its origins are mainly conceptual. In everyday life, it is common to speak about 'chemicals' or 'dangerous substances' without actually specifying what these words mean. The word 'chemical' (my personal favorite) is often used with the clear implications that there are other substances of non-chemical nature. In this context, the intended meaning of the word 'chemical' is probably a substance produced by the chemical industry, which does not occur in nature, and potentially poses some sort of a hazard. However, this book will point out that the hazards of substances are not connected to their origins at all, natural and artificial substances are not different from this point of view. For an expert, this usage of the word 'chemical' makes no sense as all substances are chemicals.

Another common misconception is that it is possible to live without chemicals. It would be quite an experiment to attempt this. Everyone with a basic competence

Fig. 1.1 The poster of DHMO.org. (Permission obtained from copyright-owner)

in chemistry knows that this goal is impossible as every substance around us is a chemical. Even if we were to accept the definition 'chemical = industrially produced substance,' leading a chemical-free life would require a return to such Stone Age levels of civilization that no environmental movement would be willing to follow (Fig. 1.2).

The benefits of chemistry and the technologies based on it for today's society are beyond any measure. Everything in everyday life has a connection to this hidden science. One of the problems may be that it is too hidden. It seems evident that society has access to colored materials, cosmetics, cleaning agents, electronics, preserved food, medicines, cars, plastics, and so on. In fact, none of these are evident. These old and new products have to be developed, studied, manufactured and continuously inspected. Only chemists have the ability to develop environmentally friendly new substances with previously unknown properties. It is the cooperation between chemists, physicists and engineers that makes the development of novel new measurement methods possible to detect substances in our environment. It is the cooperation between chemists, biologists and physicians that enables the invention of new medicines. Unfortunately, the public image of chemistry is more often determined by infamous cases of environmental pollution. While everyone seems to be willing to enjoy the fruits if chemical discoveries, the very existence of chemical industry is frequently looked upon with disgust ($\rightarrow$ **1.2**). It is like judging physics by the development of nuclear bombs, or judging biology by genetic modification only. It is impossible to lead a risk-free life, there is no free lunch. Today's comfortable lifestyle has a price tag on it. However, one need not throw out the baby with the bath water.

An American opinion poll shows that there is an inverse relationship between chemophobia and chemical knowledge: fears about chemicals decrease as the more one knows about chemistry. One should keep an open mind and listen to expert opinions as well experts rather than believing the claims of the media or environmental organizations immediately and without thinking. Experts must come forward and show the benefits and risks of chemistry and chemical technologies, and natural phenomena in their background. Chemical professionals have an obligation to make themselves heard in the media and present objective information free of fears, ignorance—or business interests ($\rightarrow$ **1.5**).

Fig. 1.2 Cartoon of Ed McLachlan. (Permission obtained from copyright-owner)

1.2 Life as a Risky Business

First, we should clarify what *danger* means. *Danger* is the possible occurrence of some unfavorable event that threatens our property, well-being, relations, health or even our life. Although everyone understands what this means, it is not very easy to define scientifically. The probability of the occurrence of the actual negative event is called *risk*. Danger and risk imply some sort of *hazard* that can be very versatile: an electric cable, hot steam, slippery floor, lightning, tainted drinking water, a speeding car, rock climbing, *etc*. The hazard itself is usually not enough for the appearance of the actual risk: if someone lives close to a busy road, but never actually goes there, the road does not pose any risk. If someone else crosses that busy road frequently, the probability of being run over by a car is much higher. This last concept is usually called *exposure*. The relationship between the mentioned concepts is given as:

$$risk = hazard \times exposure$$

From this, there is a quite logical strategy of reducing risk, that is, by avoiding hazards or avoiding exposure, or both of them. Returning to the traffic example: if someone does not drive, the risk arising from driving is zero, and avoiding crossing roads entirely will also give the same result.

This is nice, but what if we can do neither of these? Sometimes we have to drive and sometimes we also have to cross roads… In cases like these, we have to use our good old common sense: we need to compromise and find the risk that seems tolerable to us. It may be quite unnerving, but there is no absolute safety (the absence of all risks). *Safety is only temporary and relative: we can only make a choice between different risks.*

Table 1.1 Typical events with one microrisk (µR)

2500 km of travel by train
2000 km of travel by air
80 km of travel by bus
65 km of travel by car
12 km of travel by bicycle
3 km of travel by motorcycle
Smoking one and a half cigarettes
Living together with a smoker for 2 months
Drinking 0.5 l of wine
Living 10 days in a brick house
Breathing in a typical European city for 3 days
2 min of rock climbing
Suffering a bee sting within 5 years
Suffering a lightning strike within 10 years
Living at the age of 60 for 20 min

Risk assessment is used to quantify risks. The largest risk is, needless to say, losing one's life: this is called risk (R). 1 risk is an event relevant to a single human in which the human surely loses his or her life. This is too large of a unit for practical purposes, so one millionth of a risk, called microrisk (µR) is often used instead. If 1 million people are exposed to 1 microrisk independently, 1 fatality is expected as a consequence. It has long been experienced in social science that humans usually regard 1 microrisk (or often higher) as tolerable. Risk assessment is able to provide information like this based on the statistical analysis of a huge number of events (Table 1.1).

As it can be seen from the table, different events are associated with very different risks. Another important point is that some risks (transportation, breathing, ageing…) are unavoidable, where others (smoking, alcohol consumption, sport) are undertaken voluntarily. This latter type of risk is usually looked upon with forgiveness.

In efforts to avoid or lower risks, we must always consider both the cost and the benefit of the efforts. This is done in risk-cost-benefit analysis. Everything has a price tag on it.

How do chemical risks compare to others? In the forthcoming chapters, various chemical risks (industrial and domestic chemicals, environmental pollution, food additives, cosmetics, medicines, drugs) are described in detail. Here, only one example is given. The painkiller Opren is more effective than aspirin and does not cause internal bleeding. It has a side effect, though. Those taking the drug are more sensitive to sunburns, and a few elderly people suddenly died when taking this medicine. Opren was blamed for those deaths and under pressure from the mass media, it was pulled from the British market, and hundreds of thousands now take less effective painkillers. Opren is by no means an ideal painkiller. It has side effects, but let's put the risk of its use to some perspective. For example, 400 out of 25,000 patients under surgical treatment of arthritis die during surgery or because of complications related to the procedures. The use of Opren makes surgery unnecessary

1.2 Life as a Risky Business

Table 1.2 Average occupational fatality rate in different industries per 1 million person (United Kingdom, 1974–1978)

Type of industry	Mortality rate/million person
Clothes and shoe manufacturing	5
Vehicle manufacturing	15
Furniture manufacturing	40
Building material manufacturing	65
Chemical industry	85
Ship building	105
Agriculture	110
Construction industry	150
Train personnel	180
Coal mining	210
Open-pit mining	295
Other mining industries	750
Offshore oil and gas mining	1650
Deep-sea fishing (accidents before 1970)	2800

in these cases. Its side effects would kill *one* of those 25,000 patients. What would your choice be given these statistics? Opren was banned, arthritis surgery is still common...

An alternative approach to this question is based on comparing the risks of different professions (Table 1.2).

These statistical data show that the chemical industry qualifies as one of the less dangerous industries. It is revealing to compare these data with those on domestic accidents. The number of injuries because of glass windows or doors in the US in 1980 was 910 per a million people, and 3330 injuries because of a fall on the stairs. In England and Wales, falling down the stairs causes 12 fatalities per million persons each year; the same number is 108 in the group that is at most risk (women above 75).

Is there a difference between actual and perceived risks? American and British psychologists demonstrated a very interesting phenomenon when asking people about causes of deaths. In most cases, the opinion of people was moderately accurate. But there were a few causes, where occurrences were seriously over- and underestimated. Statistically rare causes (botulism, tornado, vaccination, pregnancy, and flood) were significantly overestimated, whereas common diseases (cardiac failure, stomach cancer, diabetes, stroke, etc.) were quite underestimated. Accidents and illnesses were perceived as equally likely to cause death, although, in reality, the latter is 15 times as common as the former! British students were asked similar questions. Although there were some differences (*e.g.* in assessing nuclear power plants), the main trends were the same. Students in science and engineering were tested in a similar way later. Perceived and actual death causes were closer in this case, but the probabilities of rare causes were also overestimated. These observations are explained by the theory of accessibility, which posits that causes about which information is more readily available seem more likely to people. The mass media bear a lot of responsibility, their reporting unintentionally distorts the truth.

Fig. 1.3 Labels on some products in the early twenty-first century. Zeitgeist, ignorance, or simply marketing? (Authors' own work)

Social scientists are interested in these question well beyond the approach described in the previous paragraph. The most respected of them (Mary Douglas, Anthony Giddens, Ulrich Beck), albeit with different emphasis areas, all came to the conclusion that the dynamism of our modern age poses an increasing threat to humans, and, although this may seem unintuitive, threats do not necessarily decrease with the increasing well-being of the society. Some of them (environmental pollution, industrial accidents, and food safety) might even increase. There is some truth in these thoughts. On the other hand, and this is nicely illustrated by the mentioned examples, perceived risk is very subjective. There are two typical things that could happen to someone enjoying the benefits of a modern society: the benefits are taken for granted and new concerns emerge, which would not be worth thinking about in the absence of the benefits. Some things become insignificant, whereas certain others get magnified, and this influences how risks are perceived.

1.3 Natural Products: A Delusion of Safety

You do not have to watch television for a long time to see a commercial promoting some sort of natural food or medication. The implication is clear: natural things are somehow better than their artificially produced counterparts (Fig. 1.3). It is a widely held misconception that natural materials are safe, whereas artificial ones are dangerous (→ **1.1**). This is by no means true. The psychological background of this belief will be analyzed in the next story (→ **1.4**), but to illustrate the point, let us look at some extremely dangerous, but without doubt natural products (→ **3.27**).

Food preferences are changing. A sign of these changes is the growing abundance of organic/bio/eco products in supermarkets. In addition to being more expensive, these products have the reputation of being healthier than those prepared by industrial technologies. Organic gardening itself, however, does not reduce the risk of human cancer in general, but rather in some well documented cases, it does the opposite. Bruce N. Ames (University of California at Berkeley)[1] pointed out that all plants produce toxins to defend themselves against fungi, insects and other animals (including humans). One of the most common toxins is salicylic acid, which is used in some countries in domestic preservation ($\rightarrow$ **2.18**). Laboratory studies showed that these toxins are not the least bit less poisonous than human-made pesticides used in agriculture for similar purposes. Every American citizen consumes an average of 1500 mg of natural pesticides each day. This amount is a good 15,000-times higher than the intake of artificial estimates, which the US FDA (Food and Drug Administration) estimates at 0.09 mg/day. Organic gardening often relies on new, insect-resistant varieties, which also increase the risk from natural pesticides. Ames mentions a new variety of celery which contains ten times as much (2600 **ppb**) carcinogenic materials than a more traditional one.

A common counterargument is that while humans slowly adapted to natural substances over millions of years, the same is not true for artificial substances, which pose a constant risk. Well, one cannot get used to poisonous substances no matter how long time is available. Were this statement wrong, there would be no yearly accidents because of eating the same old toxic mushrooms or fish, or snake and spider bites.

The sharp public distinction between natural (produced by evolution or created by God—which for the purposes of this story, does not matter—and human made substances is only a delusion. This assumption was once a scientific theory, it was called the *vis vitalis* **theory** and was thoroughly discredited almost two centuries ago. The modern-day variation of this theory posits that living things add some sort of extra "good nutrients" to materials, which cannot be produced in any artificial way. Table 1.3 presents a set of toxicity data (given as LD_{50} values) on natural and artificial substances. Even a cursory look on these data will reveal that natural toxins have the distinction of being among the most poisonous substances known.

The amount of the substance and mode of exposure is much more important than any assumed tolerance that might develop toward poisons. A careful look at Table 1.3 will reveal a number of natural and artificial substances listed such as paracetamol, aspirin, caffeine, sodium nitrite and nicotine, which are more or less toxic, but still used in everyday life. This might seem contradictory at first sight, but there is a marked **dose-response relationship**. A lot of toxic substances may also have positive effects and can be used if there is a large difference between beneficial and toxic doses. There is an old Latin saying: *sola dosis facit venenum*—it is the dose that makes the poison. Even very toxic botulinum toxin has an everyday use: in cosmetics, it removes wrinkles when applied cautiously on the forehead between the eyebrows. This Botox cure has an effect for more than half a year; a lot of celebrities use it instead of plastic surgeries.

---

[1] Prof. Ames (1928) is a living legend in toxicology; one of the most basic method to test for cancer causing potential was named after him.

Table 1.3 Toxicities of different substances. The number ID of the story on the particular substance is given in parentheses. *bw* body weight

Material (→ story)	Species, method of intake	LD_{50}
Sugar (sucrose) (→ **2.20, 3.8, 3.22**)	Rat, oral	29,700 mg/kg of bw
Vitamin C (L-ascorbic acid) (→ **2.1**)	Rat, oral	11,900 mg/kg of bw
Cyanuric acid (→ **2.6**)	Rat, oral	7700 mg/kg of bw
Cadmium sulfide	Rat, oral	7080 mg/kg of bw
Alcohol (ethanol) (→ **2.10**)	Rat, oral	7060 mg/kg of bw
Melamine (→ **2.6**)	Rat, oral	3161–6000 mg/kg of bw
Melamine cyanurate (→ **2.6**)	Rat, oral	4100 mg/kg of bw
Sodium molybdate	Rat, oral	4000 mg/kg of bw
Table salt (→ **2.19**)	Rat, oral	3000 mg/kg of bw
Paracetamol (acetaminophen)	Rat, oral	1944 mg/kg of bw
Arsenic (elemental) (→ **4.2, 4.6**)	Rat, oral	763 mg/kg of bw
Coumarine (from Chinese cinnamon and other plants)	Rat, oral	293 mg/kg of bw
Aspirin (*O*-acetylsalicylic acid) (→ **3.25**)	Rat, oral	200 mg/kg of bw
Caffeine (→ **2.22, 3.17**)	Rat, oral	192 mg/kg of bw
Arsenic(III) sulfide	Rat, oral	185–6400 mg/kg of bw
Sodium nitrite (→ **2.8**)	Rat, oral	180 mg/kg of bw
Cobalt(II) chloride	Rat, oral	80 mg/kg of bw
Cadmium oxide	Rat, oral	72 mg/kg of bw
Sodium fluoride	Rat, oral	52 mg/kg of bw
Nicotine	Rat, oral	50 mg/kg of bw
Lysergic acid diethylamide (LSD)	Rat, intravenous	16.5 mg/kg of bw
Strychnine	Rat, oral	16 mg/kg of bw
Arsenic(III) oxide (→ **4.2, 4.6**)	Rat, oral	14 mg/kg of bw
Arsenic (elemental) (→ **4.2, 4.6**)	Rat, intraperitoneal	13 mg/kg of bw
Sodium cyanide (→ **4.25, 4.26**)	Rat, oral	6.4 mg/kg of bw
White phosphorus	Rat, oral	3.03 mg/kg of bw
Mercury(II) chloride	Rat, oral	1 mg/kg of bw
Beryllium oxide	Rat, oral	0.5 mg/kg of bw
Aflatoxin B_1 (*Aspergillus flavus*) (→ **2.9**)	Rat, oral	0.48 mg/kg of bw
Snake poison of inland taipan or small-scaled snake (*Oxyuranus microlepidotus*)	Rat, subcutaneously	25 µg/kg of bw
Dioxin (TCDD) (→ **4.5**)	Rat, oral	20 µg/kg of bw
VX (nerve gas)	Human, oral, breathing, through skin/eyes	2.3 µg/kg of bw (est.)
Batrachotoxin (from certain species of frogs (*Dendrobatidae*), melyrid beetles, and birds (*Pitohui, Ifrita kowaldi, Colluricincla megarhyncha*))	Human, subcutaneously	2–7 µg/kg of bw (est.)
Maitotoxin (produced by *Gambierdiscus toxicus*, a dinoflagellate species and by the ciguateric fish *Ctenochaetus striatus*)	Mouse, intraperitoneal	0.13 µg/kg of bw
Polonium-210	Human, breathing	10 ng/kg of bw (est.)
Botulinum toxin (Botox)	Human, oral, injection, breathing	1 ng/kg of bw (est.)
Ionizing radiation	Human, irradiation	3–6 gray

1.4 Man-Made Commodities and Safety Issues

"Over eighty thousand (80,000) corporate-produced chemicals are currently used in the United States, and scientists estimate that over seven hundred (700) of those corporate-produced chemicals are now found within the body of every human. Many of these chemicals can be found in treated sewage sludge, 'frack' waste water from gas drilling, leachate from landfills, effluent from factories and smoke from incinerator stacks. Only a small percentage of those chemicals have ever been screened for even one potential health effect, such as cancer, reproductive toxicity, developmental toxicity, or injury to the immune system. Among the approximately fifteen thousand (15,000) chemicals tested, few have been studied enough to conclude that there are no risks from exposure. Even when testing is done, each chemical is tested individually rather than in synergistic combinations that reflect actual human exposure in the real world. One thousand eight hundred (1800) new chemicals enter the stream of commerce annually—thus entering into the waste stream, and the bodies of people, and into the air, water, soil, and food—with few of those chemicals tested for adverse impacts on human health or ecosystems. The use of many such chemicals by corporations is shrouded in secrecy as 'proprietary' ingredients. Nonethe-less, sufficient data and experience exist for a reasonable person to conclude that a significant percentage of both currently used and newly manufactured chemicals are harmful to humans, animals, and ecosystems." This quotation comes from *The Community Environmental Legal Defense Fund*, a non-profit, public interest environmentalist law firm in the USA.

In fact, the inventory of the US Toxic Substances Control Act (TSCA) of 1976 currently contains over 80,000 existing chemicals, many of which are produced or imported at low or negligible volumes. Others are polymers, which are unlikely to present significant risk concerns because of their physical size (e.g. high molecular weight) and other characteristics (insolubility etc.). By excluding low volume chemicals (~25,000 chemicals produced or imported in amounts less than 5000 kg per year) and polymers (which tend to be poorly absorbed by organisms and therefore typically exhibit low toxicity), the remaining TSCA Inventory is comprised of about 15,000 chemicals produced or imported at levels above 5000 kg per year. Of these, 3000–4000 are produced or imported in amounts over 500 t per year: these chemicals are considered by the US Environmental Protection Agency (EPA) to be US High Production Volume (HPV) chemicals. EPA has identified the subset of the mentioned 15,000 chemicals as being the broad focus "universe" of the TSCA Existing Chemicals and Chemical Testing Programs with the primary focus placed on the 3000–4000 HPV chemicals.

This book will specifically treat bisphenol A ($\rightarrow$ **4.24**), DDT ($\rightarrow$ **4.4**) and dioxins ($\rightarrow$ **4.5**) from the compounds appearing in this list.

Is the overall picture really so grave? An earlier chapter ($\rightarrow$ **1.1**) explained that there is no connection between the risk and the origin (natural or artificial) of a substance, despite widespread assumption (or advertisements) to the contrary. Natural substances ($\rightarrow$ **1.3**) can actually be quite hazardous, whereas a number of chapters in this book try to give a nuanced analysis of the risks associated with artificial materials.

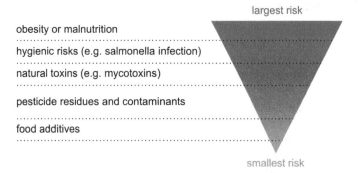

Fig. 1.4 The relative importance of food-related risks. (Authors' own work)

Why is there so much belief in the inherent dangers of man-made substances? Why do the very same dangers seem more acceptable for natural products? American science writer Michael Shermer (1954-) studied the putative connection between autism and vaccines and came to the conclusion that the anecdotal associations, particularly when amplified by stories in the media, are so strong that they tend to overwrite scientific evidence in people's minds. The human brain has evolved to care about anecdotes: a false positive association (thinking that A is connected to B, whereas there is in fact no connection) is often harmless, but a false negative association (thinking that A is not connected to B, whereas a connection in fact exists) sometimes has fatal consequences. Typically, it is associative learning through which the brain finds patterns. Belief in magic and superstition is several million years old, whereas science, which tries to avoid false positive associations, has only been practiced for a few centuries.

Food additives have a distinguished spot on the unimaginably long list of feared "chemicals". A 2005 poll in Hungary (the home country of the authors) showed that 76% of the population was worried about food additives. In a similar, but EU-wide poll based on the opinions of 26,691 adult citizens from 27 member states, many people stated that they were highly worried about chemicals remaining in fruits, vegetables and grains (the EU average was regretful enough at 31%, with Hungary scoring an embarrassing 84%).

Modern food production is impossible without food additives. The risk of their use determined by scientific methods (Fig. 1.4) is actually much lower than public opinion might ever believe. Consumers like a wide variety of products while manufacturers are interested in the minimization of costs and in implementing the simplest possible technologies. To reach these objectives, stable and easily handled auxiliary materials are necessary, which make quality assurance possible (→ **2.2, 2.3, 2.4, 2.18**).

The application of a food additive is only permitted if the necessity is demonstrated, and alternative technologies cannot be used to reach the same goal, or if the completely benign nature of the food can be proved by scientific results. The biological effects are carefully tested, new scientific results are continuously monitored, and additional testing is done if new developments make this necessary.

The opinion of independent consultant toxicologist John Hoskins could be very surprising to many: "On any plate of food the only things that can be relied on to be safe to eat are those chemicals, natural or synthetic, which are traces of pesticides or hormones or those that have E-numbers. The rest of the food must be taken on trust."

There are many consumer and environmentalist groups worldwide who watch the issue of food additives with enormous suspicion. They often create their own lists with frightening and unfounded claims about the risks of individual additives. However, it is wiser to listen to those who are specialists in this field. Authorities, e.g. the European Food Safety Authority (EFSA) in Europe, the Food and Drug Administration (FDA) in the US, the United Nations World Health Organization (WHO) and Food and Agriculture Organization (FAO) in their *Codex Alimentarius*, pay distinguished attention to the issue of food safety in most countries. Some more specific questions about food will be discussed in the next section of this book (essays → **2.1** through **2.25**).

1.5 The Cholera Pandemonium: The Blind Leading the Blind

Chemical information seldom gets noticed in everyday life as long as business goes as usual. There are unusual cases, though, which continue to attract the attention of the mass media. For experts, it is not uncommon and tragically funny to find that the information sources are totally unreliable in those cases. Decreasing public interest in science is a common phenomenon in the developed world. Lack of trust in government organizations is also rampant throughout the world, and in Central Europe, such mistrust has a long and rich history. As a result, alternative sources of information are often sought.

Civil organizations usually enjoy the reputation of being trustworthy, in contrast to government or industry. They have an important role in modern society as they provide independent checks on many things and help consumers to develop environmentally sustainable attitudes. They even receive awards for their activities. However, they are not infallible and sometimes they would benefit from some independent external reviews. Few people suffering from a toothache would turn to patient-rights advocates first. Very similarly, civil organizations often help raise awareness of environmental problems, but they are seldom the ones who can offer solutions. Unfortunately, civil organizations often show an alarming lack of scientific information with respect to chemical questions. A recent example will be described here, where Greenpeace played a very negative role.

A cholera outbreak occurred in early 1991 in Peru. The infectious disease first appeared on the coast than rapidly advanced to inland areas. There were 322,582 documented cases just in 1991, and the infection spread to two neighboring South American countries. In the four years between 1991 and 1994, 941,804 cases were recorded with 8622 fatalities. The cholera appeared again in Peru in 1998. Between

1900 and early 1991, not a single case of cholera was reported in Peru, Colombia and Ecuador. After the 2010 earthquake, a similar outbreak began in Haiti in October primarily because of poor sanitation. About 300,000 people got sick, 4500 died. The reasons behind the 1991 contagion in Peru are still not clear: poor sanitation, tainted drinking water, unwashed fruits and vegetables, consumption of raw fish, warming by El Niño, floods, *Vibrio cholerae* bacteria surviving in zooplankton, and the unusually high incidence of the 0 blood group (75%) in Peru could all have contributed. Cholera is not that difficult to fight. Dehydration of patients must be prevented to cure them and transmission can be slowed down by improving sanitation. There is a classic case in cholera studies: John Snow (1813–1858), a physician living in London, made systematic observations during a cholera outbreak in 1854 and proved that the contagion can be controlled by carefully checking the quality of drinking water delivery systems.

Fred M. Reiff, a representative of the Pan American Health Organization (PAHO) was responsible for fighting communicable diseases between 1981 and 1995. In cooperation with workers of Lima-based Centro Panamericano de Ingeniería Sanitaria y Ciencias del Ambiente (CEPIS), he studied many different possibilities of disinfection and found that chlorination is the most efficient and cost-effective method. The cost of equipment and the chemical calcium hypochlorite was about 2.50 dollars/family/year, which is by no means of secondary importance in developing countries. In his memoirs, Reiff describes that local health officials were reluctant to use chlorination during the 1991 outbreak. Most of them were concerned about the possible formation of hazardous trichloromethane derivatives. The source of this information was the scientific literature and the environmental agencies of a few developed countries. Reiff tried to explain, in vain, that the extra risk posed by these compounds is expected to cause a single extra death in a population of 100,000 people during 70 years. Greenpeace did not accept chlorination, either. Instead, it recommended the consumption of a local citrus-like fruit called toronja. This fruit provides acidic conditions, and it is true that these conditions slow down the growth of cholera bacteria. However, its effect is very poor compared to water chlorination. Greenpeace still denies its responsibility in failing to respond to the Peru cholera outbreak appropriately.

Greenpeace has lobbied the US congress quite intensely to ban *all* chlorine-containing substances. One of the arguments to back this intention was that chlorine-containing organic compounds never occur in Nature, they only appeared because of human activities. They did not mind that this is not true. Gordon W. Gribble (Dartmouth College, USA) pointed out that chlorine containing compounds are formed in massive quantities in about a dozen natural processes including by lightning-induced forest fires, volcanic eruptions or even in normal life processes of aquatic organisms. About 400,000 t of chlorophenols are formed in the moors of Sweden every year because of the decomposition of humic acids catalyzed by the chloroperoxidase enzyme. This is about 40 times as much as the amount produced by the paper industry. Biomass production is accompanied by the release of 5 million t (!) of poisonous methyl chloride gas each year, while the human contribution is merely 26 thousand t annually.

The single most important use of chlorine-containing compounds is water disinfection. About 98% of the drinking water in the US and 96% of the waste water is treated with chlorine. There are four technologies that could replace chlorination: membrane filtration, ultraviolet irradiation, filtration on activated carbon bed and treatment with ozone (**ozonolysis**). All of them are more expensive than chlorination, and none of them were studied in as much detail as chlorination was. If ozone is used, the by-products formed in the reactions of ozone with organic compounds have to be removed in a separate step using activated carbon. Overall, there is no viable alternative to chlorination today.

The Hungarian branch of Greenpeace has its own story on a red mud catastrophe which will be described later in this book ($\rightarrow$ **4.8**).

1.6 Regulating Chemicals: Maybe or Maybe Not

REACH is the acronym of the phrase *Registration, Evaluation, Authorisation, and Restriction of Chemicals*. This acronym names a piece of EU legislation introduced in 2007 after a long and tiresome negotiating process between industry and non-governmental organizations. REACH in effect aims to determine how to treat chemicals within the European Union. Perhaps it will surprise no one that the chemical industry finds it unnecessarily strict and demanding, whereas non-governmental organization criticize its leniency. Here for example is a very typical opinion from an environmental civil organization: "The lobby of the chemical industry aims to convince decision makers to allow the use of many hazardous substances even when safer alternatives are known." This may sound convincing, but a convincing statement does not necessarily make it true.

A cornerstone of REACH is that all chemicals produced in the European Union in a quantity exceeding 1 t/year must be thoroughly investigated so that its risks—not only for the producer, but for the entire cycle of use—could be assessed. This is different from earlier practice as the investigations are not the responsibility of the regulatory agencies, but the producers, so the results must be publically disclosed. There are four different ways a chemical may appear: (a) through a transformation of another chemical (this could be a useful product or waste); (b) through import for a substance that was not used in Europe earlier; (c) through isolation from natural sources (mining, harvesting, butchering, etc. and/or separation by chemical methods); (d) through reprocessing from a source (*e.g.* waste) that is not liable to REACH, but the substance is re-introduced into the product cycle.

After entering the product cycle, the following two cases are possible for a substance: *(i) It undergoes further chemical transformation*. This is called an intermediate. The life cycle of the substance is short in this case with limited and well controlled human or environmental exposure. Therefore, the requirements in REACH for these substances are somewhat relaxed. The intermediate substance can be used in two different ways: the new chemical is separated (strictly speaking, this is the only intermediate use), or it is built into a product or structure during a chemical transformation (EU does not usually recognize this as intermediate use). *(ii) It is used in its original form in some sort of application.* Usually, this use is preceded by mixing the substance in question with other substances to improve properties or effects, *etc.* This is also the end use of a substance, after which it will be considered as waste at the end of its life cycle. The substance remains in its original form in a chemical sense, there is no reaction. For these substances, there are several different layers of legislation depending on the potential exposures.

The introduction of REACH has been criticized from numerous viewpoints. Describing this would make a book rather than a short story, so only the most important reservations will be included here:

1. REACH is the most detailed legislation in the history of the European Union. The length of text exceeds 1200 pages. It is very implausible to expect a decision maker to read and understand it, this would be a daunting task even for a chemical expert. However, the devil is in the details, maybe more so in this case than in many others. The details of the legislation fundamentally determine its benefits for the citizens of the EU.
2. The use of the precautionary principle (*Vorsorgeprinzip*) is obvious in REACH. This principle may seem logical for many, but this is not proved scientifically. Actually, a cost-benefit analysis often gives a negative overall evaluation about this principle ($\rightarrow$ **4.24**).
3. REACH does not mention existing waste despite the fact that this poses a significant problem. The dream of the fathers of REACH was the omnipotence of the life cycles—the legislation is unable to handle anything outside of these cycles.
4. REACH overreaches. It contains exaggerated precautions against unlikely events. This is a disadvantage for the European chemical industry as the competitors in North America or Asia do not have to obey the same strict rules. This will disproportionately affect chemical SMEs (small and medium-sized enterprises) in Europe.
5. Lawmakers have forgotten that the chemical industry and the market are already global. European chemical companies are tempted to move outside the borders of the European Union because of overregulation. Switzerland and Norway are not members of the EU but they also have strict regulations, so moving chemical production to these countries will not mean increased risk for European consumers. The same is not true if the production moves to India or China. There are frequent scandals about chemicals and production technologies in these two vast countries, these are known to the general public as well. Developing countries often use technologies that are outdated or forbidden in more developed areas of the world. The same problems that REACH attempts to solve might very well return as imports ($\rightarrow$ **2.21**).
6. The chemical industry already introduced a number of principles before the introduction of REACH in order to fight its image problem. The most well-known of these is responsible care and product stewardship. The chemical industry is willing to do more to protect health, environment or safety than it is legally required to do (Table 1.4). These principles are based on the golden rule: "Therefore

1.6 Regulating Chemicals: Maybe or Maybe Not

Table 1.4 A new ethics of responsible action

Old ethics	New ethics
Meet the legally required minimum	Do the right thing
Be considerate	Let other people see that you do the right thing
Production is your business	Product stewardship for entire life cycles
Explain why concerns of the society are unfounded	Seek and address the concerns of the society
Share risk information only as needed	Society and employees deserve to be informed of all risks
A company fights for its own interests	Mutual help is necessary among companies

whatever you desire for men to do to you, you shall also do to them…" (World English Bible, Matthew 7:12). Of course, these principles are not held sacrosanct by every chemical company (→ **4.5**), but there is an increasing tendency to follow them.

7. Toxicological tests required in REACH have unintended consequences. The EU requires all chemicals produced on an annual scale larger than 1 t to be registered with full toxicity data by 2018. This will need 2.1–3.9 million test animals according to the EU Joint Research Centre (JRC). This estimate drew fire from toxicologist Thomas Hartung, who was the director of the member institutions of JRC, independent expert Costanza Rovida and numerous other toxicologists. Their estimate is 8–9 million (7.5–10 million according to other sources) test animals because a second generation of the animals will have to be tested to give a reasonably accurate risk assessment of possible negative effects. JRC estimates that about 92% of the data will be available from computer simulations, whereas other experts claim that this percentage cannot be larger the 50%. Reliable information will only be available from actual experiments for the other half.

Hartung and Rovida welcome the introduction of REACH and they think that it is a major step toward protecting European consumers. However, there is a major discrepancy between initial estimates and reality, and this will necessitate an overhaul of the legislation. EU estimated that 27,000 companies would file 180,000 applications for 29,000 substances in the pre-registration period before the end of 2008. In fact, 65,000 companies filed 2.7 million applications for 140,000 substances until 2008. The number of substances to be tested was probably overestimated by those who filed, the actual number is somewhere between 29,000 and 140,000. An optimistic estimate would be 68,000. For the *in vivo* tests of these substances according to current REACH regulations, 54 million test animals would be necessary between 2008 and 2018. This is 20 times the original estimated number of animals and 6 times the originally estimated cost, which could easily reach a staggering 9.5 billion €. About 90% of the animals used were planned to be tested in reproduction studies. The European Union does not have the facilities to carry out these tests. Because of the huge difference between estimates and actual numbers, Hartung and Rovida called for a suspension of second-generation tests. This may only be necessary if the first-generation offspring gives reason for suspicion. As a replacement, more developed testing methods should be considered such as high through-put

Table 1.5 Annual amounts of product, waste and environmental (E) factors in different industries

Industry or product	Scale (thousand tons anually)	Weight of waste per weight of product (E)	Overall weight of waste (thousand tons)
Petroleum and gas industry	1000–100,000	<0.1	100–10,000
Bulk chemicals	10–1000	<1–5	10–5000
Fine chemicals	0.1–10	5–50	0.5–500
Pharmaceuticals	0.01–1	25–100	0.250–100
Paper industry	325,000	250–1000	81,250,000–325,000,000*
Biotechnology products	0.0001–1	7000–15,000	0.7–15,000*

Estimated values marked with asterisks include waste water as well

methods, cell cultures, non-vertebrate animals, fish larvae, or the more extensive use of computational algorithms.

These concerns show that the legislative history of REACH is far from over, major changes and adjustments will be necessary.

1.7 Biowaste: Biotechnology in Perspective

The chemical industry as a whole suffers from a serious image problem. When non-experts hear about a chemical transformation, they most often associate it with environmental pollution or industrial accidents, because these incidents receive a lot of publicity. Some experts and most of the general public, however, hold a much more favorable view of the pharmaceutical and biotechnology industry. These industries are also based on chemical transformations, but in a more environmentally sustainable manner—at least so the word goes.

A more objective view on this question is offered by Table 1.5, which lists waste produced in different industries.

The table shows that there are major differences in the weights of product and the waste-generating potentials of different industries. The petroleum and gas industries, which are a major part of the chemical industry as a whole, manufactures vast amounts of products, but the waste-to-product ratio is quite low. This ratio is called the environmental factor (E). As the produced quantities decrease from bulk chemicals to pharmaceuticals, the E factor increases dramatically. This is not accidental. In the petroleum and gas industry, the typical processes do not usually involve more than two or three steps. On the other hand, 10-step or even 20-step processes are not uncommon in the pharmaceutical industry. The higher the number of steps, the higher the potential for generating waste.

Two of the listed industries, namely, paper manufacturing and biotechnology, use large amounts of water. Consequently, there is a further increase in the E factor, which is almost exponential. It is not an accident that paper mills are always located along rivers.

1.7 Biowaste: Biotechnology in Perspective

Fig. 1.5 Large-scale industrial fermenters used in a brewery. (Copyright-free Wikipedia picture)

Biotechnology (fermentation processes) have been used for some seven decades now to produce antibiotics, steroid pharmaceuticals and hormones (Fig. 1.5). These technologies are usually more favorable than equivalent chemical methods because the latter would be much more complicated. Fermentation of antibiotics provides an excellent example of reducing the required amount of input materials. A strain initially producing only a few milligrams/liter of the desired product can often be modified to reach much higher levels such as 1 or 10 g/l. In biotechnology, aqueous biochemical reactions are typically applied, although the product might not dissolve very well in water and the microorganisms may not tolerate high concentrations. Thus, large amounts of water are necessary. The production of 1 kg of monoclonal antibodies typically requires 7000 kg of water, 600 kg of inorganic salts (which also end up in the waste), 8 kg of organic solvents (most often alcohols) and 4 kg of other consumables (plastic pipes, filter, resins, *etc.*). Similarly, to produce 1 kg of mid-sized proteins using genetically modified *Escherichia coli* bacteria, 15,000 kg of water, 400 kg of inorganic salts, 100 kg of organic solvents (most of which qualify as hazardous waste) and 20 kg of consumable are required. Of the mentioned waste materials, the largest problem is posed by the dilute aqueous solutions containing inorganic salts. Organic waste is usually easy to burn, whereas inorganic waste can often be handled only by precipitating agents. Because of the large amounts of water used, waste water treatment and recycling are extremely important.

Even if water is excluded from the considerations, biotechnology would still have an E factor at least five times that of the pharmaceutical industry. If water is included, the ratio of the E factors is about 3000. Some improvement is possible through flow systems and further genetic modification of microorganisms in order to make them more tolerant to extreme conditions. A combination of biotechnology with chemical methods is also often a step forward. The current production of vitamin C and penicillin use such advanced methods.

The actual amount of waste produced in industry is shown in column 4 of Table 1.5. This number is the product of the numbers appearing in columns 2 and 3 in each row. The data include water, so the values shown for the paper industry and biotechnology are very high. In addition to the mere amount of waste, the type of waste is also an important factor in making a comparison. Table salt may be considered as waste sometimes, but it is a very different kind from, say, chlorinated organic solvents. Recently, attempts are made to quantify this effect by defining a quality factor Q, and using the product $E \times Q$ in comparisons.

In an attempt to find a similar comparative measure to characterize other human activities, the concept of virtual water has been introduced as the volume of water (in liter) necessary to produce 1 kg of product. Some basic food products (wheat, rice, eggs) have their numbers in the range from 1300 to 3400, but beef scores at 15,500. Other consumer products (shoes, shirts, sheets, pants) typically have values between 4000 and 10,000. Every human activity generates waste. While in certain cases (food or paper), public awareness of this fact is low, in certain other cases such as the chemical industry, society seems to be less forgiving even though their waste production compared to the weight of useful output may be much lower. Needless to say, this fact alone does not exempt anyone from the obligation to improve toward more environmentally benign practices, but a little sophistication in thinking about waste is probably beneficial for the entire society.

1.8 Ice Skating and the Science of Sliding

Ice skating, believe it or not, has long been fodder for thought for scientists, especially for a small group of people who teach a science named physical chemistry. The question seems to be centered on the issue of what makes the skate slid so effortlessly, the ice itself or liquid water produced on top of the ice by the skate.

Skates have relatively sharp blades, so someone in skates exerts an unusually high pressure on the ground. An exceptional property of water ice is that its melting point decreases under pressure (most substances behave in exactly the opposite way). This may lead people to rush to the conclusion that ice melts under a skate, and the athlete (or the beginning skater) actually slides on a thin layer of water rather than the surface of ice. However, this is far from logical: many drivers know that an icy road is more slippery than a wet one.

What is wrong with the ice-melting argument? Well, the argument only says something about the *direction* of changes, but never tries to estimate their *magnitude*. If we think a little deeper, this theory about ice melting under skates would lead to a number of consequences that are definitely not true.

1. Children would have more difficulty when skating because their weight is lower and they would exert a lower pressure.
2. The colder the temperature is, the more difficult would be to move on a skate as higher pressure would be needed to achieve a larger change in the melting point of ice.
3. On a single leg (or skate, to be more precise), it would be much easier to skate because the pressure would double compared to a pair of skates.

1.8 Ice Skating and the Science of Sliding

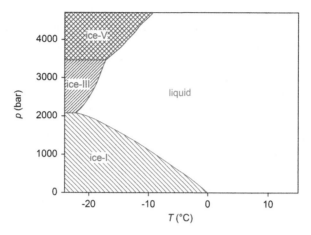

Fig. 1.6 A part of the phase diagram of water. (Authors' own work)

4. A puddle would form below someone standing on a skate on ice for a longer time.
5. Much higher efforts would be needed to achieve higher skating speeds as melting of ice is by no means an instantaneous process.

Let's estimate than how much the melting point of ice changes below a skate. To help our thinking, we can make good use of the phase diagram of water, which is shown in Fig. 1.6. In this graph, temperature (T) is displayed on the horizontal axis, whereas pressure (p) is shown on the vertical axis. One of the simplest ways to use this diagram is to decide what form of water is stable under a certain pressure at a given temperature. All we have to do is to find which part of the diagram the point falls. In Fig. 1.6, 1000 bar and $-5\,°C$ is within the area labeled liquid water, so one can be sure that the stable form of water is liquid under these conditions. At first sight, it may seem surprising that water is a liquid below its usual freezing point. As already explained, the external pressure also influences the temperature at which water freezes (or ice melts). Similarly, the boiling point of water also depends on pressure. From a phase diagram, one can read the actual melting point of ice at any pressure: all we need to do is to find where the boundary lies between ice and liquid water at that pressure.

It should not be left without notice that there are three kinds of ice on the phase diagram. Ice, similarly to many other solids, has more than just one possible crystal structures, that is it has several different ways in which the molecules can arrange themselves in a regular pattern. These crystal structures are stable at different temperatures and pressures. In fact, science knows many more form than displayed on this graph. This is also true for many other chemical substances including carbon, which is known both as graphite and diamond—the latter being thermodynamically unstable but far harder and more expensive ($\rightarrow$ **4.13**).

Finally, one can also see from the phase diagram that it is impossible for liquid water to exist below $-22\,°C$, as no part of the phase diagram below this temperature is labeled liquid ($\rightarrow$ **4.17**). If someone were to insist very much on the theory of melting ice under a skate, they would have to conclude that it is impossible to skate below this temperature. The experimental observation, however, is quite clear: ice can be very slippery even at $-30\,°C$.

How large is the pressure under a skate? This is not very difficult to calculate, but you may need to refresh your elementary physics studies. The blade of a skate may be 20 cm long and 2 mm wide, so the surface is $0.2\,m \times 0.002\,m = 0.0004\,m^2$, this doubles to $0.0008\,m^2$ for a pair of skates. If the mass of an average person is 80 kg, the weight is 800 N; the conversion factor between the two quantities is the gravitational acceleration on the surface of Earth, which is approximately $10\,m/s^2$. So the pressure exerted by the skate on the ice is $800\,N/0.0008\,m^2 = 1,000,000\,Pa = 10^6\,Pa = 10$ bar. This is barely ten times the atmospheric pressure on Earth, which is about 1 bar (or 10^5 Pa or 1 atmosphere—scientists and their units can be perplexingly diverse). In Fig. 1.6, the melting point at 10 bar is hardly any lower than $0\,°C$. Much higher pressures are needed for a significant change in the melting point.

A counter-argument for this line of thought was put forward as early as the nineteenth century. It says that the uneven surface of a skate blade makes the actual contact surface a lot smaller. Even if this is true, there is no small enough surface for skating below $-22\,°C$. Higher pressure is usually accompanied by higher friction as well, which actually acts against the movement, and could be much more significant than the melting point depression of ice.

Friction, however, offers an even deeper line of thought. Friction produces heat, which can melt ice in a very thin layer. However, friction can only produce heat during movement. Ice is known to be slippery even for objects that are stationary, there is no extra effort needed to begin the movement. And forces of friction tend to be quite small on ice—this is the very reason why it is slippery, and small forces can only generate a small amount of heat.

What could the explanation be then? First of all, the surface of ice is very even, or at least it is when conditions are ideal for skating. The even surface offers little chance for the small bumps on the surface of the sliding object to stick to anything. In addition, water molecules on the surface of ice experience less interaction than those in the middle of the substance, which means that they also move easier. It is sometimes said that the surface of ice features an extremely thin (maybe one hundred thousandth of a millimeter) layer that resembles a liquid. But this does not mean actual melting at all. This surface is always there, the presence of skaters does not make a difference.

Finally, remember that there is something called synthetic ice. Skating fields are sometimes made at places where it would be very expensive to maintain freezing temperatures. This synthetic ice is made of a special kind of plastic, the melting point of which actually increases as pressure increases. Skaters on synthetic ice do not seem to care at all.

1.9 So What's Happening to the Ozone Hole?

The ozone layer and its thinning have been extensively covered both in the daily press and the popular scientific literature. The words ozone shield and ozone hole have joined the vocabulary of many *literati*. These phrases are easy to remember and quite easily visualized. Unfortunately, they are also quite deceiving if their meaning is taken literally.

Ozone is the name of a type of oxygen molecule (chemists use the word *allotrope*) that is made up of three oxygen atoms ($\rightarrow$ **4.11**). Normal oxygen, the one everyone breaths, consist of diatomic molecules. The words ozone shield or ozone layer (the meaning is the same) are used to refer to the amount of ozone that is present in the Earth's stratosphere, which lies approximately between 10 and 50 km above the ground. Air is very thin at such high altitudes, definitely not dense enough to support human breathing. In this thin gas, ozone only makes up a tiny proportion, but it is still much more than the ozone found in the troposphere (the 10 km of air close to the ground). This may seem accidental at first sight, but if it is actually an accident, it is an extremely fortunate one. First, contrary to popular belief, ozone is in fact highly poisonous. It would be impossible for humans to live if ozone occurred in any significant quantity in the air we breathe. Second, ozone molecules absorb ultraviolet radiation, a mortal danger to life, before it reaches the surface of Earth.

So the ozone layer is by no means a continuous or dense shield: it is much more like a very thick but sparsely woven net. To make matters even more confusing, scientists often talk of the of the thickness of the ozone layer, usually given in Dobson units (DU), which was so named in memory of the Oxford meteorologist Gordon Dobson (1889–1976). The amount of ozone in the stratosphere is 1 DU if, after removing all other gases, it would form a layer 10 µm (one thousandth of a centimeter) thick on the surface of Earth. This is the result of an imaginary experiment under imaginary conditions. Normal ozone concentrations are between 200 and 300 DU, so ozone would only form a layer 2 or 3 mm thick at sea level. In reality, this amount of ozone is distributed in other gases in a mass of air that is almost 40 km thick.

The phrase ozone shield was intended to be metaphoric: it protects life from danger, which comes from the Sun as ultraviolet (UV) radiation. Other gases in the air let not only light pass, but also UV radiation. This would be harmful for living organisms including humans. Ozone happens to absorb the most dangerous part of UV radiation. There is a quite formidable irony in this: a highly toxic substance occurring high above the ground actually shields life from mortal radiation.

In the 1970s, scientists began observing a gradual fall in the amount of ozone in the atmosphere. Again, an easily comprehensible, but somewhat imprecise phrase was coined: the thinning of the ozone layer. In fact, this does not mean that the 40 km layer of air in which ozone occurs is getting any thinner. It means a decrease in the amount of ozone in the layer but that does not change its size. This amount has remarkable seasonal dependence as well, so it only makes good sense to talk about yearly average amounts. The rate of the decrease was about 4 % per decade, and even this seems to have stopped in recent years.

Fig. 1.7 The ozone hole above the Antarctic. (September 2006, the blue areas have larger transparency to ultraviolet radiation, i.e. less ozone). (Copyright-free Wikipedia picture)

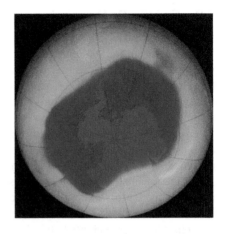

The phrase ozone hole, however, means something different. Interestingly, the loss of ozone in Earth's atmosphere is much larger above the poles than anywhere else, and even in those areas, it mainly occurs in early spring (March–April above the North Pole, September–October above the South Pole). To make matters more complicated, the two poles are quite different as far as ozone goes: the largest drop in ozone concentration was about 30% at the North Pole, whereas at the South Pole, a drop of up to 70% is not unheard of. So the ozone hole is not an actual hole in a solid shield, it is more like a large decrease in the texture of an already sparsely woven net (Fig. 1.7).

The reasons why the amount of ozone decreases in the atmosphere have been studied in excruciating detail; scientists Paul J. Crutzen, Mario J. Molina and F. Sherwood Rowland were awarded the Nobel Prize in chemistry for these studies in 1995. A group of man-made substances called freon-like compounds turned out to be the major culprits. Their main advantage was their inability to engage in any sort of chemical reactions. They were used as refrigerator liquids and propellants in spray flasks. Research showed that "freons" reside in the atmosphere for decades and they speed up (to use chemists' word: catalyze) the decomposition of ozone. Ozone both forms and decomposes in the atmosphere under natural conditions as well, but freons only speed up the decomposition, which results in a smaller overall amount. Recognition of this global problem initiated global action: the Montreal Protocol, in effect since 1989, called for freons to be phased out gradually from everyday use. Needless to say, this did not occur without much scientific debate, some of which were quite passionate: until 1988, American freon-producer DuPont Co. paid a few scientists to follow Sherwood Rowland everywhere he travelled and debate his scientific views, mostly based on reasoning motivated by financial interests.

The Montreal Protocol has been ratified by 196 countries and most freon-like substances have already been phased out. Some positive effect is already detectable: the ozone hole above the Antarctic was largest in 2006, and most recent measurements indicate that the loss of ozone in the troposphere has stopped—the trend might even have reversed somewhat.

If ozone loss had continued at the rate observed in the 1980s, the consequences would have been dire. However, few of the negative effects would have been seen in European or Northern American countries. There has never been an ozone hole above these areas. If officials or professionals warn against the dangers of sunbathing in the high summer, it has nothing to do with the ozone hole. It is just good old common sense.

1.10 Predicting the Elements: Success and Failure

Today the periodic table is displayed on the wall of every somewhat self-respecting chemistry classroom. Science history gives enormous credit to Dmitri Ivanovich Mendeleev (1834–1907) as the father of the periodic table, and not without good reason. Although a lot of scientists had been busy organizing the elements even before the work of Mendeleev, the contribution of the great Russian scientist is made unique by two important aspects. First, Mendeleev tirelessly popularized the concept of the periodic table and the periodic law behind it. Second, and more important, he made detailed predictions about elements that had not been discovered, and also gave some guidelines for their properties and possible occurrence in nature. Nowadays, textbooks often quote the successful predictions (for example, gallium and germanium) and make the reader believe that Mendeleev was infallible in this area. Nothing could be farther from the truth.

Some of Mendeleev's predictions can be seen in his first periodic table, which was published in 1869. The logic of the system dictated that there should be an element under boron ("-=44"). Mendeleev tentatively called the missing element ekaboron (the element under boron), and made predictions for its properties based on the surrounding known elements. The same happened for the element below aluminum (ekaaluminum) and silicon (ekasilicon). It took a great deal of confidence and intuition to predict two missing elements next to each other in a system where most of the elements were thought to be known. Discoveries within the next two decades confirmed Mendeleev's predictions. Ekaaluminum was discovered by Frenchman Paul-Émile Lecoq De Boisbaudran (1838–1912) in 1875 and named gallium. Ekaboron was found in 1879, the discoverer was Lars Fredrik Nilson (1840–1899) from Sweden, who chose the name scandium. Ekasilicon was identified by German scientist Clemens Winkler (1838–1904) and named germanium. So these predictions were shown to be true within two decades and element names kept patriotic feelings high in the discoverers.

Noble gases were not foreseen by Mendeleev because none of the elements were known from this group in 1869, and the concept of atomic number was also unknown before 1913. However, there was place for a new column in the system between the halogens and the alkali metals. So finding the place of noble gases in the 1890s was not much of a problem.

In Mendeleev's last periodic table published in 1904, the predicted and discovered elements were shown along with the noble metals. Throughout the years, the

Table 1.6 Mendeleev's predicted elements and the actual discoveries

Predicted element	Predicted atomic mass	Actual element (year of discovery)	Actual atomic mass
Ether	0.17	–	
Coronium	0.4	–	
Ekaboron	44	Scandium (1879)	44.96
Ekacerium	54	–	
Ekaaluminum	68	Gallium (1875)	69.72
Ekasilicon	72	Germanium (1886)	72.61
Ekamanganese	100	Technetium (1939)	98
Ekamolybdenum	140	–	
Ekaniobium	146	–	
Ekacadmium	155	–	
Ekaiodine	170	–	
Ekacesium	175	–	
Trimanganese	190	Rhenium (1925)	186.2
Dvitellurium	212	Polonium (1898)	209
Dvicesium	220	Francium (1939)	223
Ekatantalum	235	Protactinium (1917)	231.0

great Russian chemist made further predictions. Ekamanganese ("-=100") and trimanganese were still unknown at the beginning of the twentieth century, but successfully identified later as the elements technetium and rhenium. Dvitellurium (polonium), dvicesium (francium) and ekatantalum (protactinium) were similar success stories.

A closer look at the 1904 system reveals that two elements lighter than hydrogen were also predicted: these are elements x and y. Mendeleev put both of them into the group of noble gases above helium. He thought that element x was ether, which was thought to be an actual material by physicists, and referred to element y by several different names, of which coronium was the most common. In 1904, barely a decade had passed since the discovery of the first noble gas, so it was logical for Mendeleev to assume that there were still undiscovered elements in this group, and they do not necessarily occur on Earth. In addition, he had a more philosophical reason for insisting on these predictions. At that time, many scientists accepted a theory called Prout's hypothesis, which posited that all elements are eventually composed of hydrogen. From this assumption, it clearly followed that no element can be lighter than hydrogen. Mendeleev was opposed to this hypothesis. He knew quite well that atomic weights are not necessarily whole number multiples of the atomic weight of hydrogen (in his 1869 system, beryllium, aluminum and chlorine illustrate this fact), although he did not have any clue about the existence of isotopes. His opposition to Prout's hypothesis was manifested by the fact that he insisted on the predictions of elements x and y, although there were absolutely no experimental data to back up this claim.

Table 1.6 summarizes all of the predictions Mendeleev made in his life. The atomic weights of ether and coronium were estimated based on the atomic weight

1.10 Predicting the Elements: Success and Failure

Table 1.7 Mendeleev's original periodic table from 1869

	Group I - R^2O	Group II - RO	Group III - R^2O^3	Group IV RH^4 RO^2	Group V RH^3 R^2O^5	Group VI RH^2 RO^3	Group VII RH R^2O^7	Group VIII - RO^4
1	H = 1							
2	Li = 7	Be = 9.4	B = 11	C = 12	N = 14	O = 16	F = 19	
3	Na = 23	Mg = 24	Al = 27,3	Si = 28	P = 31	S = 32	Cl = 35.5	
4	K = 39	Ca = 40	- = 44	Ti = 48	V = 51	Cr = 52	Mn = 55	Fe = 56, Co = 59 Ni = 59, Cu = 63
5	(Cu = 63)	Zn = 65	- = 68	- = 72	As = 75	Se = 78	Br = 80	
6	Rb = 85	Sr = 87	?Yt = 88	Zr = 90	Nb = 94	Mo = 96	- = 100	Ru = 104, Rh = 104 Pd = 106, Ag = 108
7	(Ag = 108)	Cd = 112	In = 113	Sn = 118	Sb = 122	Te = 125	J = 127	
8	Cs = 133	Ba = 137	?Di = 138	?Ce = 140	-	-	-	- - -
9	(-)	-						
10	-		?Er = 178	Ta = 182	W = 184	-		Os = 195, Ir = 197 Pt = 198, Ag = 199
11	(Au = 199)	Hg = 200	Tl = 204	Pb = 207	Bi = 208	-		
12	-	-	-	Th = 231	-	U = 240	-	

Table 1.8 Mendeleev's last periodic table from 1904

	Group zero	Group I	Group II	Group III	Group IV	Group V	Group VI	Group VII	Group VIII
0	x								
1	y	hydrogen H = 1.008							
2	helium He = 4.0	lithium Li = 7.03	beryllium Be = 9.1	boron B = 11.0	carbon C = 12.0	nitrogen N = 14.04	oxygen O = 16.0	fluorine F = 19.0	
3	neon Ne = 19.9	sodium Na = 23.05	magnesium Mg = 24.1	aluminum Al = 27.0	silicon Si = 28.4	phosphorus P = 31.0	sulfur S = 32.06	chlorine Cl = 35.45	
4	argon Ar = 38	potassium K = 39.1	calcium Ca = 40.1	scandium Sc = 44.1	titanium Ti = 48.1	vanadium V = 51.4	chromium Cr = 52.1	manganese Ma = 55.0	iron cobalt nickel Fe = 55.9 Co = 59 Ni = 59 (Cu)
5		copper Cu = 63.6	zinc Zn = 65.4	gallium Ga = 70.0	germanium Ge = 72.3	arsenic As = 75.0	selenium Se = 79	bromine Br = 79.95	
6	krypton Kr = 81.8	rubidium Rb = 86.4	strontium Sr = 87.6	yttrium Y = 89.0	zirconium Zr = 90.6	niobium Nb = 94.0	molybdenum Mo = 96.0	-	ruthenium rhodium palladium Ru = 101.7 Rh = 103.0 Pd = 106.5 (Ag)
7		silver Ag = 107.9	cadmium Cd = 112.4	indium In = 114.0	tin Sn = 119.0	antimony Sb = 120.0	tellurium Te = 127	iodine I = 127	
8	xenon Xe = 128	cesium Cs = 132.9	barium Ba = 137.4	lanthanum La = 139	cerium Ce = 140	-	-		— (-)
9		-	-						
10	-	-	-	ytterbium Yb = 173	-	tantalum Ta = 183	tungsten W = 184	-	osmium iridium platinum Os = 191 Ir = 193 Pt = 194.9 (Au)
11		gold Au = 197.2	mercury Hg = 200.0	thallium Tl = 204.1	lead Pb = 206.9	bismuth Bi = 208	-	-	
12	-	-	radium Rd = 224	-	thorium Th = 232	-	uranium U = 239	-	

ratios of the noble gases. This table also shows that elements x and y were not the only unsuccessful predictions. The rest of the erroneous predictions (ekacerium, ekamolybdenum, ekaniobium, ekacadmium, ekaiodine, and ekacesium) are all connected to the fact that Mendeleev knew the lanthanide group, but could not correctly place them into the periodic table (Tables 1.7 and 1.8).

It may be surprising to learn that Mendeleev's success rate in his predictions was only 50%. It is quite interesting to note that while the successful predictions contributed significantly to the general acceptance of the periodic table, the failures did not have the opposite effect. These unsuccessful predictions did not lack logic, but chemistry turned out to confirm different lines of thought later. One should remember that Mendeleev did not have any idea about the concept of atomic number, neither did he have any information about the electron structure of atoms. Today, these two things make the periodic table almost trivial.

Nobody is infallible, Mendeleev was not, either. His mistakes were forgiven, possibly almost forgotten in science because his contribution to chemistry was indeed gigantic. Of all the scientists, he was the one who deserved an element to be named after him the most. This is element 101, mendelevium. The consensus on this is illustrated quite nicely by the fact that the name was not proposed by his fellow Russian scientists: it was suggested by an American group.

1.11 Vedic Wisdom: Lead and Ayurveda

Tetétlen is a small village in Eastern Hungary where a company called Greentech-Vekerd Inc. planned to invest about 20 million € to build an acid lead battery recycling plant in 2008. Originally, the company wanted to build this plant in an even smaller village named Vekerd, but the mayor of the neighboring village Zsáka raised so many objections that the company finally opted for the new location in Tetétlen. According to the plans, the recycling plant would be able to process about 25,000 t of used batteries each year. In Hungary, 28,000 t of batteries get thrown out annually, but since there is no re-processing plant, this waste is exported to Austria and Slovenia. The council of the village of Vekerd, which included one of the owners of the company, raised money to finance a trip for the mayors of the neighboring villages to Italy, where they could visit a very similar plant and ensure that it posed no significant environmental threat. The recent history of this region and its villages illustrates the difficulties Greentech-Vekerd, Inc faced in establishing the lead re-processing facility.

Vekerd and Zsáka were cut off from the natural center of their region, Oradea, in 1920 by the newly drawn Hungarian-Romanian border. About 15–20% of the active population of this region is unemployed, and industrial plants have gradually closed in this region during the last decade. Vekerd is one of the least developed villages in Hungary. When the local co-operative of farmers was disbanded in the early 1990s, many local people said this was the beginning of the end. Out of 171 inhabitants, 150 were retired at that time. Every second house is abandoned, most of them in ruins. From an agricultural standpoint, the soil is highly alkaline and less productive than the Hungarian average.

A possible solution to these problems would be finding National or European programs that support underdeveloped regions. Zsáka and its neighboring villages submitted a proposal to build an innovation park for about 500 million €, where

1.11 Vedic Wisdom: Lead and Ayurveda

researchers could study biogenetics and quantum gravity "close to nature". This lunatic proposal was rightly rejected.

Oradea and its region are among the most dynamic in Romania with an unemployment rate about half the national average, whereas unemployment across the Hungarian border near Zsáka is more than double the Hungarian average. Although new road construction was planned in this area, the plans have been continuously altered due to cross-border political differences.

Vekerd was approached by Hungarian-Italian joint venture Greentech Ltd about the possibility of building an acid lead battery reprocessing plant and creating 50 new jobs. About 13,400 t of lead could have been re-claimed from the batteries. The constituent elements would have not been decontaminated on site, they would have been transported to a hazardous waste deposit in Galgamácsa, a village about 200 km away and close to the Hungarian capital Budapest. At the Vekerd site, batteries would have been disassembled to re-capture lead, sulfuric acid and two different kinds of plastic. In a 2003 referendum, the inhabitants of the village gave their support to build the facility. The mayor of neighboring Zsáka, however, intervened. He argued that lead re-processing must not be allowed in an area protected by a program called Nature 2000, in which many farmers have already switched to organic farming. He also warned that the plant would be harmful for local inhabitants. Expert opinions, on the other hand, found no health risks to speak of in the plans and even the officials of the nearby National Park of Hortobágy approved the plans. Tomatoes are often grown next to similar plants in Italy, and ski centers operate in similar locations in Austria and Germany. In Hungary, however, an unfounded concern was enough to stop the construction in a seriously underdeveloped region. The community of Vekerd would have received about 220,000 € in taxes from the facility annually, whereas its current yearly budget is about 85,000 €. The mayor of Vekerd, however, remained optimistic. He plans to build an Ayurvedic medicinal center instead of the reprocessing plant.

Ayurveda is an ancient system of healing methods from India combining medical science and traditional wisdom: the name comes from combining the words 'ayur' (long life) and 'veda' (wisdom). There are two main principles: The first says that the human body has a spirit attached to it, and this must be kept in good condition for a healthy life. The second enumerates bodily substances as the five classical elements earth, water, fire, air and ether. These are combined in three elemental substances (*dosha*): *vata* meaning air and ether, *pitta* meaning fire and water, and *kapha* meaning earth and water.

For Europeans and North Americans, the most familiar word of Ayurveda could be *panchakarma*, which is a therapeutic way of eliminating toxic elements from the body (→ **3.15**). This mostly involves a removal of accumulated poisons in the body through a special diet, laxatives, massage, steam bath and enema. Ayurvedic medicines contain only natural substances (healing plants, animal body parts, minerals, metals and precious stones). Some of the therapeutic practices were found to be beneficial by modern scientific methods, while others (for example the use of heavy metals) are outright harmful, and not just compared to a lead re-processing plant. As we write, Vekerd has neither a lead re-processing plant nor Ayurvedic medicinal facilities.

So what is known about lead? It is one of the few elements which seem to have no biological function, and its water soluble compounds are mostly toxic. In the human body, it accumulates in the bones as insoluble phosphates. Today's human bones contain about double the amount of lead that was measured in Stone Age skeletons, yet today's values are several times lower than it was in the bones of people who lived in the last two millennia. The health effects of lead remained unknown for a surprisingly long time in human history.

The daily average lead intake is about 1 mg, but only about 10% of this reaches to the blood stream. Human food usually contains 0.001–0.025 **ppm** of lead compounds. Drinks are primarily responsible for most of the human lead exposure: old lead pipes, lead containing paint on mugs, and wine or whisky stored in leaded glass, are the main sources.

In the eighteenth and nineteenth century, many people suffered from joint gout: the American founding father Benjamin Franklin, biologist Charles Darwin, British prime minister William Pitt, poet Lord Tennison, and Methodist church founder John Wesley were all documented to have this illness. Gout is caused by uric acid crystals formed in the joints. The source of uric acid in the body is partly a diet rich in meat, and partly a metabolic disorder in purine compounds. Frequent consumption of Port wine was also often considered a possible cause, and not without good reason. Studies in the 20th century showed elevated levels of lead in about one third of the patients suffering from joint gout. In previous centuries, wine was very often stored in lead-containing bottles, so its consumption added to the risk of developing gout. The World Health Organization (WHO) proposes that the maximum tolerable amount of lead in drinking water should be 10 µg/l (0.01 ppm). Before 1995, the proposed limit was 50 µg/l.

One of the first lead compounds humans encountered was lead(II) acetate, also called sugar of lead (*saccharum saturni*), which was used to sweeten drinks, and a medication against diarrhea. In Ancient Rome, a sweetener called *sapa* (a syrup of lead (II) acetate) was produced by boiling grape juice or sour wine in lead vessels. The use of *sapa* could have caused the low fertility in Ancient Rome compared to other nations at the time, which is especially surprising because Romans were world leaders in sanitation. About 80,000 t of lead was produced each year in the ancient Roman Empire.

Soluble lead compounds exert their toxicity by inhibiting the enzyme necessary for hemoglobin synthesis in blood. This causes 5-aminolevulinic acid (ALA), the precursor of hemoglobin, to accumulate. ALA is responsible for the symptoms of lead poisoning: headache, stomach ache, diarrhea, paleness, dysfunctional intestines, kidney malfunction, infertility, miscarriage, and anemia. Lead can inhibit a number of further enzymes because its reacts with thiol groups. It is therefore, a general cell poison.

Lead in nature occurs in a number of minerals such as galena (lead sulfide), cerussite (lead carbonate) and anglesite (lead sulfate); none of these dissolve in water. The world production of lead is about 6 million t annually. If societies only used newly mined lead, deposits would be exhausted in a mere 15 years. This limited natural supply of lead and the toxicity of water-soluble lead salts is more than enough

1.11 Vedic Wisdom: Lead and Ayurveda

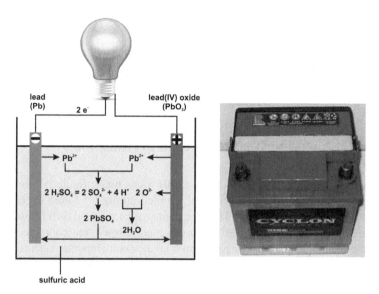

Fig. 1.8 Schematics of a lead acid battery and an electric car battery. (Authors' own work and copyright-free Wikipedia picture)

reason to make major efforts to re-cycle the metal. About one third of the world production is used in car batteries (Fig. 1.8). These are not currently considered as the most advanced batteries, but they are extremely reliable and their replacement in cars is close to unthinkable today. In the developed world, re-processing of lead batteries is among the most successful recycling technologies: 97% of spent batteries were re-cycled in the US between 1997 and 2001. The second most common use of lead is still glass making, but not for tableware. A traditional (not liquid crystal-based) computer monitor or television contains about a quarter of a kilogram of lead, and its primary function today is protecting humans from X-rays produced during the operation of these devices. Protection from radiation is also important in the nuclear industry.

American mechanical engineer and chemist Thomas Midgley made an important discovery in 1921: tetraethyl lead, an organic compound of the metal, significantly improved the performance of gasoline engines, as it inhibited the phenomenon of "knocking". Subsequently, leaded gasoline became quite common by the 1960s. Today only lead-free gasoline is sold at the pumps. Phasing out tetraethyl lead was a major step in decreasing human lead exposure. Nevertheless, the primary rationale for this was not to reduce health risk. Modern cars contain catalytic converters to preserve air quality, and the presence of tetraethyl lead in gasoline shortened the life time of these very expensive converters.

Increasing lead levels from transportation and other sources were once a major concern. An American study in 1979 found some correlation between falling IQs and increasing blood lead levels. About 400,000 children were tested in a 1993 study in the US state of New York, and 5% of them were found to have increased levels of lead in the blood. Officials of the Environmental Protection Agency, however, did not find the evidence for the connection between intelligence and blood levels convincing. In fact, these early opinions are thought to be vastly overestimating the risks. Needless to say, no one questions the toxicity of lead compounds today, so exposure to them must be kept under control.

1.12 Manipulating Weather: Ocean Fertilization

German scientist Alfred Wegener (1880–1930) was a major figure in modern meteorology, geophysics and polar research. He died during his fourth expedition to the island of Greenland. He formulated the theory of continental drift (*Kontinentalverschiebung*) in 1912, which posits that continents slowly but continuously drift around the planet Earth. Wegener could not convince contemporary scientists of the validity of his theory, so it was forgotten until the 1960s, when it was revived as plate tectonics and developed into a high impact field of science. The scientific institute Alfred Wegener Institut für Polar- und Meeresforschung (AWI, Bremerhaven) was named after him. In 2009, AWI received a great deal of attention from the media due to an ocean fertilizing experiment carried out by one of its researchers.

Eighty percent of the ocean's mass supports very little life. Different forms of life tend to accumulate in 'oases' as beautifully shown by the films of French explorer and conservationist Jacques-Yves Cousteau (1910–1997). American oceanographer John Martin (1935–1993, Moss Landing Marine Laboratories, California) once famously proved that the lack of iron salts is the main factor limiting plankton growth in the upper regions of oceans. The survival of other sea animals is dependent on planktons to a large degree. The name plankton refers to any organism in the sea that can only move with currents rather than using its own muscles. Planktons are usually classified into three distinct groups: phytoplankton (diatoms living and photosynthesizing close to the ocean surface, *Diatomeae*, cyanobacteria and green algae, *Chlorococcales*), zooplankton (small single cell and multicell organisms, eggs and larvae of larger animals, such as fish, crustaceans, and annelids that feed on other plankton), and bacterioplankton (bacteria and archaea, which play an important role in breaking down organic material deep in the water).

Martin's theory was spectacularly supported by the 1991 volcanic eruption of Mt. Pinatubo in the Philippines. The volcanic ash carried 40,000 t of iron into the ocean, increased the oxygen content, and decreased the carbon dioxide content of water. Martin assumed that adding the missing iron to oceans would slow down or even reverse global warming on Earth as the increased growth of plankton could remove substantial carbon dioxide from the atmosphere. Martin set up an experiment called IRONEX I that was carried out in 1993 to the west of the Galápagos Islands. Unfortunately, he passed away before he could see the results. The essence of the experiment was to pour ferrous sulfate into the water in an area as large as 60 km^2. The success of the experiment was spectacular: green plankton filled the water within a week, which proved that iron is indeed one of the limiting factors for aquatic life.

1.12 Manipulating Weather: Ocean Fertilization

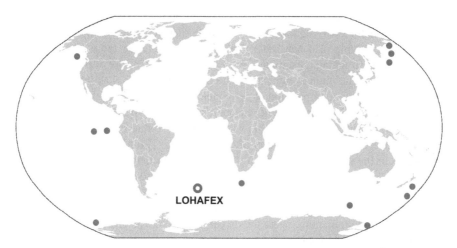

Fig. 1.9 The sites of LOHAFEX and former ocean fertilizing experiments with iron salts. (Authors' own work and copyright-free Wikipedia picture)

Fig. 1.10 Polarstern, the ship that controlled the LOHAFEX experiment. (Copyright-free Wikipedia picture)

Vast amounts of iron, about 850 million t, reach the oceans each year, mostly from rivers, with smaller contributions from glaciers. However, only a minor fraction of this actually reaches the open waters of the sea. The major source of iron is dust from deserts with about 16 million t per year. The share of southern oceans is about 1 million t, so it is not unreasonable to describe the state as a major lack of iron.

There have been 13 small scale experiments since 1993 aimed at fertilizing the oceans with iron salts (Fig. 1.9). The latest one, which drew the most media attention, was conducted by the research ship Polarstern from the Wegener Institute (Fig. 1.10). Forty-nine researchers, mainly from India and Germany cooperated in this experiment, called LOHAFEX, with the name constructed from the Hindi word *loha* (iron) and the English expression 'fertilization experiment'. Polarstern set sail on January 26, 2009 and aimed to distribute 20 t of ferrous sulfate (equivalent to 6 t of dissolved iron) in an area of 300 km^2 around Antarctic waters. There were heated protests against the planned experiment, mainly from the activists of Greenpeace.

The LOHAFEX experiment took two and a half months to complete and introduced 10 mg of iron for every square meter of water surface. The amount of iron in a 4000 m deep column of ocean water is about 10 times larger; tap water may have 100 times more iron than the amount used LOHAFEX. Therefore, it is very difficult to accept one of the main arguments of the protesters, which was that the experiment is a large scale manipulation of ecosystems.

For the LOHAFEX experiment, an area in the ocean had to be found where currents go in a circular pattern. This was necessary so that the iron would not be dispersed over a vast area too soon after deposition. The ideal site was close to Antarctica at the southern latitudes of 48° and western longitude of 16°. This was the center of a clockwise circular current of area of 10,000 km^2. During the experiment, chlorophyll concentration increased to double or triple of its pre-experiment value. This increase was detectable from space. Very small (0.2–2 µm sized) plankton called picophytoplankton were the driving engine of the growth, but they were promptly eaten by zooplankton.

In a deviation from earlier fertilization experiments on the Southern hemisphere, the silicate level of these waters was low. As a consequence, the growth in silicate and carbon consuming diatoms was negligible. So the major objective of the experiment, the significant removal of carbon dioxide, did not occur.

A key observation behind the ocean fertilizing experiments was that iron ions play a central role in photosynthesis. Estimates show that 1 kg of iron may induce the removal of 3000–83,000 kg of carbon dioxide depending on the locations and external conditions.

Opponents of the experiments (environmental groups and a number of researchers) presented arguments as to why these experiments should be discontinued. They believed that the experiments may result in unexpected and unfavorable outcomes (toxic algae, jellyfish, overgrowth of diatoms, and declining oxygen levels because of the decomposition of plankton). The opponents were also skeptical of the assumed climate-changing effect of iron salt fertilization.

Carbon dioxide has a bad public reputation ($\rightarrow$ **4.20**), based on fact that the Earth's gradual warming since the industrial revolution is almost exclusively attributed to this gas. Without doubt, the increase in the concentration of greenhouse gases is a major problem today. The essence of the greenhouse effect is that greenhouse gases reflect the infrared radiation emitted by the surface of the earth, and this energy gets trapped in the atmosphere instead of leaving it. This effect is by no means altogether harmful: the average temperature of Earth, which is about 15 °C today, is a good 33 °C higher than expected based on the Sun-Earth distance. Most of the greenhouse effect (62%) is caused by water vapor in the atmosphere. Carbon dioxide is only a distant second on this list (22%), with tropospheric ozone joining the ranks as number 3 (7%). Water vapor and cloud cover have a complex role. Increasing temperatures cause the evaporation of more water, which would increase the greenhouse effect. Increased cloud cover, on the other hand, reflects more sunshine back to space (the albedo is larger), which slows the warming down. Global warming is a complex problem (ozone, carbon dioxide, greenhouse effect, acid rain, plankton, dimethyl sulfide emission, oil spills, pesticides, heavy metals, deforestation): it attracts a lot of scientific and political attention.

With everything taken together, was the LOHAFEX experiment a failure? The evaluators, some of whom were the participants themselves, said that results of the iron fertilizing did not live up to the expectations. On the other hand, earlier experiments in waters with high silicate content actually showed highly increased removal of carbon dioxide. The LOHAFEX experiment proved that it is very important to choose the location. Both proponents and opponents of iron fertilization think there is a need for more experiments to better assess the possible benefits. The experiments thus far only covered about 0.04% of the ocean surface. In order to plan further experiments and studies, 13 oceanography institutes started a consortium named *In Situ Iron Studies* in early 2011.

Chapter 2
Food

2.1 Test Your Cranberry Pie: Vitamin C and Benzoates

Food dyes, flavor enhancers and preservatives receive the brunt of criticism in the family of food additives. Preservatives do not really deserve this bad reputation as they replaced older, traditional techniques (smoking, curing) that are actually not without dangers.

Benzoic acid and its salts are well known preservatives (E210-E213), their action depending on the inhibition of an enzyme in the citric acid cycle of microorganisms. These preservatives only work well under highly acidic conditions. They are effective against yeasts, mold, and microorganisms producing aflatoxins, but only partially against bacteria. Benzoic acid is often used in combination with potassium sorbate as this mixture is more effective than their individual constituents. Benzoic acid is not effective against oxidation or enzymatic rotting so it is combined with sulfur dioxide in treating fruits. Some people are allergic to benzoic acid; about 4 % of people suffering from asthma develop shortness of breath because of it. Many fruits (blueberries, dry plums, apples) contain benzoic acid in concentrations that exceed the limits set by authorities (Fig. 2.1). These fruits contain vitamin C as well. There is no known human risk of consuming less than 500 mg of benzoic acid daily because it does not accumulate in the body, and it is excreted quite rapidly with urination. Preserved food usually contains 0.05–0.1 % benzoic acid. It is one of the preservatives with the longest known histories, often used in canned fruit, pickle and soft drinks.

Vitamin C (L-ascorbic acid) is probably the best known vitamin, and the daily amount necessary for humans is between 50 and 100 mg depending on age (150 mg for breastfeeding mothers). It is present in many natural sources and is also a common synthetic product.

Suspicion about a possible health risk from the simultaneous consumption of benzoic and ascorbic acid arose in the 1990's in connection with soft drinks. The British daily *The Independent* picked up the line again in 2007 when it claimed that benzoic and sorbic acids may harm mitochondria, the energy producing center of cells. These claims were based on yeast cell experiments carried out by biochemist

Fig. 2.1 Different species of cranberry, especially *Vaccinium macrocarpon* (aka large cranberry, American cranberry, bearberry, shown) is rich in both benzoic and ascorbic acids. (Copyright-free Wikipedia picture)

Peter W. Piper (the name is real!) in the 1990s. There is no unambiguous connection between yeast cells and human cells. For human cells, there is no evidence of any harm done to mitochondria.

Benzoic acid and ascorbic acid are of about the same acid strength. Benzoic acid can be formed in equilibrium between ascorbic acid and benzoic acid salts. Benzoic acid can decompose to give carcinogenic benzene and carbon dioxide under *certain conditions* (Fig. 2.2). How much benzene could be formed in this reaction? There is typically 150 mg of benzoic acid in a liter of soft drink, whereas there are no legal limits for ascorbic acid. The German Federal Institute for Risk Assessment (BfR, *Bundesinstitut für Risikobewertung*) published a recommendation in 2005, which stated that this reaction depends on many external conditions, such as the concentration of the substances involved, the presence of some metal ions (copper and iron), temperature, time, and level of ultraviolet radiation. More importantly, the simultaneous presence of ascorbic acid and benzoates was not obviously connected to the formation of benzene. The amount of benzene formed increased at first when ascorbic acid was added in model experiments, but increasing the amount of vitamin C further actually slowed the reaction down. This is similar to the effect of other antioxidants (e.g. mannitol used as a sweetener or uric acid in urine). The amounts, however, were very low: the concentration remained in the **ppb** or **ppt** range. In the experiments under extreme conditions (45 °C, 20 h of ultraviolet irradiation), the concentration of benzene reached 300 ppb, but this value was much lower under physiological conditions (37 °C, pH 7.4, no ultraviolet irradiation). When experiments were done with food containing both benzoic and ascorbic acids, benzene was formed in concentrations between 0.01 and 38 ppb, but typically below 1 ppb. Just to put this into context: the World Health Organization (WHO) recommends and upper limit of 10 ppb for benzene in drinking water.

Irrational fears might result in irrational precaution. The US Food and Drug Administration published data on benzoic acid and ascorbic acid in 2007. Although no connection was established between the amount of these substances and the concentration of benzene, recommendations for the reduction in benzoate use were made for manufacturers in a few cases, which were duly followed. The British supermarket chain Sainsbury's removed sodium benzoate from 120 products because of consumer concerns.

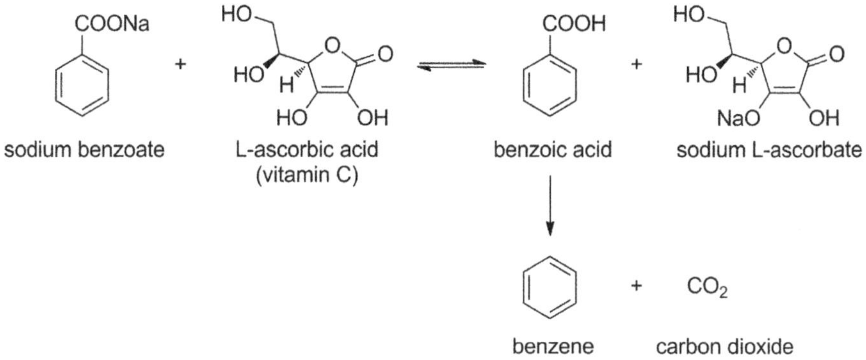

Fig. 2.2 Reaction of sodium benzoate and L-ascorbic acid (vitamin C) in a model experiment. (Authors' own work)

The majority of human exposure to benzene comes not from food but air (smoke and car exhaust). The concentration of benzene inside a car might be as high as 20 ppb, at the gas pump it may even be 1000 ppb. In the US, the average benzene concentration in the air is 2–10 ppb. Typically, 0.01 ppb is detected even in the air above the Pacific Ocean. Smoking a single cigarette produces about 500 μg of benzene, which of course makes its way to exhaled air. The British Food Standards Agency estimated that consuming about 20 l of soft drinks a day poses the same benzene exposure as other environmental sources.

The reader may rightly feel that this story is getting complicated. Or is it? In everyday life, people continuously face risks when choices are made on food, transportation, family care and so on. Society is aware of some potential hazards (e.g. transportation), but is less aware of others (food additives etc.). Naturally, there is more fear about the latter. People choose between trying to understand risks or continuing to live in irrational fear. The reader may certainly make his or her own choice.

2.2 Food Dyes: The Good, the Bad and the Ugly

Food dyes, especially the artificial ones (→1.4), are probably the most controversial group of food additives. The proven risk of allergic human reactions to these substances is seldom larger than for other substances, but their unnecessary use should no doubt be avoided: blue triple sec, green ketchup, or soft drinks containing 8–10% of orange, cherry or other fruit are made more attractive by dyes. However, natural substances may also be problematic, and some of them are banned today.

From a chemical point of view, most synthetic dyes are azo dyes, some of them from the triphenylmethane dye family, whereas there are a few whose structure does not belong to the two large, predominant groups. Why are synthetic dyes advantageous? Primarily, because synthetic dyes are more stable than natural dyes. Most

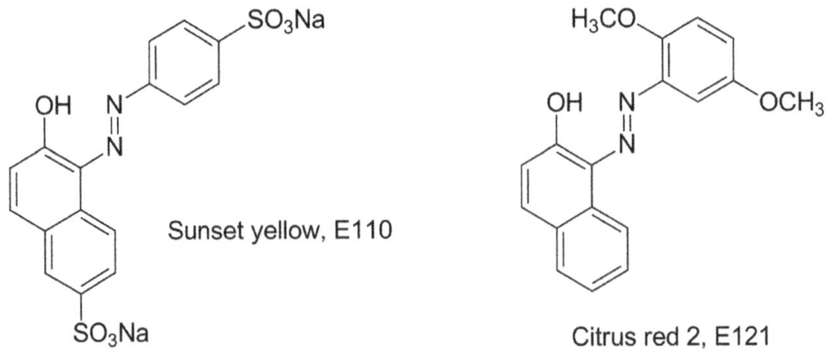

Fig. 2.3 The chemical structures of two azo dyes. (Authors' own work)

natural dyes cannot tolerate heat treatment, the oxygen in air or changes in pH. Nature did not develop them for human kitchen use. The juice of red cabbage turns an ugly grey upon pH change.

2.2.1 Azo Dyes

The largest group of synthetic dyes is called azo dyes. The name refers to the azo group, which is two nitrogen atoms connected by a double bond: $R-N=N-R'$. R and R' stand for various aromatic ring structures, which influence the acidity, alkalinity, and color, among others. The azo group maintains a continuous conjugation between the π electron systems of the aromatic rings, and this causes the vivid color of these compounds. Two examples are shown in Fig. 2.3. Sunset yellow (E110) is still in use, whereas Citrus red 2 (E121) has already been phased out.

An older name of this group is aniline dyes, which refers to the compound aniline, the simplest aromatic amine. Aniline is substance that has had a very high, but largely unrecognized impact on modern society. The names of two huge multinational chemical companies founded in Germany in the 1860s, BASF (Badische Anilin- und Soda-Fabrik) and AGFA (Aktiengesellschaft für Anilinfabrikation) are a testament to the true importance of this unsung chemical hero. Azo dyes were discovered in the middle of the nineteenth century and generated a major development in the organic chemical industry. Tens of thousands of different dyes were prepared mostly to color textile, leather or fur, but inks, paper and wall paints also benefitted greatly. One of the first commercial products introduced in the 1870s was chrysoidine, used for making yellow wool or silk. The nineteenth century dye industry was also the cradle of the later developing pharmaceutical industry. One of the first true antibacterial drugs, sulfonamide *Prontosil rubrum* (Domagk, 1935) was in fact an azo dye. Strict health and environmental criteria has only left 12–13 azo dyes for use in the modern age as food additives.

Azo dyes today are considered harmless at the concentration level of their typical usage. Although azo dyes do not occur in nature, human metabolism can cope with

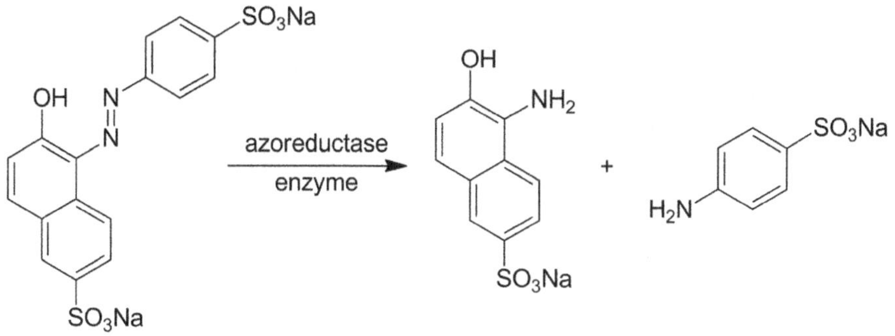

2.2 Food Dyes: The Good, the Bad and the Ugly

Fig. 2.4 The primary reductive transformation of an azo dye into the two constituent amines. (Authors' own work)

them easily. There are azoreductase enzymes in the body that reduces the azo group to produce two compounds with amino groups (Fig. 2.4).

The toxicity or the proven carcinogenic nature of some azo dyes is primarily caused by these amines or their further decomposition products. These dyes (e.g. Butter yellow or Citrus red) have been phased out. There are heated debates about the possible allergies caused by these compounds. All substances (medicines, cosmetics, even staple food included) may have this effect, typically at concentration levels of 100 or 10 **ppm**. The use of the azo dye tartrazine, which is thought to be more problematic because of allergies, is not recommended in infant formula. A 2007 report from Southampton University has received a great deal of publicity as it questioned the safety of azo food dyes in 3–9 year old children on the grounds that they cause hyperactivity. However, several scientific experts voiced serious concerns about the methodology of this study. The first counterpoint was that the reasons for selecting the dyes and their particular mixtures were unclear. In addition, measuring hyperactivity in children is by no means exact, and the statistical evaluation of the results was not rigorous.

2.2.2 Triphenylmethane Dyes

Triphenylmethane dyes are characterized by a central carbon atom which is bonded to three aromatic systems, most often phenyl rings (Fig. 2.5). There are additional basic or acidic functional groups on these rings, which change the distribution of conjugated electrons to form quinoidal structures, which in turn results in a color change. A number of these compounds are also used as acid-base indicators in chemical laboratories. The archetype of triphenylmethane dyes, magenta colored fuchsine, was accidentally discovered and purposefully patented by Frenchman François-Emmanuel Verguin in 1859 during experiments aimed at the oxidation of aniline.

It has long been known that triphenylmethane dyes with aromatic rings bearing basic, mostly dialkylamino substituents, have significant antibacterial effects. They

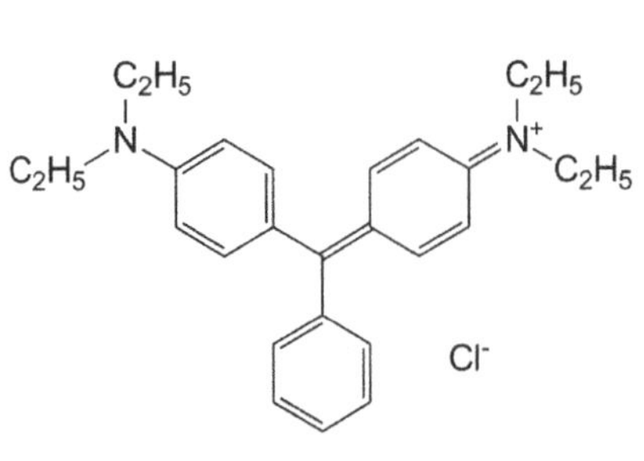

Brillant green
(antiseptic)

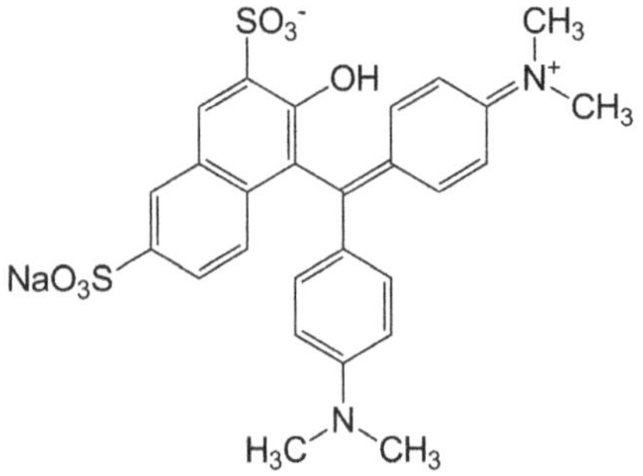

Green S, E 142
(food dye)

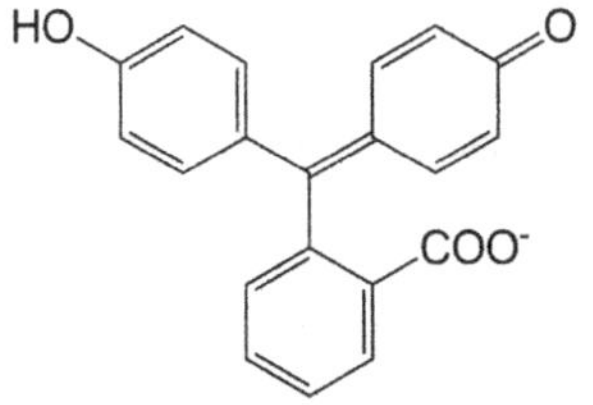

Phenolphthalein
(indicator dye, laxative)

Fig. 2.5 Examples of triphenylmethane dyes. (Authors' own work)

are also effective against fungal skin infections. Castellani's paint, which is a red solution of carbol (phenol) fuchsin, or crystal violet, are still used occasionally, but this practice is presently discouraged in developed countries. There are better and longer-lasting dyes today, such as Green S (Fig. 2.5). Significant harmful effects of triphenylmethane dyes have not been detected thus far. They may cause local tissue death when injected in concentrated solution, but this is not a reasonable health risk. Their non-toxic nature is nicely illustrated by the fact that some members of this group of dyes have been used as a laxative for more than a century. For this purpose, phenolphthalein is still one of the most reliable and safest alternatives.

2.3 The Organic Vegetable Hype

2.3.1 A Question for Eternity: Natural or Artificial Fertilizers?

This question is not that difficult to answer: avoiding too much of a good thing is an excellent guideline. Neither plants nor animals enjoy an overabundance of nutrients. Obesity and the resulting health problems (diabetes, cardiovascular diseases, and joint aches) have received copious media attention lately. Ironically, parents of overweight children are often quite proud of their offspring. Sagging and loose tissues are not only characteristic of obese people, they are found in 'overfed' plants as well. The problem is not the synthetic fertilizer, as it contains the same nutrients as the natural variety. No plant can utilize the organic components of manure directly; they need to wait for soil bacteria to change them into inorganic nitrate, phosphate and other salts. The difference lies in the speed of the chemical and biological reactions. Soil bacteria work slowly, so there is never a surplus of nitrate available and, therefore, plants cannot consume excess nitrates.

The chemical process to make saltpeter is identical to the one used in the Middle Ages: manure was mixed with straw and ash in compost piles and left alone for as long as a year with an occasional sprinkling of urine, which contains nitrogen. Bacteria slowly turned organic nitrogen into nitrate, the salts of which crystallized on the surface of the compost piles very much like soda-salts do in dry flatlands. At the end of the process, the saltpeter was collected and purified.

In the case of artificial fertilizers, there is no such natural speed-limitation; farmers can distribute as much as they wish on their lands. Excess fertilizers can cause overstimulation of growth and while the farmers make a clear profit from the increased crop, quality usually suffers. The results are similar to a person who eats constantly. The wisest course of action is to use fertilizers only to replenish nutrients in soil without creating an overabundance of them.

2.3.2 Nitrates in Vegetables

Potassium nitrate (KNO_3, saltpeter) and sodium nitrate ($NaNO_3$, Chilean saltpeter) have been known for a long time, and used in explosives, fertilizers and preservatives. They also have a long history for curing meat products ($\rightarrow$ **2.8**). Nitrates themselves are not toxic. However, nitrites, which often form from nitrates, compete in an entirely different league. On average, about 5% of nitrates are transformed to nitrites in the bowels. Poor quality drinking water may also contain nitrates and this may be especially harmful for infants. It is not particularly beneficial for adults, either. Nitrite reacts with the oxygen-carrying molecule hemoglobin in blood: it oxidizes iron(II) in hemoglobin into iron(III). The resulting compound is called methemoglobin, since it is unable to carry oxygen, the body begins choking. Sometimes this appalling phenomenon is called bluebaby syndrome.

It seems much less publicized that certain vegetables are also significant sources of nitrates in the human diet. Serious research in this field began only a few years ago. Spinach is a prime suspect as its nitrate content may vary widely from 700 to 4000 mg/kg. Is it not ironic that spinach could actually be harmful for children? ($\rightarrow$ **2.11**) Whatever Popeye might say, children might have more common sense than their parents do. Table 2.1 present the nitrate contents of a few vegetables.

Table 2.1 Nitrate contents of some vegetables

Nitrate content (mg/kg)	Plant
Very high (>2500)	Beetroot, chard, radish, lettuce, spinach, chervil, celery, cress, salad rocket
High (1000–2500)	Fennel, endive, kohlrabi, Chinese cabbage, parsley, leek, celery root
Average (500–1000)	Dill, cabbage, Savoy cabbage, parsnip
Low (200–500)	Broccoli, chicory, cauliflower, turnip, squash, cucumber
Very low (<200)	Artichoke, bean, pea, potato, sweet potato, garlic, mushroom, watermelon, onion, eggplant, green pepper, tomato, melon, asparagus

The nitrate content of a vegetable depends on the season, and even upon the time of day, the quality of the soil, and the use of fertilizers. Boiling dissolves about 70–75% of the nitrate content, so it is advisable to discard the water used for boiling after the first 2 min. The highest value (9300 mg/kg) ever found was for salad rocket (rucola). In most European countries, laws have set a highest tolerable limit for the nitrate content of certain vegetables, e.g. they are in the range from 3000 to 4500 mg/kg for beetroot. The European Commission has set a legal limit of 200 mg/kg for the nitrate content of infant formula and food specifically intended for children. The WHO (World Health Organization) recommends an **allowable daily intake** (ADI) of 3.7 mg/kg of body weight/day, which translates into approximately 200 mg of nitrate for an adult. Consuming vegetables could well be a source of larger amounts. Even if this were the case, consuming large amounts of nitrates almost never happens on a regular basis, so it is not a cause for alarm.

To sum up: vegetables are often surprisingly rich in nitrates. If surplus fertilizers are used to increase the amount of nitrogen in soils, this may further exacerbate the problem. The solution is controversial: artificial fertilizers with reduced nitrogen content must be used, and natural fertilizers should also be applied with extra care. Needless to say, these precautions will decrease the yield of the vegetables.

2.3.3 Are Organic Vegetables More Nutritious than Others?

This question was carefully studied using **meta-analysis**, in which relevant experimental reports were re-evaluated using identical and very strict criteria. All scientific articles were excluded which did not describe reproducibility, analytical techniques or statistical methods in enough detail, or presented other reasons to doubt the reliability of the conclusions. Out of 318 studies reviewed, only 55 were deemed reliable. In these well-crafted studies, nitrogen, phosphorus, vitamin C, magnesium, calcium, potassium, zinc, copper and oil contents were compared during the analysis, and the bottom line was unambiguous: organically grown products were not significantly different in any way from non-organically grown. The nitrogen and phosphorus contents of organic vegetables were a few percent higher on average, but the scattering of measured data was much larger than this difference.

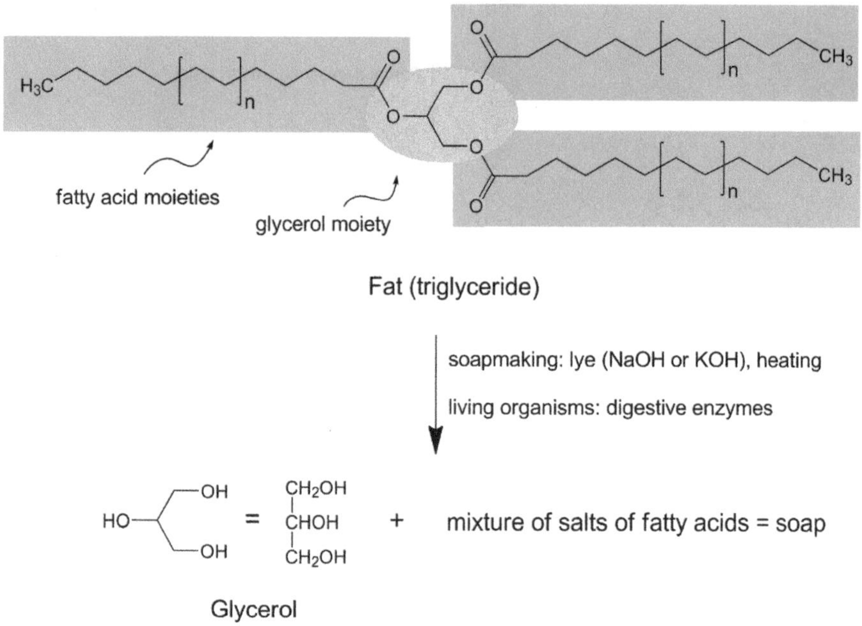

Fig. 2.6 Saponification of fats. (Authors' own work)

2.4 "You're the One for Me, Fatty": Concocted Fats

If we wish to start the story where it began, we should travel back in time to an Ancient Age when soap-making was first discovered. What is soap? If fats are boiled together with some aqueous alkali for a lengthy time, they will decompose. Chemically, they undergo alkaline hydrolysis (Fig. 2.6). Glycerol and soap are formed in this process: soap is a mixture of the sodium (or potassium) salt of different fatty acids. In an ever-popular novel, *The Mysterious Island* (*L'Île mystérieuse*), French science-fiction writer Jules Verne very nicely described that soap could be made even on a deserted island (the movies supposedly based on this novel seldom make a point of this part).

In a chemical sense, fats are triglycerides. This word may sound familiar to many as it is often used in the mass media. In blood tests, the level of triglycerides is often measured in addition to cholesterol. Triglycerides are compounds formed from glycerol (an alcohol with three hydroxyl groups) and fatty acids known as *esters*. They are fundamental substances in all living organisms, bacteria and humans alike.

This decomposition process is called saponification. It can be halted at intermediate stages to produce mono- and di-glycerides. These are compounds in which only one or two of the hydroxyl groups of glycerol form ester bonds (Fig. 2.7). In layman's terms: some of the fatty acids from the fat have been removed. In the human body, these intermediate compounds are formed when fats are broken down, and they are also formed when fat-containing ingredients are boiled for a long time

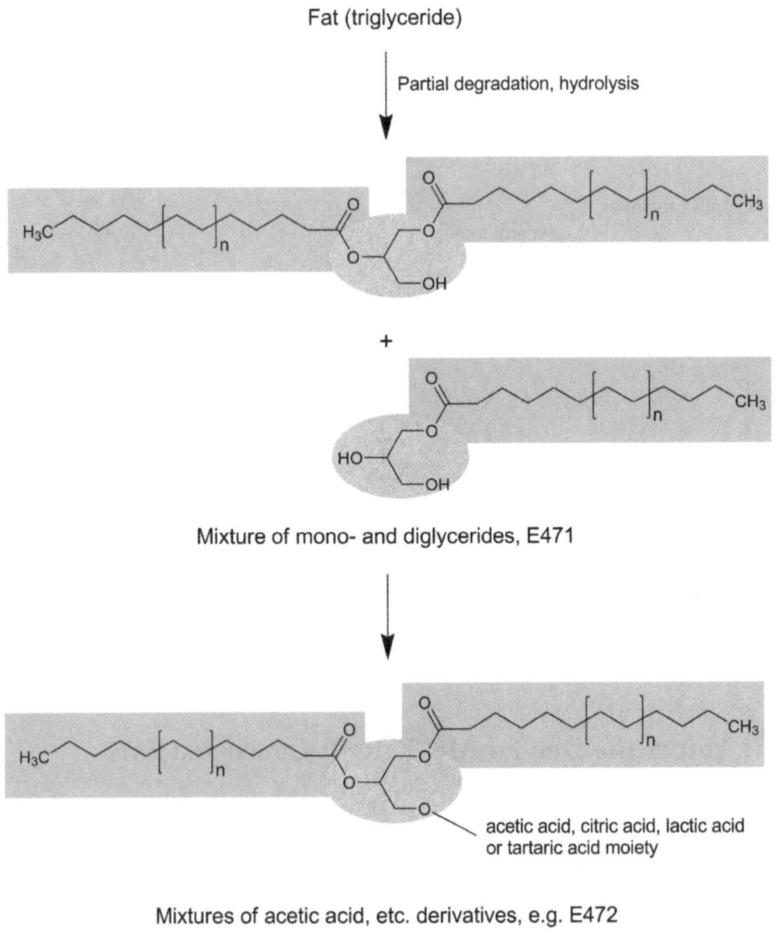

Fat (triglyceride)

↓ Partial degradation, hydrolysis

Mixture of mono- and diglycerides, E471

↓

acetic acid, citric acid, lactic acid or tartaric acid moiety

Mixtures of acetic acid, etc. derivatives, e.g. E472

Fig. 2.7 Structure of modified fats. (Authors' own work)

in the absence of alkalis. These intermediate compounds are common food additives used to blend together components of creams, shortenings, dressings, et al.

In the food additives of the E472 group, the free hydroxyl groups of glycerol are transformed back to esters, but simpler organic acids such as acetic acid, citric acid, lactic acid or tartaric acid (or their combination) are used instead of fatty acids. Sugars can also be chemically bonded to mono- and di-glycerides, these are registered as E474, and the series from E470 to E480 includes more substances of a similar chemical structure. These are mostly used as emulsifiers, stabilizers and texture modifiers in canned meat, bakery products or pudding. All mono- and di-glyceride derivatives are a type of semisynthetic fat derivatives, and they simply continue their natural breakdown process in the human body. Although they are not literally natural substances, they are comprised of natural "bricks" and have better properties than their natural counterparts. Since they smoothly take part in the natu-

Table 2.2 Food additives produced using partially hydrolyzed fats and other simple natural substances

EU Code	Name	Role	Comment
E470a E470b	Sodium, potassium and calcium salts of fatty acids Magnesium salts of fatty acids	Emulsifier stabilizer	Similar to soap
E471	Mono- and diglycerides of fatty acids	Emulsifier	Partially hydrolyzed fats
E472a	Acetic acid esters of mono- and diglycerides of fatty acids	Emulsifier, texture modifier	Compounds of mono- and diglycerides and acetic acid
E472b	Lactic acid esters of mono- and diglycerides of fatty acids	Emulsifier, texture modifier	Compounds of mono- and diglycerides and lactic acid
E472c	Citric acid esters of mono- and diglycerides of fatty acids	Emulsifier, texture modifier	Compounds of mono- and diglycerides and citric acid
E472d	Tartaric acid esters of mono- and diglycerides of fatty acids	Emulsifier, texture modifier	Compounds of mono- and diglycerides and tartaric acid
E472e	Mixed acetic and tartaric acid esters of mono- and diglycerides of fatty acids	Emulsifier, texture modifier	Compounds of mono- and diglycerides with tartaric or acetic acid
E473	Sucrose esters of fatty acids	Emulsifier	Compound of fatty acids and cane sugar
E474	Sucroglycerides	Emulsifier	Compound of fatty acids and cane sugar
E475	Polyglycerol esters of fatty acids	Emulsifier, texture modifier	Compound of polymerized glycerol with fatty acids
E477	Propane-1,2-Diol esters of fatty acids	Emulsifier	

ral decomposition process, any concern about them would also be a concern about natural processes. Some of these substances are listed in Table 2.2.

2.5 Fat Matters: Margarine vs. Butter

French emperor Napoleon III (1808–1873) liked to show off his social sensitivity. In 1869, he initiated the development of a substitute for butter which would be suitable for the Navy and low income people, had a pleasant smell, was inexpensive and would not turn rancid. The prize was awarded to French food chemist Hippolyte Mège-Mouriès (1817–1880) for his proposal of margarine. The name margarine was coined by another French chemist, Michel Eugène Chevreul (1786–1889), who synthesized a new substance from bacon in 1813 and called it margaric acid (from Greek μαργαρίτης, margarites meaning pearl-oyster or pearl). Later, it was discovered that Chevreul actually prepared a mixture of stearic acid and palmitic acid, which contain 16 and 18-membered carbon chains. Today, margaric acid is defined as a saturated fatty acid with 17 carbon atoms. This only forms about 1% of ruminant fat.

The margarine produced by Mège-Mouriès, however, was not quite what we call margarine today. He tried to mimic the butter making process by using milk, chopped cow udder tissue and the low melting part of cow tallow. He patented the product and the process, but interference from the Franco-Prussian War convinced him to sell the rights to two Dutch butter sellers, Anton Paul Jurgens and Simon van der Bergh. They founded the first margarine plant in the world between 1871 and 1873 in the Brabant territory of the Netherlands. Their efforts met with commercial success, with 46 plants in business in Germany by 1885. The composition of margarine has changed quite a lot since then, as animal fat has been replaced by plant-derived oils. One novel development was the introduction of hardening, the basis of which was developed by Paul Sabatier (1854–1941) in 1897. He used finely dispersed nickel to catalyze the hydrogenation reaction which turns unsaturated compounds into saturated ones. This method was industrialized by Wilhelm Normann in 1902, and further developed to utilize several plant and animal oils such as sesame oil, coconut oil, soybean oil, peanut oil and fish oil.

Liquid oils change into solid and half-solid fats upon hydrogenation, which are ideal for making margarine. Similar to butter, margarine is a water-in-oil emulsion, in which water drops are dispersed in oil using emulsifiers such as lecithin. The fat content is now about 80% of previous products. Modern margarines contain other additives as well as acids, vitamins, preservatives, dyes and antioxidants. In addition to low costs and successful marketing efforts, the key to margarine's rise to prominence is its favorable properties for large scale use and its increased resistance to oxidation compared to butter. Products containing hydrogenated fats preserve their original flavor for a longer time.

Ghee is another way to preserve butter: this is made from the milk of domesticated Indian water buffalos (*Bubalus bubalis* or *Bubalus bubalus,* earlier *Bos bubalus*) or cows. Ghee production removes all components from butter that go rancid quickly. Boiling the butter removes most of the water. Milk proteins and milk sugars are deposited during the treatment: the protein forms a white layer that sinks, and foam forms on the top. Removing the foam and leaving the bottom behind gives ghee, which is a much purer fat than butter.

Consumption of butter became an indicator of the well-being of a society and increased rapidly at the beginning of the twentieth century. World production was 200,000 t in 1897 and 470,000 t in 1913. Margarine, a replacement for more expensive butter, also made progress in this period: consumption increased from 90,000 to 200,000 t, with Germany being responsible for about a third of this amount. Margarine was not welcomed equally in all parts of the world. There was strong opposition to it in the United States. The state of Wisconsin only lifted restrictions on selling margarine in 1967! During World War II, when butter was hard to come by in Europe, the market share of vitamin-loaded margarine made substantial progress. In 2004, world consumption of margarine was 7 million t, and for butter, 8 million t. The 27 members of the European Union manufactured 1.8 million t of margarine and 2.1 million t of butter in 2008. In the same year, annual per capita consumption

2.5 Fat Matters: Margarine vs. Butter

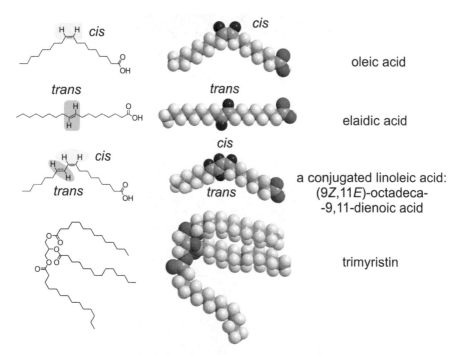

Fig. 2.8 The simplified chemical structures and 3D models of selected unsaturated fatty acids and a triglyceride. Hydrogens in *cis* and *trans* positions are *highlighted*. (Authors' own work)

of margarine was 4.33 kg, which shows a slight declining trend from the year 2000, when the value was 5.82 kg.

In order to choose between butter and margarine, some information about the differing chemical composition is probably helpful. About 99 % of natural fats and oils are usually triglycerides, which are esters of glycerol with long chain fatty acids (→ 2.4, 2.13). Saturated and unsaturated fats are both present. In the latter, there are double carbon-carbon bonds in the chain of fatty acids (Fig. 2.8). Another aspect focuses on omega fatty acids (→ 2.13).

Saturated fats are mostly (but not exclusively) found in food prepared from animals. Unsaturated fats are more abundant in plant-derived oils. The ratio of saturated and unsaturated fatty acids is the key determinant of the physical properties of oils and fats. Triglycerides with a higher content of unsaturated fatty acids have lower melting points, these are called oils. On the other hand, triglycerides with a greater quantity of saturated fatty acids melt at higher temperatures, are almost solid at room temperature, and are called fats. Fats and oils are seldom pure substances, which complicates matters further. Depending on their origin, a number of different fatty acids may form ester bonds with glycerol, and these molecules can only be understood using statistical averages. For example, milk lipids contain about 400 different fatty acids, 96–98 % of which are found in 43 main triglycerides. Fats and oils often occur in different crystalline forms. Cocoa butter is present in six different

Fig. 2.9 Premium quality chocolate. (Authors' own work)

modifications, of which the one designated β-3 is the most valuable, as it melts at 33.8 °C, and produces a very pleasant feeling in the mouth. Highly ranked chocolate manufacturers usually indicate the cocoa content on their product as this also guarantees the cocoa butter content (Fig. 2.9).

The double carbon-carbon bond of saturated fatty acids results in the possible formation of two varieties (isomers): in this *cis* isomer, hydrogen atoms are adjacent, whereas they are on opposite sides in *trans* isomers. *Cis* fatty acids have a bent molecular shape, *trans* fatty acids are more linear, and similar to saturated fatty acids. Fatty acids containing multiple double bonds also occur in fats and oils, and these offer *cis-trans* isomer possibilities at each double bond. The difference in the shape of fatty acids in different triglycerides influences crystallization properties and possible roles played in biological membranes. Food scientists agree that unsaturated fats have more favorable properties for human consumption than saturated fats (→**2.13**). Oils, on the other hand, are more sensitive to oxidation in air and turn rancid faster. Vitamins A and E in margarine are partly there to slow down rancidification.

The oil hardening process used in margarine manufacturing involves hydrogenation of unsaturated fatty acids at a pressure of 2–10 atmospheres, temperature of 160–220 °C, and the presence of 0.05–0.1 % nickel catalysts. Depending on the number of double bonds, different (partially) saturated triglycerides form in this process. *Cis-trans* isomerization and double bond movement are also possible under these conditions, which may give rise to 5–40 % of *trans*-fatty acids.

Trans-fatty acids also occur in nature, primarily as a result of bacterial action in the stomach of ruminants. Veal fat contains 2–9 % *trans*-fats. The main sources of *trans*-fats in the human diet are fast food, cooking oil and margarine. About two thirds of the *trans*-fatty acids in *trans*-fats is elaidic acid (Fig. 2.8).

The biological role of *trans*-fats was almost unexplored before the last decade. Such studies were hindered by a lack of suitable analytical techniques and the dearth of information on eating habits. Scientific results show that *trans*-fats do not influence the growth of the human body, reproduction, breast and colon cancers significantly, but they were a risk factor in cardiovascular diseases. This increased risk is caused by the increasing amount of low density lipoprotein (LDL) and a simultaneous decrease in the amount of high density lipoprotein (HDL). The 18-carbon atom monounsaturated fatty acid increases the levels of LDL-cholesterol ("bad cholesterol") and decreases the amount of HDL cholesterol ("good cholesterol"). The effects of the consumption of *trans*-fatty acids were tested in an 8-year experiment focusing on 80,000 nurses in the US. A 50 % difference was found between the groups with the highest and smallest risk of the studied population. About 3.2 % of the diet of the former group was *trans*-fats, compared to 1.3 % for the former. Statistical analysis established a connection between the increased risk of health problems and the consumption of *trans*-fats. Since the difference between the two groups in terms of actual quantity consumed was very small (3.3 g), experts have questioned the conclusive link.

Different countries have different attitudes when recommending a balanced diet. However, there is agreement about encouraging people to limit fat and oil consumption to 30–35 % percent of total calories. For saturated fats, a maximum of 10 % is proposed. As the consumption of saturated fatty acids is 5–7 times greater than that of *trans*-fatty acids, the significance of the latter should not be overestimated. To prevent cardiovascular diseases, saturated fats must be avoided, and *trans*-fatty acids are only a secondary concern. Preventing the formation of *trans*-fatty acids during food production is, of course, the responsibility of the manufacturer.

The concerns about *trans*-fatty acids have influenced methods of margarine production. Previously, the content of *trans*-fatty acids often reached 9 %. While today, the desirable quantity is below 1.5 % or even 0.5 %. This reduction can be achieved through a modification in technology. Hydrogenation in the modern process is followed by transesterification with *cis*-fatty acids in the presence of sodium methylate. There is an enzymatic technology available as well, which is suitable for infant formula production, or fats used as a replacement for cocoa butter. The US Food and Drug Administration requires that food labels show the amount of *trans*-fats in any product. This regulation has driven further improvements in technology. In modern margarine, the use of *trans*-fatty acid containing triglycerides obtained by hydrogenation was entirely phased out, and instead, plant oils and fats are emulsified with water. This change was also accompanied by a decrease of fat content. Earlier margarines contained about 80% of fat, whereas today this is typically between 40 and 60%. Margarine containing as little as 20% of fat can also be found on the shelves of supermarkets. What is then the remaining 80% in the latter product? Well, that's largely water, dear reader....

There is one group of *trans*-fatty acids that deserves special attention because of its beneficial health effects. Conjugated linoleic acids (CLA) are 18-carbon fatty acids with two double bonds with no methylene (CH_2) groups between the double bonds. This motif is different from other polyunsaturated fatty acids. The two double bonds are most often found in the *cis*-9 and *trans*-11 position (Fig. 2.8). CLA typically occurs in the fat of ruminants. Cow milk contains 0.3–3% of CLA, and the highest levels were found in kangaroo meat. The anticancer effect of conjugated linoleic acids was first noted by Michael Pariza of the University of Wisconsin in 1979 in a series of experiments on mice. CLA is also thought to lower the risk of cardiovascular diseases and inflammation. These are all results from animal experiments, and none of these effects have been proven in humans.

CLA is already used in dietary supplements because of another assumed effect: it decreases the amount of fat tissues and increases the growth of muscles. Thirty, partially contradicting studies, were reviewed in 2007 to test this effect. It was determined that there was miniscule weight-loss as a result of consuming 3.2 g of CLA a day. CLA is not toxic, and FDA approved its use in certain selected products (yoghurts, granola bars, fruit juices).

The choice between butter and margarine is primarily a question of taste and price. Concern about *trans*-fats convinced margarine producers to alter their manufacturing technology, so the risk from consumption is no longer relevant. The primary focus is fat content for both butter and margarine. When choosing a brand of margarine, it is useful to check the amount of *trans*-fats on the label. The healthiest margarines are those that do not contain any hydrogenated oils, and which are produced by the emulsification of plant oils and fats with water.

2.6 Fake Food and Kidney Stones

Milk powder tainted with melamine affected the lives of about 300,000 Chinese infants in 2008. Six thousand of these babies became seriously ill, and 150 suffered kidney failure, resulting in 6 fatalities. This was one of the worst modern examples of counterfeit or adulterated food (→ **2.25**). Melamine was added to milk to make it appear more nutritious.

Four Chinese companies sold melamine-tainted milk, and about 20 companies re-processed it. The whole case began long before it was discovered by the press. Zhang Yujun was charged with manufacturing more than 700 t of melamine for the counterfeiters and sentenced to death along with Geng Jinping and Gao Junjie, who assisted in preparing the tainted milk. Tian Wenhua, the leader of the company selling the products, received life imprisonment. Few thought this sentence was strict enough for this appalling crime. The tampering was discovered when a company in New Zealand found melamine in Chinese products it was importing, and 16 children were diagnosed with kidney stones in the Chinese province of Gansu.

The counterfeiters tried to capitalize on the widespread use of a test developed by Danish chemist Johan Kjeldahl (1849–1900) to determine the protein content of food, especially milk. With this method, nitrogen within proteins is transformed into ammonia (ammonium sulfate), and most nitrogen-containing substances can be detected *independently* of the nitrogen's origin. Melamine has a substantial nitrogen content (67%), so adding a small amount of melamine makes it appear that the food product contains more protein. Another method for nitrogen measurement was developed by French chemist Jean-Baptiste André Dumas (1800–1884). His process transforms the nitrogen content of compounds into nitrogen gas. Dumas' method is faster than Kjeldahl's, but more expensive, and is also unable to determine the nitrogen's origin. Since the milk tainting scandal, three new methods have been developed to test for melamine. The first uses electrospray mass spectrometry and can detect melamine promptly. A second method developed in China uses the interaction of luminal with melamine to cause blue light emission. The third method, developed by researchers at the British food analysis company RSSL, is based on a technique called ELISA, and uses antibodies labeled with fluorescent dyes.

People of Chinese descent cannot usually consume milk products as their bodies are unable to digest milk sugar (lactose). This phenomenon is called lactose intolerance and a common symptom after consumption is diarrhea. However, many people in China drink milk despite the prevalence of lactose intolerance, and since drinking milk can be considered a mark of prestige, milk consumption has been on the rise in modern China. Milk produced in China is used in the production of chocolate, cookies and confectionery, and the melamine scandal caused a major public outcry all over the world.

Mela*m*ine is not to be confused with mela*n*ine, which is a dark pigment in skin and hair. Mela*n*ine is formed from an amino acid called tyrosine, and it protects against the harmful effects of sunlight. This pigment is absent in the bodies of albinos. Mela*m*ine (Fig. 2.10), on the other hand, is an important raw material in the production of plastics. Melamine-formaldehyde resins have excellent mechanical and other properties (they are colorless, good insulators), and their use is very common in some household products such as plates, bowls, mugs and special papers. A compound closely related to melamine is cyanuric acid (Fig. 2.10), which is one of the longest known organic substances. Swedish chemist Carl Wilhelm Scheele (1742–1786) discovered it in 1776. Cyanuric acid and its derivatives are important

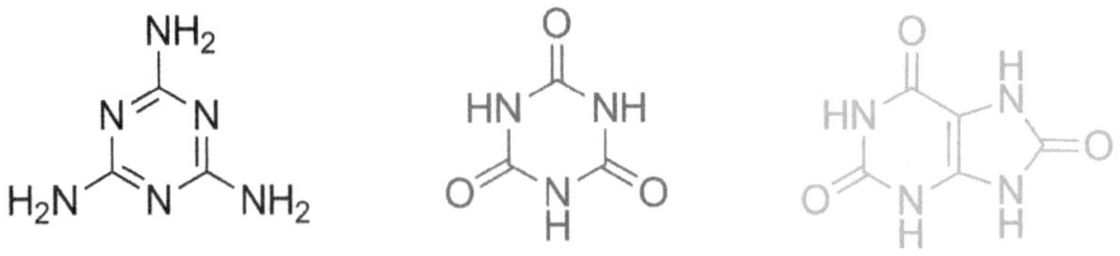

Fig. 2.10 Chemical structures of melamine, cyanuric acid and uric acid. (Authors' own work)

in the chemical industry: they are used in the production of herbicides, dyes, and disinfecting agents for swimming pools ($\rightarrow$ **4.1**). Uric acid (Fig. 2.10) is also a related compound, and was also discovered by Scheele in 1776 (although Bergman had his own claim to this discovery). Uric acid is one of the most important substances in urine.

Melamine is not particularly toxic, and its effect in rats is comparable to that of table salt: the **LD$_{50}$** in rats for ingestion is 3161 mg/kg of body weight, (with other sources showing 6000 mg/kg of body weight), data for humans could not be found. Melamine is excreted from blood plasma with a **half-life** of 3 h. Large quantities of melamine cause kidney stone formation, inflammation of the urinary bladder, a type of cell proliferation called hyperplasia, and, in male rats, urinary bladder cancer. A dose of 63 mg/kg of body weight per day (**NOAEL**, no observed adverse effect level) does not increase the risk of kidney stone in rats.

The properties of cyanuric acid are similar to melamine, with its LD$_{50}$ for ingestion in rats being 7700 mg/kg of body weight. From the human body, 98% of cyanuric acid is excreted within 24 h. Cyanuric acid causes kidney damage in animals, and the NOAEL value was found to be 150 mg/kg of body weight per day.

The 2008 melamine scandal was preceded by the death of about 1000 cats and dogs in the US, which was caused by pet food tainted with melamine and cyanuric acid. It was unclear if cyanuric acid was deliberately added to pet food, or it was a simply an impurity in low quality melamine. Individually, these substances are not very toxic, but the risk of kidney stones is much higher when the two compounds act together. The investigation in these cases showed that the stones were made of a water-insoluble mixture of melamine and cyanuric acid. It is assumed that melamine and cyanuric acid are absorbed by the body in the bowels, and they combine in the kidneys to form the stones. The details of this process are unknown. Melamine and cyanuric acid are held together by strong secondary chemical bonds called hydrogen bonds, which also accounts for the poor water-solubility of the mixture (Fig. 2.11).

Sanlu brand powdered milk was found to contain significant amounts of melamine, with the highest concentration being 2563 mg/kg of product (0.2%). The kidney stones in patients suffering from melamine were not melamine-cyanuric acid, but melamine and uric acid in 1:1.2–2.1 ratios (Fig. 2.11). Natural formation of uric acid kidney stones is not unknown, but they constitute only 5–10% of all kidney stones, especially in developing countries. Babies excrete uric acid in much

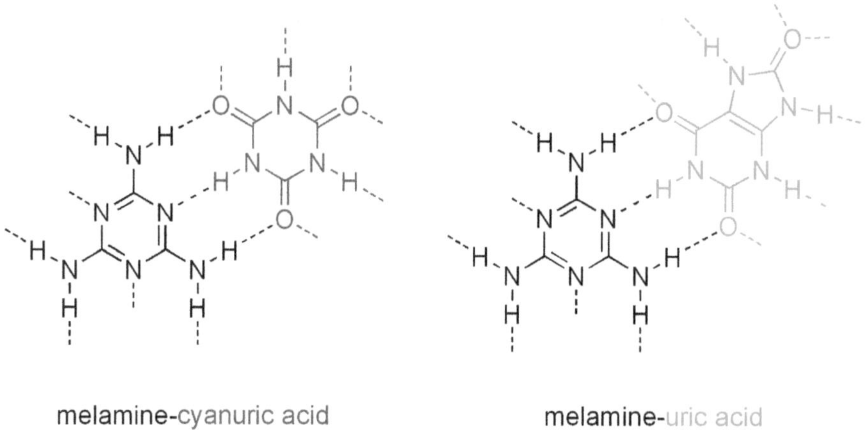

melamine-cyanuric acid   **melamine-uric acid**

Fig. 2.11 Structure of crystals of mixed melamine-cyanuric acid and melamine-uric acid. (Authors' own work)

larger amounts than children or adults do, and this was the main reason why most of the patients in the melamine scandal were very young (6–36 months). If kidney stone formation is diagnosed early, its further development is relatively easy to prevent. Stone formation occurs under acidic conditions (below pH 5.5) and treatment with the sodium salt of citric acid or sodium hydrogen carbonate (a constituent in baking powder, →2.18) can raise the pH to its normal value of 6.5–7.0, causing the stones to dissolve. Larger kidney stones, unfortunately, can only be treated by surgery.

The European Food Safety Authority (EFSA) issued a statement to consumers that they (especially children) do not face dangers if they do not consume large amounts of tainted food. The World Health Organization (WHO) says that the human body tolerates melamine in a dose of 0.2 mg/body weight a day, whereas the US Food and Drug Administration considers 0.063 mg/body weight a day harmless. Considering the fast excretion of melamine, these values almost certainly err on the side of safety and were devised to reassure consumers rather than reflecting scientific facts. Pre-scandal limits from the same organizations were much higher (WHO: 0.5 mg/body weight a day, FDA: 0.63 mg/body weight a day). The problem was not the legal limit, but the willful, and negligent, non-adherence.

2.7 The Coming Shortage: Vanilla and Menthol

The IgNobel award is a parody of the real Nobel Prize. Scientists are awarded the IgNobel for meaningless discoveries, useless inventions or simply funny ideas. The 2007 IgNobel prize in chemistry was awarded to Japanese scientist Mayu Yamamoto (International Medical Center of Japan) for developing a way to extract vanillin—vanilla fragrance and flavoring—from cow dung. The winner performed contract research for Sekisui Chemical Company, and he converted lignin-rich cow manure into vanillin using a chemical reaction (not extraction, as the official announcement implied). Lignin-based vanillin production has been known and used for decades to create synthetic vanillin, using a by-product of the paper industry as raw material.

This IgNobel prize was regarded as funny. Only a few researchers felt that there was something amiss with the award and the reasoning behind it. Many materials originate from re-processing waste from other industries. Drinking water is regularly obtained by purifying waste water. Japan has a very high population density and special environmental problems, and in this case, an attempt to solve one of these problems was seriously misrepresented.

By now it is clear that vanillin can be made from lignin, but why is synthetic vanillin needed at all? The answer is quite simple: there isn't enough natural vanillin to go round. The world currently consumes about 12,000 t of vanillin each year, only 1800 t of which is produced in vanilla plants (e.g. *Vanilla planifolia*). Vanilla originates in Central America, and when Hernán Cortés conquered the Aztec Empire, vanilla was already in use to flavor chocolate. Vanilla and cocoa arrived together in Europe in 1528 (→ **2.22**).

Vanilla is a demanding plant to cultivate (Fig. 2.12). A humid tropical climate and an elevation about 1500 m above sea level are essential along with soil rich in organic material. The vanilla plant could only be grown in Central America for a considerable length of time because its pollinator species of stingless *Melipona* bees (and maybe a few species of humming birds) exist there. Today, Madagascar is the world leader in the vanilla business with about 58% of the market share (2006). In Madagascar, vanilla is still pollinated manually, following the procedure developed by Belgian botanist Charles François Antoine Morren (1807–1858) in 1836. A single worker can pollinate about 1000 flowers a day, so the 1800 t of natural production actually requires a good deal of human labor. The vanilla bean is 10–20 cm long and does not contain vanillin, which is the main flavor of vanilla sticks. Vanillin is obtained from its precursor vanillin glucoside in a sophisticated months-long process. Vanillin crystallizes on the surface of the bean during this curing process. The cured bean is black and contains about 2% of vanillin in addition to about 20 other components. Vanillin is used in the food, cosmetic and chemical industries. Synthetic vanillin was first produced in 1874 from coniferin, 16 years after its chemical structure was discovered. The most common production method today starts with guaiacol. In addition to synthetic vanillin, which is identical to the flavor causing compound of natural vanilla, ethyl vanillin is also used. It has one more carbon atom than the original and produces a more intense flavor.

Another favorite substance for medicine and cosmetics is menthol. It is obtained from a number of herbs, peppermint (*Mentha × piperita L.*) being the most common source (Fig. 2.13). The peppermint plant contains about 1–3% essential oil, which can be obtained by a chemical separation method called steam distillation. About half of the oil is menthol and its derivatives, terpenes, and **flavonoid**s. Menthol has a pleasant odor and cooling effect, and these two properties are why

2.7 The Coming Shortage: Vanilla and Menthol

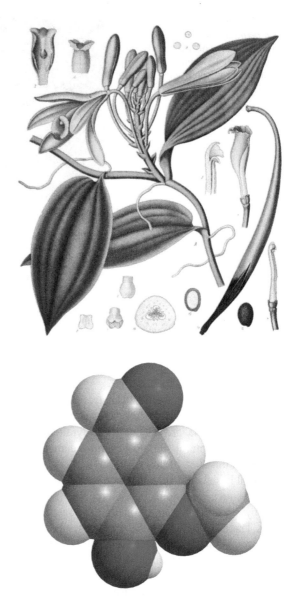

Fig. 2.12 Vanilla (*Vanilla planifolia*), a plant in the family *Orchidaceae* together with the 3D structures of vanillin (*light grey balls* represent hydrogen atoms, *mid-grey balls* carbon atoms, *red balls* oxygen atoms). (Authors' own work and copyright-free Wikipedia picture)

it is used on such a large scale in toothpaste and perfume production. Peppermint oil and menthol are used in medicine to treat spasms of the gall bladder and the intestines, gall stones, digestive problems, itches, blocked noses, headaches and as an expectorant. These substances, which have been studied in detail, relax abdominal muscles and desensitize abdominal nerves, primarily due to the menthol in the oils. Peppermint oil has no known negative effect when used in the usually prescribed amounts.

Fig. 2.13 Peppermint (*Mentha × piperita L.*) and the simplified structure of "natural" (−)-menthol. (Authors' own work and copyright-free Wikipedia picture)

(-)-menthol

Similarly to vanillin, the menthol needs of the world cannot be met from natural sources. However, the synthesis of menthol is less critical, as menthol-producing plants, unlike vanilla, are easy to grow. Synthetic (−)-menthol, identical to the substance found in natural sources but more demanding to produce than vanillin, accounts for about 30% of the world production, which is 12,000 t a year.

Menthol has eight different varieties called optical isomers. These isomers have the same sequence of atoms bonded together, but their shapes are different. This is caused by the different relative spatial arrangements of atom groups around certain centers. There are three such centers in menthol each offering two possibilities for spatial arrangements, which makes $2 \times 2 \times 2 = 8$ different molecules. Out of these eight, one dominates in nature. This variety rotates the plane of polarized light to the left, which is the origin of the notation (−)-menthol, and is simply called menthol, although some of the remaining isomers are occasionally found in nature.

It is the existence of the eight menthol isomers that makes the chemical synthesis onerous. In many procedures, more than one isomers form as a final product, and needs to be separated out at the end. These byproducts, the isomers of (−)-menthol, all have cooling and refreshing effects, but none of them is as potent as menthol itself. So the presence of these isomers lowers the quality of the final product. Another option is to look for a chemical reaction that produces (−)-menthol exclusively. This was achieved from meta-cresol by the company Haarman and Reimer, which is also a major vanillin producer, whereas a Japanese chemical company, Takasago, begins its own method from (+)-citronellal or a terpene called mircene. The Japanese company alone processes 1500 t of (+)-citronellal to produce (−)-menthol.

The histories are vanillin and menthol are, in one respect, the same old stories that keep appearing in the history of humankind. There is always something that is in short supply: raw material, fuel, food, luxuries *etc*. Human ingenuity then strives to produce them by transforming other, more readily available resources.

2.8 Red Alert: Meat Colors

Swiss art educator Johannes Itten tells this story in his noted book titled, *The Art of Color*: "A businessman was entertaining a party of ladies and gentlemen at dinner. The arriving guests were greeted by delicious smells issuing from the kitchen, and all were eagerly anticipating the meal. As the happy company assembled about the table, laden with good things, the host flooded the apartment with red light. The meat looked rare and appetizing enough, but the spinach turned black and the potatoes were bright red. Before the guests had recovered from their astonishment, the red light changed to blue, the roast assumed an aspect of putrefaction and the potatoes looked moldy. All the diners lost their appetite at once; but when the yellow light was turned on, transforming the claret into castor oil and the company into living cadavers, some of the more delicate ladies hastily rose and left the room. No one could think of eating, though all present knew it was only a change in the color the lighting. The host laughingly turned the white lights on again, and soon the good spirits of the gathering were restored." (Translated by Ernst van Haagen)

Whether we like are not, colors undoubtedly affect us very deeply. Shades of warm red color are commonly associated with liveliness, dynamism, strength, and aggression. Itten is correct to point out that "red can express all intermediate degrees between the infernal and the sublime."

The color red is most often visible in meats, fruits, vegetables and anything produced from them and these are sold on a vast scale, tens of billions of tons annually. The most common artificial dyes in the food industry are colored red and yellow, which clearly demonstrates their importance.

What is the origin of the red color of raw and cured meat? Blood in the muscle tissues would be a logical answer. Poultry meat has little blood in it, so it is pale, and red meat is often referred to as "bloody". However, the blood in animals is only 3–10% of the body weight, so proper butchering leaves very little blood in the meat. Blood in raw meat has also been known to turn black after some time or upon heating. Fresh beef or pork becomes more red where it is cut. Uncured meat loses its bright red color during cooking and turns brown, whereas cured meat retains its natural red color.

The red color of blood is caused by a substance called hemoglobin. This molecule consists of *heme*, which contains iron ions, and a protein called *globin*. In muscle tissue, a substance similar to hemoglobin is called myoglobin. It is no more than 1% of the weight of the dry meat. Both hemoglobin and myoglobin serve as

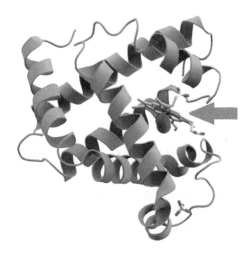

Fig. 2.14 Schematic structure of the myoglobin of the sperm whale. The *arrow* indicates the wireframe model of heme with the attached oxygen molecule (*red balls*), the *blue ribbons* represent the globin protein. (Authors' own work and copyright-free Wikipedia picture)

oxygen carrying molecules: hemoglobin picks up oxygen in the lungs, then transports it to myoglobin, which delivers it to the final destination in the cells.

A common feature of the chemical structure of the two substances is the iron-containing heme, while the difference is in the protein part. Both myoglobin and hemoglobin contain divalent iron, also called the iron(II) ion in the heme part. Hemoglobin has four protein subunits, each of which is built up of 141 or 146 amino acid building blocks. Each of these subunits binds a heme group as well. Myoglobin only contains a single heme, and the protein part is built up of 153 amino acid units in both humans and sperm whales (Fig. 2.14). In living animals, about 10 % of all iron occurs in myoglobin, whereas in properly butchered beef, this is 95 %. Since hemoglobin does not contribute much to the red color of meat, it is the myoglobin and the ambient light which are the main culprits (the more scattered the light is, the less intense the red color).

Myoglobin is a purple substance, and can only exist in its native form if there is a scarcity of oxygen. In the presence of oxygen, bright red oxymyoglobin is formed (MbO_2), and this is what causes the red color of meat. One noticeable feature of oxymyoglobin is the fact that it loses its color 50 times faster at room temperature than at 0 °C. Oxidation by air, turns it into brown metmyoglobin (MMb^+) containing iron(III) ions instead of the original iron(II), and is unable to bind oxygen. The same is true for methemoglobin, the formation of which causes a syndrome called methemoglobinemia. Metmyoglobin and oxymyoglobin are both transformed into denatured metmyoglobin upon heating (Fig. 2.15).

When raw meat is cut, a number of chemical processes are initiated. One of these is an increase in acidity, which also affects the stability of myoglobin. Below pH 5, myoglobin begins to decompose into heme and globin, and heme is readily oxidized further into hemin (Fig. 2.15).

Due to the processes described above, the food industry goes to considerable lengths to maintain the quality of its meat products both in terms of freshness (microbiological state) and color, which greatly impacts the costumer's decision mak-

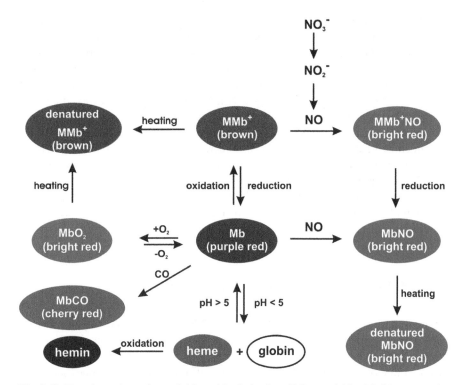

Fig. 2.15 Transformations of myoglobin and its derivatives (*Mb* myoglobin, *MMb*$^+$ metmyoglobin, *MbO$_2$* oxymyoglobin, *MbCO* carboxymyoglobin, *MbNO* nitrosomyoglobin, *MMb*$^+$*NO* nitrosometmyoglobin). (Authors' own work)

ing process. Raw meat is often kept under protective gas during transportation and storage to prevent color changes and spoilage. Carbon dioxide does the best job in inhibiting bacterial growth (mixtures containing 60% carbon dioxide and 40% nitrogen are used), whereas argon is not very efficient for this purpose. The red color of meat fades in oxygen-rich mixtures rapidly and the smell also becomes unpleasant. The stink is caused mostly by bacteria, *Brochotrix thermosphacta* or *Pseudomonas*, in minced meat.

The use of carbon monoxide seems like a bizarre idea at first as it is a well-known toxic gas. However, the use of a gas mixture containing 0.4% carbon monoxide, 60% carbon dioxide and 39.6% nitrogen poses no health risk; human exposure to carbon monoxide from transportation is much greater. Raw meat conserves its appealing color longest in this protective gas mixture and the growth of bacteria is also inhibited, because myoglobin is transformed into carboxymyoglobin (MbCO), which resists decomposition. This method has been used in Norway for two decades, and it was introduced in the USA in 2004. The European Union at present does not allow the use of this procedure as the meat retains its red color even if it is no longer fresh. Consumers in Norway and the USA are probably more used to smelling meat in addition to looking at it before purchase.

A less sophisticated method conserves the color of cured meat. Nitrates and nitrites in the curing salt oxidize myoglobin into metmyoglobin ($Mb + 2H^+ + NO_2^- \rightarrow MMb^+ + H_2O + NO$). The nitric oxide (NO) produced in this reaction preserves the red color by giving a red-colored adduct (assembly of molecules) with both myoglobin (MbNO) and metmyoglobin (MMb^+NO). MbNO is still red when it is denatured by heat treatment (Fig. 2.15). Reducing agents such as vitamin C assist in the formation of the red color by transforming nitrites into nitric oxide and metmyoglobin into myoglobin. The use of nitrites in curing procedures is condemned from time to time because carcinogenic nitrosamines are formed. However, even though their carcinogenicity was demonstrated beyond reasonable doubt in animal testing, nitrosamines decompose readily, so human exposure in a normal diet is low. Under controlled conditions (e.g. addition of vitamin C, which inhibits nitrosamine formation) the risks can be minimized. No one has yet invented an alternative to using nitrites in the curing process.

2.9 Moldy Business: Whole-Grain Cereals

Changes in eating habits occur all the time (→ 2.25). Today, it is trendy to eat products of whole grain flour. As its name implies, whole grain cereal contains the entire grain: germ, endosperm, and bran. Bran contains a lot fibers and minerals. The dietary value of fibers is not known today in sufficient detail but it does seem that they have multiple beneficial effects: they improve gut movements, increase the **glycemic index** of carbohydrates, make carcinogenic substances in food less available for the body and provide a growth medium for gut bacteria (*Lactobacterium* and *Bifidobacterium* strains).

It is not only the human diet that changes every now and then, the feed of domestic animals is also under somewhat capricious human influence. Bran was an important animal food in earlier years, but artificially prepared mixtures have gradually replaced it. Fiber-rich bran needed a new use. The marketing experts of mills took notice of the beneficial health effects of fibers and began advertising this fact heavily. The result was less bran for animals, more for people. The new-found popularity of whole grain cereal is a testament to their proficiency.

Every bean has its black, though: whole grain cereal and bran is also a source of danger from mold. Mold formation in plants is a problem long known in agriculture, it may adversely affect crop yields and the value of product. Animals usually refuse to eat moldy feed, and if they do, they often get sick or even die. In some examples, regular consumption of moldy food led to very serious human illnesses such as **ergotism**, whose first documented case occurred in 857 AD in the Rhine Valley. The reason is that some molds produce quite effective poisons called mycotoxins. Mycotoxins are among the most dangerous natural substances in food (→ **1.3**). Aflatoxins, ochratoxins and *Fusarium* toxins (4-deoxynivalenol, fumonisins, zearalenone) are the most hazardous mycotoxins (Fig. 2.16).

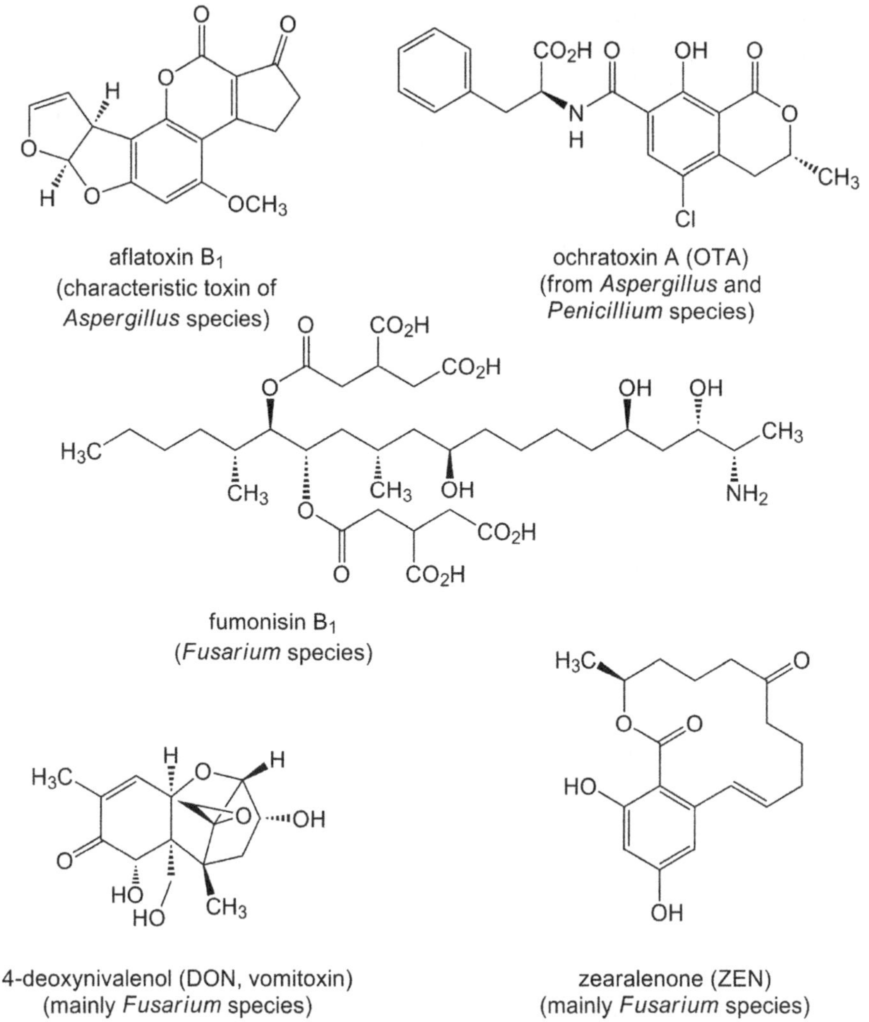

2.9 Moldy Business: Whole-Grain Cereals

aflatoxin B$_1$
(characteristic toxin of *Aspergillus* species)

ochratoxin A (OTA)
(from *Aspergillus* and *Penicillium* species)

fumonisin B$_1$
(*Fusarium* species)

4-deoxynivalenol (DON, vomitoxin)
(mainly *Fusarium* species)

zearalenone (ZEN)
(mainly *Fusarium* species)

Fig. 2.16 Chemical structures of the most common mycotoxins. (Authors' own work)

Aflatoxins are very effective poisons produced by mold of the *Aspergillus* genus. Their primary target is the liver, they harm the genetic materials in the body (genotoxicity), and also increase the risk of cancer (carcinogenicity). Consumption of highly tainted food may lead to illness and death in a short time. Lower concentrations are also carcinogenic, immunosuppressive, and harmful for the liver and the genes. Aflatoxins are not formed under moderate climate, their appearance can only be a problem in food exported from tropical countries. Aflatoxin B1, B2, G1, and G2 are the most significant with B1 being the most toxic. The products causing significant concern in human and animal health need continuous monitoring. These products are oilseeds (peanuts, sunflower seed, pistachio, other nuts), cereals, corn, soy beans, rice, dried fruits and spices (e.g. chili pepper). Aflatoxins are heat stable, they are not destroyed by cooking, only ultraviolet light decomposes them.

Ochratoxins are mycotoxins produced by molds in the *Penicillium* and *Aspergillus* genera. The most important substance in this group is ochratoxin A (OTA). OTA is highly toxic. It causes kidney damage, but is also harmful for the liver, the nervous and the immune systems. Animal tests showed it to be carcinogenic and harmful for fetuses (teratogenic). Ochratoxin A has been linked to kidney disease in humans, too. The presence of OTA has been detected in food worldwide, mainly in cereal or derived products, legumes, coffee, beer, grape juice, raisins, wine, cocoa, peanut and spices. It was also detected in blood and chitterlings. It is produced under moderate climate, so it needs attention in many countries of the world. The European Union sets limits for cereals, dried grape, roasted coffee, wine, grape juice, food for infants and babies. These are the main possible sources of OTA in the human diet.

Fusarium mycotoxins occur regularly in Europe, and are mainly produced by *Fusarium* fungi (*Fusarium graminearum* and *Fusarium culmorum*). A few of these *Fusarium* fungi produce toxins, others do not. The production of toxins continues after harvest if the product is not treated or dried appropriately. Toxins can enter the human body through the consumption of grains (corn and wheat) and products made from them. They also harm animal health and high toxin content in plants obviously causes financial damage to the producer. The *Fusarium* toxin levels in plants are influenced by weather conditions during the flowering season. Good agricultural practice can minimize the risks from *Fusarium* toxins. The European Commission Recommendation 2006/583/EC of 17 August 2006 sets the principles of the prevention and reduction of *Fusarium* toxins in cereals and cereal products. This forbids the mixing of products with high levels of toxins with others to meet the legal limitations. Zearalenone (F-2 toxin), fumonisins and trichothecenes are the most notable *Fusarium* toxins. Their effect is different. Fumonisins may be carcinogenic, zearalenone has estrogen-mimicking effects. Trichothecenes (e.g. 4-deoxynivalenol, DON, or vomitoxin, T-2 and HT-2 toxins) often cause major financial losses in pork and poultry farming as they reduce appetite and protein synthesis, therefore lower the efficiency of farming.

EU rules limit the levels of 4-deoxynivalenol, zearalenone and fumonisins in certain products. Animal tests were used to determine **tolerable daily intake** (TDI) values with a safety margin factor of 100. The TDI is 1 µg/kg of body weight/day for 4-deoxynivalenol, 2 µg/kg of body weight/day for zearalenone and 0.2 µg/kg of body weight/day for fumonisins. *Fusarium* toxins usually accumulate in the outer layers of grains, the potential risk is highest in bran and sprouts, even dinkel wheat is not an exception.

Fusariosis, the disease caused by *Fusarium* toxins cannot be entirely prevented today. Improved agrotechnical, pest control and storage techniques collectively known as good agricultural practice can minimize the risk of fusariosis. Although some industrial processes can reduce the amounts of toxins, it is impossible to remove them entirely. Rotation of crops, appropriate soil treatment, fertilization, and

Table 2.3 The UK Code of good agricultural practice to reduce fusarium mycotoxins in cereals published by the British Food Standards Agency

Good Agricultural Practice	Impact
Rotation and previous crop *Avoid maize as previous crop*	High
Crop residue management *Minimize previous crop residue on soil surface*	High
Variety choice *Choose more resistant varieties*	Medium
Weed control *Control weed populations*	Low
Insect control *Control insect pest population*	Low
Fertilizer use *Use optimum nutrient inputs*	Low
PGR (Plant Growth Regulator) use *Use where necessary to avoid lodging*	Medium
Fungicide use *Consider an ear spray to control ear blight*	Medium
Harvest and storage *Timely harvest and drying of grain*	Medium

selection of resistant varieties and use of fungicides are all important in preventing the formation of *Fusarium* toxins. The different methods have different impacts, though (Table 2.3).

Records of the Rapid Alert System for Food and Feed (RASFF), a system for reporting food issues within the European Union, show that mycotoxins were the leading cause of alerts and press releases in 2006 (31.2% of all cases), followed by pathogenic microorganisms (18.6%), and then by chemicals (17.0%). Mycotoxins kept their lead in later years as well (2007: 26%; 2008: 31%).

Consumers regularly underestimate the risks on mycotoxins ($\rightarrow$ **2.25**), whereas they tend to overestimate the risk of food additives ($\rightarrow$ **1.4, 2.1, 2.2, 2.4, 2.18, 2.21**). Old things are usually perceived safe, whereas new things are often looked upon with suspicion. Humans are slow learners....

2.10 Red Wine and the Vineyard Blues

It is becoming a commonplace today to believe that red wine is healthy. Not surprisingly, it is primarily the wine lovers who believe in the beneficial effects (Fig. 2.17). Do the facts also support this belief, or is it just spectacularly successful marketing from winemakers?

Medicinal use of wine was continuous from ancient times to the beginning of the twentieth century, but its supposed health effects kept changing. In ancient times, direct healing effects were attributed to it, whereas in the last century, the only use was as a solvent to extract or dissolve some ingredients during the preparation of certain pharmaceuticals. However, the composition of wine changes slightly each year, so the pharmaceutical industry discovered that a mixture of water and ethanol

Fig. 2.17 Red wine—to our health? (Copyright-free Wikipedia picture)

usually does a much more reliable job and, as for white wine, it has not been considered a pharmaceutical substance since the 1960s.

Scientific research on the health benefits from wine consumption began in the second half of the twentieth century primarily connected to the studying the **Mediterranean diet**. A lot of **epidemiological** and **pharmacological studies** have been published in this field. Growing evidence prompted the World Health Organization (WHO) to make a statement about the beneficial effect of moderate alcohol consumption in preventing cardiovascular diseases in 2003.

The first scientific articles connecting alcohol consumption with a lower incidence of heart disease were published about 30 years ago. The first large scale epidemiological study included 18 countries and established an inverse correlation between red wine consumption and the number of deaths from **ischemic heart disease**. This investigation is still frequently cited today to improve the reputation of red wine.

Epidemiological studies on wine consumption gave rise to something called "the French paradox" in the 1990s. The term was coined by Serge Renaud, who reached an interesting conclusion based on a 10-year WHO program analyzing data from 7 million people. He found that coronary disease, when the other known risk factors (body mass index, smoking, blood pressure, cholesterol levels) are taken into account, is significantly less common in France than in the US or the UK. Renaud interpreted this finding as a result of the higher average wine consumption in France (60–70 l per person a year) than in the other two countries (annually 5–10 l per person). In some other countries, people drink at least as much alcohol as in France, but not in the form of red wine. Heart diseases, however, are more common in these countries than in France, and this means that alcohol consumption alone is actually a risk factor.

The French mostly drink red wine, so the epidemiological effect was associated specifically with red wine rather than white wine. Other benefits of red wine have been discovered since then, some of them even proven, so a search for the active ingredients was started. Ethanol was a logical choice as it is the second most abundant component in wine after water.

A large number of studies on the ethanol component have not distinguished between the sources of alcohol, they only recorded the amount consumed. Analysis of data involving a lot of people seemed to show that moderate alcohol consumption lowers the risk of heart attacks compared to people who do not drink at all. Another study based on more than 100,000 participants concluded the same, but it also gave evidence of an increased risk when the amount of consumed alcohol was higher.

Today, the molecular mechanisms responsible for the effects of alcohol are known, which gives further support for the epidemiological studies: the benefits of red wine are higher than its alcohol content alone would justify. In small quantities, alcohol alone also seems to have some positive effects, but their practical value is largely diminished by the simultaneous negative effects.

Another study monitored people's drinking habits and health for about a decade and showed that moderate red wine consumption cut the **relative risk** of heart disease to 0.44 (1.00 is the base value for those who do not drink alcohol at all). The relative risk was 0.73 for those who drank comparable amount of beer, regular consumption of spirits had values above 1.00, which shows harmful effects. Facts from other investigations also supported the special importance of red wine. People in Northern France drink more spirits and less red wine than the average, and heart disease is more common there.

So what causes this unique effect of red wine? Part of the story is the high trace element content in comparison with beer or spirits, but this is not all. White wine and red wine are very similar as far as the main components are concerned. Red wine, however, contains about 20 times more of polyphenol derivatives than white wine. Alcoholic beverages are pro-oxidants, which means that they increase the intensity of oxidation processes. Alcohol itself is responsible for this effect. Red wine, however, is an antioxidant ($\rightarrow$ **3.31**) thanks to its polyphenol content. Antioxidants were very intensely researched in the 1980s and 1990s, which also contributed to the increasing interest in red wine. The presence of polyphenols is required for this effect, but this is still not the whole story.

The main organoleptic properties (taste and color) of wine are determined by sugars, acids and polyphenol derivatives. This last group of substances includes quite a lot of structurally diverse molecules, their chemical kinship lies in at least two hydroxyl groups connected to an aromatic ring, and this is also responsible for the antioxidant effect. Red wine has an acidic taste caused by **tannic acids**, which also belong to the group of polyphenol derivatives. Tannic acids do not have significant antioxidant effects, though.

The largest concentration of polyphenols is found in the seed and peel of grapes. The difference between the polyphenol content of red and white wines lies not only in the different varieties of grapes, but also in the preparation method: white wine is made by fermenting "must" which is obtained by pressing grapes, but it is not separated from seeds and peels in red wine fermentation. This process takes a long time, and the ethanol formed can extract a lot more of the polyphenol derivatives. Their typical concentration in red wine is 2 mg/ml, whereas only 0.2 mg/ml in white wine. Red wines (especially dry red wines) have high polyphenol contents and high antioxidant capacities. Polyphenol content decreases when wine is stored, so old wines have lower antioxidant capacity. Some of the polyphenol derivatives are formed (and not only extracted by the ethanol) during the fermentation, these also contribute to the health effects. Non-alcoholic grape juice also has significant polyphenol content, and accordingly shows some of the beneficial effect of red wine.

The beneficial effects of red wine are by no means a consequence of a single substance. Different polyphenols and the interaction between them are responsible for the effect together, which is much greater than when only a few isolated substances are used. Some main compounds have been identified in the studies, and *trans*-resveratrol seems to be a rather unique one.

The main health effect of wine and the polyphenols it contains is to slow down atherosclerosis. The key to this effect is decreasing the concentration of **LDL** cholesterol, and an additional benefit is an increase in the concentration of **HDL** cholesterol in blood.

If atherosclerotic plaques already began to form, maintaining an adequate blood supply to organs and preventing the blockage of veins and arteries is important. Olive oil (→ **2.17**) is a major ingredient of the Mediterranean diet that prevents blood clot formation, but red wine has this effect, too. Polyphenols in red wine, especially **flavonoid** type compounds quercetin and *trans*-resveratrol inhibit the aggregation of blood platelets. This effect was observed in studies of human blood coagulation, and the juice of blue grapes also works, although to a lesser extent. White wine has no such effect, which is some clinical evidence of the physiological activity of polyphenols. Orange or grapefruit also contain high levels of polyphenols but do not inhibit the aggregation of blood platelets. This finding clearly implies that not only the amount of polyphenols is very important, but also their identity. An American research group published an interesting observation: the seeds and skin of the grapes both have effects, but this effect is much stronger if the two are used in combination. Therefore, the compositions of the skin and seeds must be at least somewhat different, and there may be some synergisms between different components. Although alcohol itself also inhibits the aggregation of blood platelets, this is only of academic interest as the amounts needed are too high.

Polyphenols relax arteries and coronary arteries, which causes a decrease in blood pressure and improves the blood supply of heart muscles. This is most probably why red wine consumption lowers the risk of ischemic heart disease. A very similar effect is also observed with alcohol-free red wine, but drinking vodka does just the opposite. In Japan, a drink prepared from red wine vinegar and grape juice is sold in supermarkets. A scientific study of this mixture, which is similar to red wine as far as polyphenols go, revealed a blood pressure lowering effect in rats.

Acute (occasional) overconsumption of alcohol results in intoxication. Chronically (over some period of time), this can lead to gradual damage of the brain and the nervous system. It is an obvious but rather debated question what effect moderate alcohol consumption may exert on the brain. Are the known risks overcompensated by the expected benefits? The lifestyles of more than 1000 elderly people were monitored for at least 7 years in a related American study. Those who consumed moderate amounts of alcohol actually preserved their mental abilities better than those who opted for abstinence. But the authors of this study remarked that they would not recommend alcohol consumption for this purpose despite the positive results. The mechanism of the effect is still unknown, and another substance in the drinks rather than alcohol may very well turn out to be the crucial ingredient. It may be only speculated that the platelet aggregation inhibiting effect may lower the risk of microinfarctions in the brain, and this is why benefits for the brain arise. In **observational studies**, moderate alcohol consumption was also shown to decrease the risk of Alzheimer's disease. Mechanisms are not known here, either, but these results would warrant further studies about the effects on the central nervous system.

In recent years, wine consumption was also linked to the development of tumors—but not in a positive way this time. Overconsumption of alcohol was proved to increase the risk of cancer in multiple studies. This is thought to be the consequence of the pro-oxidant and DNA damaging effects of alcohol consumption. But moderate drinking is not quite innocent, either: it was shown to increase the risk of breast cancer (irrespectively of the source of alcohol, i.e. red wine included), especially among women with low body mass indexes. These results call for caution: although the antioxidant effect of red wine is beyond doubt, it still cannot be recommended as a preventive tool against cancer.

There have been more positive conclusions, too. Another human test showed that drinking red wine, or alcohol-free red wine and alcohol together lowered DNA damage caused by other factors, but alcohol alone did not have such activity. Again, this is not evidence of the anticancer effect of red wine or its polyphenols, but useful additional information.

Trans-resveratrol is an especially important polyphenol derivative in wine from the viewpoint of anticancer effects. The chemical structure of this compound resembles the hormone estradiol (Fig. 2.18). The substance in fact binds to estrogen receptors, it is sometimes called a phytoestrogen (→**3.24**). However, resveratrol is much more important than its minor estrogen activity would make it appear. A very intensely studied—and also hotly debated—recent question is about the enhancement of the activity of an enzyme called sirtuin. Some researchers think that resveratrol increases this activity and helps preserve healthy cell functions. Some *in vitro* studies support this theory, but human clinical tests are still missing. *Trans*-

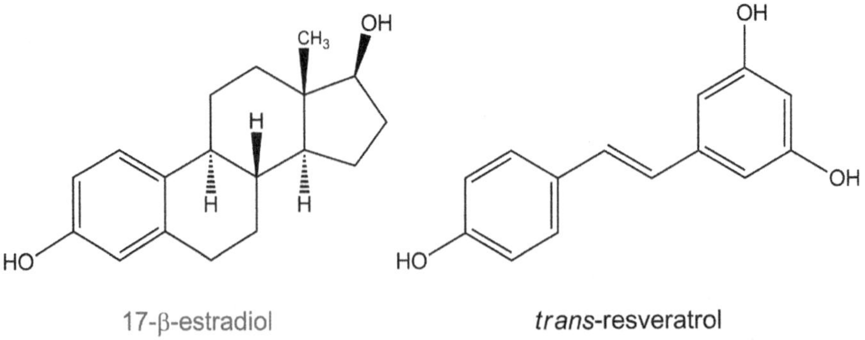

Fig. 2.18 Chemical structures of estradiol and *trans*-resveratrol. (Authors' own work)

resveratrol may be developed into a drug molecule, or its chemical structure may be the source of inspiration of other drugs. But it is also possible that resveratrol is just a fleeting star in the universe of health-related chemicals, and is not the key to the beneficial effects of red wine.

The reader might have the impression thus far that all the tests and evidence are contradictory and nothing is known for sure. Actually, this is the usual state of affairs for research in progress. What seems very likely now is that the benefits of red wine are not caused by its alcohol content. A number of experiments (even those described in the previous paragraphs) actually showed very similar effects with alcohol-free concentrates. A well prepared concentrate may have all the positive effects with much lower risks.

Yet all this does not answer the original question on red wine consumption. Is it recommended after all, or does the alcohol content ruin the whole thing? By this point, the reader may have learned not to expect a yes-or-no answer. As written in the previous paragraphs, moderate red wine consumption seems to have beneficial cardiovascular effects, but some additional studies show the same for other alcoholic drinks as well.[1] The mechanisms behind the effects of red wine are partially known, it seems better than other alcoholic drinks, yet randomized, **double blind studies** of the highest clinical quality are still missing. A report published by WHO in 2003 mentions a general cardiovascular benefit of moderate alcohol consumption without singling out red wine as the most likely reason. Even so, this report does not recommend this practice because of the known risks.

It is also not impossible that increased consumption of red wine is associated with higher status in society, and this higher status is also reflected by a more careful attitude toward health-related issues. The fact that those people who drink red wine regularly enjoy better health and higher standards of life is a statistical correlation. This says nothing about causes and effects. Another similar example: people

---

[1] Just to complicate matters further: in many of the studies about general alcohol consumption, people drinking red wine were included without distinction. It is not inconceivable that the calculated average effect was positive because this subgroup distorted the results. So the benefit, which is limited to red wine only, may have conveyed a positive general picture of alcohol.

who make frequent visits to the dentist are known to be healthier than others. But this is by no means the achievement of the dentist. Simply put, people interested in their own dental health are usually interested in their general health, too.

Although alcohol-free grape extracts showed promising results, studies on long-term (chronic) consumption have not been done yet. Drinking blue grape juice is by no means as common as drinking red wine, epidemiological studies are much more difficult to carry out. Grape consumption is very seasonal, so information about its health effects is rather scarce, too. Other fruits with high polyphenol content should also be tested in much more detail. They may even contain valuable substances not present in red wine.

Trans-resveratrol is added to several dietary supplements with the assumption that it will have the beneficial effects of red wine. This assumption has not yet been tested scientifically. Grape and wine are very complicated mixtures of chemicals, so other compounds, thus far unknown, may also be highly significant. Individual differences in human life styles may add a large amount of deviation to any observations. Resveratrol is sensitive to oxidation: in capsules, its actual amount may be much lower than the label says. In such a concentrate, resveratrol is not protected from oxidation by other components present in red wine.

It should also be noted that the benefits of red wine consumption are primarily shown in people following the Mediterranean diet. The regularity and the amount may both be important factors. Irregular red wine consumption offers no benefits, overconsumption has major dangers. It is primarily Italy and France that serve as examples to be emulated in this respect: people mostly drink red wine as part of meals (but not alone), yet alcoholism is much less prevalent than in Nordic nations. This is not simply a question of red wine: it is a question of life style.

2.11 Popeye's Spinach and the Search for Iron

The iron content of spinach gave rise to a complicated story of which any pulp fiction writer could be proud of. On the web, or even in somewhat more serious scientific literature, one can find the following story, with occasional variations:

Popeye, who was a cartoon hero before becoming a film star, gets most of his strength from spinach because this contains a lot of iron. This was a common misconception in the 1930s, so Elzie Crysler Segar, the creator of Popeye chose spinach in order to popularize healthy food among children. Under Popeye's influence, spinach consumption grew by 33% in the United States. In fact, spinach does not contain outstandingly high amounts of iron compared to other vegetables. The misconception was based on a mistake in a 1870 article published by Emil Theodor von Wolff (1818–1896), who misplaced a decimal separator. By the 1930s, the mistake itself was corrected in the scientific literature, but this fact could not stop the misconception from gaining new ground. A variation of the story blames Gustav von Bunge (1844–1920) for the misprint and places the original article in the 1890s.

On a web site, this story is listed among the seven most catastrophic misprints ever made in history. The story was repeated numerous times by numerous speakers or authors. In fact, the story is used as an example to show how important it is to find scientific data in their original source. It is a pity that the writers of these warnings did not follow their own advice. It seems that the only truth in this story is that spinach does not have an outstandingly high content of iron.

By 2010, Dr. Mike Sutton, a forensic science expert of the University of Nottingham was very bothered by the fact that he had never seen the ominous article that misplaced the decimal separator, despite the high frequency of the repetition of the story. In fact, he had never even seen a reference to this article, which means none of the story tellers cared to disclose which article in which journal was the primary source. So Dr. Sutton began what he intended to be a short investigation. It became longer and longer, and he ended up searching for weeks and finally he drew some really interesting conclusions.

There is absolutely no evidence that Wolff or Bunge ever measured the iron content of spinach, so they could not have misplaced the decimal separator. No such scientific articles were found, and no other scientific works citing these articles have turned up, either. What is more, these two scientists were never mentioned in the scientific literature before 1981 (or at least this was the case when Sutton wrote his article in 2010). However, American researchers published data in 1934 that showed an iron content of spinach that was 20 times larger than the actual value, or at least it was easy to misunderstand them in this way. The source of the error, however, could not have been a misplaced decimal separator. Other American scientists, one of whom had a German-sounding name, pinpointed this mistake, albeit in a somewhat confusing manner. There is no sign that the erroneous information received any significant publicity, though.

In the Popeye cartoons, there is absolutely no sign of a connection between Popeye's preference for spinach and the iron content of the vegetable. Indeed, Elzie Segar wanted to influence children's diet, but the reason for this was quite clearly vitamin A and not iron. Popeye first ate spinach in a cartoon picture published on July 3, 1932, which was 2 years before the 1934 article with the erroneous data. Even at this time, he did not forget to mention vitamin A. None of the original cartoons made any connection between spinach and iron.

Spinach consumption was on the rise in the Unites States even between 1915 and 1928. For this reason, it is not quite convincing that Popeye played any role in the further increase that occurred during the 1930s. American authorities recommended the consumption of this vegetable together with meat, some even said instead of meat.

Finally, at that time, Sutton found no evidence that this Popeye-iron-spinach story had ever been written down before 1981, when an editorial by Terry J. Hamblin told the cautionary tale in the *British Medical Journal*. The article contains literature references, but it gives no clue as to the origin of the story. Sutton even contacted Hamblin personally, who said that he could not give the source of the story because of an editorial decision and later forgot what it was. Sutton concluded that Hamblin must have made up the story of thin air.

Table 2.4 The iron content in some food products

Food	Iron content (ppm)
Curry (powder)	295
Sugar cane molasses	170
Chicken liver	120
Dehydrated peach	41
Lentil	33
Spinach	27
Garlic	17
Onion	14.8
Red cabbage	8.0
Broccoli	7.4
Tomato	2.7
Carrot	2.0

A year later, in a blog entry, Sutton apologized to Humblin as he found an earlier printed version. He traced the story back to an inaugural lecture given by Professor Bender in 1972, in which Bender attributed the discovery of the decimal error to a professor Schupan. He repeated this claim in a later article published in the Spectator. Yet, no references to any work published by a Professor Schupan have been found. Bender himself acknowledged "Professor den Hartog of Holland for tracing the possible origins of the belief", which proved to be another dead end for the investigation.

Table 2.4 shows the iron contents in a number of food products. Among vegetables, spinach contains a relatively high amount of iron, but it is by no means outstanding. From 1935, it is also known that the human body can utilize very little of the metal because of the oxalic acid ($\rightarrow$ 4.15) in spinach. The authors of the related scientific publication actually made a funny reference to Popeye, but they did not say that Popeye ate spinach because of iron, and also did not mention the alleged mistake with the decimal separator.

It is quite interesting that a story about the importance of checking scientific references became its own example. Will there ever be a separate word for misconceptions about misconceptions?

2.12 The Perplexing Spice: Curry

Curry is a popular spice mixture with easily recognizable taste and odor. On the other hand, the name curry also occurs on the label of a small, green spice plant that is sold in large retail stores. Are these two connected? And after all, what is curry?

The title question does not seem to have much relevance to chemistry. Yet, this spice is worthy of a little more attention as it gives an example how the same word may refer to different substances in different ages and cultures.

The origins of the word curry and the food curry is not entirely the same. There are several explanations. Some think that 'curry' derives from the Tamil word *karil*,

Fig. 2.19 The title page of the cookbook entitled 'The Forme of Cury', which was published in 1780, but originally compiled in 1390. (Copyright-free Wikipedia picture)

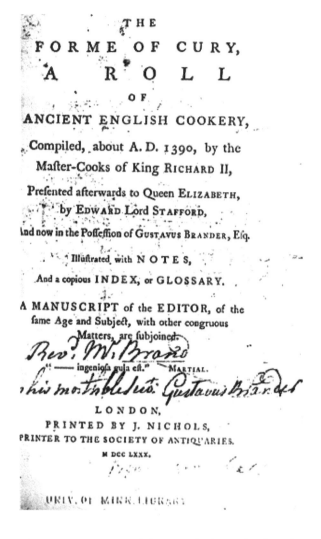

which is a dish with sauce on it. Others think the origins must be in England, as the word cury was already used in the title of the fourteenth century cookbook (*The Forme of Cury*, Fig. 2.19), which originally was borrowed from French and meant cooking. A third opinion derives the word from Hindi *karahi*, which is used for a specially shaped cooking dish.

Curry is actually not a spice, but a type of food. It can be made using meat, vegetables or fish, the common element is the spicy sauce on them. Curry dishes come from Asia and they exist in countless local varieties. It is hardly surprising that the spices used to flavor curry also vary greatly. India has a dominating influence on curry sold in Europe, the basic ingredients of which are turmeric (*Curcuma longa*), cumin (*Cuminum cyminum*) and coriander (*Coriandrum sativum*). Curry powder in Europe is yellow and has a recognizable odor, which is caused by turmeric. In many parts of Asia, this spice as actually not used in curry at all. From India to Korea, curry powder may contain green cardamom, cinnamon, black pepper, bay leaves, mustard seed, fennel, ginger, nutmeg, cloves, anise, garlic, fenugreek, and a number of other spices, which are completely unknown in Europe and do not even have names in English. Chili pepper, which is often thought to be an indispensable ingredient in curry in the Western World, is in fact uncommon. It is by no means a traditional spice in India as it only arrived there in the sixteenth century. Curry is as diverse as the spices in it. Chefs and traditional families often have their own, carefully guarded special curry recipes. In India, the mixture of spices is called *masala*. *Garam* (hot), *tandoori* and *chaat* (sweet and sour) *masala* are known even in Europe.

Recent studies showed that curry spice has some desirable health effects (anti-inflammatory, antioxidant, anticancer and anti-Alzheimer), which is mainly caused by turmeric. Other ingredients only have positive effects on digestion. The effects are primarily caused by the color-causing components in turmeric, called curcuminoids, but the medicinal benefits have not been proven beyond reasonable doubt.

Today, curry may be one of the traditional *British* foods. This influence of Indian cuisine is obviously of historical origins. The Netherlands had monopoly on the trade (and price, as the reader can imagine) of black pepper in the sixteenth century. To combat this monopoly, the famous British East India Company was founded. The original objective was to establish a safe transportation route to Indonesia, but it was soon recognized that trade with India offered more profit. Trade interests got mixed with imperial ambitions and as a result, India became a British colony for several centuries until 1947. This period had a major effect on both Indian and British culture.

Soldiers and civil servants working in India often moved back to Britain after fulfilling their duties, and they took home the recipes of food they became fond of in India. This was how curry was brought to Europe, and it was understandably influenced by British taste. In a cookbook written by Hannah Glasse in 1747, the recipe of 'currey' calls for the use of coriander and pepper. Turmeric, the most important current ingredient, only appeared in the fourth edition.

The city of London saw the opening of the first Indian restaurant in the beginning of the nineteenth century. Curry became immensely popular in Europe after World War II, when mass immigration from Asia began. Inexpensive restaurants, mostly run by owners of Bangladeshi origins, sold 'Indian' food adjusted to British tastes. It was in the melting pot of London where the Indian food sold throughout Europe was born. Indian people may very well see this food as unfamiliar. *Chicken tikka masala*, although widely regarded as Indian food, is in fact a British invention based on Indian spices and was successful enough to become increasingly popular in the Indian subcontinent. An even more local food is *Currywurst*, a cooked sausage in Germany covered by turmeric sauce: it even has its own museum.

Although curry is primarily a dish, many people also use the word for the spices used in curry. In addition, two plants also have their own claims for this name. The first is the curry tree, and not without some justification. The leaf of the tree *Murraya koenigii* is actually used as a spice in some parts of India, it is one of the ingredients used in curry. Curry tree alone does not make a tasty curry. Neither does the plant that is sold under the name of curry in many European countries. This 'curry plant' (*Helichrysum italicum*) has south European origins. It has yellow flowers, a pleasant odor and it can certainly be used as a spice in soups, salads or vegetable dishes. In Asia, however, its use in curry is unknown.

2.13 Bigger the Better? The Alpha and Omega of Fatty Acids

It is no small irony that consumers can be misled by true statements in advertisements. These true lies only work on those who do not have all the correct information. As far as unsaturated fatty acids are concerned, advertisers capitalize on the usual "more is better" attitude of consumers: on some products, the amount of omega-9 fatty acids is also displayed in addition to omega-3 and omega-6. Does anyone in the general public have a reasonably balanced picture about omega fatty acids? This sort of information is seldom taught in school.

Advertisements seem to pretend that everyone knows about omega fatty acids and their highly beneficial properties. Despite these benefits, most chemistry books are not interested in these substances at all. The misunderstanding lies in the fact that the letter ω is mostly used in food science: chemistry books refer to the same molecules under different names. The structure and some of the properties and of fatty acids need to be reviewed first in order to understand the importance of the number behind the omega.

Fatty acids are saturated or unsaturated carboxylic acids with unbranched carbon chains (Fig. 2.20). These molecules have an even number of carbon atoms in plants and animals—this property is caused by the method the body uses to build them up. Fatty acids occur in a form chemically bonded to glycerol through **ester** groups. Glycerol is a common component in all fats, differences are solely in the fatty acids. All of these have a single carboxylic acid group, the other end of their carbon chain is closed by a methyl group. Natural fatty acids contain at least eight carbon atoms. If there is no double bond in them, they are called saturated. Monounsaturated fatty acids have a single double bond, polyunsaturated fatty acids feature several double bonds. The double bonds in natural fatty acids usually have *cis* configuration (→ 2.5). There are common abbreviations, used even in languages other than English such as MUFA (monounsaturated fatty acid) and PUFA (polyunsaturated fatty acid). The methyl end of a fatty acid molecule is called omega (ω), the second carbon atom after the carboxylic acid end is called alpha (α). In food science, numbering usually begins at the ω end, whereas chemists prefer to start at α. Since the significance of these materials is much larger in food science than in chemistry, it is the omega notation that found its way into the advertisements. The number after the letter omega means the location of the first double bond as counted from the methyl end.

One of the most common fatty acids in plants is oleic acid with a chain of 18 carbon atoms, in which the first (and only) double bond is between carbon atom 9

2.13 Bigger the Better? The Alpha and Omega of Fatty Acids

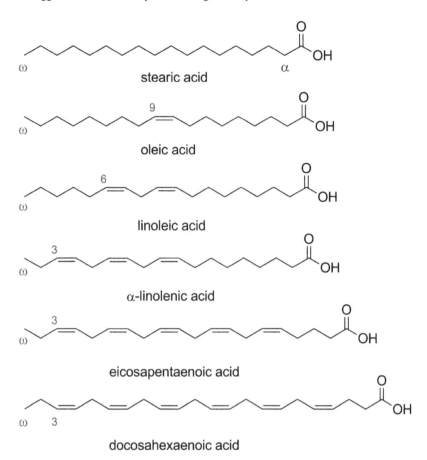

Fig. 2.20 Chemical structures of saturated stearic acid, monounsaturated oleic acid (ω-9), and polyunsaturated linoleic (ω-6), α-linolenic (ω-3), eicosapentaenoic (ω-3), and docosahexaenoic (ω-3) acids. (Authors' own work)

and carbon atom 10: this is called an ω-9 fatty acid. Significant polyunsaturated fatty acids belong to two big groups: ω-6 and ω-3. An example of an ω-6 fatty acid is linoleic acid, whereas α-linolenic acid is an ω-3 acid. Both of them contain 18 carbon atoms (Figs. 2.20 and 2.21).

The human body does not have the enzyme to make ω-3 and ω-6 fatty acids. Therefore, an appropriate amount must be consumed in the food to maintain health. Linoleic and α-linolenic acids were considered vitamins (vitamin F) at a certain point in the history of science, but the term 'essential fatty acid' is the accepted version today. There are no such essential compounds among monounsaturated or saturated fatty acids.

In fact, it is not linoleic and α-linolenic that are essential, but their derivatives. In the body, they are transformed to longer fatty acids (20–22 carbon atoms) with more (4–6) double bonds. In turn, these are used in the body to build fatty acid derivatives

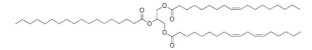

Fig. 2.21 A triglyceride containing oleic, stearic and linoleic acids. (Authors' own work)

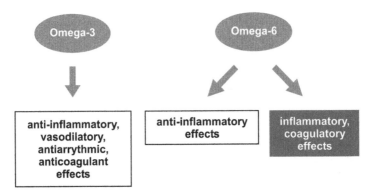

Fig. 2.22 Major effects of compounds formed from ω-3 and ω-6 fatty acids. (Authors' own work)

(prostaglandins, leukotrienes, thromboxanes). Although there are a number of common enzymes in the separate ω-3 and ω-6 pathways, their final products and their physiological significances are different.

With a little bit of oversimplification, ω-3 can be declared much more favorable to human bodies, although ω-6 fatty acids are also essential (Fig. 2.22).

Large quantities of linoleic acid are present in plant oils (sunflower, corn, soya, peanut). It is first transformed into γ-linolenic acid (GLA), which contains three double bonds. Sometimes there is a shortage of the enzyme catalyzing this process in the body, then the insufficiently low levels of γ-linolenic acid lead to skin symptoms. This acid is only present in very few plants, so these symptoms can only be treated with the exceptional oils (borage, evening primrose and blackcurrant seed).

The most significant ω-3 fatty acid of the plant kingdom is α-linolenic acid: it is only present in a small number of plants used for food. Linseed oil is the most significant source, about 50% of this is α-linolenic acid. However, linseed oil has an unpleasant smell and is seldom eaten. Rather, it is used to protect the surface of wood. Linseed oil is a so-called drying oil, exposure to air and light induces polymerization to a solid form through the reactions of the double bonds. In addition to the unattractive taste, linseed oil has a further disadvantage: only a small percentage of the ω-3 fatty acid α-linolenic acid is transformed into useful molecules in the human body. It is better to consume fatty acids that are further down in the biosynthesis chain such as eicosapentaenoic acid (EPA, 5 double bonds) or docosahexaenoic acid (DHA, 6 double bonds). These fatty acids are not found in plant-derived oils, but they are present in fish (Fig. 2.23).

Fig. 2.23 The 3D structures of a saturated (stearic), monounsaturated (oleic), triunsaturated (α-linolenic) and a hexaunsaturated (docosahexaenoic) fatty acid. (Authors' own work)

Interest in EPA and DHA does not only stem from the molecules formed from them. They have direct effects as well: they lower the levels of fats in blood. They are also found in brain tissue, so there is good reason to assume that they have an important function in the brain. Many other additional benefits are described as well, the most well documented is a decrease in the risk of cardiovascular diseases. This was recognized as early as the 1920s, when studies of the usual diet of Inuit people highlighted the connection between consuming a lot of ω-3 fatty acids and the absence of heart diseases in their society. Inuit who did not follow their traditional diet did not experience similar protective effects.

The scientific results about materials present in commercial products are often portrayed in a highly exaggerated fashion. Fish oil lowers cholesterol levels and inhibits blood clot formation—these are proven facts. Other advertised effects are less unambiguous. Some products are claimed to increase the IQ of children and lower the incidence of attention disorder. The scientific background is that DHA is necessary for the development of the brain and it is present in large amounts in the membranes of nerve cells. The number of double bonds influences the spatial structure of a molecule quite significantly, this is why specific fatty acids are needed to build up membranes. Although the significance of DHA is clear, the conclusions presented are often sensationalist. Current understanding is that excess DHA consumption does not help brain or nerve development during pregnancy or infancy. It is true that a shortage of DHA can slow down development—especially during pregnancy. Regular consumption of fish is recommended for pregnant women to avoid a shortage.

Eating habits have changed so much recently that increasing attention should be paid to ensuring an adequate intake of ω-3 fatty acids. For several thousand years, about one fifth of the overall calorie intake was from fats in the typical human diet, 7–8% of which were unsaturated oils and fats. The ω-6 / ω-3 fatty acid ratio was about 4:1. Since the industrial revolution, meat consumption and the use of margarine have become favored, which increased the ratio of saturated fatty acids. Recent decades have seen growing nutrition awareness, but the plant oils only supply ω-6 fatty acids. So the current ω-6 / ω-3 fatty acid ratio is about 15–16. Consumption of fish or fish oil products is necessary to bring this ratio close to the optimal value of 5. About 1 g of EPA+DHA needs to be consumed daily to achieve this goal. This does not necessarily mean fish oil, eating fish on a regular basis might be just as

Fig. 2.24 The "Man of Bicorp" collecting honey from a beehive as depicted on an 8000-year-old cave painting in Cuevas de la Araña (Spider Caves) near Valencia, Spain. Note how the bees attack the gatherer. (Copyright-free Wikipedia picture)

good. There are several grams of ω-3 fatty acids in 100 g of various sea fish (tuna, herring, salmon, mackerel, pilchard). Freshwater fish have not enjoyed such positive image in this respect, but in fact, several of them (trout, perch, freshwater herrings) are just as good as sea fish.

Finally returning to the question asked in the title: most of the unsaturated fatty acids are healthy, but only the ω-3 and ω-6 are essential. The ω-3 group receives more attention because of the favorable health effects. The ω-9 group does not have any known special properties that would warrant extra attention.

2.14 Sweet as Birch: Xylitol

The craving for tastes is an inborn characteristic of the human being and the sweet taste has an outstanding role in the history of mankind. For a long time during the evolutionary process, sweet foods were not available in limitless quantities. Yet some common sources of sweetness in ancient times would have been found in fruits, berries and honey (Fig. 2.24). This situation changed, however, with the cultivation of sugar cane, which was native to Polynesia and taken first to China then to India around 700 BC. The culture of growing this plant slowly moved westward, it finally arrived to the warmer parts of the Mediterranean through Persia in the early Middle Ages (ca. 800 AD). Muslims have been especially eager to cultivate this

2.14 Sweet as Birch: Xylitol

crop and many traditional Muslim foods rely heavily on the sugar obtained from sugar cane.

Sugar reached England as late as 1319 AD, but it was very expensive and used as a medicine in the beginning. It took a while until the inhabitants of Northern Europe, around the seventeenth century, could start enjoying this novelty from less expensive imports arriving from the Caribbean.

Sugar cane is a 2–3 m (6–9 ft) tall jointed reed that grows for up to 10 years in a rich and wet soil. Sugar cane can only be cultivated in an area with warm climate, where it is harvested twice a year. Sugar cane shoots were taken to the New World by Columbus during his second voyage in 1494 as the West Indies proved to be ideal for sugar cane cultivation. However, it took almost two centuries to begin the export of this valuable crop back to the Old World. The processing of ripe sugar cane to obtain refined sugar was a further step in the process but it proved to be labor intensive before mechanization and it required a lot of slaves. This fact certainly belongs to the dark side of European colonial history.

The sugar cane industry had grown considerably until the Napoleonic Wars, but the continental blockade of France by England dramatically changed the situation by cutting the access to cane sugar produced in Haiti. The shortage of sugar had to be remedied quickly. French scientists commissioned by Napoleon evaluated the results of Andreas Marggraf (1709–1782), a German chemist from Berlin, who extracted sugar from sea beet. Selective breeding of this plant resulted in today's sugar beet which was higher in sugar content and ideal for growing even in the colder climates of Europe. Thus, dependence on imported cane sugar had been eliminated and a new sugar industry, based on sugar beet, was born in Europe over a 100 years ago. Beet sugar, however, is more expensive than cane sugar and the recent worldwide food crisis seriously influences the economic status of common sugar.

What is sugar exactly? Sugars belong to the larger class of substances known as carbohydrates. The name carbohydrate suggests that they have something to do with carbon and water. Indeed, the composition of many carbohydrates is close to the chemical formula $C_n(H_2O)_n$. Carbohydrates are the most abundant organic matter on Earth and make up about three fourths of the biomass produced annually. This is a huge amount, ca. 30 t of biomass is produced per capita every year on Earth!

Carbohydrates are very complex substances, they can be classified into simple carbohydrates (or mono-saccharides) and complex carbohydrates (oligo- and polysaccharides). The difference between these saccharides comes from the fact that oligo- and polysaccharides are built from monosaccharide units. For example, the common table sugar (also called saccharose or sucrose, Fig. 2.25) is a di-saccharide made up from glucose (grape sugar) and fructose (fruit sugar). Starch and cellulose are both polysaccharides and made up of thousands of glucose units, but these units are attached to each other in different ways. In general, mono- and lower oligosaccharides tasting sweet are called sugars. The sweet taste, however, is not an exclusive characteristic of sugars. In addition, many sugars are less sweet than common sugar: the disaccharide called gentiobiose is actually bitter.

The role of carbohydrates in our life is also very complex. They supply us with energy day by day and play an important role in the cell to cell communication (the

Fig. 2.25 The three-dimensional structure of saccharose or sucrose and birch sugar (xylitol). (Authors' own work)

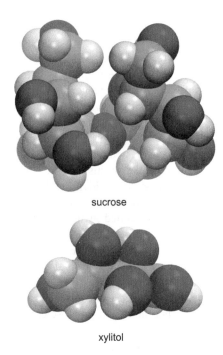

sucrose

xylitol

difference between blood groups is attributed to different carbohydrates attached to the cell surface). Carbohydrates also make up a part of the human hereditary material and the cell wall of bacteria and plants provide structural materials (e.g. wood).

From the point of view of our essay, the role of sugars in our food is an important issue. Glucose, also known as 'blood sugar' constitutes 0.1 % of human blood and our body strives to maintain this level constant. Glucose, derived from our food by means of different enzymes in our stomach, passes directly to our blood stream and is delivered to different body parts where energy is needed. When used, glucose breaks down into simpler molecules and finally carbon dioxide and water are formed. It may sound ironic, but these latter two substances are actually the starting materials for the plants to produce carbohydrates (and oxygen) in the reverse process, which is called photosynthesis:

$$carbon\ dioxide + water + light\ energy \rightarrow carbohydrate + oxygen$$

When someone has more glucose in her or his blood than necessary (e.g. after a heavy meal), excess glucose is turned into the polysaccharide glycogen and stored in the liver. If the glucose level of blood falls, glycogen is turned back into glucose. An average adult may have 350 g (¾ pounds) of glycogen stored in her/his body, enough to keep it going for a day. When the glycogen reserve is full, the body turns excess glucose into fat, which is a more economical form of energy storage than glycogen. This reserve stores enough energy to keep an individual alive for

2.14 Sweet as Birch: Xylitol

Table 2.5 Sweetness of bulk sweeteners (some with E numbers) compared to sucrose (100%)

Fructose	120%
Xylitol (E967)	100%
Erythritol (E968)	75%
Glucose	70%
Galactose	60%
Maltose	50%
Sorbitol (E420)	50%
Mannitol (E421)	40%

a month. The long-term storage of glucose in the form of fat turned out to be an evolutionary advantage in ancient times due to uncertain food supplies, but it poses problems nowadays as more and more people became overweight. Obesity is a serious threat to our health, and there is a constant need for sweeteners, which taste pleasantly sweet but do not have as high a calorific value as common sugar.

Brown sugar varieties are becoming more popular but most of these varieties nowadays are made of refined sugar colored by molasses ($\rightarrow$ **2.20**). Many people who praise honey over table sugar simply forget the fact that honey is also a concentrated solution of glucose and fructose (70%). Honey also contains the disaccharides sucrose and maltose (10%), whereas the rest is water ($\rightarrow$ **3.8**). The characteristic aroma and flavor of honey is due to some small amounts of other substances, but in any case, the calorific value of honey is not less than the equivalent amount of table sugar.

The advent of cheap sugar (compared to the earlier times) also changed the European gastronomy: the salty and sweet tastes have been separated and the dessert has appeared as a separate dish. The world production of raw sugar now amounts to over 160 million t every year. Current sugar consumption varies between 3 kg/person/year in Ethiopia to around 40 kg/person/year in Belgium (oh, those fine Belgian chocolates, Monsieur Poirot!).

Sweeteners can be classified into bulk and intense sweeteners ($\rightarrow$ **2.16**). Bulk sweeteners need to be used in relatively large amounts to achieve the desired sweetness in contrast to intense sweeteners that are much sweeter and needed only in tiny amounts. The sweetness of individual bulk sweeteners is compared in Table 2.5. Sweetness is subjective. It is usually measured by preparing a 10% solution of the compound in water followed by a request to a panel of people to taste it. It is then diluted and tasted again, diluted and tasted again, and so on until the panel declares that the solution is no longer sweet. The perception of the sweet taste in food also depends on concentration, acidity, temperature, and the presence or absence of additives. But just as sweetness is affected by other factors, so it is that sweetness influences the perception of fruit flavors, sourness and bitterness as well.

Current bulk sweeteners include sucrose, starch syrup, glucose (grape sugar), high fructose corn syrup, fructose (fruit sugar), lactose (milk sugar), sugar alcohols, and invert sugar (the mixture of glucose and fructose obtained from sucrose, honey is largely invert sugar). Sugar alcohols are produced by the addition of hydrogen to sugars, in a process called reduction. The most common sugar alcohols include xylitol, sorbitol and mannitol. Thus, xylitol is indeed a sugar, which serves as an unexpected single-word answer to the question asked in the title.

Xylitol (E967, Fig. 2.25) is a five-carbon sugar alcohol, small amounts of which occur naturally in yeast, lichen, fungi, vegetables, fruits, in birch sap (hence its older name birch sugar). In the human body, 5–15 g xylitol a day is processed as part of glucose metabolism. Its economical production (ca. 30,000 t/year) involves the extraction of xylan (an oligosaccharide composed mainly of xylose) from plant materials (birch wood, corn cobs and straw). Xylan, once transformed into its xylose building blocks, is then hydrogenated to obtain xylitol.

Xylitol is just as sweet as sucrose (Table 2.5), but it has some 30% less calories. It is also non-cariogenic (tooth-friendly), which makes it ideal for use in confectionery, chewing gums, toothpastes, mouthwashes, throat lozenges, multivitamin tablets, cough syrups and diabetic products. In dental health, xylitol has the additional advantage of preventing the proliferation of certain plaque-forming bacteria (*Streptococcus mutans*), reducing the growth and adhesion of dental plaque and a possible favorable effect on the balance between the de- and re-mineralization of dental lesions. Consuming xylitol produces the sensation of freshness on our tongue because its dissolution requires heat, which is taken away from the tongue. Xylitol is absorbed slowly from the small intestine, the remaining portion is a nutrient for bacteria in the large intestine, where it is degraded to short-chain fatty acids and some gas (hydrogen, methane, carbon dioxide). Common baker's yeast cannot ferment xylitol, thus, the dough will not leaven properly with yeast in the presence of xylitol.

Xylitol is also beneficial in ear and upper respiratory infections. Xylitol prevents the growth of bacteria in the ear, not only in the mouth. When bacteria enter the body, they adhere to the tissues using a variety of sugar complexes, which is much more difficult in the presence of xylitol, which forms many different sugar-like structures that interfere with the ability of many bacteria (e.g. *Staphylococcus* species) to adhere.

Xylitol has no aftertaste. It has also a very low **glycemic index** (GI) of 7 (glucose has a GI of 100). The glycemic index provides a measure of how quickly blood sugar levels rise after eating a particular type of food. Although not completely perfect, the value of GI gives an idea about which food influences the blood sugar content more. High-GI carbohydrates are associated with increased risk of obesity.

Xylitol has no toxicity in humans, it enjoys the rare status of 'generally recognized as safe' (GRAS). In one study, the participants consumed a diet containing a monthly average of 1.5 kg of xylitol with a maximum daily intake of 430 g (which is a huge amount of a bulk sweetener!) with no apparent adverse effects. One side effect of consuming large quantities of xylitol is its laxative effect, bloating and flatulence, mainly because, like other sugar alcohols, it is not fully broken down during digestion and it feeds intestinal bacteria. The laxative effect shows high individual variability. In a study of 13 children, four experienced diarrhea when consuming over 65 g per day, which is again quite a large amount. Studies have reported that adaptation occurs after several weeks of consumption. An alternative of xylitol is the four-carbon sugar alcohol erythritol (E968), ca. 90% of which is absorbed from the small intestine and excreted unchanged in the urine. Therefore, its laxative effect is much smaller compared to other sugar alcohols. Erythritol, however, is much more expensive than xylitol.

In contrast to humans, dogs are very sensitive towards xylitol. Dogs that ingested foods containing xylitol (greater than 100 mg of xylitol consumed per kilogram of bodyweight) showed low blood sugar levels (hypoglycemia), which can be life-threatening. Low blood sugar level can result in a loss of coordination, depression, collapse and seizures in as little time as 30 min. The intake of higher doses of xylitol (greater than 500–1000 mg/kg of body weight) can be fatal for dogs, and liver damage (hepatic necrosis) often appears. Other sugar alcohols (sorbitol and mannitol) have little to no effect on blood glucose concentrations or insulin secretion in dogs, although over-ingestion may result in an osmotic diarrhea. Artificial sweeteners, such as saccharin, aspartame ($\rightarrow$ **2.16**), and sucralose, are generally regarded as safe (GRAS) and should not cause significant illness even if large amounts are ingested by dogs.

What makes a substance sweet? The perception of tastes is both an art and science. Regarding the sweet taste, it was established in the 1960s that the sweetness-tasting buds at the tip of our tongue can be activated by so called weak interactions between certain parts of molecules and the taste bud. It was found that nitrogen or oxygen atoms attached to hydrogens (N–H or O–H groups) at one location, whereas nitrogen or oxygen atoms (N, O) at a second location and a water-repelling group at a third location at a given distance should be present to induce the sweet taste. The validity of this model is proven by the fact that the amino acid phenylalanine can occur in two forms: the right-handed one tastes slightly sweet, while the left-handed one tastes bitter because they fit differently into the above-mentioned triangular "pocket". Later, it was found that five more regions are also present in this pocket. Thus, the current model includes eight molecular fragments in our tasting buds that are responsible for the perception of sweetness. The sweet substances can interact with at least two or three such fragments: the more fragments interact, the more intense sweetness is perceived. This is the reason why substances of so different structures (glucose, saccharin, aspartame, amino acids, certain proteins, terpenes etc.) taste sweet.

Inventing good sweeteners is a challenging task. One has to consider nutritional, physiological, processing, chemical and physical properties, quality and intensity of sweetness (eventual aftertaste etc.), effect on teeth (cariogenicity), behavior during cooking/baking, consumer attitude and economic impact. Most often for different purposes different sweeteners are recommended as long as our desire for sweet food does not cease…

2.15 Bittersweet Effects: Grapefruit Seeds

Grapefruit is one of the most widely known tropical fruits. In recent times, its extract, which is prepared from the seeds, is marketed and advertised for medicinal purposes as well. Grapefruit seed extracts are claimed to have antibiotic, antifungal, immunostimulant activities and are very popular among consumers who prefer natural remedies instead of synthetic drugs. This preference is triggered by several articles on the Internet, with claims such as "it is effective in killing over 800 bacterial and viral strains, 100 strains of fungus, and a large number of single and multi-celled parasites. No other naturally-occurring anti-microbial can come close to these results. However, as a non-toxic, natural remedy, it had none of the side effects of the other treatments!". However, common phytotherapy textbooks do not list grapefruit as a medicinal plant. Something is amiss here. Does grapefruit really possess such extraordinary antimicrobial effect, which makes this plant one of the most prominent medicinal plants?

Grapefruit (*Citrus paradisi*) is a tropical citrus species originating from the islands of the Caribbean Sea. It is presumably the natural hybrid of sweet orange (*C. sinensis*) and pomelo (*C. maxima*) and was described first on the island of Barbados in the eighteenth century. With a height of 5–6 m, the grapefruit tree is very similar to the pomelo, but the fruits are smaller. Grapefruits are valuable constituents of the diet as they contain a significant amount of water-soluble vitamins and flavonoids. The latter compounds are also responsible for the orange color of the fruit. The diversity of color (from pale yellow to pink) is attributed to the remarkable carotenoid content. The bitter limonoids are further characteristic compounds in grapefruit (Fig. 2.26). They are concentrated in the seeds and furocoumarins, which can be found in the fruit peel.

A large number of grapefruit-containing dietary supplements are available on the market; the majority of them are based on grapefruit seed extract. Taking into account the known components of the seeds, the rationale of the application of these products is rather questionable. However, it cannot be ruled out that ethnopharmacologic data may support the therapeutic value of this plant part, which may be confirmed by future research. Unfortunately, there are no data from folk medicine supporting the widespread and specific medicinal application of grapefruit seeds, and scientific studies on the plant are fairly limited. The majority of scientific papers deal with the liver enzyme inhibiting effect of grapefruit juice. This effect is attributed to the furocoumarins of the pulp that block certain isoenzymes of the liver, and decrease the metabolism of several medicines and other compounds as a result. This may lead to the augmentation of pharmacological activity and to the increase of frequency and severity of side effects at the same time. This phenomenon results in a contraindication for grapefruit juice drinking in cases when certain medicines are applied.

Grapefruit juice, peel and seeds have antioxidant effects because of certain components present in them. This activity is one of the main reasons for the reduction of cardiovascular risks described in regular grapefruit (or grapefruit juice) consumers. However, such effects have not been confirmed and seem to be irrelevant for grapefruit seed extracts.

The majority of scientific papers report the antimicrobial effects of different parts (peel, seed) of the fruit. Grapefruit peel extracts are effective especially against Gram positive bacteria. Similar activity has been reported for the limonoids of the seeds. In one experiment, a product containing grapefruit seed extract proved to be effective against several bacterial strains. This activity was ascribed to some qua-

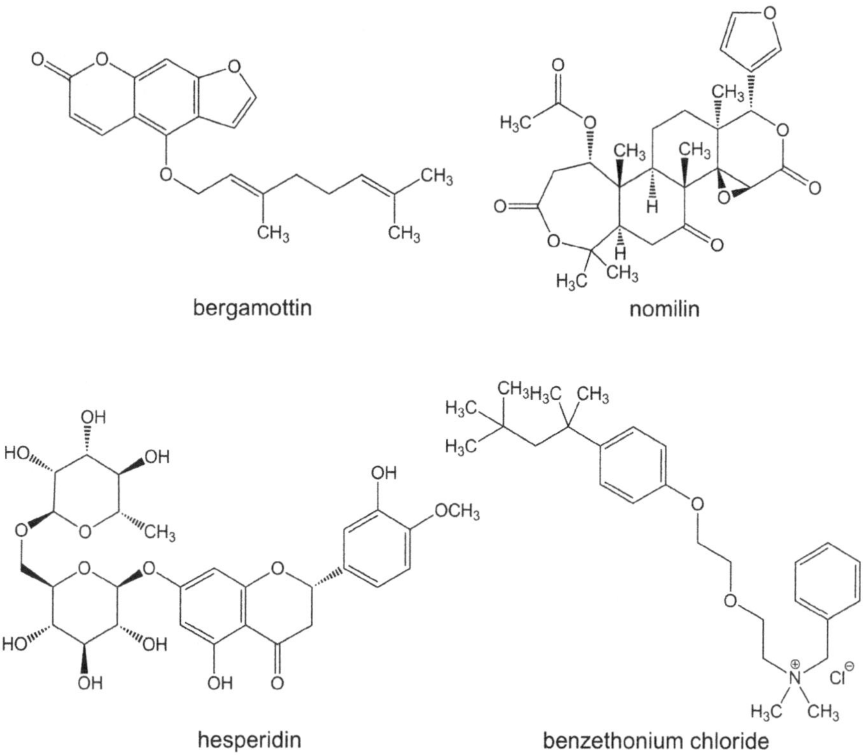

2.15 Bittersweet Effects: Grapefruit Seeds

Fig. 2.26 A characteristic furocoumarin (bergamottin), a limonoid (nomilin) and polyphenolic compound (the flavonoid hesperidin) described in grapefruit and the synthetic benzethonium chloride—no chemical relationship. (Authors' own work)

ternary ammonium salts detected in the product. These compounds (benzethonium and benzalkonium chloride) are not genuine natural products, and presumably were present as preservatives in the grapefruit seed-containing dietary supplements. In a further experiment, the compositions and activities of six grapefruit seed extract-containing products were analyzed. Five of the six preparations exhibited marked antibacterial effect and significant amounts of benzethonium chloride were detected in all the effective products. The only dietary supplement free of benzethonium was inactive and the seed extract prepared by investigators from an authentic seed sample also lacked antimicrobial activity.

Based on all the available data, it seems feasible that *the antibacterial effect of grapefruit seed preparations is based largely on the presence of preservatives.* And to top it all, not even the "fortified" products are able to kill the bacteria causing systemic diseases. Although quaternary ammonium salts have local antibacterial effect, this is by no means the same as antibiotic efficacy. To achieve such an effect, the concentration of antimicrobial compounds should reach a value that is sufficient for killing bacteria in blood plasma and tissues as well. In the case of benzethonium and benzalkonium salts, there is no evidence for a systemic antibacterial (antibiotic) effect. Only one, not very convincing human trial is available concerning the therapeutic antibacterial activity of grapefruit seed. In a case study carried out in Nigeria, four patients with urinary tract infection were treated with grapefruit seeds (participants had to consume 5–6 seed every 8 h for 2 weeks). Although the therapy was reported to be successful, the methodological weaknesses (first of all, there was no control group) of this study question the validity of the results.

Despite the lack of evidence for therapeutic efficacy, grapefruit seed extract is very popular and a plethora of products is available on the market, which are promoted as natural antimicrobial agents for both internal and external use. There are books devoted to this material, which is claimed to be able to treat various diseases such as eczema, acne, sore throat, athlete's foot, gastrointestinal infections, gastric and duodenal ulcers, allergies, and parasitic diseases in a safe and efficacious manner. Although there is no scientific or empirical evidence to support these indications, it cannot be ruled out that grapefruit seeds may be useful under certain conditions. However, grapefruit seed extracts may become part of modern medicine only after their efficacy is confirmed. Good quality products are also required. At the moment, much remains to be done to fulfill the latter criterion. A quality control study carried out in Hungary revealed that the majority (4/6) of the tested products did not contain grapefruit seeds; one of the products with inferior quality contained significant amounts of furocoumarins indicating the presence of grapefruit peel, and the main component of another product was benzethonium chloride. According to some manufacturers, grapefruit seed extract is prepared by a proprietary manufacturing process that involves the conversion of polyphenols in grapefruit seed to a quaternary ammonium salts. Unfortunately, this explanation contradicts the rules of organic chemistry….

2.16 Sweet Dreams Without Sugar: Artificial Sweeteners

Hearsay and hoaxes about artificial sweeteners are among the most widely discussed aspects of food additives. In addition to natural sugars and sugar alcohols (sugar substitutes), artificial sweeteners make up a considerable part of this category of additives. Natural sugars (sucrose, glucose, fructose, etc.) have their own problems as we have already discussed ($\rightarrow$ **2.14**): although Winnie the Pooh and other animals like honey, human metabolism has not evolved to deal with the immense amount of sugars (and other simple carbohydrates) regularly consumed today. The results of this chronic sugar poisoning are diabetes, obesity, and cardiovascular problems, to name just the most important issues. The use of other type of sweeteners has been presented to consumers as a possible solution to reduce the overconsumption of sugars. Table 2.6 lists a number of common artificial sweeteners (the last three entries are of natural origin!).

Unfortunately, the public views these compounds with deep distrust. This essay will summarize two seminal stories about artificial sweeteners.

2.16 Sweet Dreams Without Sugar: Artificial Sweeteners

Table 2.6 Summary of artificial sweeteners and their properties. Maximum levels are allowed only in products consumed in very low quantity such as chewing gum or breath-freshening microsweets—no sane person would eat 1000 pieces of chewing gum daily. Allowed maximum levels may slightly differ in some countries

Name	Relative sweetness, saccharose (sucrose)=1	Typical levels used in food	Maximum level
		(mg/kg or mg/l)	
Saccharin	300–400	80–200	3000
Cyclamate	30	100–250	3000
Aspartame	180	350–500	2500
Neotame	~5000	20–60	250
Acesulfame	200	250–350	6000
Sucralose	600	200–400	3000
Thaumatin (natural)	~2000–5000	50	400
Neohesperidin DC (modified natural)	1500	30–50	400
Stevia (natural glycosides)	250	100–250	3300

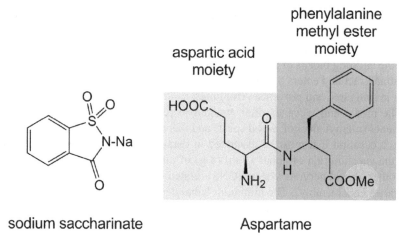

Fig. 2.27 The structure of saccharin (as sodium salt) and aspartame. (Authors' own work)

2.16.1 The Saccharin Case in a Nutshell

The restrictions on performing human experiments were not anywhere near as strict 80–100 years ago as they are today. Such tests revealed no problems on consuming 5–10 g (!) doses of saccharin (Fig. 2.27), and no serious toxic side effects were observed. Therefore, it was a great surprise when, in 1977, Canadian researchers reported that saccharin was found to be carcinogenic. The testing conditions were quite specific: only male specimens of certain species of rats displayed malignant bladder tumors, having consumed an unrealistically high level of saccharin in their food for more than a year (in human measures, nearly a thousand bottles of soft drinks daily for a year or so!) The use of saccharin was immediately banned in North America and a few other countries. However, most countries were more cautious in their response to the report and did not follow suit because of the questionable scientific proof and methodology. Since then, a huge number of independent investigations have been performed (for example, follow-up experiments for more than ten generations in animals) resulting in the complete rehabilitation of saccharin, which was declared safe for consumption once more. Investigations have shed light on the carcinogenic action of saccharin in rats. More recent results showed that its mode of action is physical rather than chemical. Owing to the special micro environmental conditions in the bladder of rodents, saccharin salts crystallized and the continuous long-time mechanical damage caused by these microcrystals eventually led to tumor formation.

2.16.2 Aspartame and Methanol

The sweet taste of aspartame was serendipitously detected in 1965. The great advantage of aspartame is that this simple dipeptide comprises two common amino acids: phenylalanine and the methyl ester of glutamic acid (Fig. 2.27). During the rapid digestion of aspartame in the gut, it is split into its component parts, which are also the building blocks of proteins. Aspartame itself *never* absorbs from the gastrointestinal tract and enters the body, it cannot be detected in the blood. This fact has been investigated and proven several times.

The media and websites often marked in red letters that aspartame contains methanol (methyl alcohol, wood spirit) and therefore is extremely dangerous. This is an accusation that must be discussed in detail. Methanol is really highly toxic and the consumption of about 10 ml (8 g) of pure methanol may cause blindness and other central nervous system (CNS)-related symptoms (individual susceptibility varies considerably). The minimum lethal dose is 0.3–1 g/kg body weight. Not regarding industrial exposure, acute toxicity may most often occur through the consumption of illegally concocted spirits (unfortunately, serious mass accidents happen every year). It is not widely known that *methanol is present in our diet*, it occurs naturally in fresh fruits and their juices in the form of methyl esters of different fatty acids. These are the so-called fruit esters with distinctive fruit-like odors. Additional methanol sources are the methylated polysaccharides such as pectin, the natural gelling agent in jellies and jam. Gastric juice does not act enzymatically on pectin. However, anaerobic digestion by several bacteria in the colon can yield methanol in nearly stoichiometric amounts. In addition, a tiny amount of methanol is always the by-product of natural fermentation processes. Thus, it also occurs at low levels in alcoholic drinks. The origin of methanol in drinks is also related to the pectin content of fruits. Last but not least, the human body also produces endogenous methanol (Table 2.7).

2.16 Sweet Dreams Without Sugar: Artificial Sweeteners

Table 2.7 Possible methanol sources

Endogenous methanol production by the human body:	0.3–1.2 g/day
From dietary sources 1 l of fresh 100 % fruit juice	About 0.14 g
1 kg of apples	About 0.5 g (from the bacterial degradation of the pectin content)

Table 2.8 The maximum allowed methyl alcohol content of some selected spirits. Actual values vary considerably in different products

Categories	EU max. allowed level expressed in	
	Grams per hectoliter of 100 % vol. alcohol	Grams per liter of 33 % vol. alcohol
Grape marc spirit (produced by distillation of fermented grape marc)	1000	3.3
Fruit marc spirit (produced by distillation of fermented fruit marc)	1500	5.0
Fruit spirit (produced by distillation of fermented fleshy fruits or must)	1000–1350 (depending on fruits)	3.3–4.5
Cider spirit and perry spirit (produced by distillation of fermented fleshy cider or perry)	1000	3.3
Brandy, wine spirit (produced by distillation of wine)	200	0.66
Vodka (obtained by fermentation of potatoes and/or cereals followed by distillation), whisky	10	0.03
Gin (made from ethyl alcohol of agricultural origin)	5	0.015

Tiny amounts of methanol pose no problem at all: the intracellular chemical factory called metabolism gets rid of it without any problems. The capacity is not very high, though. Larger amounts of methanol lead to a toxic state as its rate of elimination is about 7–8 times slower than that of ethyl alcohol.

Table 2.8 shows the maximum allowed methanol contents of several spirits as it is regulated by the European Union. They are officially expressed in grams per hectoliter of 100 % volume alcohol, but they are included in the table in the more user-friendly grams per 1 l of 33 % by volume drink as well. Note that fruit-based distilled drinks have considerably higher methanol content than grain-based ones (vodka, whisky and alike).

And what about the methanol from aspartame? For a 60 kg human, the **allowed daily intake** (ADI) of aspartame is 2.4 g. The degradation of aspartame yields 11 % methanol, which is 0.26 g in this example. 1 l (about 1 quart) of an average low-calorie soft drink may contain max. 0.6 g of aspartame, this will produce 0.066 g of methanol. Compare this value with those presented in the two tables above. It is obvious that *the consumption of a few apples or swigs of fruit spirits might cause a much higher methanol burden than that originating from aspartame used to sweeten low-calorie soft drinks or yoghurt or anything else!*

Fig. 2.28 Oil trees live for several centuries. (Copyright-free Wikipedia picture)

2.17 An Ointment for Your Throat: The Secrets of Olive Oil

Olive oil is becoming increasingly popular even far away from the Mediterranean Sea. Everyone seems to know that it is healthy, but its actual benefits are seldom mentioned. There are numerous brand names in the stores and the prices also vary greatly. Why is olive oil special? Is it truly special? Is the fatty acid composition better than in other oils? Its price is about double the price of common sunflower oils. Is olive oil a good value for the money, or is it just very successful marketing?

There is ample evidence showing that people living in Mediterranean Europe enjoy better health on average than those inside the continent. The main reason is the **Mediterranean diet**, whose ingredients (fish, vegetables, fruit) are present in other regions as well. The use of olive oil, however, is—or at least used to be—a unique feature in this region. Various scientific studies have proved that olive oil helps prevent cardiovascular diseases, cancer and old-age dementia. Even in scientific articles, it is common to equate the population on a Mediterranean diet with people consuming a lot of olive oil. This may be an oversimplification, but olive oil is clearly a dominating component (marker). Let's see if the composition of olive truly lives up to the hype.

Olive oil is produced from the fruit of olive tree (*Olea europaea*) native in the Mediterranean (Fig. 2.28). It is greenish yellow, it has a characteristic, pleasant taste and odor. There are several different types of olive oil, the difference is the method of pressing and the growing area. Virgin olive oil is made by pressing ripe fruits, the residue remaining (in Italian, *pomace*) can be pressed further and extracted with an organic solvent to obtain more, but lower quality oil. This lower quality oil is called by several names: *pomace, sansa*, and even a few other names, but the word virgin is not permitted on the label. Even among virgin olive oils, there is a premium group called extra virgin. Refined olive oil is also sold.

As far as fatty acids go, there is no basic difference between the different types of olive oils: there are large amounts of monounsaturated and polyunsaturated fatty acids ($\sim 70\%$ oleic acid and $\sim 10\%$ linoleic acid, $\rightarrow$ **2.13**). The fatty acids are mostly bound in mono-, di- and triglycerides in olive oil, the relative amount of glycerides

2.17 An Ointment for Your Throat: The Secrets of Olive Oil

can easily reach 99%. Extra virgin olive oil has the lowest content of free fatty acids, no more than 0.8%. In virgin olive oil, this is below 2%. These two types of olive oils are the most enjoyable, and also the most expensive. *Pomace* and refined olive oil have more neutral taste. The latter is prepared from virgin olive oil that has a high content of free fatty acids or an unpleasant taste.

In comparison with other vegetable oils, olive oil does not seem special at all. There is a higher content of healthy fatty acids in quite a few vegetable oils (corn oil, grape seed oil, peanut oil, sunflower oil). Considering the fatty acid content alone, it is difficult to understand why communities using these other vegetable oils do not show the benefits seen for olive oil.

The answer must be in the minor components of olive oil. Some of these (vitamins A and E, squalene, phytosterols) also occur in other oils, so they cannot be responsible for the special effects of olive oil. The odor and the aroma are caused by volatile compounds in the oil, with *trans*-2-hexenal being the most significant of these. Chlorophyll, pheophytin (chlorophyll without magnesium) and carotenoids are responsible for the color of olive oil, which is often regarded as exquisite. The color, however, is not connected to the beneficial effects, although greener oil obtained during the first pressing is more valuable than that obtained later during processing. Phenolic compounds such as tyrosol and hydroxy-tyrosol also contribute to the taste and antioxidant effects. Clinical studies in the late 1990s and early 2000s proved that these phenolic substances increase the protecting effect against cardiovascular diseases. Olive oil richer in these compounds lowered cholesterol levels more significantly.

A combination of favorable fatty acid content with the presence of phenolic compounds is still not enough to understand the health benefits of olive oil. American scientist Gary K. Beauchamp published an important paper in the prestigious scientific journal *Nature* in 2005, which helped unravel this mystery. The study reported the isolation of an anti-inflammatory compound from extra virgin olive oil. The discovery was prompted by the off-hand observation that extra virgin olive oil has a taste and a throat-irritating effect similar to that of the non-steroid anti-inflammatory drug ibuprofen. Gary K. Beauchamp worked at the Monell Chemical Senses Center, Philadelphia. This institution studies the organoleptic and pharmacological properties of different substances, so the discovery was anything but accidental. The taste of the oil focused attention on studying anti-inflammatory components in olive oil. A dialdehyde later named oleocanthal was isolated as a result (Fig. 2.29). It was also determined that the level of oleocanthal is somehow connected to the 'pungency' of the taste of olive oil. Its influence on inflammatory enzymes was also studied. The effect of oleocanthal is similar to ibuprofen: they inhibit the cyclooxygenase-1 and -2 (COX-1 and COX-2) enzymes. Oleacanthal is actually more effective than ibuprofen: at a concentration level of 25 µmol/l, the inhibition of COX enzymes was 56%, compared to 15% from the synthetic drug.

In Spain and Italy, the average annual consumption of olive oil is 14 l/capita, which is 40 ml a day. In Greece, the average is even higher: 26 l/year/person (70 ml/day/capita). Extra virgin olive oil has an oleocanthal concentration about 200 µg/

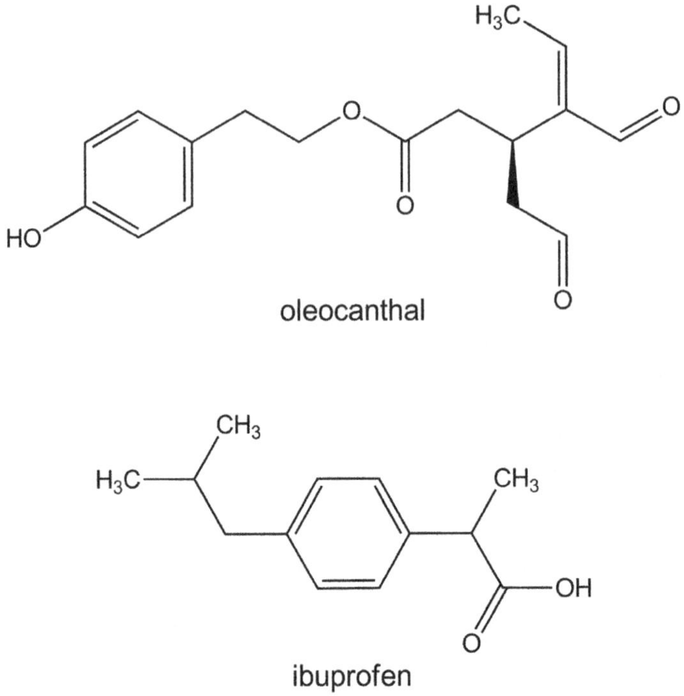

Fig. 2.29 Structural formulas of oleocanthal and ibuprofen. (Authors' own work)

ml, which translates into absorbing 9 mg of olecanthal from a rough average of 50 ml of olive oil a day. This has about one tenth of the anti-inflammatory potential of ibuprofen used in therapy. This amount is insignificant to treat acute inflammations or to relieve pain, still, the continuous exposure may have noticeable long-term effects. It is well known that anti-inflammatory drugs in low doses decrease the risk of heart attack as they inhibit blood clot formation. Acetylsalicylic acid, better known as aspirin, is used for this purpose in a daily dose that is about one fifth the amount used for anti-inflammatory therapy (→ **3.25**).

So the anti-inflammatory effect of oleocanthal implies that this dialdehyde may contribute significantly to the beneficial effect of olive oil, in addition to the favorable fatty acid composition. Accordingly, the effect of olive oil may be similar to the long-term effect of aspirin. A significant difference from the effect of aspirin is that oleocanthal also inhibits the COX-2 enzyme. A recent review article analyzing 91 earlier epidemiological studies concluded that regular use of non-steroidal anti-inflammatory drugs lowers the risk of certain types of cancer (mostly in the digestive system) by 36–73 %. Some studies showed that ibuprofen, which inhibits COX-2 in addition to COX-1, has a higher anticancer potential than aspirin, which is selective for COX-1. Anticancer effects for oleocanthal have not been demonstrated yet, but the discovery of this compound gave new vistas for research on the Mediterranean diet.

The importance of this substance may grow further as evidence is accumulating about the role of non-steroid anti-inflammatory drug in preventing Alzheimer's disease. New results give reason to assume that some chronic inflammatory processes also contribute to the development of this disease. The anti-inflammatory effect is supplemented by an inhibition of amyloid plaque formation for certain drugs in this group (e.g. ibuprofen). This effect is proven in animal tests, but human studies are not yet complete. Such an effect for oleocanthal is only assumed at this point, but its detection would explain why Alzheimer-related dementia is less common in the Mediterranean then elsewhere in Europe.

It is difficult to imagine that the isolation of oleocanthal is the final word in studying the benefits of olive oil. This discovery re-energized research in this area and provides a long-sought explanation for the observed effects. Thus far, what seems to be clear is that the benefits do not come from the free fatty acids or the triglycerides, but from the minor components. Their amount is highest in the virgin olive oil obtained during the first pressing, so this type is hailed as healthy with some justification. Other types of olive oils cannot as yet be assumed to be unhealthy, but their benefits may not exceed those from other vegetable oils such as sunflower oil.

2.18 To Add or Not to Add? Food Additives

An entire book could be written in an attempt to answer this question. A really unbiased report on this topic, however, would no doubt conclude that very few of the food additives give reason for any sort of concern. Unfortunately, public opinion is often shaped by sensationalism rather than by confirmed facts. Here, three examples will be presented to show how relative the words *dangerous* and *safe* could be in the public eye.

Many people proudly claim that they check the number of food additives (in Europe designated with the so-called E-numbers) in any food product before purchase. These claims usually go further to say that if the number is higher than two (or three or four, depending on the person), that is a strong reason against buying. This intention is not altogether incomprehensible, but totally misses the point.

The only way in this claimed procedure to ensure consistency would be to memorize the meaning of all E-numbers or keep their list always at hand. Reading about additives E160, E260, E300 and E410 on a food label will most probably repel the customer. Seeing the names carotene, vinegar, vitamin C and Carob gum is much more likely to have a positive impact. Too bad the actual substances are the same in the two cases. Similarly, additive E262 might not invite much trust from a consumer. The name of this substance, sodium acetate, is at best unfamiliar for most people. If only they knew that this is exactly the substance formed by mixing vinegar with baking soda. A well-known cooking trick is to get rid of too much vinegar by neutralizing it with the addition of baking soda.

2.18.1 Baking Powder

Baking powder, which is a chemical leavening agent, was discovered by several investigators simultaneously in the nineteenth century: German pharmacist August Oetker (1862–1918) is the best known in Europe, whereas Joseph C. Hoagland (1841–1899) and Cornelius Hoagland (1828–1898) are credited in the USA. There are several different types of baking powder. It is most often a mixture of baking soda (sodium hydrogen carbonate) and an acid salt of either phosphoric acid or tartaric acid (calcium hydrogen phosphate, disodium hydrogen phosphate, potassium hydrogen tartrate and so on) and it is quite stable in dry state in a well closed sachet. Baking soda produces carbon dioxide gas when it reacts with acid and this gas makes dough lighter and more airy. The reader can even try mixing some vinegar and a speck of baking soda: gas formation will be evident. The same reaction, albeit in a somewhat tamed version, occurs when baking powder does its own trick upon heating.

Baking powder is not precise enough of a name for regulators, the exact chemical composition of the stuff has to be given on food labels. The identifiers E500 (sodium hydrogen carbonate), E341 (calcium phosphate), E336 (potassium tartrate), and others are often found on the packages. Does this fact make baking powder hazardous? It is the same stuff our great-grandmothers used for cooking.

2.18.2 Salicylic Acid

The story of salicylic acid (Fig. 2.30 or its sodium salt called sodium salicylate) is strange. This substance has been known and used in the kitchen for a long time in certain countries. Despite the fact that it has proven health risks, public opinion is much more willing to accept its use than more recent alternatives, which scientific studies showed to pose much less of a danger. Salicylic acid occurs chemically bound in willow bark. This was the source where it was first identified and prepared in a pure form back in 1838. This substance has a quite strong decomposing effect on biological tissues, hence its use in skin lesion removal (keratolytic therapy). Also known for a long time is its ability to ease pains and reduce fevers, but its side-effect, stomach irritation, seriously limits its medical uses. More sensitive patients can even suffer headache, buzzing in the ears or allergy symptoms. This is why acetylsalicylic acid was invented as a medicine as early as the 1890s: it is popularly known as aspirin.

In addition, salicylic acid can also be used to kill fungi: it has long been used as a food preservative ($\rightarrow$1.3). To prevent the formation of mold, our great-grandmothers threw a little salicylic acid on top of jam or other stuff meant to be stored for a long time. However, because of the known and dangerous side effects, the food industry is not allowed to use it in any country of the world. It does not even have an E-number! It is more than just a bit ironic that salicylic acid can be sold freely in supermarkets in many countries around the world. Benzoic acid is a lot less dangerous

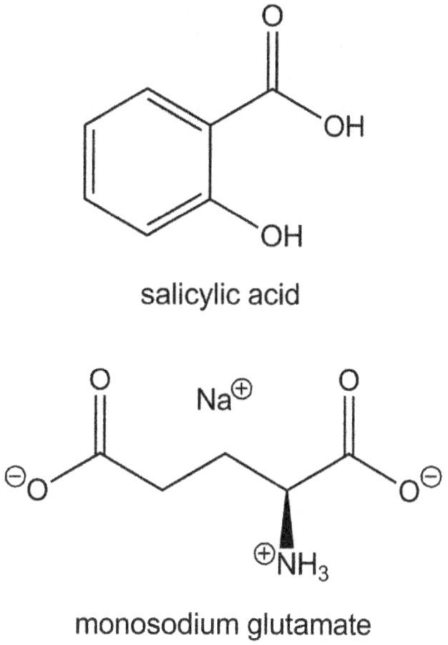

Fig. 2.30 The chemical structure of salicylic acid and monosodium glutamate (*MSG*). (Authors' own work)

salicylic acid

monosodium glutamate

than salicylic acid is, sorbic acid does not even have any known side effects. Yet the public views them with suspicion because they are "artificial" food preservatives (actually both benzoic and sorbic acids do occur in Nature, →2.1).

2.18.3 The Chinese Restaurant Syndrome and Monosodium Glutamate

Monosodium glutamate (MSG) is the sodium salt of glutamic acid, which is one of the major constituents of natural proteins present in all living creatures. In addition, it is also an essential component in higher organisms: a mediator in the central nervous system, involved in signal transmission from one nerve cell to another. Glutamic acid also happens to be a flavor enhancer. Soya sauce, fish sauce, oyster sauce and other similar additives have been use in many Asian cuisines for centuries, sometimes even for millennia. In western countries, Asian cuisine has gained widespread recognition relatively recently. Nevertheless, in Poland and Slovakia, soup flavors prepared from cereal flour have been known and used for a long time. These are made of fermented or hydrolyzed proteins, and a major component in them is MSG formed in the treatment process. The scientific study of these flavors led to the discovery of the flavor enhancing effect of glutamates in Japan in the early twentieth century. This is now understood to be a main component of the fifth primary taste (there are five of them!) called *umami*. Food additives from E626 to E635 have similar flavor enhancing effects: these are nucleosides and their sodium salts, which also occur as the ubiquitous components of RNA and DNA in all known forms of life.

Even the name of glutamic acid has its origin in wheat. The substance *gluten* occurs abundantly in high quality wheat. Glutamic acid was first prepared in a pure form from hydrolyzed gluten, hence its name.

In spite of its immense popularity, the view that high amounts of glutamate in food are dangerous and may cause headache seems to be quite generally held ("Chinese Restaurant Syndrome"). Nothing could be farther from the truth. A lot of scientific tests have been carried out, and none of them even hinted at any such possible risk when glutamate was used at the "kitchen" level. Properly controlled double blind studies have failed to establish a relationship between the "Chinese Restaurant Syndrome" and ingestion of MSG, even in individuals claiming to suffer from the syndrome. What is more, there are a number of other types of food (for example beans, tomato), which contain high amounts of glutamates in free or chemically bound form. Edible kelps and seaweeds contain outstandingly high amounts of free glutamic acid. The yeast-based food and bread spreads Marmite and Vegemite and especially Parmesan cheese also feature a high amount of glutamic acid (1.2% by weight), yet the scientific world is still awaiting the first reports of the "Parmesan syndrome".

2.19 There's Salt, and then There's Salt

Supermarkets all over the world sell two kinds of salt: sea salt and mined salt. In landlocked countries, sea salt is often more expensive than the alternative. It is difficult to understand why, although a widespread opinion claims that sea salt is of higher quality, or even "healthier" than mined salt. Frankly, this is absolutely rubbish, which may only serve the financial interests of sea salt producers.

The properties of a substance depend on its chemical composition, but not on its history. The reason is quite simple: materials do not have memory, they do not contain any information about what happened to them earlier. A careful reader may discover some contradiction here: modern forensic methods, which are often highlighted even in television crime series, can readily identify the origin of a wide variety of substances (glass or paint, for example). In fact, these methods are based on the analysis of components in these substances that are present in small amounts but quite revealing as far as the origin of the substance goes. In principle, it is not impossible to forge these clues, this even happens form time to time, but it needs a lot of know-how and money. Salt sold in supermarkets contains at least 99% sodium chloride no matter where it comes from, and the fundamental properties of this substance are independent of the other components.

Sea salt is produced from sea water, which is widely known to be salty, through the process of evaporation, most often from natural solar energy. What remains after the evaporation is usually not pure salt, some more purification is necessary. What is not called sea salt on the food labels usually comes from mines. However, eventually all salt comes from the sea. Salt deposits in mines usually originated in ancient seas or salt lakes, from which water gradually evaporated during geological ages.

2.19 There's Salt, and then There's Salt

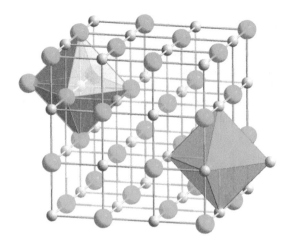

Fig. 2.31 The crystal structure of sodium chloride. (Copyright-free Wikipedia picture)

So mined salt comes from the sea, the only difference is that evaporation occurred tens or hundreds of millions years ago. This fact may even favor mined salt: seas are somewhat more polluted today than they used to be, so sea salt is more likely to contain man-made pollutants than mined salt. Incidentally, this does not mean pollutants may be present in harmful amounts in any of the salts sold in supermarkets.

The essence of table salt is sodium chloride (Fig. 2.31). For our taste, it simply does not matter whether it was crystallized from the sea a few months or a much longer time ago. What actually matters is the size of the salt crystals. If salt is put on food without letting it dissolve in water, larger crystals usually taste saltier. This phenomenon is also called salt burst in the mouth. Fine salt has no such effects. When salt has to be dissolved, for example for cooking soup or pasta, it does not matter anymore how large the crystals were before dissolution. The only thing that matters is the amount of salt used. Again, substances have no memory: once the salt crystals dissolve in water and dissociate into sodium and chloride ions, they do not remember how big their crystals were.

For both sea and mined salt, it is worth watching the amount one consumes. There is no doubt, salt is essential for the human body. However, experts cannot yet agree on exactly how much is needed for a healthy life, figures between 3 and 10 g a day can easily be found in different sources (the World Health Organization recommends 5 g per day). What they do agree on is that European or North American people usually consume much more than necessary. Scientific results linked sodium in the salt, or at least excessive amounts of it, to an increased risk of high blood pressure. Too much of a good thing can actually be bad. Too much salt can cause serious health problems or even death. This may be surprising at first, but is actually well known from the tragic examples of people who drank seawater in their ultimate desperation. It is no small irony that the high salt content in seawater actually causes the body to dehydrate seriously. This apparent paradox is caused by a natural phenomenon called osmosis. Cells are separated from the body fluids by a membrane through which water permeates following the laws of osmosis. These laws ensure that water always flows from a more dilute solution into a more concentrated solution so that the two different concentrations change toward a common value. If someone drinks seawater, the concentration of salt in the body fluids will be much larger than within the cells, so water will flow out of the cell and the cell dehydrates sooner or later.

There are salt mines and salt pans open to tourists all over the world—a visit to one of them may be a day of fun.

2.20 Sugar for Your Tea? White or Brown?

The shelves of grocery stores usually offer a sweet choice between brown and white sugar. Brown sugar is often more expensive than the more common white variety. This may very well be the fundamental cause of the belief that brown sugar is also healthier. Producers of brown sugar may not be involved in spreading this belief but they do not try to stop it, either. In fact, as far as health effects go, there is no difference between the two varieties.

What we call sugar in everyday life is a compound in the carbohydrate family of substances. It has the chemical name sucrose, and belongs to the group disaccharides, which means that its molecule is composed of two simpler sugar molecules, D-glucose and D-fructose, connected to each other through a chemical bond (Fig. 2.32). When it is important to distinguish sugar from yet another common type of sugar, grape sugar (chemical name D-glucose, a monosaccharide, a simple sugar molecule), it is usually called cane sugar or beet sugar - depending on the local climate. In areas with moderate climate, the most common source of sucrose is sugar beet, whereas tropical places usually grow sugar cane. Both of these plants contain 10–20% of sucrose.

In countries with moderate climate, cane sugar is usually also sold in grocery stores, but its price is much higher than the price of beet sugar because of the transportation costs. Cane sugar and beet sugar are chemically the same. Why some people experience a sweeter taste for a more expensive sugar is a very interesting scientific question, but currently it is psychologists who try to answer it.

Sugar production is nothing else but removing everything from sugar cane or sugar beet except sucrose. The 'everything else' is called molasses for both plants. A slight difference here is that the molasses from sugar cane is suitable for human consumption, whereas the same is not true for sugar beet. The color of brown sugar is caused by small amount of cane molasses. In principle, brown sugar could be prepared by stopping somewhere during the white sugar preparation process, but it is more common to prepare white sugar first and then add some molasses back in. Brown sugar can be produced from sugar beet by first making white sugar, then adding some sugar cane molasses. The reader may have already realized that mountains of molasses must pile up during sugar production, but their exact fate will be left in darkness here.

2.20 Sugar for Your Tea? White or Brown?

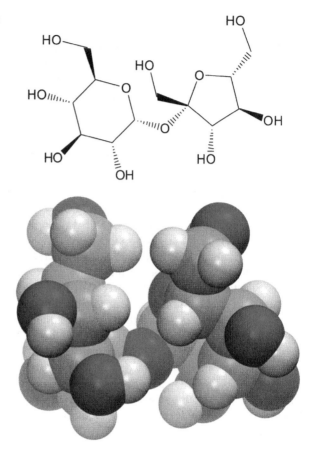

Fig. 2.32 Chemical formula and three-dimensional molecular model of sucrose

It does not take a star detective to conclude that brown sugar is nothing else but dirty sugar. Experience shows that a little bit of dirt does not threaten much harm to humans. It can even be somewhat beneficial because it prepares the immune system for more serious attacks. Many doctors say that allergies have become very common during the last two or three decades (at least) partly because of too much cleaning: the human body does not learn how to distinguish between dangerous pathogens and harmless other stuff, therefore it overreacts to the latter. Nevertheless, it cannot be concluded that brown, that is dirty, sugar is healthier for this reason than the white variety. White sugar is very pure sucrose, whereas brown sugar is just a little less sugar with a range of other substances. Purifying something cannot make it less healthy.

To support claims for a healthier brown sugar, it is often said to contain minerals (manganese, copper, iron) and vitamin B^6. This is not untrue as these components are present in the molasses of sugar cane. However, in line with this argument, eating dirt would be even more desirable as it contains more minerals. Another fact to consider is that brown sugar contains more minerals than white sugar, but it is still not much more. If someone were to eat enough brown sugar to supply his or her body with the necessary amount of minerals, it would be so much sugar that it would definitely be incompatible with human health.

From a health perspective, it does not matter which sugar you use. If someone likes the dark color or the taste of molasses in brown sugar, he or she can lead just as a healthy life as a more conservative person with a preference for white sugar.

2.21 The Thickening Stuff: Guar Gum Gumbo

Some of the readers may recall that unacceptably high levels of pentachlorophenol and polychlorinated dibenzodioxins (e.g. TCDD→**4.5**) were found in guar gum (also called guaran) sold by India Glycols Ltd. in 2007. The pollutants seemed to have an environmental origin, so other Indian suppliers were placed under suspicion, too. Through a trading company in Switzerland, tainted goods reached nine EU countries, and the European Commission ordered all member countries to take necessary action, which included a recall of food products (e.g. yoghurts) containing guar gum. In 1999 in Belgium, a large quantity of food had to be destroyed because of high dioxin levels, and the resulting shortage of food culminated in calling early elections. Dioxin surfaced again in Germany in 2011: high levels were found in human and animal food including eggs.

Natural gums are usually powders that remain after liquids from plants are dried up or they may in some cases be flours. They contain polysaccharides similar to starch. Unlike starch, they form a real solution in water rather than a **colloidal sol**. The solutions are highly viscous and thixotropic, which means that they resemble gels, but will liquefy upon mechanical disturbances. Gums serve as thickening agents, gelling agents, and emulsifying agents in the food industry (E410: locust bean gum or carob gum, E412: guar gum, E413: tragacanth, E415: xanthan gum, E418: gellan gum). They bind water quite strongly, this property is also important for their uses in the cosmetics, paper, paint and oil industries.

Guar gum is a white powder (Fig. 2.33) milled from the seeds of the plant guar bean (*Cyamopsis tetragonolobus*). Chemically, about 80% of guar gum is a polysaccharide containing D-mannose and a branching D-galactose at every second position on average (Fig. 2.34). Branching prevents the formation of a crystalline structure in guar gum, but these side chains can soak up a lot of water. Guar gum, water and a little bit of borax make an excellent and safe adhesive for children. In addition to the uses in food, guar gum is also used in medication to control digestion and to increase lactation. Guar gum decreased the levels of fats, D-glucose and insulin in a clinical test of patients suffering from type 2 diabetes. It can be used to treat both diarrhea and constipation. The purified product is also used as a lubricant in eyes.

Locust bean gum has a structure that is very similar to that of guar gum, but it has a different ratio of D-mannose and D-galactose (about 3.5:1). Traditional interpretation says that St. John the Baptist survived in the desert by eating locust

2.21 The Thickening Stuff: Guar Gum Gumbo

Fig. 2.33 Guar beans (*Cyamopsis tetragonolobus*) and the powder called guar gum. (Copyright-free Wikipedia pictures)

beans of the carob tree (Matthew 3:4), but there is some confusion as the original Greek word used here could also be taken to mean the locust insect. All 22 other occurrences of this word in the Bible clearly refer to the insect, so the modern view in this question is that John had the less attractive survival strategy of feasting on insects, which are still commonly eaten in Arabia. The uniform, small carob seeds were used as a unit of measure in jewelry. The seeds were called *qīrāṭ* (طاريق) in Arabic, and *kerátion* (κεράτιον) in Greek, which is the origin of the name of the still used unit carat. Another possible origin of the unit carat is the seed of African coral trees (*Erythrina* species) called *kurara* seeds. The seeds of the two plants did not weigh exactly the same, so old records may disagree on the size of historical diamonds depending on the seed used for comparison. Metric carat (exactly 200 mg) was introduced in Europe and the US in 1907. Locust bean gum can bind a lot of water, about 50–100 times its own weight. It can be used to treat diarrhea, especially in children who find its taste attractive. Locust bean gum has been eaten for a long time, it has no known risks. The food industry uses it in cream cheeses, ice creams and various meat products.

Tragacanth is a natural gum obtained from the dried sap of several species of Middle Eastern or Mediterranean legumes (e.g. *Astragalus gummifer*). Tragacanth is used as a food additive in crèmes and toothpaste and also has medicinal application as a binder in tablets. It is a mild laxative. Xanthan gum is produced by fermentation of D-glucose with the microorganism *Xanthomonas campestris*. It is one of the most important microbial polysaccharides; world production topped 50,000 t in

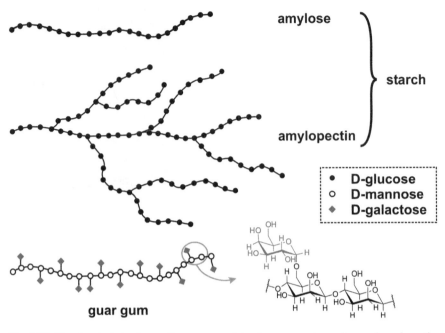

Fig. 2.34 The chemical structures of guar gum and starch components amylose and amylopectin. (Authors' own work)

1995. The food industry uses it in salad dressings, ketchup, mayonnaise, and other sauces. Gellan gum is another microbial gum made by fermentation using the bacterium *Sphingomonas elodea*.

Guar gum and other natural gums are not poisonous at all, they might cause some slight allergy if the protein residues are not removed properly. But how could pentachlorophenol and dioxin make it into guar gum? Globalization comes with the side effect that even though a certain product might be produced in one place, it can very likely be packaged in a second and sold in a third. There may be public opposition to this, but it cannot really be helped as long as society wishes to enjoy the benefits of globalization (larger choice of less expensive food). However, as a result, outdated and polluting technologies may also be moved to less developed countries, where they are still legal. This was an unexpected (and most certainly unintended) side effect of the introduction of REACH, a comprehensive European policy ($\rightarrow$ 1.6) on dealing with chemicals. In this case, developed countries were successful in decreasing the dioxin levels in pesticides, but they could not require developing nations to make the same efforts. Today, it seems impossible to decide whether pesticides with abnormally high dioxin levels were used on guar beans in the 2007 incident or whether the dioxin originated from some other source of environmental pollution.

No one doubts that dioxins are toxic, especially after the industrial accidents in Seveso and elsewhere ($\rightarrow$ 4.5). However, risk assessment is a much more complex issue. Today's legal limits were set very cautiously and are probably unnecessarily low. The public may even be shocked by the opinion of noted dioxin toxicology specialist Christopher Rappe, who says that more people make a living out of dioxin than suffer from it. Finnish dioxin specialist Jouni T. Tuomisto pointed out that a balanced risk assessment would be important (albeit not only for dioxins). As an example, a risk calculation was made for farmed salmon, which may accumulate dioxins in their fat. In the European Union (387 million residents), this probably causes 40 extra cases of cancer a year. However, the beneficial effects of fish oil consumption on the cardiovascular system prevent about 5200 fatalities annually. Life is complex, but people often prefer to oversimplify it. The hysteria surrounding the 2007 guar gum case was probably induced more by chemophobia ($\rightarrow$ 1.1), the general aversion of society to chemicals and the chemical industry, then by any factual or objectively analyzable risks.

2.22 Is Caffeine Free of Risk?

Every civilization needs stimulants, and caffeine is one of the best known. It is a significant ingredient in about 60 different plants, with coffee, tea and cocoa being the most common sources. These plants began their European career as exotic herbs. An 1671 treatise titled *Usage du caphé, du thé et du chocolate* published anonymously in Lyon listed the virtues attributed to coffee: "It dries up all cold and dump humors, drives away wind, strengthens the liver, relieves dropsies by its purifying quality; sovereign equally for scabies and impurity of the blood, it revives those who have stomach ache and have lost their appetite; it is equally good for those who have a cold in the head, streaming or heavy.... The vapor which rises from it [helps] watering eyes and noises in the ears, sovereign remedy also for short breath, colds which attack the lungs, pains in the spleen, worms; extraordinary relief after over-eating or over-drinking. Nothing better for those who eat a lot of fruit."

The coffee shrub probably originates in Ethiopia, cocoa comes from Central America, whereas tea is native to China. Muhammad ibn Zakariyā Rāzī (865–925) Persian Muslim scholar left the first written record of coffee consumption. The habit moved to Europe through Muslim and Venetian tradespeople in the seventeenth century but its real success began in Paris. Coffee consumption increased dramatically in the eighteenth century as Europe began to organize production. France imported 38,000 t of coffee in 1787, 36,000 t of which was re-sold to other countries. About 300 million people began drinking coffee in the eighteenth century. London was home to about 2000 cafés around 1700, eighteenth century Paris had 700–800 of them, and thirsty gentlemen had a choice of 500 such places even in Budapest, Hungary by 1900. Cafés were more than just places for enjoying coffee, they became place of lively society life. Artists, authors and publishers met in cafés, they were job fairs, clubs and substitute homes at the same time. Global culture owes a lot to cafés in big cities.

Fig. 2.35 The Mayan sign for kakaw(a) (cocoa) (**a**) and the Chinese character for tea (**b**). (Authors' own work and Copyright-free Wikipedia pictures)

Coffee was first a drink of the aristocracy, but later its consumption became more widespread in society and advanced a notion of a sober society that refrains from excessive drinking. Coffee replaced alcoholic drinks, beer in the north, wine in the south. The English began drinking coffee as a morning drink instead of ale in the eighteenth century. Coffee overtook beer as the most popular drink in Germany by 1979, the annual average consumption is 187.8 l/person.

Tea was first brought to Europe by Portuguese, Dutch, and English travelers from China, where it had already been popular for the previous 10–12 centuries. The first cargo of tea reportedly arrived in Amsterdam in 1610. Tea was successful in regions where grape did not grow: England, Netherlands, Northern Europe, Russia and Muslim countries, and vice versa, grape was unknown in tea-producing regions. Canton exported 7000 t of tea to Western Europe in 1766 in addition to about 500 t to Russia. Europe was unable to gain control of tea production for a long time: the first tea shrub was planted only in 1827 in Java, and as late as 1877 in Ceylon, after the coffee plantations there were ruined. The home of tea is still Asia today: about three fourths of the 2006 world production of 3.6 million t were grown in this continent (→3.17).

Cocoa was already known in Mexico as early as 1500 BC. Christopher Columbus brought the plant to Europe after his fourth American voyage in 1502, but the drink made from cocoa was first tasted only in 1528 by Spanish conquistador Hernán Cortés. Chocolate drink first appeared in Spain in 1544, cocoa trade began between Seville and Veracruz in 1585. Therefore, caffeine in cocoa made its first European appearance before caffeine from coffee or tea did (Fig. 2.35).

The main growing areas of coffee and cocoa shifted from their native lands as history moved on. Today, most coffee is grown in central and South America, whereas cocoa is mainly farmed in Western Africa (Ivory Coast alone is the origin of one third of all the cocoa on Earth). The annual world production is about 7.8 million t for coffee, and 4.1 million t for cocoa. Coffee trade is worth $ 7 billion a year, which ranks as number four after coal, oil and cereal grains.

The cultivation of all three plants has had major effects on the economies of the producing countries. Monocultures of plants in West Africa, Central America, the Caribbean and Brazil played a decisive role in history and many of the current social and economic problems originated in the age when plantations were founded.

2.22 Is Caffeine Free of Risk?

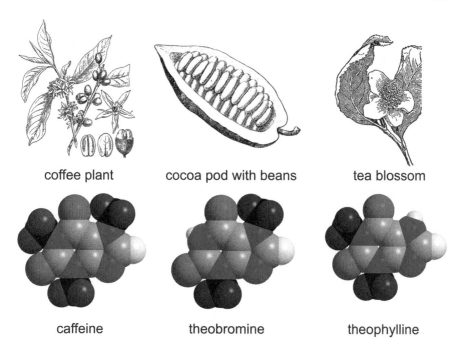

Fig. 2.36 The coffee shrub and its berries, cocoa beans and tea flowers. Shown below are the 3D molecular models of caffeine, theobromine and theophylline. The difference between the chemical structures is the number and position of methyl groups (highlighted by *dark shading*). (Authors' own work and copyright-free pictures from the book Bessette, Alan E., Chapman, William K. (eds.): Plants and flowers. 1761 Illustrations for artists and designers. Dover Publications, Inc., New York, 1992)

The dominating active ingredients in the three plants are xanthine alkaloids caffeine, theobromine and theophylline (Fig. 2.36). Raw coffee contains 0.9–1.4% (Arabica) or 1.5–2.6% (Robusta) of caffeine. Tea leaves contain 3.2–4% of caffeine, but the different preparation method results in about half as much caffeine in the drink compared to coffee. The main active ingredient of cocoa is theobromine, there is only 0.2% caffeine in it, but this small amount is still responsible for the stimulating effect.

Caffeine is one of the most studied substances in science. It stimulates the central nervous system: it counteracts adenosine in the brain, thereby inhibiting its effect causing a feeling sleepiness. Caffeine also increases the level of adrenaline (a hormone) and dopamine (a neurotransmitter). Caffeine, contrary to some beliefs, does not wake people up, it only inhibits the effects of adenosine, increases the rate of heartbeat, dilates certain blood vessels and narrows some others, and makes it easier for muscles to contract. Caffeine also boosts pattern separation memory in humans. In addition to this general stimulating action, side effects may include irritability, nervousness, headache or insomnia, especially if unreasonably high amounts are consumed. Caffeine has been linked to a number of negative effects as well (heart, liver and kidney disease, cancer, osteoporosis, ulcers, premenstrual syndrome, low sperm motility and fertility, hyperactivity, mental dysfunctions, sports performance decrease, etc.), but none of these have been justified by detailed studies. The only advisable precaution is to avoid consuming more than 200 mg of caffeine (equivalent to 1 or 2 cups of coffee) a day during pregnancy in order to lower the risk of abnormal prenatal development.

Caffeine can be toxic: the lethal dose is about 10 g for an adult of average weight. A cup of coffee usually contains 80–180 mg of caffeine, so 55–125 cups would make a lethal amount. Caffeine pills are more of a danger in this respect: consuming 2 g of caffeine may result in hospitalization.

High levels of caffeine (more than 12 mg/l) in the blood of athletes qualified as an anti-doping rule violation before 2004. German swimmer and European Champion Sylvia Gerasch was penalized in January 1994 because the caffeine level in her blood was 16 mg/l. This roughly means drinking 6 cups of coffee 30 min before the competition. The World Anti-Doping Agency removed caffeine from the prohibited list in 2004, but the substance is still monitored in order to detect patterns of misuse in sport. The 2010 and 2011 monitoring programs did not reveal global specific patterns of misuse, though a significant increase in the consumption of athletes was detected.

Theobromine and theophylline have milder effects on the central nervous system, their physiological effect is to relax smooth muscle in the lungs and act as a diuretic. These effects are used in asthma treatment. Horses are more susceptible to the effects of caffeine and theobromine than humans, so eating chocolate before a race is a sort of doping for them. Theobromine is toxic for cats, dogs and birds, so pets should not be treated to chocolate. But humans have no problem, and they can also enjoy the beneficial effects of a daily dose of caffeine.

2.23 The Biofuel Dilemma

An aircraft of Virgin Atlantic Airways flew 356 km from London to Amsterdam in 2008 using 22 t of fuel, 1 t of which was biofuel. This was a major step toward introducing biofuel into commercial aviation, but a shadow was cast upon this achievement by the fact that 140,000 coconuts were needed to produce that single ton of biofuel.

The crude oil production of the world is about 4 billion t annually, half of which is used in transportation. Aviation alone requires 340 billion l of jet fuel per year. If the entire amount is to be replaced by biofuel, a huge land area must be converted into a coconut plantation—its size would be about the combined area of France, Spain and Germany, or the largest US state of Alaska…

These data illustrate the dilemmas surrounding biofuels. Biomass is produced on a vast scale each year—about 50 times the amount of crude oil, so it makes sense to consider this as a possible source of energy for mankind. But how?

The basic needs of society include energy, raw materials and food. There is no telling which is more important than the others. Let's begin with energy here. En-

2.23 The Biofuel Dilemma

Fig. 2.37 A Mærsk Triple E class container ship (Mærsk Mc-Kinney Møller), one of the biggest ships in the world. An ice hockey rink, a football pitch, and a tennis court would fit on the deck simultaneously. Danish business conglomerate Maersk contracted Daewoo Shipbuilding to build 20 of these huge ships each of which can carry 18,000 containers. As of September 2014, eleven of the ships were ready. Some earlier supertankers were longer and heavier, but none of them are in service now. A highly energy efficient ship like this burns 100 t of Diesel a day. Cargo ships are responsible for about 4% of global air pollution. (Copyright-free Wikipedia picture)

ergy is valuable for mankind because of the services it provides. Heating, electricity and transportation all need energy (Fig. 2.37), but the requirements are very different. The problem of energy is a complex one, in which science is only one factor, so chemists actually do not have much influence on the decisions made. Yet their expertise is indispensable.

About half of plant-derived biomass is cellulose, one quarter is hemicellulose, one fifth is lignin, the remaining 5% is shared between oils and fats, proteins, nucleic acids and alkaloids. Cellulose, hemicellulose and lignin are difficult to break down: this is why they mostly serve as structural materials or supports in nature. Materials from plant-based foods (sugar, starch, oils, etc.) are easier to access, but only waste should be used for energy production. Otherwise, there is a serious conflict between food and energy supply. There are already telltale signs: the worldwide food price index increased by 65% between 2006 and 2010, which was partially caused by the application of food raw materials as energy sources.

The annual biofuel production was about 40 billion l in 2006, 90% of which was bioethanol, the rest in the form biodiesel. Some properties of these alternative fuels are described in the next paragraphs.

Bioethanol Ethanol as a fuel made a lot of progress in the US in the 1970s and peaked in 1986: about one and a half billion liters of gasohol was sold, which typically contains 10% ethanol and 90% gasoline. Altogether, 120,000 t of ethanol was blended with gasoline in that year. In 2002, 20 million t of bioethanol was produced. About 6 million US cars use E85 fuel today (85% ethanol and 15% gasoline), but the number of petrol pumps selling this fuel is relatively low.

Ethanol, commonly called simply alcohol, can be produced from virtually any plant by fermentation. The most economical raw materials are sugar cane (Brazil), corn (US), wheat, potato and sugar beet (Europe).

Brazil is home to the most advanced bioethanol technology today. About 77 t of sugar cane can be grown on a hectare, which yields 4800 l of distilled bioethanol. Waste is used as animal feed, fertilizer and combustible material. The project was very carefully planned: even the soil removed from the fields on the wheels of trucks is transported back. These fields are far away from the Brazilian rain forests and represent about 1% (3.4 million ha) of the arable land there. Crop rotation (with peanut, for example) is used to replenish the nitrogen content of the soil. The return on energy investment is quite generous: eight times the energy used up in the production process can be gained from the resulting bioethanol. About 45% of all Brazilian vehicles use ethanol or ethanol-gasoline mixture. The current plan is a threefold increase in bioethanol production by 2020. Doubling of the area of the plantations will be necessary to achieve this goal.

In the US, about 100 biorefineries produced 35 billion l of bioethanol in 2008 based on corn. The official goal is to increase this amount to 220 billion l by 2030. This would require doubling the current percentage (20%) of corn used for bioethanol production. Corn yields about 400 l of ethanol per ton.

Ethanol could be obtained from cellulose-containing waste as well (360–450 l/t of waste), but the process itself is more difficult to carry out as cellulose first needs to be broken down into sugars by a combination of physical, chemical and biological processes. The sugar formed can then be further digested by yeasts. There is very intense research to increase the efficiency of ethanol production and to find alternative technologies. Some plants can be grown in areas otherwise less than ideal for agriculture. *Miscanthus* species (such as "elephant grass"), switchgrass (*Panicum virgatum*), and hybrid poplars are all possible choices as energy crops. Another possibility is using new enzymes to break down cellulose. The Dutch company Royal Nedalco, currently part of the American multinational corporation Cargill, converts 160,000 t of food waste annually into bioethanol using a technology built on a fungus isolated from elephant droppings. A very advantageous feature of this method is that it can process not only cellulose, but hemicellulose as well, which has a high content of a sugar called xylose.

For the past decades, there have been heated debates about the energy balance and environmental impact of bioethanol production. In a scientific study, David Pimentel (Cornell University) concluded that bioethanol is "energy negative", meaning it takes about 30% more energy to produce than is contained in the final product. Critics of this work say that any production of energy requires energy input in some other form. Other calculations have shown bioethanol "energy positive" by

2.23 The Biofuel Dilemma

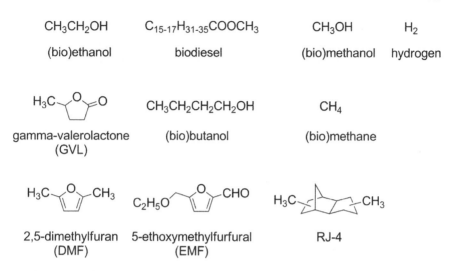

Fig. 2.38 Fuels of the future? (Authors' own work)

27–120 %. The exact number heavily depends on factors that are not usually considered scientific (energy independence, price of crude oil, national security issues and political factors).

Biodiesel Biodiesel is composed of fatty acid methyl esters (Fig. 2.38). These can be produced by the chemical modification of animal fats or plant-derived oils. Methanol is needed as a raw material in the process, and the by-product glycerol forms. Biodiesel is very similar to petroleum-derived Diesel as far as fuel applications go, but its waxing temperature is usually higher. This means extra care is needed in cold weather. Its transportation is not ideal through pipelines as it tends to stick to the pipe walls and may contaminate other fuels. This would be especially dangerous for jet fuels, so biodiesel is almost exclusively transported in tank cars or trucks. Producing biodiesel is definitely energy positive, the gain is about 93 %.

The world uses about 500 million t of Diesel each year. The annual production of plant oils is only 120 million t, only 6 million of which is used for fuel production. These numbers show that replacing Diesel with biodiesel on a large scale would be a huge problem. The most common raw materials used for producing biodiesel are sunflower, rapeseed, coconut, palm, or jatropha oils. Rapeseed oil is very popular, about 3600 l can be grown on a hectare. Animal fats can also be used: a pilot plant in the US processes 1 million t of waste chicken fat into 1 billion l of biodiesel.

Algae could produce oils about five times more efficiently than plants, sometimes even under extreme conditions. Yet, the technology needs to be developed further to be competitive economically.

The Finnish oil refining and marketing company Neste Oil has developed a notable alternative technology for processing animal fats and plant oils. The raw materials are reacted with hydrogen, the fatty acids are transformed into hydrocarbons similar to Diesel or paraffin wax (renewable Diesel), whereas glycerol is transformed into propane gas. The company finished its third such plant a Singapore in 2010. Currently, it is the largest renewable Diesel plant in the world, with annual production of 800,000 t.

Biomethanol California-based American–Hungarian Nobel Laureate George Olah (born Oláh György) is probably the most noted supporter of methanol economy. The core idea is to produce methanol from virtually all carbon-containing material (biomass, natural gas, coal, glycerol obtained as a by-product of biodiesel production, carbon dioxide in the air, carbon monoxide produced as a by-product of steelmaking). In the case of carbon monoxide, the chemical equation $CO + 2H_2 = CH_3OH$ would be the essence of production. Further chemical transformation of methanol could yield hydrocarbons similar to those obtained in petroleum refining. The technology to use these as fuels or as feedstock for the production of further materials is well known. There is a drawback to methanol, however: it is somewhat toxic. Drinking small amounts causes blindness, whereas consuming larger amounts will result in death. George Olah often makes the counterpoint that he has never ever seen anyone drinking gasoline at a pump, but the problem is a little more complicated as methanol smells a lot like the considerably less dangerous ethanol, which is the alcohol in alcoholic drinks. Accidental death caused by methanol consumption is not unknown in industry even today…

About 33 million t of methanol is produced today annually, mostly from natural gas. China has a huge reserve of coal, and the idea of methanol use has been particularly attractive there. China already uses 7 billion l of methanol as a fuel today and is planning to build 200 (!) plants to produce methanol from coal in the near future.

Biohydrogen Hydrogen is beyond doubt the cleanest of all fuels. Burning it yields nothing else but water: $2H_2 + O_2 = 2H_2O$. This combustion process can also be used for direct electricity production in fuel cells. On the other hand, there are quite formidable obstacles—both theoretical and practical—to using hydrogen as a fuel. Hydrogen is the lightest of all gases, it only liquefies at $-253\,°C$, its energy density in unit volume is only one eighth of gasoline even when hydrogen is compressed to a high pressure of 200 atmospheres. Its storage, transport and use pose special safety problems, which can only be handled in the aerospace and aviation industries today (rocket fuel). It is highly unlikely that hydrogen-powered vehicles will make a conquest on the roads any time soon.

About 40 million t of hydrogen is produced annually as a byproduct of chlorine production, which—incidentally—consumes a vast amount of energy. Establishing a meaningful hydrogen economy would require at least 8 times this amount. Economical hydrogen production from other sources (biotechnology, solar power, nuclear energy) has not been solved yet.

Gamma-Valerolactone (GVL) GVL is a non-toxic liquid with high boiling and low melting point, nice odor, and a research octane number (RON) of 130. It occurs in some food and seems to be an ideal fuel. Hungarian-born chemist István Tamás Horváth (currently working in Hong Kong) is credited for developing an economical way of producing GVL from cellulose. GVL can also be used as a chemical feedstock.

Other possible fuels (Fig. 2.38) include biobutanol, biogasoline, biomethane (biogas), various furans (2,5-dimethylfuran, 5-ethoxymethylfurfural) and terpene derivatives (e.g. RJ-4). All of these have advantageous properties and some economic potential in certain application areas.

The initial enthusiasm about the use of biofuels seems to have died down and the problems are much more clearly seen today. The EU production of bioethanol was 2 billion l in 2008, but this is not a profitable business. Germany made 3 billion l of biodiesel, but it came with a high price tag. A number of experts, including those in the Netherlands Environmental Assessment Agency propose that a more economical strategy could be burning biomass (about 3% of which is currently used) for electricity production, and using that to produce liquid fuels. The overall efficiency of cars today is about 30% (hybrids can score as high as 60%). If these numbers do not change, biofuels are unlikely to meet the needs of road transportation. Investing in improving fuel economy seems to be a much better strategy today. The airline industry cannot make long-term plans based on biofuels, either: hydrogen could be a viable alternative, but a method to produce it economically is still badly needed.

2.24 Perfect Timing: Egg Cooking

Worldwide egg consumption was 61 million t in 2006, with Asia alone responsible for 37 million t. Eggs are considered staple food, and boiling them is one of the simplest bits of cooking. Most cookbooks recommend boiling for 3–6 min to prepare a soft-boiled egg and 8–10 min for hard-boiled eggs. Hard and soft refer to the egg yolk, the egg whites are (or at least should be) solid in all boiled eggs. Overcooked eggs have rubber-like whites and crumbly yolks with a thin green layer on the outside. Cookbooks seldom specify the temperature needed for boiling an egg, the universal assumption is that boiling hot water is meant, which is close to 100 °C (small differences may be caused by elevation and weather conditions). A scientific mind, however, enjoys looking at still further questions. Does size (of the egg, or even of the pan) matter? Does initial temperature (fridge-cold or ambient egg) make a difference? Is it better to put the egg into cold or hot water? If the water is cold in the beginning, when should the stopwatch be started? Is the volume of the water important? How about the power of the burner? These questions have received a surprising amount of attention from egg-loving scientists. In order to find some answers, the reader should first get familiar with the structure of an egg.

About 10% of the mass of an egg is the shell, the yolk is about 30%, and the remaining 60% is the white (Fig. 2.39). The shell is mostly made up of minerals: calcium carbonate, or more precisely, calcite dominates with small amounts of magnesium carbonate and various phosphates. The shell also contains about 3.3% proteins in a form called mucopolysaccharide complex, which is a network of globular and fibrous protein bits which hosts a large number of calcite crystals (about

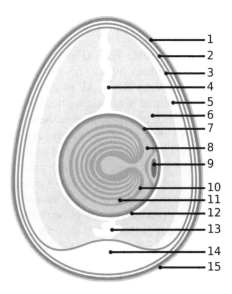

Fig. 2.39 The schematic structure of an egg: *1* eggshell, *2* outer membrane, *3* inner membrane, *4* Chalaza, *5* exterior albumen (outer thin albumen), *6* middle albumen (inner thick albumen), *7* vitelline membrane, *8* nucleus of pander, *9* germinal disk (blastoderm), *10* yellow yolk, *11* white yolk, *12* internal albumen, *13* chalaza, *14* air cell, *15* Cuticle. (Copyright-free Wikipedia picture)

50 times its own weight). The egg shell has four distinct layers. The outermost is called the cuticle, then the palisade layer followed by the mammillary layer and the innermost part is a porous membrane. The cuticle is 10–30 μm thick, it is made of a transparent, mucous protein that also has carbohydrate parts and makes the shell water-resistant. The palisade layer is responsible for about two thirds of the entire thickness and is responsible for the strength of the eggshell. Its main components are calcite and the protein network already mentioned. The mammillary layer looks like a collection of knobs connected to the palisade layer outside and sticking to the membrane on the inside. The shell membrane is divided into an inner and an outer layer (48 and 22 μm thick), both of which are mostly composed of interconnected protein-polysaccharide fibers. The air cell is located between these two layers

The number of pores in an egg is about 6000–10,000, the diameter of a pore is between 10 and 20 μm. These pores are present throughout the entire eggshell. The cuticle is permeable for gases but partially blocks the pores, which provides a limited protection against microbial pathogens. Eggs with damaged cuticles (by friction, for example) pose a higher risk of bacterial infections. The cuticle penetrates the palisade layer to different degrees in the eggs of different birds: quite deeply in the eggs of guinea fowls, less so for chicken eggs. The cuticles of duck eggs barely cover the pores, so these get infected quite easily and can only be stored for short times. If the pores are blocked artificially, e.g. with sodium silicate (also called soluble glass), bacteria cannot enter the eggs and water cannot evaporate. Eggs preserved in this way can be stored for almost a year, but may explode on boiling because the formation of water vapor may increase the internal pressure. To avoid this phenomenon, a small hole must be pierced on the shell with a needle. This may be useful even if the egg is not preserved with silicates: it will prevent the eggshell from cracking. As time goes on, the water content of the egg white evaporates through the shell and air fills up the remaining space. Air expands when the egg is

2.24 Perfect Timing: Egg Cooking

Table 2.9 A. Denaturation temperatures of egg constituents. B. Times needed to prepare honey-like egg yolks

A. Denaturation temperature of egg constituents		B. Temperature and time necessary to cook egg yolks to the consistency of honey	
Proteins	Denaturation temperature (°C)	Temperature (°C)	Time (min)
Egg white		60	310
Ovotransferrin	61	61	200
Ovomucoid	70	62	125
Lysozyme	75	63	75
Ovalbumin	84.5	64	55
Globulin	92.5	65	45
Egg yolk		66	40
LDL	70	67	26
HDL	72	68	25
Alpha-livetin	70		
Beta-livetin	80		
Gamma-livetin	62		
Phosphovitin	>140		

heated, so old eggs crack more readily than fresh ones. The volume of the air cell indicates how fresh the egg is: very fresh eggs tend to sink in salty water, old ones float instead.

Egg white is in aqueous solution of proteins, the concentration is about 10%. The solutes include fibrous ovomucin and various globular proteins (ovalbumin, conalbumin, ovomucoid, lisozyme, ovoglobulins), and very small amounts of other substances. The difference between jelly-like and free-flowing egg white and is the ovomucin content—it is four times larger in the former. The pH of the white of a very fresh egg is about 7.6–7.9 (slightly basic), dissolved carbon dioxide makes it look opalescent. With passing time, the pH increases and carbon dioxide escapes. Egg yolk is an oil-in-water type emulsion: about half of the mass is water, one third of the rest is proteins and the remaining two thirds are various lipids. After the addition of salts, the egg yolk can be divided into fraction by an ultracentrifuge: lipovitellinin and livetin in the plasma, making up 78% of egg yolk, in addition to lipovitellin, phosphovitin and low density lipoprotein (LDP) in the granules. Vitamins (A, B_2, B_3, D, E) are more abundant in the egg yolk than in the egg white, the same is true for phosphorus, manganese, iron, iodine, copper and calcium. The pH of the egg yolk is about 6.0, and it does not change much with time as there is no carbon dioxide present.

The structure of an egg is quite complicated, so the changes induced by the increase of temperature cannot be simple, either. A hard-boiled egg is easily made, but soft-boiled eggs leave much more room for error. The proteins in eggs denature at different temperatures (Table 2.9, column A), as evidenced by less successful attempts to prepare soft-boiled eggs. Professional and amateur cooks, some of them well established scientists, have devoted time and energy to this problem and claim that 63–65 °C is sufficient to make a soft-boiled eggs, while hard-boiling needs at least 85 °C. Some experts (e.g. physicists Peter Barham and Charles D. H. Williams) swear that cooking time is the most important factor, whereas others (e.g.

chemist Hervé This) put temperature first. Common sense tells that cooking time must be important as it takes time for eggs, which are poor thermal conductors, to reach the temperature of the cooking medium. The denaturation of proteins changes the thermal conductivity, which makes things even more complicated. The green outer layers in overcooked egg yolks are caused by the combination of smelly hydrogen sulfide, formed from decomposing proteins, with iron ions. This does not make the yolk inedible, but the color and the smell of hydrogen sulfide are not exactly attractive.

What is the ideal way of preparing soft- or hard-boiled eggs then? Scientists César Vega and Ruben Mercadé-Prieto went into painful detail in order to answer this question. Time and temperature are both important, they claim, as there is no "63 °C egg" or "65 °C egg". They noted that at between 60 and 68 °C, the denaturation of egg yolk is inversely proportional to time. Different combinations of time and temperature are suitable to prepare egg yolks of a given consistency (condensed milk, mayonnaise, honey, nutella or toothpaste), things get difficult to control above 66 °C (Table 2.9, column B). Two cooks added their own expertise: Chad Galliano treated egg yolks at 63.8 °C for 90 min, David Barzelay at 70.0 °C for 17 min to obtain similar end products that could be formed into a layer. But what about egg whites? To solidify egg whites, 6–7 min of cooking at 85–90 °C is sufficient. The still-liquid egg white conducts heat and its properties may alter the state of the egg yolk somewhat, but this is not very significant if the yolk is cooked at low temperatures. Japanese cuisine actually makes use of egg cooking at relatively low temperatures. They refer to it as *onsen tamago*, which means egg in a hot spring.

In summary, cooking at 100 °C is not really necessary to make a boiled egg. However, this temperature is practically constant and easy to identify even for those who do not have a suitable thermometer in the kitchen.

Finally it is time to ask the most important questions of all. Who devoted so much time and effort to study these questions in so much detail? These are enthusiastic amateurs, adventure-cooks and curious experts who are fans of the scientific field now called molecular gastronomy. This term was coined by Hungarian-born physicist Nicholas Kurti (1908–1998, born Kürti Miklós), who was an expert in low temperature physics at the University of Oxford and had food science as a hobby. The methods were further developed and popularized by French chemist Hervé This (pronounced Tis). Today, noted Catalan and French chefs Ferran Adrià and Pierre Gagnaire prepare surprisingly diverse dishes using the principles of molecular gastronomy.

2.25 Food Fraud: Then and Now

Long gone are the good old times when everyone produced food for the family in their own garden or bought it from someone who was a direct personal acquaintance. However, farmer's markets and self-production are not entirely relics of the

Fig. 2.40 Advice worth following in food safety: *CLEAN* wash hands and surfaces often. *SEPARATE* separate raw meats from other foods. *COOK* cook to the right temperature. *CHILL* refrigerate food promptly. (http://www.fda.gov/food/resourcesforyou/consumers/ucm188807.htm)

past even in the developed world: French people grew about one fourth of the potatoes they ate in 1990. This percentage is obviously higher in the developing world. Although many people in cities purchase an extensive amount of ready-to-eat food and self-farming shows a declining trend, production for self-use, even in urban areas, will probably never stop entirely.

People are inclined to look at the past with nostalgic feelings and often forget that "the good old times" can only be described credibly by those who lived in it. Food production has always faced huge challenges in history. Ensuring steady output and maintaining the quality of agricultural products were a constant problem, and every age contributed to solving it. Increase in the crop yield (crop rotation, fertilization, herbicides, new varieties) and preservation of products (smoking, salting, curing, cooling, freezing, preservatives, freeze-drying) are among the most significant developments which enabled the production of food in increasing amounts and quality, which led to the growth in the human population of Earth. Food has never been a simple matter of trade, in addition, economic and public health issues have also been involved, especially as they have impacted emotions and trust (Fig. 2.40).

The most common hazard factors for food today include natural toxins, chemicals used in plant protection and animal farming, random pollutants, and substances formed normally or by malfunction in food processing technologies. Risk assessment for different factors is characteristically different for experts and consumers. Consumers tend to underestimate the hazards posed by mycotoxins ($\rightarrow$**2.9**), whereas the case against food additives is regularly overstated ($\rightarrow$**1.4, 2.1, 2.2, 2.4, 2.18, 2.21**).

Another issue in food safety is intentional foul play, primarily counterfeit food. In her book titled *Swindled. The dark history of food fraud, from poisoned candy to counterfeit coffee,* British food writer and historian Bee Wilson presents the cultural background and the history of food forging. This practice is as old as humankind itself: there are very few types of food the forging of which has not been attempted in some form or other. Bans on adding water to beer and wine are known from Biblical times. Since then ham, eggs, olive oil, margarine, mustard, ketchup, sardines, cheese, bread, rice, milk, saffron, tea, coffee, chocolate, paprika, and countless other products have joined the ranks of the food forgery victims.

London-based German chemist Friedrich Christian (Frederick) Accum (1769–1838) was an outstanding figure among the whistle-blowers on food forging. His 1820 book titled *A Treatise on Adulterations of Food and Culinary Poisons* was a wake-up call for the British and American societies and pointed out that many supposedly perfect foods were vulnerable to forgery. Another researcher, Arthur Hill Hassall (1817–1894) published a book titled *Food and its adulterations* in 1855, in which he reported from his microscope findings that 74–96% of the coffee sold in London was forged to some extent by diluting it with chicory, roast wheat and caramelized sugar. There has been some improvement since then. In 1993, instant coffee sold in the UK "only" showed a 15% rate of forgery. Today's coffee forgery is more sophisticated, though: forgers add less expensive Robusta coffee to the premium quality Arabica. In addition to tasting, the difference can be discovered by analytical studies that determine the ratio of chlorogenic acid and caffeine.

One of the easiest and most common forms of food forging is to add a harmless (and dirt cheap) substance to the original food to increase its mass, e.g., mixing powdered sugar with flour. Another form is more dangerous: harmful substances are sometimes added to products in order to make them look more desirable (e.g. adding lead-containing red pigment minium to paprika to improve its color, or adding melamine to milk in order register higher protein content on standard tests, →2.6).

Eating habits keep changing. German statistician Ernst Engel (1821–1896) made an interesting observation in 1890: the higher the income of a family, the smaller the portion of this income spent on food. Indeed, the family of a bricklayer in Berlin around 1800 spent about three fourths of its entire income on food, about a half of that on bread. An average French family spent 42% of its income on food in 1950, about 5–10% lower than at the beginning of the twentieth century. This ratio dropped by a further 30% in the last 50 years, which opened up possibilities for families to spend more freely on other things such as travel, domestic appliances and hobbies. The composition of food related spending also changed as times changed. About 22% of the money used to purchase food was spent on meat in an average worker's family in the Belgian society of the 1920s. The corresponding figure was 26% in the 1940s, 31% around 1960, and 26% in the early 1970s. Today, it is less than 30%, so meat has lost some of its appeal and the consumption of fruits and vegetables has been on the increase. Some food was replaced by most upper-class varieties. Belgian workers began buying pastry products, croissant and raisin bread in the 1950s. Only wealthy people consumed these earlier. Yoghurt began to supplement milk, pre-cooked and frozen French fries took some of the place of raw potato, and people started opting for bottled mineral water instead of tap water. It is by no means certain that this trend of declining food spending will continue indefinitely. The consumer price index for food jumped 65% (!) between 2006 and 2010. There may be some speculative business in the background, but the future is as impossible to predict as ever.

To counter the culture of mass consumption, the international movement named *Slow Food* was started by the Italians in 1986. The movement aims to resist the global standardization of food, and promotes variability, originality and something called "ecogastronomy". The high efficiency, internationalization and businesslike conduct of the food industry symbolized by fast food in McDonald's was challenged by the founder of *Slow Food*, which was, surprisingly, mostly portrayed as new and innovative rather than conservative. *Slow Food* renewed interest in local recipes, food, tastes and technologies. The movement is criticized for its blind respect for traditions, its highly elitist and staged pursuit of originality, and, most importantly, for ignoring the simple fact that traditional processes cannot supply the entire world with enough food. International academic and freelance writer Rachel Laudan called the activity of *Slow Food* "Culinary Luddism". Paradoxically, the slow food movement is a meeting point for the pursuit of originality in modern societies and the traditional food preparation methods of the developing world.

Slow Food is only one of the movements built on criticizing the modern consumer society. Several consumer organizations were founded after World War II. They accepted new technologies, but they sought to control the development of mass food production and distribution. *Consumentenbond* in the Netherlands has significant monitoring responsibilities and already had 275,000 members in 1970.

A change of eating habits in the developed world has led to the gradually worsening problem of obesity. This was originally an American phenomenon, but about a third of the population in South Africa, Hong Kong and Morocco was overweight in 2004. There is a huge contradiction between this phenomenon and the mass of 1 billion people who starve every day on Earth.

Changing food supply chains present new problems of food safety. As the population of Earth has grown, so has the amount of food produced. The attraction of the extraordinary large scale is a temptation for careless manufacturing and food forging as exemplified by dioxin-related scandals ($\rightarrow$ **2.21**), or the 2011 diarrhea epidemic in Germany. Globalization has some unpleasant by-products: food supply chains are increasingly segmented, long-range transportation is often necessary to maintain a competitive selection of foods and to decrease costs. However, although free trade or globalization did not create the motivation for food forging, they certainly have created new opportunities. Government authorities make major efforts to monitor food of different origins continuously. Sophisticated analytical methods provide the means of efficient testing, but the ultimate responsibility lies in the consumer to buy or not to buy a certain food after a sufficient search for information. The dangers seem to have multiplied: marketing experts have achieved considerable sophistication in their quest to convince consumers to purchase certain products, and they often use information that is portrayed as scientific with or without justification.

Chapter 3
Medicines

3.1 Are Generic Medicines the Same as the Original?

No doubt, healing does not only depend on the active ingredients in medicines, but it is influenced by a number of mental factors as well. Could it be possible that two medicines containing the same substance have different effects? This question is becoming quite important today as generic drugs flood the market.

Pharmaceutical research is probably the most expensive business in the world. The development of a new drug takes long years and the costs may easily top 1 billion (one thousand million) €. This investment is regained by the developers through patent protection, which is valid for about 20 years in most countries. In this period, only the patent holder is allowed to sell the drug. However, patenting is done very early during product development, so the developer can only sell the new drug without competitors for 7–10 years. Just to give the reader some idea about the size of the market: the total retail earnings (paid by patients and insurance companies) in the pharmaceutical market of the European Union was 214 billion € in 2007, that is about 430 €/person/year. Most of the profit is made by blockbuster drugs, which have yearly sales above 1 billion € under the patent protection period.

When the patent protection period is over, other companies can start selling the same active ingredient. This is rather advantageous for patients, as competition drives down the prices. Drugs containing the same active ingredients are called generic drugs. The category 'generic' was first used in the USA in the 1980s, with the intention of helping pharmaceutical companies to compete in making drugs whose patents expired. The 1984 Drug Price Competition and Patent Term Restoration Act was signed into law by President Ronald Reagan, who made the following comment: "The legislation will speed up the process of Federal approval of inexpensive generic versions of many brand name drugs, make the generic versions more widely available to consumers, and grant pharmaceutical firms added incentives to develop new drugs."

The essence of the legislation was that pharmaceutical research was helped by lengthening the patent protection period, whereas generic drug makers were helped by introducing a faster drug approval process. Europe followed suit quickly. Generic drugs are not tested as extensively (laboratory, animals, healthy people, patients) as original drugs. The applicant must only file the results of the original approval process and show that the generic version is identical to this original in a number of respects (mass of active ingredient, method of application). Of course, the generic drug will be less expensive than the original. Generic drugs do not have the same medicine names as the originals because the names are usually trademarks, whose protection never expires. Generic drugs can easily be identified by checking the chemical name of the active ingredient and the dose.

The introduction of the generic version of an original drug into the market typically occurs within 1 or 2 years of the expiry of patent protection and results in a price drop of 25–50%. In addition, the yearly price increase of a generic drug is usually smaller than the rate of inflation, whereas the opposite is true for patent protected drugs. Generic drugs therefore save a lot of money for the society as a whole. On the other hand, the competition in the generic drug market results in lower profitability. European pharmaceutical companies spend huge amounts of money (typically 17% of their entire revenue) on developing original drugs, which keeps profit margins high for a few patent-protected drugs. With generic drugs, the primary means of gaining profits is marketing, which also requires huge amounts of money. When generic versions of a drug become available, intensive marketing often causes an increase in the amount of drug sold and, somewhat paradoxically, the overall money spent on the drug by patients and insurance companies may also increase even despite the falling prices.

As pointed out earlier, fewer tests are sufficient to approve the introduction of a generic drug. However, bioequivalence must always be demonstrated. Although it seems logical that the same active ingredient given in the same form and the same dose should have the same effects as the original, bioequivalence must be tested by measuring the concentration of the active ingredient in the blood. These experiments are done on healthy volunteers and the concentrations are measured after blood tests. Figure 3.1 gives an example of the time-dependence of the concentration of a drug in blood plasma. Bioequivalence requires that the area under the curve (AUC) in this graph and the maximum concentration (c_{max}) should not be more than 20% lower or 25% higher than the original drug. If the reader wonders why the two numbers are different, consider this nice little statistical oddity: a 20% drop in a value must be followed by a 25% increase to get back to the original value. So these numbers were set to ensure that the test results do not depend on which drug is the original and which drug is the generic.

Bioequivalence also assumes that the two products are therapeutically equivalent, so a therapy started with one drug can be continued with the other at any time. A demonstration of therapeutic equivalence in not required for approving a generic drug, as this would involve much higher costs and the price reducing effect of the introduction of generic drugs would be less pronounced. Although the requirement for bioequivalence is that the difference between the bioavailability should be lower than 20–25%, practice shows that generic drugs are typically within 3% of the original. Such small differences render therapeutic equivalence studies a waste of money: the expected benefit from them would be very small, and the costs extremely high.

3.1 Are Generic Medicines the Same as the Original?

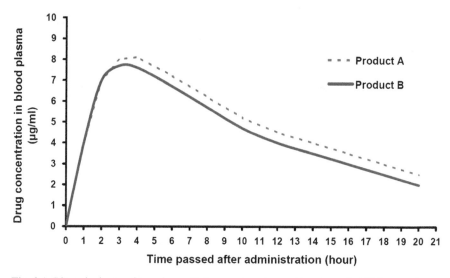

Fig. 3.1 Bioequivalence of two drugs. Reference drug A caused a somewhat higher maximum concentration in blood plasma (c_{max} = 8.4 µg/ml, AUC = 125 µg h/ml), but drug B satisfies the bioequivalence criterion as its bioavailability is better than 90 % (c_{max} = 7.8 µg/ml, AUC = 115 µg h/ml). (Authors' own work)

How is it possible that patients sometimes report a change in effectiveness when their doctor prescribes a generic version of a well-known brand name drug? This phenomenon mostly has psychic origins. A patient is used to the name of a drug and trusts it, the new name may cause suspicion and finally a feeling of reduced effect, even if this cannot be proved in any objective manner. It is also typical to report more side effects. Sometimes it may even happen that the patient trusts the new drug more (something new must be better than the old) and reports it as more effective. These are very similar to the placebo or nocebo effects, which are described in another chapter of this book (→ **3.6**). Other factors might be significant, though. A generic drug may sometimes contain the active ingredient in a different salt form, which may have some influence on efficiency. A well-documented case is the antidepressant paroxetine hydrochloride, which is formulated as paroxetine mesylate in one of the generic forms. A patient used to the hydrochloride reported a new side effect and decreased efficiency when the generic mesylate was used instead of the original.

Negative effects of switching from the original to a generic drug are more common for drugs whose therapeutic range is narrow (there is a small difference between harmfully high and uselessly low doses). This has been documented for some epilepsy drugs. The quite minor differences in blood plasma concentrations between original and generic drugs actually caused an increase in the frequency of epileptic seizures.

Under certain circumstances, generic drugs may differ from originals. But this is very rare and is limited to certain illnesses. Ideally, the doctor or pharmacist is aware of the possibility and pays special attention to the possible side effects or lack of therapeutic effects. It may also happen that a generic drug is marginally more effective than the original. Generic drugs should not be confused with counterfeit medicines: in contrast to the latter, these are perfectly legal and appropriately tested. Counterfeit drugs are usually sold only on the Internet in developed countries, but more widely in the developing world. These are illegal, their efficiency and safety is not tested.

3.2 The Vitamin that Never Was: B_{17}

Mr. Jason Vale, an executive of the New York based company Christian Brothers Contracting Corporation, was sued by America Online in 1998 for allegedly sending over 20 million "spam" e-mail messages to its subscribers. In peak time, Mr. Vale sent about 100,000 messages an hour to advertise the products of his company, which were alternative cancer medicines (apricot seeds, Laetrile, "vitamin B_{17}", and amygdalin). A federal judge awarded AOL US $ 631,585 in damages in 1999. The Federal Food and Drug Administration (FDA) got a court injunction in 2000 to stop Mr. Vale from selling these products and advertise them as a cure. Mr. Vale seemed to comply, but in fact only changed his marketing strategy: he asked his relatives to maintain websites and telephone services. He was arrested and, in 2003, convicted of criminal contempt of the injunction and sentenced to 5 years in prison.

This was not the first court case in the history of "vitamin B_{17}" or Laetrile. There is a good chance that further cases will see the inside of courtrooms in the future. Dr. Irving J. Lerner, a noted medical researcher at the University of Minnesota described Laetrile as "the slickest, most sophisticated, and certainly the most remunerative cancer quack promotion in medical history."

Internet search engines return a lot of websites relevant to "vitamin B_{17}". The National Cancer Institute (NCI) of the US tested its efficiency in 23 animal cancer models and on humans as well. No therapeutic benefit was detected, but the side effects resembling cyanide poisoning were significant. Figure 3.2 shows the chemical structures of the mentioned substances. In Mexico, "vitamin B_{17}" or Laetrile usually means amygdalin (Fig. 3.3). The cohort of believers uses these words to refer to two different substances.

Natural amygdalin, occurring in the seeds of stone fruits of the *Prunus* genus (for example apricot or bitter almond), and synthetic "vitamin B_{17}"/Laetrile belong to the group of compounds called cyanogenic glycosides. Both of these substances are broken down into monosaccharides (simple sugars), benzaldehyde and hydrogen cyanide by beta-glucosidase enzymes. The hydrogen cyanide formed in this process is responsible for the side effects. Beta-glucosidase enzymes are not present in the human body, but food may contain them and the breakdown process is also aided by acids, such as ascorbic acid (vitamin C).

Why do people not die after eating a small amount of apricot seed, almond or marzipan? The eventual fate of amygdalin depends on how much beta-glucosidase

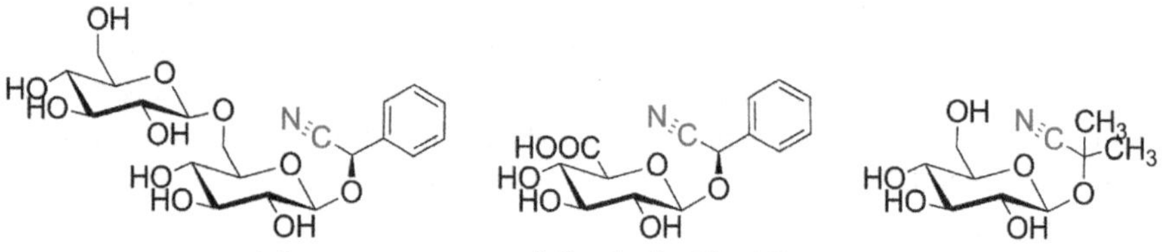

3.2 The Vitamin that Never Was: B_{17}

Fig. 3.2 The structures of some natural and artificial cyanogenic glycosides. (Authors' own work)

Fig. 3.3 A Mexican B_{17}/amygdalin injection. (http://vizionizator.hu/forum/pics/Amigdalina%203g%2010ml%20-%20500.gif, copyright-free picture)

enzyme the gastrointestinal tract already contains at the time of ingestion. If there is only a low amount of this enzyme present, amygdalin goes through unchanged and is excreted with feces. If there is a lot of enzyme, some risk of cyanide poisoning is created. The **LD_{50}** value of amygdalin in rats is 880 **mg/kg of body weight**. Taken together with beta-glucosidase enzyme, 600 mg per kg of body weight causes almost instantaneous death. The amount of amygdalin is lowered significantly by heat treatment, which is a usual step during marzipan making. Another cyanogenic glycoside, linamarin (Fig. 3.2) is present in cassava (*Manihot esculenta*, also called manioc, yuca, balinghoy, mogo, mandioca, kamoteng kahoy, or tapioca), which is a major staple food in the world. Without heat treatment, the root is toxic, linamarin obtained from it is used as an arrow poison.

The story of "vitamin B_{17}" is a complex one. Let's begin at the beginning.

Ernst Theodore Krebs, Sr. (1876–1970)—not to be confused with German-born British Nobel Laureate Hans Adolf Krebs (1900–1981)—began his career in San Francisco, where he became medical doctor in 1903. He sold parsley extract during World War I, supposedly effective against Spanish flu, but the FDA deemed this ailment fraudulent. Later, Krebs Sr. tried promoting the enzyme chymotrypsin under the trade name Mutagen as a cancer remedy. He and his son also patented pangamic acid or "vitamin B_{15}" to treat heart disease, cancer and a number of other serious illnesses. The vitamin-like properties of pangamic acid have still not been proved.

Ernst Theodore Krebs, Jr. (1911–1996), considered the father of Laetrile, was often quoted as Dr. Krebs, but never in fact received a doctorate. His education was not quite a success story: he ultimately received a Bachelor of Arts degree from the University of Illinois after failing science courses at various colleges. His only doctoral degree was an honorary one received to acknowledge presenting a single hour of lecture at American Christian College in Tulsa, Oklahoma, which was not accredited to award any advanced degrees.

The story of Laetrile exists in several variations. Krebs, Sr. first encountered apricot seeds when he was preparing odors for smuggled whiskey. He isolated a substance and attributed anticancer effects to it. Krebs, Jr. modified his father's method in 1949 and called the product Laetrile (as an abbreviated form of chemical name laevo-mandelonitrile beta-D-glucuronoside). However, Krebs, Sr. recalled a different sequence of events during an FDA interview. In this version, he gave 1951 as the year of inventing Laetrile, with the probable intention of evading FDA regulations. Father and son patented the preparation method of Laetrile in 1961.

Krebs, Jr. started the John Beard Memorial Foundation in 1945 in the memory of Scottish embryologist John Beard, to revive a theory of cancer formation. Krebs, Jr. argued that cancer cells have an abundance of an enzyme that breaks amygdalin down to hydrogen cyanide. The poison formed is able to kill the cell, but healthy cells remain unaffected as they do not contain the enzyme. This was the time when regulators began attacking the medicinal use of Laetrile. Krebs changed his theory: he considered amygdalin a vitamin, the lack of which causes cancer. None of these claims have been proven to this day. Krebs first advertised Laetrile as a cancer ailment, later the slogan changed to "controlling" cancer, then the vitamin era brought the buzzword of prevention into the campaign.

Canadian businessman Andrew R. L. McNaughton—not to be confused with his father, Canadian scientist, army officer, cabinet minister, and diplomat Andrew George Latta McNaughton (1887–1966)—gave new momentum to the clinical use of "vitamin B_{17}"/Laetrile in 1956. McNaughton was a test pilot during World War II, and made a fortune after the war by selling surplus army supplies at a huge profit. For example, he sold weapons to Israel and the Cuban Batista government, which somehow ended up in the hands of supporters of Fidel Castro. His services won him honorary citizenship in communist Cuba. McNaughton started a foundation named after himself, and later a company named International Biozymes Ltd., in order to advance Laetrile in Canada. McNaughton had serious financial trouble in the 1970s. In 1973, he was charged by the Italian police with having taken part in a $ 17 million swindle. In 1974, a Canadian court found him guilty of stock fraud involving a company named Pan American Mines. McNaughton was fined $ 10,000 and sentenced to serve a single day in jail. He seems to have refused to pay the fine and left Canada without serving the sentence.

McNaughton was not only responsible for producing and selling "vitamin B_{17}"/Laetrile, but was also in charge of marketing efforts. He recruited surgeon Dr. John A. Morrone and freelance journalist Glenn Kittler to write articles and even a book about the benefits of the product. These were not unsuccessful. Dr. Ernesto Contreras, a former Mexican army pathologist, set up a clinic just across the US border in the Mexican city of Tijuana. He mostly treated American cancer patients with "vitamin B_{17}"/Laetrile. This turned out to be a brisk business and a foundation (International Association of Cancer Victims/Victors and Friends) was soon established to aid patients residing in a nearby motel in California and who commuted to Tijuana. Dr. Contreras claimed that the clinic treated 100–120 new patients every month in 1974 in addition to returning customers. A month-long Laetrile treatment cost as little as $ 150. In the 16 years leading up to 1979, this Tijuana clinic was visited by 26,000 patients. Dr. Contreras estimated that 30% of these responded dramatically to the drug. Yet when the FDA asked him for case histories, he could only come up with 12 full records: six of these died of cancer, one was treated by a conventional method, one died of an unrelated illness, another one still suffered from cancer, the remaining three could not be located.

Legal problems soon became daunting in the US. In 1961, Krebs, Jr., and the John Beard Memorial Foundation were indicted for selling the unapproved drug pangamic acid. Krebs was fined $ 3750 and sentenced to a prison term. The sentence was suspended when he agreed to a 3-year probation, in which neither would manufacture or distribute Laetrile unless the FDA approved its use for testing as a new drug. In 1962/63, the California Cancer Advisory Council reviewed more than 100 cases and found no evidence to support the anticancer therapeutic efficiency of "vitamin B_{17}"/Laetrile. The Council recommended banning the drug. The Krebs family returned to court several more times. In 1965, Krebs, Sr., pleaded guilty to a contempt charge for shipping Laetrile in violation of injunctions and received a suspended 1-year sentence. In 1974 Ernst, Jr., and his brother (Byron) were fined $ 500 and given a suspended sentence of 6 months for violating the California state health and safety laws. Ernst, Jr., violated the terms of his probation and spent 6 months in a county jail in 1983.

Meanwhile, Howard H. Beard (not a relative of Scotsman John Beard), developed three tests that were alleged to diagnose all cases of cancer early. In 1963 Krebs, Jr., stated that the "scientific implementation" of Laetrile relied upon Beard's test. During the early 1960s, the California Cancer Advisory Council assessed the reliability of these tests by a randomized method. It turned out that the method is unable to distinguish between urine from cancer patients and healthy individuals. Furthermore, it was also demonstrated that the test depended mainly on the amount of lactose in the urine. Beard was indicted by a federal grand jury in Texas and given a 6-month suspended jail sentence with a 1-year probation.

Another interesting player of the story was Dr. John Richardson, a.k.a., the "metabolic" doctor. He was a general practitioner whose practice was not exactly flourishing. He met Krebs, Jr. in 1971, and promptly became a self-promoted cancer specialist and an ardent believer in the use of "vitamin B_{17}"/Laetrile. Richardson's practice boomed as a result. He charged patients $ 2000 for a course of Laetrile. Income tax returns show that Richardson grossed $ 2.8 million from his practice between 1973 and 1976. Michael Culbert, an editor at the Berkeley Daily Gazette and prominent Laetrile promoter estimated that Richardson had treated 4000–6000 patients by 1976, and his actual total income could have been between $ 10 and $ 15 million. In spite of these alleged "dramatic improvements," Richardson admitted that most of his cancer patients died. In an attempt to overcome this problem, he increased the Laetrile dosage, advised patients to observe a vegetarian diet and also prescribed huge doses of regular vitamins. He even coined the phrase "metabolic therapy" to refer to this combination measures. Richardson was arrested and convicted in 1972 for violating California's Cancer Law. After a lengthy legal battle, his California medical license was revoked in 1976. Later he worked at a Mexican cancer clinic, then practiced under a homeopathic license in Nevada.

Dr. Richardson's arrest received a lot of publicity and triggered the formation of the Committee for Freedom of Choice in Cancer Therapy (CFCCT). The group's founder and president was Robert Bradford, a former laboratory technician at Stanford University. Michael Culbert was also quite instrumental in it and published two books promoting Laetrile: *Vitamin B-17: forbidden weapon against cancer* (1974) and *Freedom from cancer* (1976). CFCCT's activities were coordinated with the John Birch Society, where Richardson, Bradford, and Culbert were all members.

In addition to the charlatans and profiteers, two experts with recognized academic background also appeared in the "vitamin B_{17}"/Laetrile story. These are biochemist Dr. Dean Burk of the NCI and Prof. Harold W. Manner, chairman of the biology department at Loyola University in Chicago. Loyola University officials soon became upset with Manner's new activities, so he had found a new job at a Tijuana clinic. Movie star Steve McQueen (*The Magnificent Seven, The Towering Inferno*) was treated with Laetrile at another Mexican clinic, but died shortly afterward.

During the mid-1970s, 27 states passed laws permitting the sale and use of Laetrile within their borders. But federal law still forbade interstate shipment of Laetrile, so the state laws had little practical consequence. In 1977, a US Senate subcommittee chaired by Senator Edward Kennedy held hearings on Laetrile. Dr. Richardson claimed that the FDA, NCI, American Medical Association (AMA), American Cancer Society, Rockefeller family and major oil and drug companies had all conspired against Laetrile. However, he and Krebs, Jr. were unable to agree on the formula for Laetrile, which was the source of considerable amusement in the subcommittee. Senator Kennedy concluded that the Laetrile leaders were "slick salesmen who would offer a false sense of hope" to cancer patients

Responding to public and political pressure, the NCI carried out two major studies on "vitamin B_{17}"/Laetrile. The first was a retrospective analysis, in which 455,000 health professionals were asked to provide case reports about the use of the drug. Various pro-Laetrile groups were asked to provide information as well. Although it had been estimated that at least 70,000 Americans had received treatment, only 93 cases were submitted for evaluation, 68 of which gave sufficient documentation. Two of these showed complete remission of disease, four displayed partial remission, there was no measurable response in the remaining 62 cases. Although the NCI did not ask for negative case reports, 220 physicians submitted data on more than 1000 patients who had received "vitamin B_{17}"/Laetrile without any beneficial response. In 1980, the NCI undertook a clinical trial of 178 cancer patients who received Laetrile, vitamins and enzymes at four prominent cancer centers. Since proponents were unable to agree on the formula or a testing protocol for Laetrile, NCI decided to use a preparation distributed by the major Mexican supplier, American Biologics and the dosage was based on the published recommendations of Krebs, Jr., and the Bradford Foundation. The results of this trial left no ambiguity: not one patient was cured or even stabilized. The average survival rate was about 5 months, and tumor size had increased in those still alive after 7 months. In addition, several patients suffered from cyanide toxicity. Bradford and American Biologics filed three different lawsuits against the National Cancer Institute, alleging that as a result of the study, they had sustained serious financial damage. All three suits were thrown out.

In 2002, British scientists developed a recombinant protein that was able to break down linamarin (Fig. 3.2) to hydrogen cyanide in cancer cells selectively (antibody-guided enzyme nitrile therapy, AGENT). However, the protein has major disadvantages and is unlikely to be useful as a medicine.

The case of vitamin B_{17} has a broader context as well. Today, the market of developed countries is flooded with dietary supplements and alleged vitamin products, which do not need to be approved by medical authorities. Many of these products share one characteristic: using them is known to result in no health benefits at all, or as in the case of "vitamin B_{17}", they are outright harmful. When will society learn to trust its own education system and screen out profiteering charlatans?

3.3 Joint Efforts: Glucosamine and Chondroitin

Osteoarthritis is a degenerative joint disease (see Fig. 3.4) involving the degradation of articular cartilage and subchondral bone. This disease seriously affects the quality of the life of patients, and often leads to muscle atrophy and pain. Treating it may even require surgery. The disease typically occurs over the age of 60, but some athletes develop similar symptoms much earlier in life. In the US only, about 27 million people suffer from some form of osteoarthritis, which may be an ultimate cause for more than 10 % of the visits at family doctors.

In osteoarthritis patients, the tissue that covers and protects the ends of the bones (hyaline cartilage) is gradually degenerating, giving way to small bone outgrowths. This degeneration can proceed to the point where the ends of the bones come into direct contact, which is highly painful. The disease mostly attacks the hips, knees, hands and feet. Typical bone outgrowths called Heberden's nodes or Bouchard's nodes appear mostly on the hands. Being overweight is a major risk factor in developing osteoarthritis as the weight that needs to be supported by the joints is much larger. In treatment, any loss of weight could alleviate the symptoms, especially in the knees, but a change of life style and getting more physical exercise is often difficult for elderly people. The pain in joints can be relieved by various painkillers called NSAIDs (non-steroidal anti-inflammatory drugs), for example meloxicam, diclofenac, and ibuprofen. Unfortunately, NSAID treatment is not without side effects. It involves risks of toxicity and cardiovascular diseases. There are also reports of NSAID-induced toxic effects on cartilage in joints.

Cartilage in the body mainly contains proteoglycans and glycosaminoglycans (GAGs), which are large molecules composed of proteins and sugar derivatives. There are 40–100 repeat units in GAG, all of them sulfates of nitrogen-containing sugar derivatives. Most well-known in this family of compounds are chondroitin sulfate, dermatan sulfate, heparin, hyaluronan and keratan sulfate. The most significant proteoglycan is aggrecan, which is the major chemical component of cartilage tissue and is also called cartilage-specific proteoglycan core protein (CSPCP). The sugar and protein parts of aggrecan can absorb a huge amount of water, which makes the cartilage tissue an excellent natural lubricant.

Fig. 3.4 The X-ray image of the hand of an osteoarthritis patient. The *arrow* shows directly contacting bone surfaces. (Authors' own work and copyright-free picture)

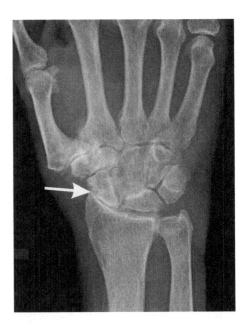

Osteoarthritis involves the weakening or full degradation of cartilage tissue in addition to bone deformation. To treat this condition, a seemingly obvious solution is to make up for aggrecan loss by administering chondroitin sulfate, keratan sulfate or the latter's building block, glucosamine (Fig. 3.5). Chondroitin sulfate can be produced from animal cartilage tissue and glucosamine is found in the shells of crabs in large quantities. Using these materials as medicines has a long history. One of the most careful reports was published by an independent nonprofit organization called *The Cochrane Collaboration*. This analysis evaluated data from 20 different clinical studies, which altogether involved 2570 patients (this kind of study is called **meta-analysis**). Non-Rotta brands of glucosamine sulfate had no effect on the joint pain or joint function. In contrast, the Rotta brand was better compared to placebos: they eased pain and the movement of joints. On a subjective test, the improvement was 21 % after treatment. In terms of drug safety, there was no difference between glucosamine sulfate and placebos.

This conclusion was further strengthened by a European study in 2008, where 318 osteoarthritis patients were treated with acetaminophen (the active ingredient of Tylenol), glucosamine sulfate and placebos. Tylenol and placebos resulted in no medical improvement, but glucosamine sulfate treatment caused 20 % decrease in pain.

In contrast, a large scale American study (Glucosamine/Chondroitin Arthritis Intervention Trial, GAIT) between 2006 and 2010 led to a fundamentally different conclusion. In this, 1500 patients were treated with glucosamine hydrochloride, chondroitin and placebo in this period, each with a dose of 500 mg of glucosamine hydrochloride and 400 mg of chondroitin three times a day. The treatment was no

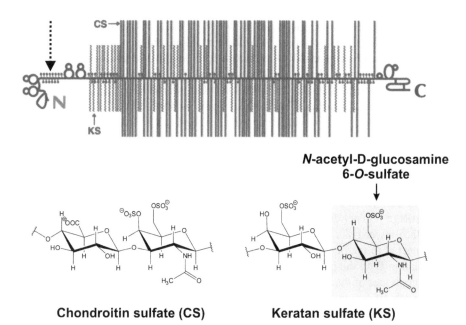

Fig. 3.5 Schematic structure of aggrecan and the sugar components in it. The *blue horizontal line* in the middle represents the protein components from the *N*-terminus to the *C*-terminus. Aggrecan is sensitive to break-up (proteolysis) where the *dotted arrow* indicate, this is responsible for the degradation of cartilage tissue. (Authors' own work and http://glycoforum.gr.jp/science/word/proteoglycan/PGA00E.html, permission obtained from copyright-owner)

better than the placebo except for a minor improvement in patients who suffered from the worst pain.

The results of GAIT were further supported by findings of some scientists in Bern, Switzerland. They analyzed 10 clinical trials, which involved a total of 3803 patients, and concluded that glucosamine and chondroitin, either alone or in combination, did not outperform placebos in preventing joint damage. Their results left no room for doubt: "Coverage of costs by health authorities or health insurers for these preparations and novel prescriptions to patients who have not received other treatments should be discouraged."

The latter results are hardly surprising. Chondroitin is a delicate macromolecule, which does not survive the harsh conditions in the stomach and intestines, and is unable to reach the joints intact through the blood stream. Glucosamine is easily formed from glucose, which is quite abundant in the body and is also continuously replenished by food. So the availability of glucosamine in the body is excellent, and its insufficient quantity cannot be the cause of cartilage degradation.

Is there any hope for developing another medicine, not based on chondroitin and glucosamine? Well, hope dies last, but giving birth to a medicine is not simply a question of hopes. One hopeful candidate is called methylsulfonylmethane (MSM or dimethyl sulfone), was actually a common ingredient in earlier preparations containing chondroitin and glucosamine. MSM contains sulfur, and it is important in the biosynthesis of a number of sulfur-containing proteins including those in cartilage. Horse breeders usually have a high opinion of MSM treatment, they claim it protects horse joints. However, there is no evidence for any human effect yet. In 2000, the US Food and Drug Administration (FDA) warned a prominent MSM promoter, Karl Loren, to stop making therapeutic claims for MSM, reminding him that the marketing of drugs without the approval of the FDA is illegal. S-Adenosylmethionine (SAM, SAM-e) is another important sulfur compound found in the human body. Some experts are convinced that SAM plays a significant role in cartilage regeneration, but they have no clinical evidence to support this claim.

The key to treating osteoarthritis might partly lie in human diet. Vitamin C has a confirmed role in collagen and proteoglycan synthesis, which can be beneficial for cartilage regeneration. People who consume at least 150 mg of ascorbic acid daily face lower risks of developing osteoarthritis. The health of cartilage tissues is connected to the health bones supporting them, so a proper amount of calcium and vitamin D in the diet can actually go a long way in protecting the bones and cartilage tissues of the body.

The role of fats might be even more important. Omega-3 fatty acids (those in fish, soya, linseed and rape oil) have anti-inflammatory effects, whereas omega-6 fatty acids (those in the oils of corn, safflower, cottonseed and sunflower seed) are associated with causing inflammations ($\rightarrow$ **2.13**). The healthy ratio of these fatty acids is about 1:5, whereas in the food currently consumed by average Americans, it is more like 1:10. To make the omega-3:omega-6 ratio healthier, one should eat less meat and processed food, and more fish, vegetables, and fruit. Daily physical exercise also seems important. Would it not be simpler just to take a pill and not care about life style?

3.4 A Man's Drug of Choice: Testosterone

After the closing ceremony of the 2008 Beijing Olympics, the Russian media was the first to report that the testosterone levels of Belarussian athletes Vadim Devyatovskiy and Ivan Tikhon, silver and bronze medalists in men's hammer throw, were found to be suspiciously high. Of course, this was important news for the athletes finishing fourth and fifth in the final, because sanctioning the two athletes for doping violation would have resulted in their automatic promotion to medal winners. A considerable part of the international media reported this case by saying that testosterone was detected in the body of the two Belarussian athletes. Maybe the author of this sentence thought that this was already a doping violation. In fact, this observation itself would only confirm the otherwise undisputed fact that the athletes are living men. Testosterone is a hormone that is indispensable for normal body functions. Despite this, it is on the list of prohibited substances in sports, but a doping violation exists only with abnormally high levels or proved to be of external origin.

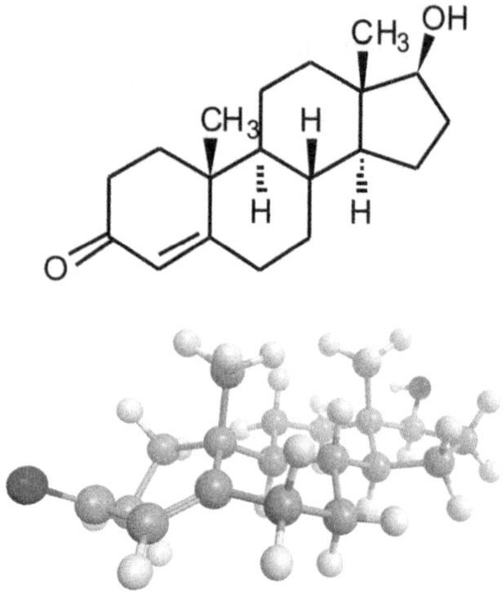

Fig. 3.6 Chemical structure and three-dimensional molecular model of testosterone. (Authors' own work)

In the case of Devyatovskiy and Tikhon, the International Olympic Committee deliberated at length before making a decision. On December 10, 2008, the two athletes were finally sanctioned by retraction of their medals and suspension. The Court of Arbitration for Sport in Lausanne, Switzerland, on the other hand, ruled in favor of the two athletes because of discrepancies in the drug testing. The court emphasized that it did not clear them of suspicion and the verdict should not be taken as exoneration. The extended legal battle is not accidental. For substances that do not occur in the human body, detection in a sample is clear evidence of guilt. Hormones, on the other hand, are produced in the human body by normal physiological processes. It is difficult to establish a case of doping violation because differences need to be established between endogenous (within the body) and exogenous (outside the body) origin.

Testosterone is the archetype of anabolic steroids (Fig. 3.6), which promote the muscle-building processes of the body. Its use is proven to improve results in some sports if applied during the training period before the competition. Ironically, a scientifically rigorous study of these effects is a side effect of the state-funded doping program of the former East Germany. The use of testosterone in sports doping has a long history: as early as 776 BC, a few athletes in the ancient Greek Olympiads were reported to consume sheep testicles, which contain high levels of the hormone. For the 1936 Berlin Olympics, Germany, governed at that time by Adolf Hitler, also prepared its athletes by giving them an extra dose of the hormone.

The most infamous case is that of American cyclist Floyd Landis. The story begins on July 20, 2006, when Landis won section 17 of the Tour de France. He was required to undergo a doping test following the race. The level of testosterone in urine is determined by a method called GC-MS (Gas Chromatography—Mass Spectrometry), and is often referred to as the T/E test. Here, T means testosterone, whereas E stands for epitestosterone, which is a variant of the hormone that occurs in the body but has none of its effects. In most humans, the T/E ratio is about 1. The rules of the World Anti-Doping Agency (WADA) declare that a T/E ratio in excess of 4 may indicate a doping violation, but further tests are necessary to establish an unambiguous case. The wording of the rules leave a lot of room for legal bickering: a single sentence mentions the cut-off ratio of 4 and occupies nine lines of text. In the legal case of Floyd Landis, a three-member court declared the cyclist guilty of a doping violation in a 2:1 split decision on 20 September, 2007. The appeal of the cyclist was rejected by Court of Arbitration for Sport in June 2008. Throughout this entire ordeal, Landis maintained that he was innocent. However, on May 19, 2010, he made a detailed confession. He admitted using a number of forbidden substances and also stated that this is a common practice in cycling. He still denied the use of testosterone in 2006, and said that he was using another hormone, GH (growth hormone, somatotropin by its chemical name) at that time.

3.5 Doped Up and Ready to Win. Maybe

It appears that doping is becoming more and more important in modern sports, or at least the number of athletes sanctioned for doping violations and the time devoted by the international media to these cases are on the rise. In a few high-profile cases, however, doubts also surface about the scientific background of prohibiting certain substances or the reliability of the methods used to detect them.

According to the classical definition, doping is the introduction of an exogenous substance into the body during a competition in order to improve results. This definition has been out-of-date for several decades. First, substances present naturally in the body can also be used for doping. Second, a substance that is used during the training period, but is no longer present during the competition can also improve the sports results of an athlete. Nowadays, doping has a legal definition: it is the use of a substance that the World Anti-Doping Agency (WADA) prohibits. It should not go unnoticed that this list is called a list of prohibited substances, and not a list of doping agents.

Improving someone's sports results is by no means prohibited in general. Its most conventional method is called training (what a surprise!). In addition, there are a few natural or artificial substances, which may have positive effects without being banned. The use of vitamins, great quantities of special energy drinks or special protein concentrates as dietary supplements is allowed.

The use of the substances on the list is banned for two major reasons. The first is to provide the conditions of fair play. This is a very naive and idealistic thought. Does a Jamaican bobsleigh team has the same opportunities as its counterpart in Switzerland? Another, and more down-to-earth reason is the protection of the health of athletes. A common property of the prohibited substances is that their use involves substantial health risks. Doping agents usually act, if they act at all, by mobilizing the reserves of the human body. The body of a trained athlete can do this without the doping agent, so any further mobilization is at the expense of health.

3.5 Doped Up and Ready to Win. Maybe

Table 3.1 Groups of prohibited substances in WADA rules

Prohibited in- and out-of-competition	1. Anabolic Agents
	2. Peptide hormones, growth factors and related substances
	3. Beta-2 agonists
	4. Hormone and metabolic modulators
	5. Diuretics and other masking agents
Prohibited in-competition	6. Stimulants
	7. Narcotics
	8. Cannabinoids
	9. Glycocorticosteroids

Extreme believers in human rights may even say that every individual can risk his or her own life as he or she wishes.

Even WADA readily admits that some of the prohibited substances do not have any effect on sports results. Table 3.1 shows the current categories on the prohibited list. Two of the categories in the rules (hormone and metabolic modulators, diuretics and other masking agents) are declared to have no improving effect. Most of the substances in the category hormone and metabolic modulators are so-called anti-estrogenic substances, and these are only prohibited for men. The reason is that the legal use of these drugs is exclusively for women, whereas men using these substances are extremely likely to use anabolic steroids and take additional anti-estrogens to fight the side effects. Diuretics and other masking agents are banned for both men and women because they make it more difficult to detect other doping agents—at least using today's detection methods. A memorable case in the international media was that of Hella Böker, a senior German woman athlete, who was 'caught' using a diuretic called hydrochlorothiazide in 2006. This drug is a component in many over-the-counter antihypertension medications. The only inexplicable part of this story is why bother performing doping tests in senior competitions at all. Here, the fame or money that can be gained by winning is insignificant (how many of the readers can recall the name of any senior world champion in high jump?), and renders the major motivation of risking someone's health by doping irrelevant. Why test those who have no interest in cheating?

For most of the substances in the remaining seven categories on the prohibited list, the performance improving effect of the drug has never been proven in scientifically rigorous studies. But the opposite (the absence of these effects) has not been proven, either. This lack of information is mostly caused by the fact that such physiological effects can only be proved by very costly studies, which would normally be ethically questionable. Experimentation on humans is only permitted for very good reasons, and testing the doping effects of a substance is not one of them. One of today's supposed star drugs, EPO (erythropoietin or erythropoetin), was only investigated in four studies. All of them show that the use of EPO may increase the oxygen uptake of the body, but only for a very short time. Therefore, it was deemed highly unlikely that this could cause any significant effect in sports results. The health risk, on the other hand, is quite clear, as EPO is used as a medicine and this question had to be studied in painstaking detail.

There is only a single group of substances which have a proven sports performance enhancing effect: these are the anabolic steroids. It is no small irony that the by-product of the former East German doping industry provides unquestioned evidence.

Most of the prohibited substances are banned because their reliably known physiological effects make it likely that they could be used to gain an unfair advantage in sports. However, some of the prohibited substances such as heroin, morphine or hashish are incompatible with any sport activities. There is still an important rationale in banning these: someone under the influence of these drugs could be a danger to themselves, their teammates or opponents.

It is a popularly held belief that prohibiting a substance is often based on anecdotal rather than scientific evidence. Others claim that adding a new substance to the list also guarantees that someone will try it. Finally, one urban legend should also be busted here. Reports in the media often imply that one or more very rich athlete use undetectable, highly advanced doping methods to achieve their success. This cannot be true—even if the athlete believes it to be so (without publically admitting). The most that can be true is that a highly unethical doctor, using reasoning and tests that appear scientific, misleads a sports star with considerable fame but without a good sense of judgment. As the East German example shows, it is not impossible to develop a truly successful doping program. However, the scientific and medical background necessary is enormous. Keeping such a program secret for a long time requires ruthless efforts: in East Germany, the secret police (Stasi) was heavily involved. In addition, a program like this would be an economic disaster. It should be remembered that the former East Germany did not start this infamous program for financial gain. The main reason was state ideology.

3.6 Can Placebos Heal?

The development of a new drug usually takes several hundred million € and a decade of time, but some of the candidate drugs fail one innocent-seeming test despite the major research effort: their effect is not significantly better than the effect of an 'empty' capsule without any active ingredients. How is it possible that a drug that appears effective on animals in laboratory tests is ineffective on humans in clinical trials?

A placebo is defined, with some simplification, as a medicine without an active ingredient. The use of such medications dates back to ancient times, albeit without the knowledge of the patients or the doctors as well! Several medicines used in the Ancient or Middle Ages are now understood to not contain any active ingredients at all. Most of these 'placebos' were widely considered effective for centuries. Conscientious people are appalled by the news that rhinos are being driven close to extinction by the belief that their horns have healing powers. It is

3.6 Can Placebos Heal?

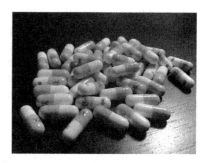

Fig. 3.7 Good placebo looks exactly like a real medicine. (Copyright-free picture from http://www.flickr.com/photos/59334544@N00/2322167178/)

good to remember that even European medicine had a collection of its 'rhinos' in the not very distant past.

Today, the successful medical use of a material without active ingredients is known as the placebo effect. The essence of this phenomenon is that a patient's positive expectations have healing powers. Modern placebos are actually more than just an empty capsule. To be effective, it is absolutely necessary that the patient believes the placebo to be a real drug. Therefore, a good placebo mimics real medicine: it is available in capsules, tablets and injections (Fig. 3.7).

The placebo effect was intentionally used for the first time to heal in the eighteenth or nineteenth century. Hooper's Medical Dictionary, published in 1811, defined "placebo" as: "an epithet given to any medicine adopted to please rather than to benefit the patient." The name placebo itself reflects the meaning: it is the first person singular future form of the Latin verb *placere* (like), so it means 'I shall please'. The essence of placebos has not changed in the last two centuries, but the role they play today is quite different. Originally, they were used to provide the appearance of a medical treatment (in order to calm the patient or because no effective treatment was known). Today, the primary role is in drug development.

The father of the modern placebo movement is Henry Beecher, who defined the use of placebos as a medical treatment supplementing the use of an active substance in a 1955 article. He became interested in the placebo effect when he worked as an army doctor in the battles of World War II. Severely injured soldiers were sometimes injected with a solution of salt instead of morphine, which was often in short supply, yet they reported a decrease in pain and felt generally better. Beecher proved similar benefits can be achieved in the treatment of a number of different illnesses.

Modern medicine is based on the effects that active ingredients have on the body. At a glance, it is difficult to understand how a medicinal drug without an active ingredient can still have a positive effect. First, symptoms often improve or illnesses even heal without any treatment. More importantly, mental issues play a central role in the placebo effect. Patients who believe they have received an effective treatment more easily accept symptoms, and have a more positive view of the changes in their health. If they trust their doctors, they try to live up to the expectation of the treatment. An empathic and attentive doctor can actually strengthen the placebo effect a great deal. There are a number of other significant factors, too. The color of the drug matters: blue capsules are best for tranquilizers, yellow works well for antidepressants, white is the preferred color of antacids, whereas red color helps the effect of heart medicines. Unsurprisingly, more expensive placebos have been shown to be more effective. Cultural factors are also important in the placebo effect: blue tablets do not generally work well as tranquilizers for Italian men, probably because the Italian national soccer team plays in blue T-shirts. Placebos taken more frequently have more benefits in general, but the effect is also influenced by the brand name and the shape of tablets. This is not just theory: the pharmaceutical industry makes good use of the potential of the placebo effect.

In most of the cases, the effect of placebos is primarily psychogenic. In contrast, pain relieving effects have a physiological background. The opioid receptors in the brain have been proven to increase their activity upon placebo treatment in a fashion similar to the real effect of morphine. Using a morphine-antagonist (a substance inhibiting the effect of morphine) also stops the pain relieving effect of placebos.

The placebo effect depends obviously on the type of illness and on the method used for diagnosing it. The symptoms that can be relieved best by placebos are those with primarily psychogenic backgrounds. Clinical trials with an insufficient number of patients or inadequate design can also show high placebo effects. It is not unusual to experience 15–60%, or even as high as 70%, placebo effect in certain psychiatric syndromes. It is very difficult to develop truly effective drugs for diseases in which the psychogenic effect may be more important than the physiological.

Placebos have side-effects, too. These are usually digestive or psychogenic symptoms (constipation, flatulence, diarrhea, anxiety, insomnia), and may occur in 10–20% of the cases. More frequently taken placebos cause side effects more often. The side effect of placebos is not to be confused with the nocebo effect. The nocebo effect is at work when someone, primarily because of prejudice, reports negative health effects of a substance that is known to cause no harm. The effect is very real as was demonstrated in a study of asthmatic patients. Patients were asked to inhale water vapor and several of those who believed the water vapor contained an allergen experienced a shortness of breath.

Today, placebos are not used in medicine, and this is not because of the side effects, but because ethical obligations require doctors to use the best available treatment, and placebos are never the best of the known options. In drug tests, though, the placebo effect plays a very important role. When a new drug is tested, it must be determined if it works better than the psychogenic effects induced. Drug tests using placebos as a comparison became common in the 1960s after Beecher published his results. To exclude subjective elements as much as possible, modern drug tests randomly assign patients real medicinal drugs or placebos, which look identical. Neither the patients nor the doctors diagnosing the effect know which group a given patient belongs to. This arrangement is called a **double blind study** (by no means the same as blind leading the blind). Double blind, randomized, and placebo controlled clinical tests represent the state of the art drug testing today. If a substance is proved to be effective by such a test, its medical usefulness is beyond doubt.

The objective of pharmaceutical research is to continuously improve drugs, so comparison with drugs already in use is a very important aspect of the research process. Since using placebos is ethically questionable, drug comparison studies are expected to gain more significance in the future, and when a new drug is in development, placebo studies will continue to be indispensable.

3.7 Synthetic Drugs vs. Natural Herbs

Modern medicines typically contain a single active ingredient which is produced by the pharmaceutical industry using chemical methods. The amount of the active ingredient is carefully controlled in such a medicine, and other substances are also used at optimum levels. Is there a place for old-fashioned herbs in modern medicine?

The question seems simple enough, but a reasonably precise answer is anything but straightforward. Synthetic and herbal medicines cannot be easily distinguished due to the overlap between the two categories. Herbs are not only used in herbal teas, but also serve as sources of active substances in the pharmaceutical industry and the molecules in herbs provide a source of chemical inspiration in drug development. In some cases, traditional herbs are better for therapy than any of the known synthetic alternatives. In the following paragraphs, a few examples will be described to illustrate the role herbs play today in medicine and the pharmaceutical industry.

Herbs enjoyed a sort of monopoly in medicine during the early ages of history. These applications were sometimes based on observations (e.g. the behavior of sick animals occasionally providing good guidance), but such treatments were often without benefits and sometimes outright damaging. Accumulated evidence was passed on from generation to generation, and some plants became herbs used against specific illnesses and their use was documented in increasing detail. Typically, tea prepared from the herb, its alcoholic extract or the finely chopped plant, was used for medical purposes. Some of these observations were later proven using the scientific method. Since the Age of Enlightenment, the value of an increasing number of non-rational (or mystic) applications has been questioned by researchers.

In the beginning of the twentieth century, synthetic medicines began replacing herbs. There was plenty of optimism at that time because the medical advances were extremely fast and many people hoped that effective medicines would be available against all illnesses within a short time. The initial success was marked by the discovery of Salvarsan (Fig. 3.8), which was an effective treatment against formerly incurable syphilis. Somewhat later, the introduction of modern antibiotics (Fig. 3.9) became a real milestone in the history of medicine (→ **3.23**). Unfortunately, drug development has slowed down. Many serious illnesses (certain types of cancer, rheumatism, Alzheimer's disease) are still without treatment. In the 1980s, combinatorial chemistry offered new hope, as it provided for the simultaneous testing of a large number of potentially active compounds in a short time. However, these hopes were not translated into significant success, and there is currently only a single medicine developed using the principles of combinatorial chemistry, namely, the anticancer drug called sorafenib.

Fig. 3.8 Arsphenamine or Salvarsan was developed by Paul Ehrlich and Sahachiro Hata in 1909 and was effective against syphilis. An improved version of this drug named Neosalvarsan was introduced 2 years later. (Copyright-free picture from http://www.paul-ehrlich.de/img/Salvarsan1.gif)

Fig. 3.9 Alexander Fleming accepts the Nobel Prize from Gustaf V of Sweden in 1945 for discovering penicillin. The first antibiotic was a natural compound produced by a fungus called *Penicillium notatum*. (Copyright-free Wikipedia picture)

Medical research has never abandoned the study of naturally occurring compounds. Materials from plants (or other natural sources) often have physiological effects, and studying these effects and the plants is much more likely to lead to new medicines than preparing new materials synthetically. This is not simply because of traditional knowledge from human society. Plants, bacteria or fungi produce many compounds in order to protect themselves from environmental harm. Evolution has had a long time to establish active ingredients that can also channel human drug development. Today, the primary value of plants in the pharmaceutical industry

3.7 Synthetic Drugs vs. Natural Herbs

is not their role as raw materials. Exploring the chemical structures of the active ingredients often reveals information about the way they act and points the way for synthetic chemists. Natural compounds often have complicated structures, but simpler analogs that possess the same properties may be readily synthesized.

In the last several decades, most of the newly developed drugs were connected to plants somehow: the idea of the chemical structure may have originated there, or at times, they were the actual raw material. Among 1200 new drug molecules introduced in the past 25 years, about 30% was purely synthetic. The remaining 70% had human proteins, neurotransmitters, bacterial, fungal or plant compounds acting as primers.

Some of the well-known drugs occurring in herbs are still important in therapy, and their structures are still a source of inspiration for research. Digoxin is one of the plant-derived active ingredients: it is produced from foxglove (*Digitalis* in Latin) extract by chromatographic methods and used in the treatment of heart failure. The first drug obtained in a pure form from plants was morphine, and this tale goes back to the beginning of nineteenth century. Despite being known for more than two centuries, morphine is still regarded as the benchmark drug to relieve severe or agonizing pain and suffering. Studying the chemical structure of morphine gave rise to an entire family of painkillers. Other notable compounds found in the poppy plant are codeine, papaverine and their derivatives, which are used to relieve cough, suppress spasm, and treat hypertension.

The large family of medicines known as nonsteroidal anti-inflammatory drugs (NSAID, → **3.3**) today also owes its existence to primitive observations about the effect of the willow (*Salix* in Latin). In this case, it was not the original compound, but a chemical derivative that became a blockbuster drug. The taste of willow bark is bitter, and similar to the taste of the bark of a South American plant *Cinchona*, which is an effective treatment against malaria. As well as the taste, the effect was also similar: willow bark seemed to possess an antifever property. Today, it has been established that salicin is the active ingredient, but a simpler derivative named acetylsalicylic acid proved to be more potent (→ **3.25**), and is currently sold under the trade name Aspirin.

The most intense research is currently aimed at developing anticancer agents, and plant-derived drugs have been outstandingly successful in this field. Vinblastine and vincristine, the alkaloids extracted from Madagascar periwinkle (*Catharanthus roseus*) were first isolated in the 1950s, and their anticancer potential against different types of leukemia was recognized as a result of systematic screening. Their semisynthetic derivatives, called vindesine and vinorelbine, were developed two or three decades later, and are used against breast cancer. All four of the above-mentioned drugs are still very important in cancer treatment.

One of the most successful antitumor compounds is paclitaxel, whose structure is among the most complicated (Fig. 3.10). An anticancer screening study carried out in the US during the 1960s revealed that the extract of the bark of pacific yew (*Taxus brevifolia*) shows very marked antitumor activity. The compound responsible for the effects, paclitaxel, was duly isolated, but since the quantity in the bark was miniscule, its chemical structure was way too complicated for any financially

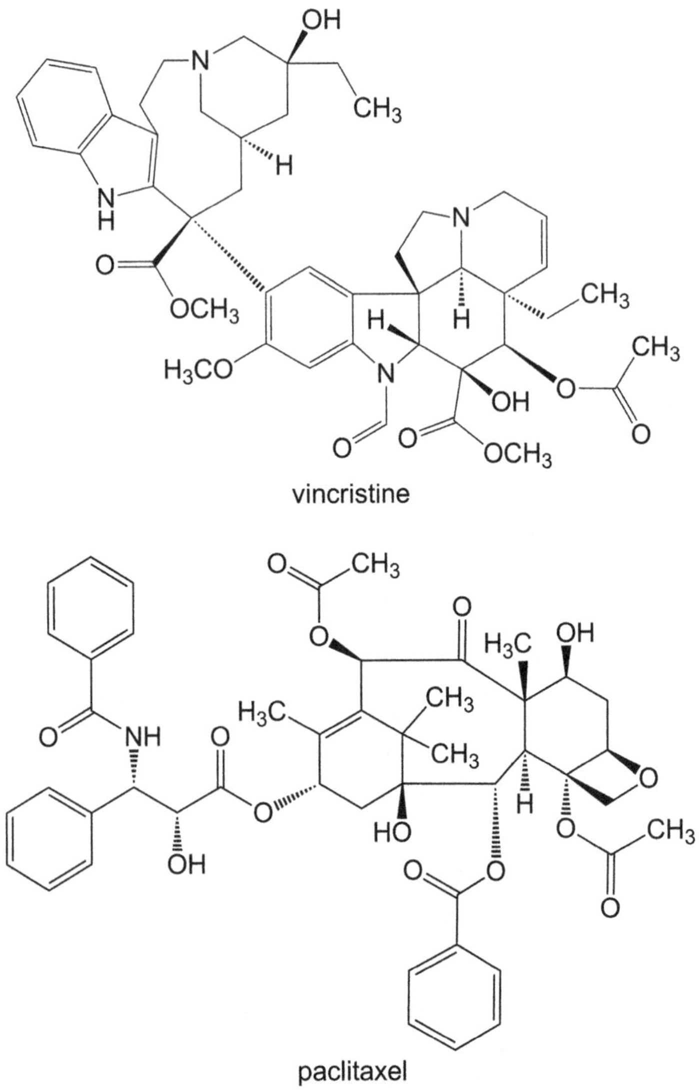

Fig. 3.10 These complicated chemical structures can only come from plants—the molecular formulas of vincristine and paclitaxel. (Authors' own work)

reasonable synthetic process. So, despite its extreme efficacy, paclitaxel was not exactly a promising drug candidate. Decades of research and huge financial investments finally yielded a solution: a compound was isolated from a related (and quite common) *Taxus* species, from which paclitaxel could be obtained by a relatively

3.7 Synthetic Drugs vs. Natural Herbs

short synthetic procedure. Among cancer drugs, paclitaxel is the one with the largest annual sales. There is still a great deal of research related to this drug, and one recent development has been that paclitaxel was identified in a cell culture of common hazel (*Corylus avellana*). This finding could open up an entirely new method of synthesis which could make the drug less expensive.

Older plant-derived drugs sometimes lose their importance when modern, synthetic alternatives are developed, but many of them still have no significant competitor. Some of the plant products in widespread use do not only contain a single active substance, but a full plant extract, or even a concentrate. Some of the stories in this book describe the efficacy and safety issues connected to these herbs ($\rightarrow$ 3.27, 3.30).

The most traditional application of herbs is in the form of herbal tea, many of which are still relevant. Two examples are given here to prove this point. The first is "nursing tea" used to help lactation in breast-feeding women. Although the hormones and neurotransmitters important in the process of lactation are all known, nursing tea is still second-to-none in efficiency. Pharmacological studies of the common ingredients (anise and fennel) appear scientifically support the observations made in the course of centuries about their lactation-inducing effects. Another group of herbal teas, for digestive problems, provide much more versatile ways to treat mild digestive problems (lack of appetite, stomach ache, flatulence) than synthetic medicines.

The list of herbs used today has not stopped growing. Most of these "new" plants have already been used in remote regions (South America or Africa) in the past, and after the necessary testing, some of them may be successful in the medicinal markets of Europe or North America. One example is devil's claw (*Harpagophytum procumbens*), which has a unique effect on joint inflammations.

The usual objective of product development is to prepare drugs with the highest possible purity and concentrations. Nevertheless, surprises still await those who are involved in drug development. A few years ago, a drug named Nicosan was introduced. It was based on a combination of African plant extracts and successfully prevented **sickling** medical crises caused by **sickle-cell anemia**. As no similar drug was on the market, the European Union classified Nicosan as an orphan drug, one which is specifically designed to treat a rare disease. The legal status 'orphan drug' is very important for manufacturers, as it provides financial incentives to develop and market drugs that would otherwise be unprofitable.

Although Nicosan is a rare exception, plant extracts with certified (standardized or carefully quantified) composition are still important in modern medicine. Many synthetic drugs are also connected to plants in some way, and it is almost meaningless to try and distinguish between natural and synthetic active ingredients. The significant difference is between high quality, properly researched medicines, and unreliable products promoted by charlatans.

3.8 Bitter, Stronger: Herbal Tea Sweeteners

Articles about herbal teas often warn that they should not be sweetened or flavored in any way because this would lower or even completely destroy their beneficial effects. In less strict texts, only sugar is banned, while honey is portrayed as a viable possibility. To check the validity of these warnings, the interaction between sugar and honey should be investigated, and some information about tisanes and their preparation methods is also worth pursuing.

First, a small technicality should be put to rest. Both food labels and the literature use the word 'tea' to refer to very diverse products. Strictly speaking, the word 'tea' should be reserved for drinks only made from the leaves of the tea plant (*Camellia sinensis*). Drinks prepared with other plant extractions and hot water should be called *tisanes*. Nevertheless, the catch-all term herbal tea is often used to refer to both the drinks and the plants they are made of, so it is useful to read the food labels.

Little is known about the early history of tisanes, but it is likely that they were already prepared in ancient times. Specific healing effects were recognized for a few plants, and it was also discovered that hot water enhances the extraction of the active ingredients.

The composition and use of herbal teas are mostly based on traditions. Although medicines underwent a complete overhaul in the last century, there was little change in the world of herbal teas save perhaps for the introduction of electric kettles.

Different varieties of herbal tea continue to be highly regarded despite the availability of modern medicines, and for some uses, even dominant (*e.g.,* to increase appetite or lactation → **3.7**). They have a pleasant taste and drinking them induces positive feelings, which explains their lasting popularity.

The fruit, shoot, leaf or blossom of a plant is needed to make a tisane. The use of underground parts (root) is less common but not unheard of. All plant parts may contain substances with medicinal effects, but usually it is the segments above the ground that deliver a tasty tisane. Some herbal teas are sold in teabags to control the dosage, but it is more typical to pack them in a finely chopped, homogenized bulk form. In this case, the user must measure the necessary quantity for making the drink.

Usually, hot water must be used to extract the active substance in herbal tea, and this *infusion* is prepared by keeping the tea immersed in hot water for 10–30 min. This ensures an effective extraction of the active ingredient(s) and flavors in the tea, and also attempts to regulate heat exposure in order to minimize any high temperature decomposition. If the active ingredient is particularly sensitive to heat (for example vitamin C in rosehip), warm or cold water should be used for extraction. Cold extraction is also convenient if the active ingredient dissolves at low temperature and the presence of other, non-desirable flavors or substances needs to be avoided. An example is bearberry (*Arctostaphylos uva-ursi*), whose cold extract is quite effective against urinary tract infections (contains hydroquinone derivatives active as disinfectants), but is free of unpleasant flavors that would be present if dissolved at high temperatures. A less common method of preparing a tisane is boiling the tea for

3.8 Bitter, Stronger: Herbal Tea Sweeteners

Fig. 3.11 Is sugar harmful for the effect of tea? (Copyright-free Wikipedia picture)

some time—which results in a *decoction*. This is necessary if the active ingredients are difficult to dissolve (from bark or root), but otherwise not recommended as high temperature may cause decomposition processes.

Banning sweeteners in herbal tea would be reasonable if sugar or compounds in honey reacted with the active ingredients or inhibited their physiological effects (Fig. 3.11). The chemical properties of cane sugar and the main components of honey render the first possibility very unlikely. Cane sugar (sucrose) is a disaccharide built up of one molecule each of glucose and fructose bonded to each other through a glycosidic bond. Upon heating with acid, sucrose hydrolyzes to give the two constituent molecules. The resulting mixture contains D-fructose and D-glucose in a 1:1 ratio, and is also known as invert sugar. Honey is about 70–80 % invert sugar, which causes the sweet taste. When bees collect honey, the enzymes in their stomachs hydrolyze the sugars in the nectar into invert sugar (Fig. 3.12).

Invert sugar is chemically more reactive than sucrose, as glucose contains an aldehyde functional group that makes the molecule reducing. In a basic solution, both glucose and invert sugar reduce silver ions to metallic silver. In the form of the silver-mirror test, this reaction is used to identify reducing sugars.

Even with this reducing character, sucrose and invert sugar are not very reactive in the tea making process. Herbal teas contain numerous chemical compounds, none of which react with sucrose, glucose or fructose. In addition to invert sugar and about 20 % water, honey also contains a complex mixture of compounds (minerals, pollen, substances produced in the bodies of bees). The possible reaction of these active ingredients with herbal tea has not been studied in detail yet. However, experience shows that honey does not interfere with the efficacy of herbal teas.

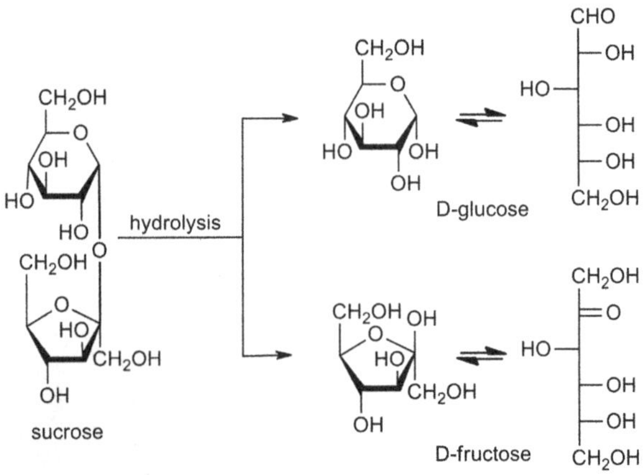

Fig. 3.12 The hydrolysis of sucrose to yield D-glucose and D-fructose. (Authors' own work)

There are exceptions, though, and adding sweetener actually does make one group of herbal teas less efficient, but this has nothing to do with chemical reactions. The bitter taste of herbal teas used to increase appetite (common wormwood, holy thistle, great yellow gentian) is essential for their effect. The intensity of the taste is the most important factor, and the identity of the actual substance causing bitterness is almost irrelevant. The sensation of bitterness is transmitted through the vagus nerve and increases the formation of saliva, gastric juice, pancreatic juice and bile. This leads to an increase in the level of the protein-decomposing enzyme pepsin, and the consequent formation of oligopeptides accelerates the formation of stomach acid. Bitter substances also activate peristalsis in the intestines, which further aids digestion. So the increase in appetite is an indirect, positive effect brought upon by better digestion. The bitter taste is the catalyst of the effect, so sweetening herbal teas that are recommended for increasing the appetite is not a very good idea.

Sweetening may have some advantageous effects as well, provided that honey is the sweetener. The minor components of honey have a number of favorable medicinal effects including anti-inflammatory and antibacterial activity. Consequently, it is wise to use honey as a remedy for a sore throat. In summary, sweetening herbal tea usually does no harm, and may even be useful in the case of honey. Using too much sweetener is not advisable as herbal teas usually have pleasant tastes even without sugar. Obese people or those suffering from diabetes are especially advised to be careful when adding sweetener to their tea.

3.9 Homeopathy: No Active Ingredient? No Side Effect?

Homeopathy is one of the most debated medical methods. Proponents claim that it is effective and—unlike conventional medical methods—has no side effects. Skeptics say that homeopathy is nothing but a placebo, as the substances used in treatments do not have any active ingredients. It is not easy to do this topic justice here, but trying to do so is not entirely without merit.

Homeopathy originated at the end of the eighteenth century, with its founding father, German physician Samuel Hahnemann (1755–1843), who rejected contemporary medical practices that caused great amounts of suffering to patients (bloodletting was one of the least intrusive). He attempted to find cures that alleviated symptoms with lower-intensity or fewer unpleasant effects. He conceived of the idea of homeopathy when he was translating a medical book, which stated that the bark of cinchona tree could be used to treat malaria. Hahnemann began experimenting on himself and found that cinchona caused malaria-like symptoms in his otherwise healthy body. This experience was the root of why Hahnemann believed that testing medicines on healthy people was extremely important. His healing principle was: "that which can produce a set of symptoms in a healthy individual, can treat a sick individual who is manifesting a similar set of symptoms." So homeopathy means healing by similarity, 'like cures like'. The word itself comes from Greek: hómoios- (ὅμοιος-) means "like–" and páthos (πάθος) means "suffering".

Later on, Hahnemann began testing what effects different substances produced in humans, a procedure that would later become known as "homeopathic proving". His theory postulated that the symptoms of "overdosing" can be removed by lowering the dose, but the therapeutic effects can be preserved. To reach this goal, he developed the method of "potentization", whereby a substance was diluted in a number of successive steps (Fig. 3.13). Dilution steps with a factor of ten, one hundred and fifty thousand were labeled D (decimal), C (centimal), and Q (quinque millessima = 50,000). The number of dilution steps is given as a digit after the letter. The dilution is done with water for liquids, whereas lactose is used for solids. The process of "dynamization" or "potentization" has to follow certain rules (vigorous shaking, striking against an elastic body) in a process homeopaths call "succussion". The objective of these rules is to increase the efficacy of the substance, so believers assert that the final product is much more than simply a dilute solution or powder.

Homeopathy often uses extremely diluted substances (which are conversely said to be of high potential). A C30 homeopathic remedy is prepared by 30 successive hundredfold dilutions, such that the final solutions contains the "active ingredient" diluted by a factor of 100^{30}. This means such a product is very unlikely to contain even a single molecule of the active substance. As an example, a quite sweet solution of sucrose with a concentration of 342 g/l or 1 **mol/dm³** contains 6×10^{23} sugar molecules in a liter. Dilution with a factor of 10^{24} will produce a solution that is unlikely to contain even a single molecule of sugar in a liter. With homeopathic wording, this would be D24 or C12 dilution. Homeopathic practice in fact uses products with even higher potential, whose content of active ingredients is also 0 molecules. Opponents often point out that such a product is nothing but water or lactose, so the effect is actually the same as the placebo effect. In the following paragraphs, some examples will be given to paint a more detailed picture about the problem of active ingredients in homeopathy.

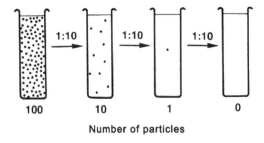

Fig. 3.13 How do 100 particles decrease to zero in three steps? The principle of decimal potentization in a solution that contains 100 particles in the beginning. (Authors' own work)

About two thirds of the substances used in homeopathy come from plants. It is often the entire above-ground plant, but occasionally only a few select parts. The raw material is first used to prepare a "mother tincture", which is further processed. Another large group of homeopathic products are based on minerals, including metals such as gold or silver and some inorganic elements such as sulfur. Animal-derived substances (from insects, snakes) form a much smaller group, but are not insignificant, either. Homeopaths also use treatments called "nosodes" (from the Greek 'nosos', νόσος, disease) made from diseased or pathological materials such as fecal, urinary, and respiratory discharges, blood, tissue, bacteria, viruses, fungi, parasites or other pathogens.

Homeopathic products have a unique status in the health marketplace. They are marketed as drugs in many countries. In the US, the 1938 Federal Food, Drug, and Cosmetic Act, which passed Congress with the active support of a senator who was also a homeopathic physician, recognized all substances in the Homeopathic Pharmacopeia of the United States as drugs. The Food and Drug Administration (FDA) of the US does not require the same standards from homeopathic products as from other drugs. In many countries, the legal situation is very similar. The quality of the products is usually carefully controlled, but even if they contain only a single active ingredient at high dilutions, it is not necessary to prove medical efficacy. As a consequence, most of them do not have officially accepted therapeutic recommendations.

Despite the general trends, a small percentage of "homeopathic" medicines actually have such recommendations. In specific cases, these uses must be supported by tests on patients in the same way as for standard drugs. Many of these exceptional homeopathic drugs share the characteristics contained in their mother tinctures or their low dilutions, e.g. D1 potentized solutions (tenfold dilution). There is often more than one active ingredient in such solutions, and the labels also include information about possible side effects.

However, there are relatively few clinical tests regarding the effectiveness of homeopathic products. One of the most extensive databases for controlled clinical trials is the *Cochrane Library*, which has registered the results of 641,000 trials, 350 of which were done on homeopathic products. These trials typically investigated solutions with low potentization and more than one active substance.

The old debate about the efficacy of homeopathic drugs cannot be simply wrapped up with a yes or no answer. There are a few published analyses based on available clinical evidence, but they arrived at contradictory conclusions. Proponents and opponents of homeopathy can both pick their favorite studies and ignore the others. Without dwelling on details, the most recent summaries usually show that homeopathy has no more effect than a placebo. However, this is actually the wrong question. Homeopathy uses a large number of methods, often based on very different materials, and it only makes scientific sense to evaluate the individual products, not homeopathy as a whole.

The action-mechanism of homeopathic products is just as unclear as in the time of Hahnemann. Lack of knowledge about these mechanisms does not necessarily mean the absence of an effect. Many traditional drugs today have verified therapeutic effects, but understanding the mechanism is fragmentary at best. However, there is a serious ethical issue present: a product should only be officially declared to be a medicine if there is evidence to support this claim. Otherwise, citizens are misled.

To screen out the placebo effect as a possible source of error in judging therapeutic efficacy, placebo-controlled clinical studies are necessary. The *Cochrane Library* registered 150 studies utilizing this methodology, and some of them showed positive conclusions. This is a very small number, especially compared to the huge variety of homeopathic products sold, but in fact still larger than many of the opponents would believe.

Surprises also await a curious mind in the world of homeopathic remedies that have effects proven by placebo controlled, randomized, clinical **double blind studies**. Plant-derived products containing a mother tincture or a solution of D1 potentization often feature detectable amounts of active ingredients. Chasteberry (*Vitex agnus-castus*) is used in both homeopathic and conventional (allopathic) medicines. The two 'different' kinds of products contain just about the same amount of the same active ingredient (called casticin; Fig. 3.14 shows analysis results) and their therapeutic recommendation is identical (managing the premenstrual stress syndrome).

In cases like this, the effects are not surprising at all and are fully explained by traditional medical theories based on receptor binding or the influence of enzyme activity. Hahnemann's intention was to lower doses in order to avoid side effects, but not all of today's homeopathic products follow this principle, which is not illegal. The legal definition of homeopathy only specifies the method of preparation, not the dilutions or concentrations. It would not be surprising if some of the products advertised as 'homeopathic' were shown to be effective even by the standards of conventional drugs, but it is unlikely that such therapies would follow Hahnemann's principle of concentration 'lowering', and would most certainly have possible side effects as well.

3.10 Aromatherapy: Nice but Useless?

In a nice environment, people feel better. This is hardly a discovery, but some experts have actually proved this connection using scientific methods. Even the symptoms of sick people feel less severe if the surroundings are pleasant, and the scents

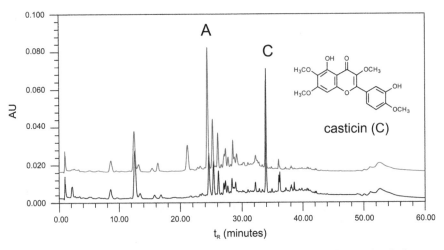

Fig. 3.14 *High Pressure Liquid Chromatograms* of an allopathic (*upper blue* trace) and a homeopathic (*lower black* trace) medicine. The peaks labeled with A correspond to agnuside, C to casticin, respectively. (Authors' own work)

of essential oils can contribute to this. Aromatherapy certainly improves the emotional states of people, but does it actually heal?

Aromatherapy is medical treatment with essential oils. Essential oils are volatile liquids obtained mostly by distillation (occasionally by pressing, extraction with a solvent or supercritical extraction) from plants. Some families of plants have a high content of essential oils and have a scent similar to that of the plant.

Although the name essential oil and its variation 'ethereal oil' both refer to the volatility of these compounds, this is not always accurate since the strong scent is in fact caused by evaporating molecules even if they are less volatile than e.g. water. Most components in essential oils are hydrocarbons with 10–15 carbon atoms. Their composition is very different from those of fixed (or fatty) oils (sunflower oil, olive oil), which are glycerol esters of fatty acids. These fatty oils are not volatile and they do not have pleasant (or other) scents, either.

A number of ancient cultures used essential oils or the plants containing them. The first distillation of suitable oils was found in today's Pakistan and is estimated to be about 7000 years old. China and Egypt have had a written history of essential oil use for several thousand years. Aromatic plants and their oils or extracts played an important role in both medicine and religious ceremonies. The first cosmetics were mixtures of essential oils, which is true even today, although some components are produced synthetically. Scented incense was also commonly used. This is the origin of the word 'perfume' as the Latin preposition *per* means 'through or along', and *fumo* means 'to smoke'.

Plants producing essential oils were used in medicine, both externally and orally. Ancient Greek physician Hippocrates of Kos (c. 460 BC—c. 370 BC) recorded that scented plants were burned during a plague in Athens. Today, it is clear that several essential oils are active against bacteria, but the plague was not stamped out because of this effect. It is more likely that the strong scent was too much for rats, which are the hosts of the fleas spreading the plague pathogen.

Essential oils are evaporated today for medicinal purposes in the same manner as in ancient and medieval times. Yet aromatherapy does not truly have such long traditions. Today's method was invented less than 100 years ago. The creator was French chemist René-Maurice Gattefossé (1881–1950), who burned his hand very badly in a 1910 laboratory accident and later claimed he treated it effectively with lavender oil. His pain was promptly relieved and the burn healed much faster than expected and without notable scarring. After this "miraculous recovery", Gattefossé devoted his life to studying essential oils. In 1937, he published his findings in a French book titled *Aromathérapie: Les Huiles Essentielles, Hormones Végétales*. This was also the first time the word aromatherapy appeared in print. The book is still considered the Bible of the method, which became very popular, especially in the French-speaking world. An English translation was only recently published in 1993.

Due to a kind of Renaissance in alternative medicine today, aromatherapy is experiencing a golden age. Numerous books and innumerable (often self-proclaimed) traditional healing experts recommend aromatherapy, but as is often the case, the real value and limitations of aromatherapy remain unknown to patient and healing expert alike.

Aromatherapy relies on interviewing the patient in order to select the suitable essential oil. The application of the oil follows, and this can be done by evaporation and inhalation, but more often it means skin massage. For massage, the essential oil is always diluted with a neutral fatty oil such avocado, olive, or peanut oil, because the essential oil alone often has an irritating effect on the skin. The massage takes one or two hours. The personalized treatment and the interview usually build up trust between patient and therapist, which has psychological advantages. The essential oil massage provides a nice, calming effect on most people, so it is hardly surprising that most patients report feeling better and relief of some symptoms.

Books do not only recommend aromatherapy to simply relax people, they claim real physiological effects as well against acute (ear ache, bronchitis, migraine) and chronic symptoms (hypertension, eczema). An examination of the clinical trials reveals that very few of the claimed effects were scientifically tested. The *Cochrane Library* database contains 121 studies related to aromatherapy. Most of these focused only upon distress and depression, and the selection of essential oils studied was also quite narrow: lavender oil seems to be immensely popular among scientists.

From these scarce data, it is clear the most practices in aromatherapy are not based on science, and to make a clear distinction, several experts have proposed that experimentally proven methods should be called *aromachology*.

Lavender, as the flagship of aromatherapy, deserves more explanation. The relaxing effect of this essential oil has been studied and confirmed by a number of animal tests and *in vitro* (outside living organism) experiments. The physiological effect resembles that of some known medicines, and one possible mechanism relies upon inhibiting the binding of glutamic acid to a receptor. Glutamic acid is an essential

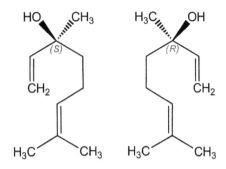

Fig. 3.15 Enantiomers (mirror image molecule pairs) of linalool, a major ingredient of lavender oil: (*S*)-(+)-linalool and (*R*)-(−)-linalool. The former has a sweet smell, the latter is characteristic of lavender. (Authors' own work)

amino acid and is known to cause stimulation of the nerve cell. Another contributing aspect might be the enhanced receptor binding of sedative gamma-aminobutyric acid. The active ingredients of lavender oil are lipophylic, which means they can exert their effect directly upon the brain after uptake through the mucous membrane of the nose (inhalation) or by skin absorption (massage). A particularly valuable study monitored the blood levels of lavender components. The two most significant ingredients, linalool (Fig. 3.15) and linalyl acetate, were detectable both in blood and brain tissue 15 min after inhalation. Similar results were obtained when lavender oil was absorbed through the skin in a massage. Although the connection between blood concentrations and clinical effects was not studied, these data alone show that the effect of lavender must be more than just a placebo with a pleasant scent.

The relaxing and sedating effects of orally taken or inhaled lavender oil have been proved in a number of animal tests showing lengthened sleep times or decreased movement intensity. Pure linalool or linalyl acetate had effects similar to those of the oil. Linalool was even studied in human tests. Only (*R*)-(−)-linalool was active and this represents more than 90% of the linalool in lavender. The other enantiomer, (*S*)-(+)-linalool did not have a similar effect. This is also evidence of some tangible therapeutic effect as receptors are expected to bind the two enantiomers (which do not have the same scent either) in different ways.

Paradoxically, the pleasant aroma of lavender makes the scientific study of its effects quite difficult. It is next to impossible to carry out a placebo-controlled test, as the characteristic scent makes the presence of the active ingredient obvious. The effect of lavender oil in bath water or massage oil was also investigated in a handful of studies, and these also turned out to be positive. In addition, EEG tests have also proved the oil's effects on the central nervous system.

In contrast to lavender oil, scientific data are very rare for other essential oils used in aromatherapy. Some of them seem to have real pharmacological effects on the central nervous system, but the evidence is not as strong as in the case of lavender oil. Joint and muscle pains can be relieved by massage-aromatherapy, but available evidence does not demonstrate anything more than just the result of improved blood circulation. As scientific studies are seldom carried out on other essential oils, their application has no confirmed role in modern medicine. As this is a case of absence of evidence, and not evidence of the absence of effect, further and more detailed studies may change the current professional zeitgeist.

3.11 Innocent or Guilty? Chamomile

Chamomile is a particularly well known herb and its allergic effects, including the phenomenon of cross-sensitivity, are discussed in several books and articles. Although the plant can in fact cause allergy, the symptoms are usually anything but dramatic.

German chamomile or wild chamomile (*Matricaria recutita*) is a member of the *Asteraceae* plant family and owes its popularity to various properties, which include confirmed anti-inflammatory, antibacterial and antispasmodic activities. As a consequence, it is used against the common cold, stomach aches and skin inflammations. Despite its widespread use, very few adverse effects have been reported, and only some of these side effects are allergic in nature. So then, why is there so much talk about chamomile and allergies?

First of all, it should be noted that the origins and symptoms of allergies vary quite a bit: modern medicine classifies hypersensitivity reactions into five types (I–V). All of these involve an exaggerated immune response: the body tries to defend itself against something that is certainly not dangerous and the body's defense reaction is exaggerated.

In type I (or immediate) hypersensitivity, the **allergen**, which is the substance responsible for the immune response, initiates the formation of a special immunoglobulin in the body, which in turn causes some special immune cells to produce substances causing inflammation, smooth muscle contraction, and blood vessel dilation. The major symptoms concern the constriction of the airways and are similar to asthma. In more serious cases, an **anaphylactic shock** may develop that can be life threatening. In modern classification, the use of the word allergy is restricted to this type I reaction, and this is the mechanism for common pollen allergy: some proteins in pollens act as antigens. Common ragweed (*Ambrosia artemisiifolia*) is quite famous for causing pollen allergies. Thus far, 22 allergens have been identified in its pollen, with the one denoted Amb 1 as the most significant (Fig. 3.16). The molecular weight of this protein is 38,000, and about 95% of people suffering from allergy are sensitive to it.

Some other plants may produce type IV hypersensitivity. In this case, the immune reaction takes a longer time to develop as there is no **antibody** matching the allergen. The cells of the immune system identify the allergen in this case, after which the body starts producing the substances which cause the symptoms. **Contact dermatitis**, which is a skin irritation after direct contact with an allergen (including some herbs), is a type IV hypersensitivity. Plants belonging to the plant family of chamomile (*Asteraceae*) contain sesquiterpene lactones, which often have beneficial anti-inflammatory effects but may also play a role in a type IV hypersensitivity and cause skin irritation, rashes, or even inflammation. Their effect depends on contact time, dose and the actual compound present. The worst allergens contain a CH_2 (methylene) group outside the oxygen-containing five-membered ring, whereas other common derivatives featuring epoxide or cyclopentenone motifs are less harmful. Compounds like these occur in common ragweed and other plants, but not in chamomile. Anthecotulide, another sesquiterpene lactone, was singled out as a

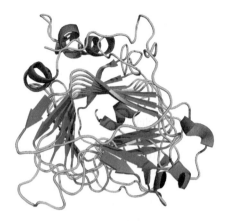

Fig. 3.16 The ribbon structure of antigen Amb 1. (Picture kindly provided by Ovidiu Ivanciuc)

possible allergy-causing component in chamomile (Fig. 3.17), but this substance was undetectable in European chamomile varieties. This is not the end of the story because the compound does occur in another related plant species called stinking chamomile (*Anthemis cotula*) that is very similar to chamomile. So the "chamomile" not collected by experts may actually contain the allergen.

Globalization cannot be stopped, and may have a pronounced affect upon the utility of herbs as well. Cheap chamomile imported from remote countries may have different amounts of allergens than local products. One of the leading producers of chamomile is Argentina, and the anthecotulide content of its product is much higher than in European countries. Different environmental conditions and genetic factors may account for this variation. However, even in Argentine chamomile, the concentration of anthecotulide is about 0.003–0.01 %, which should not cause significant problems. A scientific article once claimed that this concentration can be as high as 7 % in some chamomile varieties, but it turned out that the authors misidentified the plant species and were actually studying *Anthemis cotula*...

The same compound (or structurally very similar varieties) may occur in several related species, which explains why allergy is usually not specific to a single plant. This is called cross-reactivity, and since chamomile belongs to the same family (*Asteraceae*) as notorious allergenic common ragweed, cross-reactivity cannot be excluded.

Allergy to common ragweed is a very widespread problem, whereas chamomile allergy is not a significant concern. In fact, sesquiterpene lactones are not the primary concern for ragweed (or many other plants), but instead, the pollen is much more dangerous, as the allergens are proteins. For someone sensitive to the pollen of a chamomile-related plant species, cross-reactivity does not automatically arise as the pollens may be quite different. In chamomile pollen, none of the typical strong allergens can be found, so ragweed allergy does not automatically imply any sensitivity to chamomile.

Sesquiterpene lactones only cause allergy on contact with skin, with no oral effect being observed. These molecules are relatively small, and they can only cause hypersensitivity when they bind to proteins in skin.

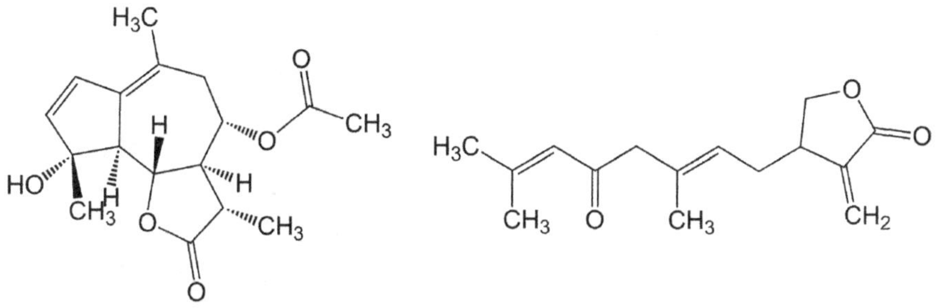

Fig. 3.17 The chemical structures of sesquiterpene lactone matricin, an anti-inflammatory compound in chamomile and allergen anthecotulide. (Authors' own work)

There are many misconceptions about chamomile allergy: some people think that it is similar to pollen allergy and affects the airways, while still others seem to believe that it can be life threatening. On the other hand, all reliable reports describe type IV hypersensitivity, which means entirely different symptoms, so in most of the known cases, it is quite possible that the allergen did not come from chamomile, but from another plant. A post-publication analysis of 51 case studies found that 46 of these did not identify the plant causing the allergy, and chamomile was most likely not a culprit in at least half of them.

Some reports of type I hypersensitivity to chamomile were also analyzed scientifically. Again, it was almost always highly doubtful if chamomile had anything to do with the observed symptoms. Chamomile is used by millions of people every day, and the number of allergy reports compared to this usage is extremely low.

3.12 Depressed: Antidepressant Side Effects

Proponents of conspiracy theories often claim that many drugs can only be sold because of the political clout of the pharmaceutical industry (often called Big Pharma). This story usually goes on to say that most of these drugs have more risks than benefits, and some of them even lack positive, medicinal effects. Antidepressants are the most frequent targets of these attacks, and skeptics say that they not only lack any beneficial effects, but also make patients suicidal.

Depression should not be confused with sorrow or intense mourning, which are natural and healthy emotions in certain situations (Fig. 3.18). These emotions are by no means abnormal unless they continue for an excessively long time. Depression, on the other hand, often lasts longer than its cause would predispose or it does not have a readily apparent cause at all.

Estimates show that about 20% of women and 12% of men experience states of low mood (actually depression) in their life that would warrant medical treatment. Although it has easily recognizable symptoms, depression should be diagnosed

Fig. 3.18 Depressed patient in the painting of Vincent van Gogh entitled *Sorrowing Old Man* (aka '*At Eternity's Gate*'). (Copyright-free Wikipedia picture)

by a trained doctor and the treatment must be adjusted to its severity. Depression has significant social, economic and societal effects, and a high risk of suicide is an unfortunate attribute of this mental state. From 15 to 25 % of deaths related to depression among untreated or mistreated patients are caused by suicide and more than half of all suicides are committed by depressed people. Depression has a severe impact on society as well—the patient's ability to work decreases, which leads to economic consequences. It is clear that depression requires medical treatment, but this does not necessarily mean a drug prescription: psychological methods aimed at identifying and ameliorating the cause can replace or at least supplement the use of medicines.

Several factors (inheritance, childhood experience, stress) influence the development of depression, but the physiological background is related to abnormal levels of certain neurotransmitter substances which transmit signals from one nerve cell to another. The levels of norepinephrine (noradrenaline) and/or serotonin are usually low in the central nervous system of people suffering from depression. Other neurotransmitters such as dopamine or gamma-aminobutyric acid (GABA) also play significant roles. Medicines are primarily used to increase the levels of these substances. Some of them increase serotonin levels, others influence norepinephrine, but numerous medications have different mechanisms involving different neurotransmitters.

In contrast to the claims that antidepressants lack beneficial effects, clinical studies showed that these drugs are significantly better than placebos at relieving the symptoms of the disease. As doctors must always use the best available treatment, the use of some older medicines was halted after the introduction of improved versions. Drugs available in pharmacies today certainly meet modern pharmaceutical criteria.

The myth of useless antidepressants stems from the multitude of causes of the disease. If the medical treatment targets the wrong neurotransmitter (the level of which is not abnormally low), the drug will appear to have no effect. Unfortunately, the level of neurotransmitters cannot be directly measured in the space separating nerve cells, so predicting which of the drugs will benefit a certain patient is next to impossible. Therefore, depression treatments do not have a high success rate, which is also mirrored by clinical studies. In practice, a trial-and-error method is adopted: after some period of time spent trying one drug without improvement, the treatment will continue with another one (preferably with a different mode of action).

Unsuccessful treatments do not ameliorate the symptoms, but may very well make them worse, which could cause a higher rate of attempted suicides as well. Opponents of the use of antidepressants claim that the suicidal behavior is caused by the treatment itself and advocate abandoning the use of these drugs altogether.

Reality, unfortunately, is much more complicated. Several studies showed that the risk of suicide increases at the beginning of the treatment, especially for young patients, but special care must be taken when drawing conclusions. This phenomenon of increased suicide risk may be caused merely by the unsuccessful treatment, and not the drug. Another possibility is that the drug creates more neuro-chemical activity long before it improves the mood of the patient. Most of these antidepressants need several weeks to work even when they are effective because the levels of neurotransmitters can only be changed slowly. An attempted suicide a few days after the beginning of the treatment cannot be rationally blamed on a drug which had yet to develop its physiological outcome.

On average, a successful treatment does not increase the risk of suicide. A 27-year study even reported a 20% drop in the suicide rate among treated patients. Another analysis based on 295 clinical trials including 77,000 patients found that those receiving a placebo had a somewhat higher suicide rate (0.72%) than those actually treated (0.62%). Patients older than 30 years showed a significant decrease, but it should also be noted that patients younger than 24 showed an increase in suicide rate upon treatment. A third, short-term test focusing on young people found some decrease in the frequency of attempted suicides upon depression treatment. However, the positive effect was less pronounced or even disappeared for patients who suffered form an additional psychological symptom (psychosis or personality disorder).

In summary, treating depression requires accurate diagnosis and careful therapy. Successful treatment relieves symptoms and decreases the suicide rate, but the risk cannot be reduced if other psychiatric conditions accompany depression, especially for the young age group. Cases known to involve increased suicidal risk require special care from people close to the patient. The cooperation of family members is not only necessary to avoid suicide attempts, but is also inevitable in curing the depression itself. In any case, drug therapy involves lower risk than no treatment at all.

3.13 Pre-Emptive Action: Vitamin C

A household method for treating the common cold is to take tablets of vitamin C or food that is rich in this vitamin (rosehip tea, lemon). Does this practice withstand scientific scrutiny?

Vitamin C is one of the most popular products of the pharmaceutical industry. Its career began in the first decades of the twentieth century. Soon after the discovery of ascorbic acid, Hungarian physiologist Albert Szent-Györgyi (1893–1986) won the Nobel Prize in Physiology or Medicine in 1937 for studying this substance. By definition, a vitamin is a material that the body cannot prepare but needs for its health, and most vitamins were discovered at the beginning of the twentieth century. Although vitamin C is often thought to be identical to ascorbic acid (L-ascorbic acid, to be more precise), in fact its reduced version (dehydroascorbic acid) and one of its stereoisomers (D-isoascorbic acid) have similar, but weaker vitamin effects. These two substances are called vitamers of L-ascorbic acid. D-Isoascorbic acid is an epimer of L-ascorbic acid, a stereoisomer that differs in the configuration of only one stereogenic center (Fig. 3.19). The enantiomers of these compounds have no vitamin-like properties.

Although L-ascorbic acid itself was unknown, the symptoms of vitamin C deficiency have been documented for many centuries. In the eighteenth century, it was finally recognized that the illness scurvy, which was common among seamen, was caused by their diet, or more specifically, their inadequate diet. During long journeys, sailors typically ate a steady diet of salted fish and hard tack. They often became sick and died of an illness, the symptoms of which were malaise and lethargy, followed by formation of spots on the skin and spongy gums. The cause was unknown, and it was a Scottish surgeon in the Royal Navy, James Lind (1716–1794), who first began a scientific study of scurvy. He divided the scorbutic sailors in his ship into six groups, each of which consumed different food supplements in addition to their usual food. One group ate oranges and lemons, while members of other groups consumed cider, elixir of vitriol, vinegar, medicinal plants or seawater. Those (and only those) eating fruits showed signs of recovery. This would not be unexpected today, as these fruits are known contain a plentiful source of vitamin C. Lind's findings were largely ignored by contemporary physicians and the breakthrough only came later. Legend has it that one of the advantages of the troops commanded by Admiral Lord Nelson at the Battle of Trafalgar (21 October 1805) was that they had to regularly consume lemon juice. Even so, many English soldiers died of scurvy caused by a vitamin C deficiency five decades later in the Crimean War.

Lind had discovered that scurvy can be *treated* with lemons, but he did not recognize that the illness was actually *caused by* an insufficient intake of vitamin C. This discovery is credited to Norwegian researchers Axel Holst (1860–1931) and Theodor Frølich (1870–1947). They experimented with guinea pigs and got lucky. Scurvy occurred in animals when their feed consisted solely of various types of grain, but the symptoms were prevented when the diet was supplemented with

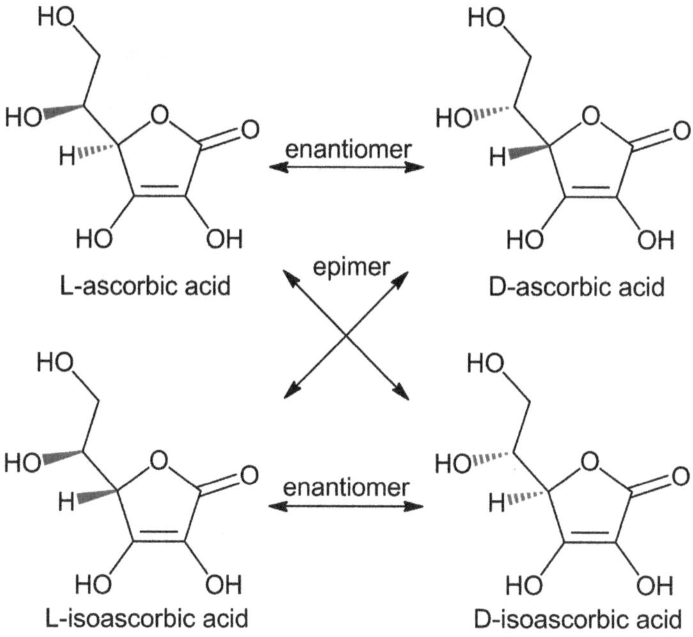

3.13 Pre-Emptive Action: Vitamin C

Fig. 3.19 L-Ascorbic acid and it stereoisomers. (Authors' own work)

known antiscorbutics like fresh cabbage or lemon juice. This was quite a fortunate coincidence: guinea pigs are now understood to be one of the very few animals that, similarly to humans, cannot produce L-ascorbic acid, so this substance is vital for guinea pigs as well. The findings were published in 1907, and it was proven that scurvy is caused by nutritional deficiency. In animal tests, several other foods were found to show anti-scurvy properties, and it was also clearly shown that heating or boiling considerably reduces the effect. The unknown active ingredient was named vitamin C in 1920, by the "godfather", distinguished British biochemist Sir Jack Cecil Drummond (1891–1952).

The substance responsible for the effects of vitamin insufficiency was isolated by Albert Szent-Györgyi at the end of the 1920s (Fig. 3.20). As the chemical structure was still unknown, he first called the compound ignose, after combining the Latin verb *ignosco* (I do not know) with the chemical suffix -ose usually used in sugars. The editor of the journal publishing the article did not like this name, so Szent-Györgyi first changed it to godnose (God knows what kind of sugar…), but the editor did still not appreciate the humor, so finally the prosaic name of hexuronic acid was used. It was difficult to prepare the substance in sufficiently large amounts, and since the first unsuccessful sources were animal adrenal glands, they simply contained too little ascorbic acid. Citrus fruits did contain high amounts of the vitamin, but numerous other sugars present made the isolation very difficult. Szent-Györgyi, who received his PhD from Cambridge in 1927, accepted a position

Fig. 3.20 Nobel laureate Albert Szent-Györgyi in lecture. (Copyright-free Wikipedia picture)

at the University of Szeged in 1930. He tried local paprika as a source of vitamin C, and his efforts were finally successful. He was awarded the Nobel Prize in 1937, and in the same year the British chemist Sir Walter Norman Haworth (1883–1950) received the Nobel Prize in Chemistry for working on the synthesis and structure determination of vitamin C.

Why is ascorbic acid considered a vitamin? Although it may have an indirect role in preventing the common cold, its value goes well beyond this. L-Ascorbic acid (together with the already mentioned dehydroascorbic and D-isoascorbic acids) is necessary for the proper functioning of a number of essential enzymes in the human body. Its main role is to keep iron and copper ions bound to these enzymes in their reduced state (Fe^{2+} and Cu^+) such that the vitamin effect is actually an antioxidant effect. From a purely chemical standpoint, other substances could serve the same function, but the physiological conditions are quite unforgiving, and apparently, nothing can replace ascorbic acid. Through essential enzymes, vitamin C is significant in the biosynthesis of carnitine and norepinephrine, and the amidation of various proteins. However, its main role is in collagen synthesis in the connective tissues. A deficiency of vitamin C has therefore a negative impact on bone, joint and muscle formation, and causes scurvy.

The potential of vitamin C to prevent the common cold was illustrated by another Nobel Laureate, American chemist Linus Pauling (1901–1994). For his scientific work focused on chemical bonding, Pauling was awarded the Nobel Prize in Chemistry in 1954. His peace activism and opposition to nuclear weapons won him the Nobel Peace Prize in 1962. He is the only person to be awarded two unshared Nobel Prizes. Only three other individuals have won more than one Nobel Prize (Marie Curie, John Bardeen, and Frederick Sanger) with Pauling and Mme Curie being the only two who were awarded Nobel Prizes in different fields.

Linus Pauling often suffered from bad colds, so he was quite receptive to the advice of one of his biochemist friends, who proposed consuming several grams of ascorbic acid to prevent the common cold. Pauling found the method effective and began popularizing it. He even wrote a book that advised readers to consume 1–18 g of vitamin C each day. Later on, he studied the anticancer and psychiatric effects of ascorbic acid, but his assertions regarding these effects have remained unproved to date. Those who cite Nobel Prize winner Linus Pauling's authority in relation to vitamin C seem to forget the fact the his Nobel prize recognized an entirely different sort of scientific work. The same is true for Albert Szent-Györgyi. In a common, but apocryphal belief, he took several grams of ascorbic acid each day. Both Szent-Györgyi and Pauling lived to be 93, but this fact by no means proves that vitamin C would be the secret of a long life.

Many a book has been written about the secrets of long life, but the present authors' ambitions are far more limited, and only concern the putative effect of vitamin C to prevent the common cold. Without doubt, vitamin C is necessary for a properly functioning immune system. In its absence, some white blood cells cannot operate fully, so this could increase the risk of certain diseases. Ascorbic acid is excreted from the body quite rapidly, which means that regular consumption is important. But this regular consumption does not necessarily have to be in the form of tablets: a balanced diet does contain enough of the substance. A number of fruits contain large amounts of vitamin C. For some nations (e.g. the British), the main general dietary source of ascorbic acid is the potato, which does not contain it in high concentrations but the large amount eaten makes up for the relative scarcity. Orange and lemon contain about 50 mg of ascorbic acid in 100 g of fruit, while the number is 10–30 mg for an apple, 1000 mg for rosehip, 10–30 mg for a potato, and 125–200 mg for paprika. The recommended daily intake for adults is 75–90 mg. This amount is already sufficient for the vitamin's purpose, and only a small percentage of this is excreted, which means that most of it is utilized by the body. It seems quite obvious that a healthy diet needs no supplements on the vitamin C front.

The tablets sold in pharmacies, which are used to prevent or treat colds, usually contain much larger amounts (200–1000 mg) of ascorbic acid. Could this high dose be harmful? Does it make sense to exceed the physiologically necessary dosage or to take megadoses of vitamin C? A general answer to the first question is given in a separate story (→ 3.14). For present purposes, it suffices to state that the approved ascorbic acid products are safe even if they contain several times of the recommended daily intake of vitamin C.

The debate about preventing the common cold by consuming vitamin C is old and not without its passions even today. If we restrict our thinking to scientific evidence, the review published by *The Cochrane Collaboration* in 2006 seems to be a good starting point. This summarized the results from 56 different placebo controlled tests, each of which involved a dose of at least 200 mg of vitamin C a day to treat or prevent the symptoms of the common cold.

Of the reviewed tests, 30 were about prevention. Their analysis did not show any general decrease in the number of cases of cold developed as a response to the treatment. However, a subgroup analysis revealed that among those exposed to extreme physical stress (marathon runners, skiers, and soldiers in the Polar regions), the decrease was quite notable—about 50%. In the 5 tests yielding positive results, the applied doses were not particularly high (1 g on average). Consumption of vitamin C somewhat shortened the time of the sickness. This was statistically significant at 8–13 %, but in real life, this would translate into 11 days instead of 12 as the average length, which seems marginal to most patients. The cold symptoms were somewhat less serious among people taking vitamin C. Altogether, 26 of the 30 tests found some benefit to the use of vitamin C.

In tests using ascorbic acid as a treatment only during the sickness, no general benefits were found. There were exceptions, though. Two of the studies used extremely high doses (4–8 g) and showed that recovery was somewhat faster, although the symptoms were just as bad as in the control group. However, the design of these studies is not entirely beyond criticism, and they have not been repeated in independent experiments since the 1970s.

Despite these partial successes, the results have not provided unambiguous evidence of the benefits of vitamin C in preventing or treating the common cold under normal conditions. The statistically significant differences found in comparison with the placebo group usually do not translate into real advantages for patients. On the web, there are some claims of benefits from consuming 30–150 g (!) of vitamin C a day. These are extreme doses, at which even unexpected negative consequences might be seen. The believers of vitamin C mega-dosing usually ignore the proven fact that a dose in excess of 400 mg a day does not increase the amount of ascorbic acid in blood or in the cells of the immune system. The extra ascorbic acid is simply excreted with the urine.

The current scientific position is that vitamin C should only be taken in tablets by persons whose diet does not contain a sufficient amount due to eating or digestive disorders or a strong preference against eating fruits. The recommended daily amount in these cases is about 100 mg. There are no scientifically proven benefits of taking larger doses.

3.14 Vitamin Megadosing: the Real Truth

Vitamins are materials that are needed for the proper functions of human (or animal) life, but the body cannot synthesize them in isolation; they have to be consumed with the diet. Today, it is well known that both insufficiently low and unnecessarily high amounts of vitamin intake may result in health problems. There are widely accepted recommendations for the daily intakes of all vitamins, but a sizable group of people do not seem to trust this expert advice. Believers in consuming megadoses of vitamins basically say 'more is better', and even the term 'alternative vitamin therapy' has been coined. Are there any scientific facts to support this view? Could too much of a good thing be bad in this case?

The beginning of the twentieth century saw major discoveries about the roles of vitamins: it was recognized that consuming food with sufficiently high caloric value was not enough to maintain a healthy diet. Some substances, later called vitamins, were found to be indispensable in small amounts. Their deficiency was assumed to cause long known diseases such as scurvy (defective connective tissues, bleeding)

3.14 Vitamin Megadosing: the Real Truth

and rickets (softening of bones). Later, these assumptions were proved correct: scurvy is caused by vitamin C deficiency, whereas insufficient sun exposure and vitamin D intake cause rickets. The first substance recognized as a vitamin is now called vitamin B1. The discovery dates back to 1906 and was followed by a few decades of similar findings, many of which were significant enough to win Nobel prizes for the dominant researchers involved. The letters of the alphabet were initially used to refer to different vitamins, but this practice is changing, and the chemical name is more frequently used today in the scientific literature. Known vitamins are listed in Table 3.2. In some cases, such as for the bioflavonoids initially called vitamin P, later research showed that these substances are indeed important in the diet, but their deficiency does not cause a disease. Occasionally, several different substances (called vitamers) have almost identical effects. For example, cholecalciferol (vitamin D_3) and ergocalciferol (vitamin D_2) both have very similar functions, the term vitamin D refers to these substances collectively.

Today, there are no doubts about the necessity of vitamin consumption. However, a debate is still in progress about the amount necessary and the form of intake. A healthy diet does provide sufficient quantities of all necessary vitamins, so some experts do not recommend taking vitamins as a supplement to a normal diet. A few diseases (mostly gastrointestinal) and even pregnancy may necessitate the intake of certain vitamins in concentrated forms as capsules. Some claim that a modern diet is an unhealthy one, so taking extra vitamins is the only way to ensure a healthy life. More extreme opinions even posit that the nutritional experts usually recommend amounts that are way too low, and much more is in fact necessary.

How do the experts arrive at their recommendations? These are results of quite complicated calculations, which take into account minimum amounts necessary in the biochemical processes of the body, results of pharmacological tests, findings of epidemiological analyses based on large populations and a number of other sources of information that would be too boring to describe here. The EAR value (estimated average requirement) is usually an important starting point and this means the dose that would provide the vitamin needs for half of all humans and leads to an RDA (recommended daily allowance) value. RDA is an amount that provides a sufficient amount of vitamin for most adults. These declared RDA values are usually higher than EAR values because they must also be suitable for people whose vitamin requirements are larger than the average. New scientific results may lead to the modification of RDA values: a recent example is the increase of the recommended dose of vitamin D. Over the past decades, mean blood vitamin D concentrations declined. This is likely due to simultaneous increases in body weight, reduced milk intake, and greater use of sun protection. At the same time, there is increasing evidence for the preventive role of vitamin D in several diseases. This led to the modification of the RDA values of this vitamin (the new values are several times higher than the previous ones). However, an RDA value does not mean that this amount must be consumed every day: this is an average over a longer time period. Several vitamin products on the market can be evaluated or compared using the RDA values as a reference. Most of these products are sold as dietary supplements, and regulations ensure that the daily dose in them cannot be lower than 15% of the RDA value.

Table 3.2 Milestones of vitamin research from recognition to Nobel Prize

Vitamin	Recognition	Isolation	Structure determination	Synthesis	Nobel prize (C: chemistry, M: medicine or physiology)
Vitamin B_1	1906	1926	1932	1933	Christian Eijkman, Frederick G. Hopkins (M, 1929), Paul Karrer (C, 1937)
Vitamin C	1907	1926	1932	1933	Albert Szent-Györgyi (M, 1937), Walter N. Haworth (C, 1937)
Vitamin A	1915	1937	1942	1947	Paul Karrer (C, 1937)
Vitamin D	1919	1932	1936	1936	Adolf Otto Reinhold Windaus (M, 1928)
Vitamin E	1922	1936	1938	1938	
Vitamin B_3	1926	1937	1937	1867	
Vitamin B_{12}	1926	1948	1955	1970	George H. Whipple, George R. Minot, William P. Murphy (M, 1934), Alexander R. Todd (C, 1957), Dorothy Crowfoot Hodgkin (C, 1964), Robert Burns Woodward (C, 1965)
Biotin	1926	1939	1942	1943	
Vitamin K	1929	1939	1939	1940	Henrik Dam, Edward A. Doisy (M, 1943)
Pantothenic acid	1931	1939	1939	1940	
Folic acid	1931	1939	1943	1946	
Vitamin B_2	1933	1933	1934	1935	Paul Karrer (C, 1937), Richard Kuhn (M, 1938)
Vitamin B_6	1934	1936	1938	1939	Richard Kuhn (M, 1938)

Calculation methods and, therefore the values of RDA, may vary in different countries, even if the populations are similar. The European Food Safety Authority (EFSA) has already begun reviewing all available data, and there is hope that there will be a single Europe-wide recommendation that will replace the current multitude of national approaches.

Requiring a minimal dose serves to prevent misleading the consumers. A legal maximal dose also exists and this ensures the safety of the product. Some of the vitamins are water-soluble and are excreted easily from the body. Others are only soluble in fats and oils, and these can accumulate in tissues possibly resulting in hypervitaminosis with symptoms resembling those of poisoning. Vitamins A, D, E, and K are not water-soluble, it is important to avoid exceeding the maximum doses. The maximum amount is called UL value (tolerable upper intake level), this is determined by scientific risk analysis and is considered safe. No product can contain any of the vitamins in amounts that are higher than the UL value (Fig. 3.21).

Unfortunately, there are not enough reliable scientific data to determine the UL value for some vitamins, but some sort of guidance is also given for these. These values are not legally binding and they only serve to inform the public. These values are based on clinical trials that showed a certain dose of the vitamins safe: they

3.14 Vitamin Megadosing: the Real Truth

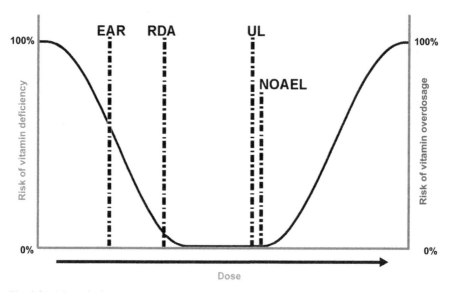

Fig. 3.21 The safe dose range of vitamins is found at the minimum of this curve. Doses smaller than the RDA involve a risk of deficiency, whereas does larger than the NOAEL (no observed adverse effect level) may result in side effects or toxic symptoms. (Authors' own work)

usually do not imply that the stated amounts are the highest safe doses (Table 3.3). The Food Standards Agency (FSA) of the United Kingdom published the following values: vitamin K_1: 1000 μg; vitamin B_1: 100 mg; vitamin B_2: 40 mg; vitamin B_5: 200 mg; vitamin B_{12}: 2 mg; vitamin B_7: 0.9 mg; vitamin C: 1000 mg.

As already stated, a common characteristic of vitamins is that they need to be consumed in order to maintain normal body functions. Preventing or treating a deficiency is the usual purpose of using these substances in medicine. Of course, such cases may well warrant doses different from the RDA value, but this is up to physicians to decide. Some vitamins are occasionally used in doses higher than the RDA for purposes that are not connected to the vitamin function. Vitamin A (mostly in the form of its synthetic analogs) is used to treat psoriasis and acnes, vitamin C is thought to be effective in preventing or treating the common cold (→ **3.13**), whereas vitamin B_6 has long been publicized as a cure for the premenstrual syndrome. These applications have some scientific background, whose levels of certainties differ. Other practices use very high doses of vitamins: usually, there is no scientific evidence to support these.

Proponents of alternative vitamin therapy claim that the amount of vitamins the human body actually needs is several times the RDA. In addition, therapeutic effects are attributed to these high amounts. An argument to support this view is that modern agricultural practices have decreased the amount of nutrients in soil, and cultivating new plant varieties has caused the vitamin content of common foods to fall substantially in the past decades. Some results showing a decrease in the vitamin contents of vegetables are often cited to support these views. However,

Table 3.3 Recommended daily intakes of vitamins for adults

Vitamin, unit	RDA	UL
Vitamin A, µg	800	3000
Vitamin D, µg	15	100
Vitamin E, mg	12	300
Vitamin K, µg	75	–
Vitamin C, mg	80	–
Vitamin B_1 (thiamine), mg	1.1	–
Vitamin B_2 (riboflavin), mg	1.4	–
Vitamin B_3 (nicotinic acid, nicotinic amide), mg	16	900
Vitamin B_5 (pantothenic acid), mg	6	–
Vitamin B_6 (pyridoxine), mg	1.4	25
Vitamin B_7 (biotin), µg	50	–
Vitamin B_9 (folic acid), µg	200	1000
Vitamin B_{12}, µg	2.5	–

even if some measurements seem to be in agreement with this trend, the problem is much more complicated. The vitamin content of any plant is heavily dependent on external factors such as the variety of plant used, the time of harvest, and even differences in the method used for analysis may seriously distort the results. A common characteristic of the statistics shown in advertisements is the lack of these vital details. Without these, no scientifically valid conclusion can be drawn.

In the absence of evidence, anecdotes are a primary means of promoting alternative vitamin therapy. The stories of Nobel laureates Albert Szent-Györgyi (1893–1986) and Linus Pauling (1901–1994) are often recited: they consumed several grams of vitamin C daily to preserve their health. It is rarely mentioned that it was not this custom that earned them Nobel prizes. Another emblematic figure is physician Abram Hoffer (1917–2009), who prescribed huge doses of vitamin B_3 for his own mother to keep her mind young. The anecdote claims that this was not only effective against old-age dementia, but also cured arthritis, improved the vision of the elderly woman, lowered her cholesterol levels, and returned her earlier vitality. She lived for 20 more years and wrote two books about her childhood. Hoffer and her followers saw an outstanding success for the vitamin therapy in this case.

Proponents of this vitamin therapy often understand the risks of vitamin overdose. The scientific literature reports several well documented cases, sometimes with fatal consequences. Fat-soluble vitamins are of the highest concern in this respect, both acute (a single extremely large dose) or chronic (from continuous large doses) toxicity is known for them. It is not an accident that three of the four fat-soluble vitamins have widely accepted UL values, whereas water-soluble vitamins rarely have reliably known upper limits. Studies of both hypo- and hyper-vitaminosis contain enough scientific literature to fill libraries, and this short story cannot even scratch the surface. Table 3.4 displays a few vitamins with reliable UL values and lists the symptoms of overdose. Care must be taken, as the toxic dose is usually lower for children than for adults, and the symptoms in kids may be more serious.

3.14 Vitamin Megadosing: the Real Truth

Table 3.4 Symptoms of vitamin overload (hypervitaminosis)

Vitamin	Symptoms of vitamin overdose
Vitamin A	*Acute toxicity*: abdominal pain, decreased appetite, nausea and vomiting, headache, dizziness, blurred vision, loss of muscular coordination, skin discoloration
	Chronic toxicity: hair loss, loss of appetite, nausea, liver damage, reduced bone mineral density, muscle pain and weakness, loss of muscular coordination, fever, blurred vision, headache, insomnia, menstrual disorders
Vitamin D	Decreased appetite, decrease of calcium levels in bones, increase of calcium levels in blood, muscle pain and weakness, constipation, vomiting, kidney stone
Vitamin E	Gastrointestinal problems, nausea, increased hemorrhaging for patient treated against epilepsy
Vitamin B_3	Itching, headache, decreased appetite, liver damage, jaundice, irregular heartbeat
Vitamin B_6	Nervous disorders, skin abnormalities
Folic acid	Allergy

The risk of vitamin A overdose is exceptionally high as it has a relatively narrow safe range (the UL value is barely four times the RDA). Although adults are not expected to show symptoms of toxicity below one-time doses of about one hundred times the RDA, the regular consumption of 8000 μg/day already involves a significant risk of developing the symptoms of toxicity. Children are especially sensitive. Overdose in pregnant women may also lead to problems in the prenatal development of the fetus.

Children are also more susceptible to vitamin D overdose than adults, who face noticeable risks if they consume amounts about 50 times the RDA value. Taking vitamin E is much safer: regular consumption of 2–3 g seems to cause virtually no side effects. The vitamers of vitamin B_3 are different in this respect: excess nicotinic acid seems to cause many more problems than large doses of nicotinic amide. Several hundred milligrams of vitamin B_6 is known to cause some (reversible) nervous disorder and skin abnormalities, but these are not serious. In the case of folic acid, it is not the possible side effects that give the reason for maximizing daily intake, but the fact that large amounts may make diagnosing a deficiency of vitamin B_{12} difficult.

Analysis of a typical megadose vitamin product shows that the manufacturer does not violate the regulations directly, but is more than happy to mislead the customers. One capsule does not contain significantly higher amounts of vitamins than the RDA values. A compulsory warning about "not taking more than x capsules in a day" (where x is usually a not very high number that depends on the particular label) is usually also displayed. However, there is a recommendation for a "superstrong" cure with a big number 15 displayed, which means that only 15 capsules a day will provide the benefits of a megadose. Following this advice will obviously provide much higher amounts than the RDA. In the case of vitamin A (4500 μg instead of the 3000 μg limit), vitamin E (800 mg instead of the 300 mg limit) and vitamin B_6 (200 mg instead of the 25 mg limit), the intake will be much larger than the tolerable upper limit.

When the pros and cons of vitamin megadoses are discussed, a common argument is that overdose-related diseases are actually very rare. Vitamin consumption is very high in the United States and yearly sales top $ 25 billion. The US Toxicology Data Network registered 65,000 vitamin-related reports in 2007, but only 17 of these involved serious symptoms with a single case of a fatality. So vitamin overdose is by no means an imminent threat today. Yet taking megadoses would invite the side effects without any demonstrated benefit. Mild as the side effects may be, they still outweigh the expected positive effect (which is often zero).

3.15 Detox Hoax: Poison Removal from Your Body

Detoxification seems to be all the rage, at least in the tabloids and lifestyle magazines, especially in the spring. This is the season of re-birth, so it is high time to get rid of the toxins that have accumulated in the human body. According to the advertisements, this is such an important project for everyone that investing major financial resources into it is absolutely necessary. So why do schools refuse to teach this practice in science class? Do scientists think that detoxification is useless but does no harm? Or is it in fact harmless?

Proponents of detoxification claim that toxins continuously enter human bodies with food where they accumulate and eventually cause illnesses. This is why annual (or even more frequent) detoxification is necessary to help remove the toxins. Herbs are often advised for this purpose, but some of the suggestions are more counterintuitive (to put it mildly).

Detoxification in alternative medicine is an approach that claims to remove "toxins" from the body, which accumulate there during normal life. To make matters more confusing, the very same word is also used in a legitimate medical context, where it means the physiological or medicinal removal of specific toxic substances (alcohol, narcotics, and poisons) that have entered the body as a result of exceptional events. Even taking a charcoal tablet to treat gastric problems is a sort of detoxification.

Food always contains smaller or larger amounts of substances that the body does not need. Yet this fact seldom has any noticeable consequence on human health because "detoxification", which here means removing the unnecessary substances, never stops in a healthy person. Actually, the body has a very sophisticated system that ensures the rapid excretion of preservatives, flavor enhancers, drugs or mycotoxins to avoid any possible damage. Part of the job is done by enzymes, which identify unnecessary substances (xenobiotics) and transform them into water soluble forms. Those taken up from the gastrointestinal tract are usually excreted in urine. A smaller percentage of these substances make their way into bile, with an even smaller part to other body fluids (saliva, sweat, and breast milk).

If the metabolized product is non-polar, it dissolves in fats and oils, and is reabsorbed in the kidney into the blood. If the solubility of the metabolized product is better in water, excretion is more efficient. An interesting calculation was made

3.15 Detox Hoax: Poison Removal from Your Body

for the sedative named pentobarbital. Without an increase of water solubility during metabolism, this drug would remain in the blood stream and exert its effect for 100 years (!).

The metabolism of vitamins mostly occurs in the liver, where various enzymes oxidize or reduce the substance, or cause some other chemical transformation. Conjugation is also possible in the liver: this mean chemical attachment into a glucoronide (a sort of sugar) or sulfate derivative. These forms usually find their way into the bile, where they are more soluble in water and cannot be re-absorbed.

If the amount of xenobiotics surpasses the detoxification capacity of the body, the symptoms of poisoning may appear, and damage to the body may follow. In this case, induced or artificial detoxification seems to make sense. The question is: do the usually recommended methods help the process? Another question is whether it is sufficient to do this annually or it is better to detoxify continuously.

Well, the detoxification advertised in lifestyle magazines or tabloids is by no means the same as the process described in the previous paragraphs. The popular detoxification does not help the enzymes in the body. Rather, it increases the rate of metabolism using diuretics and laxatives. Most of the recommended plants have one of these two effects. Unfortunately, neither of these is beneficial in the long run. Regular defecation might lower the risk of certain gastrointestinal cancers (to be more precise, frequent constipation increases this risk above normal), but this is by no means a reason to take laxatives. It is better to eat healthy amounts of healthy foods. A sufficient amount of fiber in the diet is usually enough to ensure regular defecation. Plants with strong 'detoxifying' agents usually induce diarrhea and spasms, which is actually quite harmful in a long term. Instead of using diuretic plants, drinking sufficiently large amounts of water (or other liquids) is more advisable. This will increase the volume of urine without dehydrating the body, which also minimizes the risk of stone formation in the urinary tract. Neither laxatives nor diuretics specifically increase the excretion of xenobiotics. In other words, there is no 'detoxifying' effect at all. On the contrary, this treatment may even be harmful. The short-term negative effects are seldom serious: the body uses less of the nutrients in food, but the water and ion balance of the body may be disturbed. For those who think that leading a healthy life is equivalent to suffering, the need for an occasional run of unpleasant diarrhea and the consequent loss of body weight might sound convincing. People wishing to lose weight also view detoxification in a positive light. Few of them know that their health does not improve, the promised detoxification never occurs, and the loss of body weight is anything but permanent.

Some plants used in detoxification actually have some positive effects. Milk thistle (*Silybum marianum*) is known to protect the liver against certain poisons. Flavonolignans in the seeds of this plant protect liver cells from damage and speed up their regeneration. The extract of the plant is the active ingredient in a number of medicines as well. One of the compounds isolated from milk thistle is often used in infusions to treat mushroom poisoning. Milk thistle products should only be used when the liver is already somewhat damaged or the risk of damage is very high.

Plants that increase bile production (artichoke, turmeric) actually speed up the excretion of xenobiotics, and help detoxification. There is no evidence to prove that this effect would improve the general health of anyone—unless they suffer from biliary diseases.

Less scientifically sound is the use of plants with antioxidant effects ($\rightarrow$ **3.31**). Although antioxidants may decrease the levels harmful free radicals in the body, this is not real detoxification. A diet with high levels of antioxidants (a lot of vegetables and fruits) is healthy, but concentrated antioxidants (high doses of vitamins or fruit extract) do not necessarily help improve human health.

For those who do not fashion plant-based cures, a detox foot-bath may be an attractive alternative: this is called Ionic Foot Detox Bath or Detoxification Foot Spa or Aqua Chi or Energizer Detox or Cell Spa Foot Detox or Chi Detox or Bio Detox or anything a creative advertiser can think of. The machine itself does not seem very different from a regular electric foot wash device, but a strange phenomenon may be observed during its operation. The bath water is becoming darker and darker, which is attributed to the toxins removed from the body. A typical description is like this: *"The real reason for not healing is your body running out of energy! The Aqua Chi Machine is a water energy system used to re-balance and amplify your body's bio electric field enabling the body to heal itself, by creating specific electromagnetic frequencies and harmonics, which are transmitted to the body through the water medium. When a person bathes in the Aqua Chi bath, the body absorbs the energy provided to use as needed. We believe the order of events in getting well first starts with the bio-electric field, (our cells have different vibratory rates), the second is oxygen (darkfield studies have shown the machine improves oxygen levels), and the third is nutrition and foods. The Aqua Chi balances the electric body and increases oxygen levels."*

Sentences like these are scientific mumbo jumbo: they are skillfully phrased to deceive non-specialists by appearing scientific. The changes during the operation of the device may in fact be connected to electric current, but the process itself is called electrolysis. Most baths contain common salt (sodium chloride), whose electrolysis affords hydroxide ions and hydrogen on the cathode with simultaneous dissolution of the anode material iron.

The electrolysis of a sodium chloride solution on iron anode is represented by the following equations:

$$\text{cathode reaction}: \quad 2\,H_2O + 2e^- = 2\,OH^- + H_2$$
$$\text{anode reaction}: \quad Fe = Fe^{2+} + 2\,e^-$$

Iron(II) ions combine with the hydroxide ions formed on the cathode to produce iron(II) hydroxide [$Fe(OH)_2$], which is transformed into iron(III) hydroxide [$Fe(OH)_3$] by the oxygen in air. The color of the bath darkens because of the iron(III) hydroxide formed. Anyone with a healthy dose of common sense would try operating such a device without immersing his or her feet into the water. The same color will most probably develop. Who would conclude that it is possible to detoxify feet without even immersing them into the bath? The electrodes are slowly consumed as the device operates, and the amount of salt used also varies. The appearance of the colored material depends on these factors, so different colors do not mean different levels of toxins in the body. A reader who is too busy to conduct such an experiment may also read the full documentation of a similar citizen-science-experiment on the website Herb Allure Studies. The conclusion was. "As a result of this experiment, Herb Allure decided not to manufacture its own brand of foot spas, despite the considerable potential financial gains that would likely result from such a venture."

Detoxification of the body is an illusion, maybe even a delusion. Maintaining such illusions is certainly good business for a number people, but does not improve the health of any users. Readers of tabloids and lifestyle magazine should not forget that physicians take the Hippocratic Oath, salesmen do not. Instead of occasional detoxification, maintaining a healthy diet is much more useful to avoid health problems.

3.16 Cloudy Dreams in Green: Absinthe

The golden age of absinthe was at the end of the nineteenth century, but its name is still widely recognized today. Its popularity owes much to the myths surrounding it. Absinthe is often said to change the mental state of the consumer, which is different from the effect of alcohol. Do scientific methods support this view?

The origins of absinthe are unclear. Most probably, it was first made in the eighteenth century, and the recipe used then is still followed. The above-ground parts of grand wormwood (*Artemisia absinthium*), the fruit of anise (*Pimpinella anisum*) and fennel (*Foeniculum vulgare*) are extracted with concentrated (more than 80%) alcohol, and then distilled. This distillate is sometimes used to extract other herbs as well, so absinthes may vary somewhat in color and taste. The essential oil of wormwood is already pale-blue, and the typical green color is developed during the second extraction as a result of dissolving chlorophyll. Colorless and red varieties are also known (the latter contains hibiscus extract), with the alcohol content between 45 and 75%.

Absinthe was supposed to be a brain stimulant, an aphrodisiac and a hallucinogen. These effects attracted a long list of artists including Henri de Toulouse-Lautrec, Vincent van Gogh, Émile Zola, Paul Gauguin and Paul Verlaine. Oscar Wilde described the effect as follows:

> After the first glass of absinthe you see things as you wish they were. After the second you see them as they are not. Finally you see things as they really are, and that is the most horrible thing in the world. I mean disassociated.

This supposed mental effect was the reason why absinthe was in high demand among artists. A drink of absinthe was traditionally prepared by placing a sugar cube on top of a specially designed slotted spoon, and then placing the spoon on the glass which has been filled with a measure of absinthe. Iced water was then

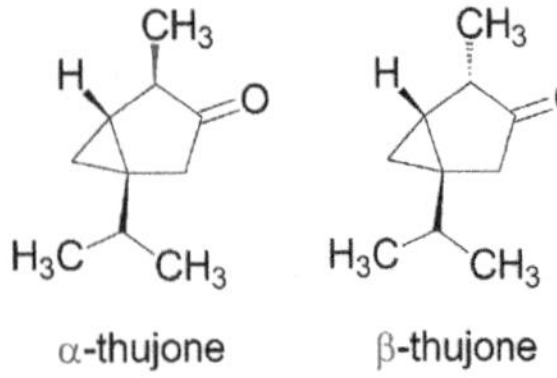

Fig. 3.22 Chemical structures of α-thujone and β-thujone. (Authors' own work)

poured over the sugar cube. As water diluted the spirit, the drink became cloudy because of its high content of essential oil not miscible with water. This was also a test of the quality of absinthe. From the end of the nineteenth century, the press connected a number of negative effects to absinthe consumption: murders, epilepsy, and delirium. Some even say that van Gogh cut off parts of his own left ear under the influence of the drink. These concerns led many countries to ban absinthe: the USA did so a few years before the Prohibition. Other nations also prohibited the sale of absinthe by the 1910s.

Although absinthe contains the extract of several plants, the essential one is wormwood, which is also the origin of the word (*grande absinthe* means grand wormwood in French). This plant is a known herb whose bitter taste increases appetite. Many related plants have been used in folk medicine around the world. In the western world, only four species were used widely: grand wormwood, common wormwood (*Artemisia vulgaris*), annual or sweet wormwood (*Artemisia annua*) and tarragon (*Artemisia dracunculus*)—the last one mostly as a spice. The bitter taste of wormwood dominates in Vermouth, but that is most often Roman wormwood (*Artemisia pontica*). High-quality absinth is not bitter: the sesquiterpene alkaloids (primarily absinthin) causing the stringent taste in the plant are not volatile and do not make it into the distillate. One of the main ingredients in the essential oil is thujone whose physiological effects are studied intensely today. Two stereoisomers, (+)-3-thujone (or α-thujone) and (−)-3-thujone (or β-thujone) are present in the oil (Fig. 3.22), and their ratio is determined by the genetic properties of the plant and the growing conditions.

Thujone is thought to be responsible for the psychoactive properties of absinthe. The neurotoxic and spasmodic effects of thujone have been proved in animal tests, but the mechanism remained a secret for a long time. Speculations about the mechanisms based on a chemical similarity with tetrahydrocannabinol, the active ingredient of marijuana, were falsified by several studies. The key to the effect of thujone was found in 2000 by Höld and his coworkers: thujone was proved to block the $GABA_A$ receptor, which explained the stimulating and epileptic activity (α-thujone is 2–3 times more potent than β-thujone). Activation of GABA (gamma-aminobutyric acid) receptors usually causes relaxation or sedation, whereas inhibiting them has the opposite effect. So the consumption of absinthe, at least in theory, combines the sedative effect of alcohol with the stimulating effect of thujone: this would certainly make a special experience.

As always, reality is a little more complex. Absinthe became popular at a time when *phylloxera* plague destroyed almost all vineyards in France. The new drink meant strong competition for the recovering wine producers. A battle against absinthe began by coining the term 'absinthism', which referred to a set of symptoms identified by famous French doctors. Valentin Magnan (1835–1916) described how seizures developed in rats treated with the essential oil of wormwood, and also observed a higher incidence of epilepsy and a tendency toward hallucinations among absinth drinkers. Today, Dr. Magnan's evidence would be seen as rather weak, but it was an excellent *raison d'être* for the movement calling for a ban of absinthe. Needless to say, French wine producers were more than happy to join this movement. Today, growing evidence show that absinthism is actually the same as alcoholism (Fig. 3.23).

Nevertheless, when the toxicity of absinthe is assessed, substances in addition to alcohol and even thujone must be considered. Copper sulfate, antimony trichloride and copper acetate were sometimes used to enhance the color of the drink, and methanol was a concern too. The concentration of thujone in typical absinthe was estimated as 200 mg/l a century ago. Realistically, this number was way too high. 2.5 kg of wormwood was used to make 200 l of absinthe, which means that a single liter of the final drink could not have contained more than 90 mg of the essential oil. Assuming a very high, 70 % thujone content in the oil, it still leads to an estimate of 60 mg/l for the concentration in the final product.

Recent measurements have not shown high thujone contents, either. Modern absinthe made by a traditional recipe has an average thujone concentration of 1.3 mg/l. Absinthe samples preserved from the pre-prohibition period did not show levels higher than 10 mg/l. A study of 147 commercial products showed that 55 % contained less than 2 mg/l of thujone. European Union regulations today set the limit for the thujone content in drinks at 35 mg/l, as experience shows that the neurotoxicity of this substance is negligible at this level. In the light of scientific results, the prohibition of absinthe was lifted in most European countries.

The Council of Europe has set a **tolerable daily intake** (TDI) of 10 μg/kg of body weight a day per thujone based on a study of female rats, where no adverse effects were found for a dose of 5 mg/kg of body weight (this dose is called the **NOAEL**, no observed adverse effect level). The recommended TDI represents a safety factor of 500 compared to the NOAEL. The TDI would mean about 700 μg/day for a 70 kg human, which is present in about 2 cl of EU-compatible drinks with the highest thujone levels. On the other, hand, average products typically contain 5 mg/l, which means that even drinking a full liter a day would only amount to 1 % of the NOAEL.

It is still unclear whether absinthe has any special effect on the mind. Many consumers drink it in the hope that it does. However, this is highly unlikely if it is legally bought absinthe in moderate amounts. Heavy absinthe drinkers are a different case, but they are alcoholic at the same time and the neurotoxic effects of thujone and alcohol are difficult to separate. Some short-term studies were carried out, one of which showed significant mood changes and attention disorders in comparison with the effects of alcohol only. Very high doses (15 mg) were necessary, and since this is only present in about half a liter of legal absinthe, the alcohol content would wreak havoc with the drinker.

Fig. 3.23 An empty-faced couple drinking cloudy drinks of absinthe in the painting titled *L'Absinthe* by Edgar Degas. (Copyright-free Wikipedia picture)

Woodworm is not to be blamed for van Gogh's lost ear, as the deviant behavior of the painter was also caused by excessive alcohol and turpentine consumption, as well as (most probably genetically determined) psychological disorders. Similarly, it would be very difficult to prove that absinthe, and not its alcohol content, is responsible for any of the effects described as absinthism. Vermouth has very low levels of thujone, which does not pose any extra danger in addition to alcohol. Legal absinthes do not involve much extra risk, either. However, in home-made absinthe, the thujone content may occasionally be high enough to be dangerous.

3.17 Is Green Tea Caffeine-Free?

Tea is becoming more and more popular around the world. Teahouses seem to open everywhere and ice tea consumption is also on the rise. Green tea, originally only known in Asia, arrived to Europe some decades ago. It is supposed to have a long list of favorable effects, which often include the lack of caffeine. Other sources say that green tea contains a compound called thein, whose properties are much healthier than those of caffeine.

Before going into more detail, some conceptual clarification about the liquids served in teacups may not be out of place. Tea is the drink made of the leaves of tea plant (*Camellia sinensis*), but the word is often (mis)used to mean other drinks made by extracting various plant parts in hot water. Technically, this other drink should be called herbal tea or tisane ($\rightarrow$ **3.8**). Tea itself is diverse enough, and this can only be appreciated in a specialized tea shop. Capsules and tablets containing tea extracts are also marketed today, and there are many misconceptions around these issues.

The tea plant is an evergreen shrub native to Asia. A legend says that Buddha was the first to try it when a tea leaf fell into his cup. Another, better documented and earlier story credits Chinese emperor Shennong with the discovery, who again got a tea leaf in his cup of hot water accidentally in 2737 B.C. No matter what the origins, tea consumption has been a part of Asian culture for millennia, and it has probably contributed to preserving the health of those who consumed it.

Today, the leading tea producing nation is India (Fig. 3.24, $\rightarrow$ **2.22**), but the plant is also grown outside Asia, most notably in Africa and South America. Again, the fact that the word tea is used for drinks not made of the tea plant may be misleading. Real tea is made of young tea leaves, the most valuable part being the tops of buds. The highest quality tea is marketed in bulk, whereas tea bags typically contain cheaper varieties. Black tea is common in Europe, China and India, whereas green tea is dominant in Japan. Tea has always been a valuable commodity. The Boston Tea Party (1773) is an excellent example of its historical significance, and disputes about tea trade were also a major contributing factor leading to the Opium Wars (1840–1860).

In Oriental medicine, tea was used against various illnesses. To begin with, drinking hot tea was a means of disinfecting water. But this is not the whole story. The main ingredients of tea are caffeine, theophylline and theobromine ($\rightarrow$ **2.22**). There is about 1–5 % of caffeine in dry tea leaves and this is mainly responsible for the stimulating effect. The other two notable alkaloids are present at much lower levels. The diuretic effect of tea also comes from caffeine and the essential oils in leaves give the pleasant flavor and aroma.

Today, tea leaves are mostly sold as black or green tea. The plant is the same, and the different aroma and color are caused by the different processing technologies. The leaves are fired on charcoal or dried in an oven when green tea is made, which destroys the enzymes in the leaves and stops most changes. Black tea, on the other hand, is fermented, which causes some of the original plant materials to decompose. Polyphenolic **tannin** materials react with the polyphenol oxidase enzyme liberated from the crushed leaves, the amount of tannins decreases, and gold brown oxidized compounds (theaflavins and thearubigins) form, which are responsible for the tea's characteristic color. The differences between black and green tea are not limited to aroma: green tea contains 25–30 % tannins, black tea only 5 %. The tannin content

Fig. 3.24 Tea plantation in the island of Java. Tea leaves may end up as green or black tea. (Copyright-free picture from http://i.images.cdn.fotopedia.com/flickr-3865034541-original/Java/Tea-Java-original.jpg)

of *oolong* tea, which is fermented but not as long as black tea, is somewhere in between. White tea is made by drying the leaves in the sun—no other process is used. *Pu-erh* is obtained by pressing fresh tea leaves together and then fermenting them for years, sometimes even for decades.

The aroma, and even the effect of tea, is greatly influenced by the preparation method. There are special procedures, even "tea rites" in traditional tea-drinking nations such as China or Japan. Entire books are written about the methods. Compared to them, the European tea making method looks positively unsophisticated. Extraction time may vary from 1 to 10 min: typically shorter (1–3 min) for green and white tea, and longer for *oolong* and black tea. Black and oolong tea usually require boiling hot water, while green and white tea are best made at about 90 °C. Some high quality tea brands may be made at even lower temperatures.

The amount of caffeine in the drink is influenced by both the preparation method and the caffeine levels in the fresh tea leaves. Different hybrids of the tea plant and tea types may vary greatly in this respect (Table 3.5).

Although green tea usually contains less caffeine than black tea, it is by no means caffeine-free. A cup of black tea may contain as much as an espresso (about 100 mg). A cup of green tea typically contains about one third of this amount. High tannin content in green tea slows down the take-up of caffeine and moderates the stimulating effect. Recent research shows that the presence of an amino acid named L-theanine is also significant (Fig. 3.25).

L-Theanine is structurally related to essential amino acid L-glutamine. Its presence in tea was known about 60 years ago, but research on its effects has only begun recently. It enters the brain and intensifies alpha waves there, which is manifested in an increasingly relaxed state after theanine consumption. Theanine inhibits the stimulating effects of caffeine. The mechanism of this effect is not known exactly, with current opinion emphasizing the role of the increased level of the neurotransmitter gamma-aminobutyric acid (GABA). Tea fermentation reduces the levels of L-theanine, which is also part of the explanation for the stronger stimulating effects of black tea. The presence of theanine, the slower uptake

3.17 Is Green Tea Caffeine-Free?

Table 3.5 Caffeine content in a cup of tea

Black tea	23–110 mg
Oolong tea	12–55 mg
Green tea	8–36 mg
White tea	6–25 mg

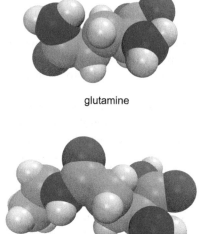

Fig. 3.25 Space filling models of glutamine and theanine. (Authors' own work)

glutamine

theanine

(tannins inhibit absorption) and the smaller amount of caffeine are the critical bits of this story.

Caffeine-free tea is available in the supermarkets today. This is made by removing caffeine with an organic solvent. Biotechnology already works on genetically modified caffeine-free tea, not without some success.

Many people think that a short extraction time will ensure the low caffeine content of tea, but this household tip did not stand experimental scrutiny. In a series of tests, tea prepared by 1-minute extraction contained 20–30 mg of caffeine, 3 min brought 34–35 mg, whereas 5 min resulted in 30–50 mg. So shortening the time of extraction does not decrease caffeine levels significantly, but it does decrease the general quality of the drink.

Finally, a few words should be offered on the mysterious substance of thein. Actually, this is the same as caffeine, and articles attributing special effects to thein have no scientific credibility. Thein was identified by a researcher named M. Oudry in 1827 as the ingredient responsible for the stimulating effect of tea. However, later it was shown to be identical to caffeine, which was also discovered in the 1820s. German Nobel laureate Hermann Emil Fischer (1852–1919) explored the chemical structure of caffeine at the end of the nineteenth century, and also synthesized it from much simpler compounds. This was one of the achievements for which he was awarded the Nobel Prize in chemistry.

3.18 Red Rice Remedy: Cholesterol Relief?

As cardiovascular diseases are among the leading causes of death in the developed world, prevention receives a great deal of attention. Lowering cholesterol levels is one of the most important means of achieving this end. Many modern drugs are available, but patients prefer supplementing (sometimes even replacing) medical treatment with natural cures. One of the latest wonder foods is red yeast rice, which is considered very effective and safe.

High cholesterol levels increase the long-term risk of the thickening and hardening of arteries, which is often caused by the deposition of the oxidized and insoluble form of this cholesterol on the walls of blood vessels. This deposition decreases the diameter of arteries and veins. As a consequence, the flow of blood to body parts deteriorates, and this may cause damage over time. If the deposition is formed in the heart, this organ is unable to function properly.

Cholesterol lowering medicines usually work by inhibiting the uptake of this compound or slowing down its synthesis in the liver. Herbs used for this purpose exert their effects in the same way. A decrease of 20–50 % in cholesterol levels is the usual result of medical treatments, but this can be improved by a low-cholesterol diet and the use of some herbs and vegetables (artichoke or garlic). Red yeast rice is quite heavily advertised as a very effective and natural substance, and millions of people purchase these products. Red yeast rice is sold in dietary supplements and is not a listed medicine, which—most ironically—just adds to its popularity (Fig. 3.26).

Red yeast rice has been known for a long time. Its first application was described in China during the Tang dynasty in the eighth century, but, needless to say, not as a cholesterol drug. Red rice was used as a dye, spice and preservative in food, and during the production of rice wine to speed up fermentation. In Southeast Asia, red yeast rice is still a common food. Records of some medical applications were found from the Ming dynasty (fourteenth to seventeenth century) from a medicine book that recommended it to help blood circulation and digestion.

Red yeast rice is prepared by fermentation of rice with fungi. The red color of the marketed product is caused by pigments produced by microscopic fungi. The most common Chinese names are *Ang Khak* and *Hong Qu*, whereas in Japan it is known as *Beni-Koji*. Chemical testing of red rice began in 1979, when Akira Endo reported the isolation of a compound named monacolin K from Thai red yeast rice. This substance was proven to have strong cholesterol lowering potential in animal tests. The discovery was by no means an accident: it was part of a systematic search. Eight thousand samples were tested, and the microbe producing the compound, *Monascus ruber*, was identified. The project also found eight additional compounds similar to monacolin K. Fermented rice contains about 0.4 % monacolin derivatives,

3.18 Red Rice Remedy: Cholesterol Relief?

Fig. 3.26 Chicken, spinach, red rice—a healthy meal? (Copyright-free Wikipedia picture)

about half of which is monacolin K. Sterols and unsaturated fatty acid also present in the product may contribute to the physiological effects somewhat. The pigments of the fungi include yellow, orange and red dyes. The fermentation of rice wine is aided by the enzyme amylase present in red rice, which breaks down starch in rice to glucose, which in turn ferments into alcohol.

Prior to the discovery of monacolin K, Endo isolated a compound called mevastatin from grain ferment. This is the "parent compound" of most modern cholesterol drugs. Some of its derivatives can be used more favorably, but mevastatin is still a useful medicine. In 1980 Lovastatin (or mevinolin), a mevastatin-related compound isolated form of the fungus *Aspergillus terreus* (Fig. 3.27), had been shown to be identical to monacolin K. This discovery was not particularly useful for some time.

Lovastatin proved to be an effective and very popular drug, but red yeast rice was not used for medical purposes for a long time. It was only "discovered" in the 1990s, at the dawn of the age of dietary supplements. Red yeast rice is relatively inexpensive. Marketing it as a dietary supplement rather than a medicine makes the approval process simpler and faster, and distribution is also more convenient than for drugs which are only available in pharmacies by prescription.

However, lovastatin and its derivatives may cause adverse effects. The most common side effect is muscle pain, and in the most severe cases, muscle damage may also develop. The monacolin K content of freely available red yeast rice products may lead to abuse (overdosage) resulting in health damage. This was the reason why all food supplements containing red yeast rice were removed from the USA market by the Food and Drug Administration in 2001.

Nevertheless, the business has never been better. These natural products can be sold without significant restrictions in many parts of the world (for example in the European Union), despite their confirmed content of active substances that also have serious side effects. A daily dose of a typical product is 1.2–2.4 g, which may contain 5–10 mg of monacolin K. This amount was proven to lower cholesterol levels in clinical tests. The cost of the treatment can vary greatly for both prescription drugs and dietary supplements, but average costs per day are similar. From a strictly scientific point of view, red rice products do not have advantages over medicines, but they may be more attractive to patients, especially those who are prejudiced against medicines. However, the amount of active substances in dietary

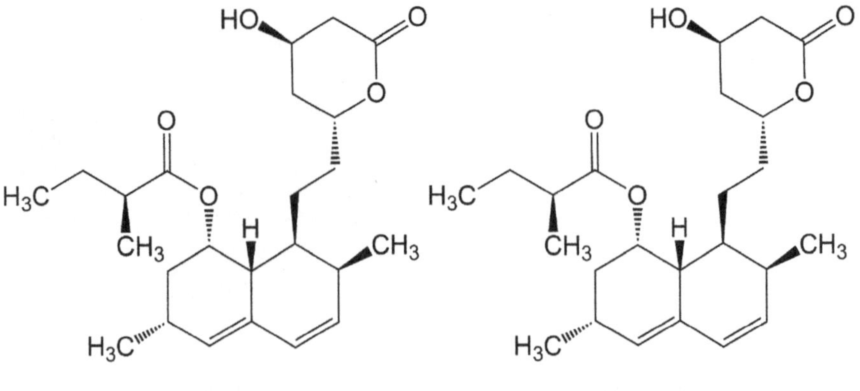

Fig. 3.27 Can you spot the difference between monacolin K and lovastatin? (Authors' own work)

supplements is generally unreliable. In a study of red rice products, the dose of monacolin K in a single capsule was shown to vary between 0.1 and 10 mg. In addition, some of the ingredients may not be favorable to human health, occasionally even outright harmful. One of these ingredients is a fungal toxin called citrinin, which was proven to cause kidney damage in animal tests. Human data are not available, but the fact that more than 1 µg was detectable in 70 % of the products in one study, may warrant some caution.

Tests showed that although side effects of red rice products are not more common or more severe than those of competitor medications, safety issues must be considered very carefully. Red rice has been tested under controlled conditions on fewer people that the pure active substances in medicines. Another cause for caution is that the known side effects of drugs have to be registered by authorities, whereas no similar obligation exists for dietary supplements. Side effects may very easily remain hidden or totally unrecognized, especially if they only develop in the long term, which makes establishing causality difficult.

3.19 What Exactly Does Aloë Cure?

Aloë vera (*Aloë barbadensis*) is a frequently used alternative medicine, and manufacturers of its related products certainly run an exceptionally successful advertising campaign (Fig. 3.28). However, the amount or intensity of marketing seldom correlates with the medicinal value of the product advertised. Leaflets or magazine articles about *Aloë vera* are next to useless as sources of credible information. Their claims are often ridiculously inflated, or even scientifically absurd.

Readers of such sensationalist articles or product promotion leaflets may have the feeling that one of the dreams of Middle age alchemist has finally come true

3.19 What Exactly Does Aloë Cure?

Fig. 3.28 *Aloë vera* plant and flower. (Copyright-free Wikipedia picture)

as *Aloë vera* seems to be a modern day "panacea". This word was originally used in Greek mythology to refer to a goddess of Universal remedy (Greek Πανάκεια, Panakeia). The roots of the Greek word go back to 'pan' (παν, everything) and 'akeomai' (ακέομαι, "to heal" or "to cure"). The golden age of panacea was in the sixteenth century as many an alchemist sold his or her potions prepared through the use of secret recipes. Despite a tremendous advance in science and technology, there seems to be a rather disgraceful renaissance of panacea in the twenty-first century. Even with all the education in sciences, many still believe that one or another plant can cure every illness. This is wishful thinking, fueled by intense marketing, but is also a testimony to the credibility of printed words. Even 50 years ago, printing a text (a book, magazine article, or even a leaflet) was quite costly and the author of the text had a great deal of incentive to ensure that the information appearing in print was correct. This is no longer the case. Printing and copying are dirt cheap now, so some printed texts are close to dirt in value as well. Furthermore, some experts of reasonable scientific stature are lured by big money into lending their credibility to questionable advertising campaigns. The primary target of marketing is the consumers, not their health. A fool and his money are soon parted—this saying has never been more relevant than in our age.

A myriad of ways have been developed to mislead customers, and some of them are, in fact, quite sophisticated. It is typical for a label to describe the properties of the plant instead of the product. Explicit medical recommendations are often missing, but there is a wealth of information about what consumers typically use the product for, which is not exactly the same, and—needless to say—much less verifiable. Patients with positive experience are often quoted and some independent-looking experts are cited in magazine articles. Digging a little deeper into these claims often reveals that the experts are not independent (for example, they were paid by the manufacturer) or they were not even experts, and patients are usually difficult to locate for a verification interview.

Listing a large number of putative medicinal uses makes the impression that a product is a registered medicine. In fact, there are no *Aloë vera* containing medicines for internal use in the European Union, and even if there were, their recommended uses would be a lot narrower than advertising would make it appear. In Europe, advertising of medical effects is legally limited to registered medicines. Most of the available *Aloë vera* products are dietary supplements and not medicines. Many of the labels of *Aloë vera*-containing products (food supplements or cosmetics) or their advertising campaigns border on illegal. It would take skillful lawyers and a large dose of word-twisting to prove that they do not violate the laws. This is quite unfortunate, as some of the heavily-advertised plants actually have some health benefits.

Although *Aloë vera* is by no means a panacea, it is a notable herbal medicine. The yellow exudate (juice) drained from the leaves (and not the advertised gel!) of a related species, Cape Aloë (*Aloë ferox*) has been used as a purgative medication for centuries. Several laxatives contain the extract of this plant. The active ingredients are hydroxyanthracene derivatives, but their only known effect is to make the feces less solid. The popular *Aloë* gel does not show this laxative effect at all, which may be surprising as it is also prepared from the leaves of an *Aloë* species. This apparent paradox is easily explained: the yellow juice is derived from the pericyclic cells of the vascular bundles located just beneath the thick green rind of the leaf. The transparent gel is found in mucus containing cells in the middle of leaf. The exudate contains active hydroxyanthracene derivatives, whereas the gel is about 1% polysaccharide and 99% water. Some *Aloë* species used in laxatives also produce gel, and primarily gel-prone species may contain laxative compounds. However, the gel and juice do not cross-contaminate each other if modern technologies are applied during processing. The gel is obtained by cutting the middle part of the plant out, whereas the juice simply pours out of the leaves when they are incised (Fig. 3.29). The gel and the liquid are easy to distinguish from each other: the gel is transparent, but the liquid is milk-like and yellow-tinted and darkens when it dries.

With such high popularity, researchers could not ignore the medicinal claims and thus began systematic studies of the plant. The laxative effect of the juice is already sufficiently well documented, so recent studies focused on other areas. Acemannan is the polysaccharide of the gel that may be responsible for health effects: this is a complex sugar built up of β-$(1\rightarrow 4)$-D-mannopyranosyl residues and about 5–30% of glucose (Fig. 3.30). The prefix 'ace' refers to the fact that mannose unit are partially acetylated at C-2 and C-3 positions. The human body does not have the enzymes to break this polysaccharide down, and some biological effects also arise from its special structure.

Hundreds of scientific articles have been devoted to *Aloë vera* gel. Most of them have yielded evidence of anti-inflammatory, immunological and wound-healing properties, which is expected if the gel is used directly on the skin (externally). Some studies also found beneficial effects in treating other skin problems. The use of *Aloë vera* gel in cosmetics seems to be justified. The World Health Organization (WHO) published a monograph on the plant more than a decade ago, which only mentions the dermatological use, but finds the evidence weak.

Most of the miraculous effects are expected through the oral use of the gel. Peptic ulcers are influenced positively by *Aloë* gel, which may not be very surprising in the light of the confirmed anti-inflammatory effects. One of the most advertised effects is the potential to lower blood sugar levels. This was also supported by animal

3.19 What Exactly Does Aloë Cure?

Fig. 3.29 Succulent plants, such as *Aloë vera*, store water in their fleshy leaves. (Copyright-free Wikipedia picture)

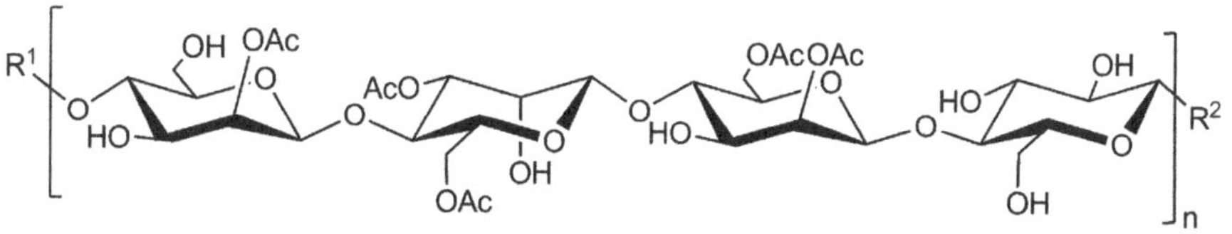

Fig. 3.30 The structural formula of the repeating unit of acemannan (Ac=CH$_3$CO). (Authors' own work)

tests. However, hardly any human studies have been carried out, and more evidence is needed before claims to treat diabetes can be substantiated.

The anticancer effects of *Aloë vera* have also been investigated in a number of studies. Again, the majority of the studies found positive effects on cell cultures and animals, but there are still diverse opinions about the mechanism of the action. Human studies have not been done yet in sufficiently large numbers to claim any therapeutic gain. The picture is quite similar when the antibacterial and antivirus (for example, AIDS treatment) effects are assessed: initial results seem encouraging, but much more work needs to be done. The following example illustrates the paradox, which should not be forgotten about the necessity of human studies. Sodium hypochlorite is an excellent antibacterial, antivirus and anticancer agent. Yet nobody wants to test it on humans as its toxicity is so well known—it is a major ingredient of the disinfectant sold as chlorine bleach.

For *Aloë vera*, toxicity is not the major problem: it is the long and often tortuous procedure needed to prove any human effects. Another important point is that the benefit must be shown not only with the plant, but also with an extract whose pharmaceutical quality is carefully controlled. Until this occurs, *Aloë vera* is just one in a long line of chemically interesting and medically promising raw materials.

3.20 Island Pleasure for Businessmen: Noni

Noni would most probably end up among the best known herbs in a public opinion poll. This new-found popularity is quite surprising as the plant was unknown outside its native range in some Pacific islands a mere 25 years ago. Advertisers have made major efforts to introduce noni and its supposedly unique active ingredient, xeronine, into modern life.

Noni (*Morinda citrifolia*) is a tree native to Southeast Asia. Its seeds were carried by ocean currents, birds and ancient seafarers to most of the Pacific islands, so today it is found in a huge area (Fig. 3.31). Polynesian cultures have used it for healing since ancient times, and Captain James Cook (1728–1779) was the first to bring information about noni to Europe. Traditional healing practices rely mostly on the leaves, but medical application of the bark, root and unripe fruit has also been documented. Noni has been applied externally to treat skin abnormalities, cuts or other wounds. The taste of ripe fruit is not very pleasant and island dwellers only eat it if there is little other choice.

The popularization of noni began with an article in a lesser-known scientific journal titled *Pacific Tropical Botanical Garden Bulletin*, in which author Ralph Heinicke reported finding the active ingredient of the fruit and called it xeronine (Fig. 3.32). The compound is present in the plant as proxeronine, which is chemically transformed in the stomach to xeronine by the acid and the digestive enzymes found there. Xeronine is responsible for the beneficial health effects of noni: it lowers blood pressure, relieves spasms, joint aches and depression, improves digestion, and prevents atherosclerosis, in addition to a long list of other effects claimed by the author. The article also identified xeronine as an alkaloid, but was not detailed enough to describe its chemical structure. Heinecke patented xeronine and its versatile applications, but the quite lengthy patent specifications still do not include any specifics about the chemical structure. The production and investigation of xeronine is vividly described in a number of popular books, articles and websites, but authentic scientific articles on this topic seem to be totally missing. On the web, one can find a putative structure for xeronine, but the source of this information and its connection to Heinecke's work remain mysterious (Fig. 3.32).

It does not happen every day that an article published in an obscure journal changes the fate of a plant, but this seems to be the case for noni. A whole industry was set up to process the fruit of noni in the years following the first report. Noni juice became popular in the US first, but not just on the continent. In Hawaii, where noni was actually known and had a traditional use, people began using it for the purposes recommended by Heinicke. The application has widened in the last 25 years considerably: many people attempt to use it as a weight loss drug, a blood cleanser, an immune stimulant, and to fight nervous problems or even cancer and AIDS.

The success of noni also required a company that was devoted to selling it. Morinda, founded in 1996 as Tahitian Noni International, is a multi-level marketing company that sells products made from the noni plant, often flavored with other fruits to make them more appealing to costumers. They also made extensive

3.20 Island Pleasure for Businessmen: Noni

Fig. 3.31 The fruit of the noni plant. (Copyright-free Wikipedia picture)

attempts to market noni highlighting its putative health effects, but these finally resulted in a 1998 court order that banned such advertisements because there was no scientific evidence to back up such claims.

This legal setback did not affect the financial success of noni much. Earnings continued to increase as the annual sales reached US $ 33 million in 1999, and exceeded 250 million in 2005. Today, the size of the noni market is estimated to be $ 1.3 billion, with the share of Morinda close to 500 million (they do not publish any of the earnings publicly).

Scientific studies on noni seem to paint an entirely different picture. Dozens of articles have reported the identification of more than 200 compounds in the fruit of noni, but none of them was able to find the active ingredient described by Heinecke. It is not possible to determine where he made a mistake. As attempts to reproduce the original work failed, the claims today are not simply dubious but discredited. It is also curious (and, in hindsight, understandable) that a biochemist should publish results with such high perceived significance in a lesser-known botanic journal instead of a highly reputed scientific venue. None of the known ingredients of noni

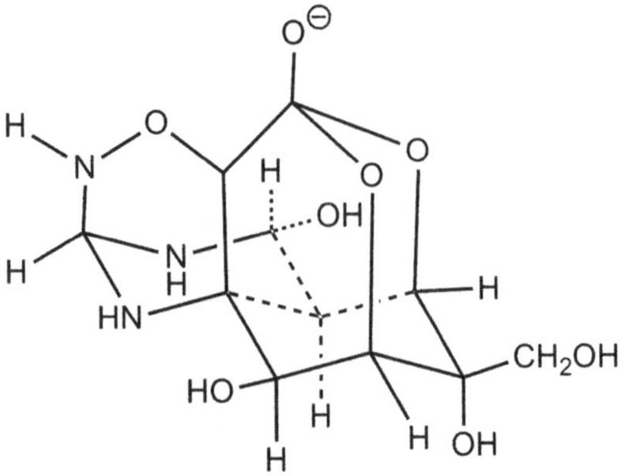

Fig. 3.32 The putative structure of xeronine—a mythical treasure? (Authors' own work)

has the miraculous effects attributed to xeronine. These ingredients are by no means special in any respect as they are found in many other fruits and herbs.

Thus, xeronine is very unlikely to exist. Despite this fact, noni could be a plant with substantial healing potential, but an objective look at the uncovered facts does not support such a classification. The lack of evidence obviously does not prove the lack of effect, but detailed efficiency studies must always precede widespread medical use. In the case of noni, no such positive results have been reported yet. It is already known that noni is not toxic even at high doses (80 ml/kg of body weight), but its medical efficiency was only tested in five previous studies, only one of which is connected to the versatile use of the plant. This study was started in 2001 and followed cancer patients. Although no significant change in the size or number of the tumors was detected, the patients reported more positive attitudes to life. More than a decade has passed since this study began, but no final results have been released yet.

Noni has been available for sale in the European Union since 2003. The advertisements in the media on the web certainly give the impression of a successful business. Sellers and consumers alike seem to ignore the fact that the unproven health benefits of noni are attributed to a non-existent substance. There is another lesson to be learned here. The fact that authorities issue a patent to someone means that the patent proprietor is legally entitled to enjoy the financial benefits of the claims, and does not guarantee that the claims are in fact true.

3.21 Castor Oil: A Painful Story

In some areas of the world, castor oil and its laxative effect are frequently mentioned in jokes. Although castor oil is not much used for this purpose any more, it has found an even bigger role in modern life.

The castor oil plant (*Ricinus communis*) is a subtropical, perennial species of flowering plant in the spurge family, which can be grown under moderate climatic

3.21 Castor Oil: A Painful Story

Fig. 3.33 Castor beans. (Copyright-free Wikipedia picture)

conditions (Fig. 3.33). The oil is obtained by cold pressing the seeds of the plant, which contains 40–50% oil, in addition to other compounds, two of which are extremely toxic. The protein called ricin and the alkaloid ricinin are often to blame for deadly poisoning from castor beans. Ricin is among the most toxic proteins. A dose of 50 mg (10–20 beans, oral) may cause the death of an average adult ($\rightarrow$ **4.26**). Castor beans are very similar to some normal bean varieties: it is very dangerous to grow this plant in places easily accessible to children (e.g. kindergarten yards).

In some regions, castor beans are consumed in large amounts without any toxic effects to speak of. The main ingredient in the Nigerian food *ogiri-igbo* is castor beans, treated with 18 h of cooking and several days of fermentation. The level of toxic substances falls to very low levels during this procedure.

Military use of ricin as a toxin was considered by both the USA and the Soviet Union, but it was not weaponized. Secret services were less picky: Bulgarian agents wounded dissident writer and anti-communist critic Georgi Markov with a small ricin-loaded projectile, and the poisoning grew deadly within a few days ($\rightarrow$ **4.26**).

Castor oil is not toxic: it contains neither ricin nor ricinin (Fig. 3.34), which are insoluble in the oil. Most of castor oil is glycerol triricinoleate. Digestive enzymes (lipases) in the intestine break down this molecule into glycerol and ricinoleic acid, and the latter possesses a laxative effect. Despite the fact that the use of castor oil in case of constipation dates back centuries, the mechanism of the action is not fully understood yet. About 5–10 g of castor oil is usually quite efficient with the onset of the action after 6–8 h. A larger dose may do its job within a shorter time. Despite certain jokes, the effect is never instantaneous.

The use of castor oil as a laxative is declining. Its drawback is that the consumption of a relatively large amount is needed and it also has an unpleasant aftertaste. Additionally, it causes painful cramps. More specifically, it cramps uterine muscles, and was earlier utilized in inducing abortions. For this same reason, the usage of castor oil is not recommended for pregnant women.

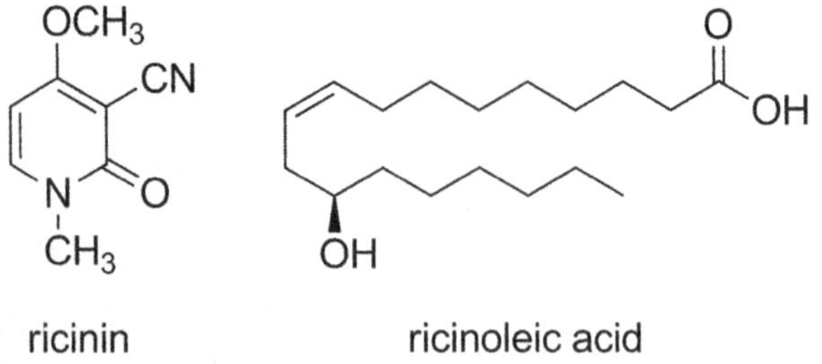

Fig. 3.34 Chemical structures on ricinin and ricinoleic acid. (Authors' own work)

ricinin

ricinoleic acid

Medical reasons were not the only ones for using castor oil as a laxative. Secret police used it as a means of interrogation in Mussolini's Italy. In more than just a handful of cases, the diarrhea combined with the cramps resulted in the death of the prisoner. The Italian language still remembers this era: *usare l'olio di ricino* (use castor oil) is an idiom still used to satirize patronizing politicians, or the proponents of unpopular legislation.

Castor oil is also used in the cosmetics industry. It has the advantage of being miscible with alcohol, which few other vegetable oils feature. The molecular background of this property is the presence of the hydroxyl group in ricinoleic acid, which intensifies the interaction with polar solvents. It dissolves in other oils as well, and also in water to some degree, so it is ideal for "dissolving" (the more precise term is emulsifying) fatty substances in water. The pharmaceutical industry also takes advantages of this rare property. If the active ingredient of a medicine dissolves in oil rather than water, the use of castor oil containing polyethoxylated ricinoleic acid is a standard procedure to prepare it for injection or intravenous infusion. The ability of substances to lower surface tension can be characterized by the ratio of their solubilities in water and oil. The quantity itself is called HLB value (hydrophilic-lipophilic balance), and its scale goes from 0 to 20 with the higher end meaning a substance soluble in water. Polyoxyethylene glycerol triricinoleate has a HLB value of 12–14, which is usually suitable for emulsifying fat-soluble substances in water. Hydrogenated castor oil can be used for the same purpose, with the HLB value depending upon the extent of hydrogenation.

If one reads cosmetics labels carefully, the name of castor oil will become familiar very soon. It is not only used to facilitate dissolution, but for its skin softening effect. Some Web-based sources attribute more dramatic effects to the oil: regular treatment is claimed to make eyelashes longer and denser, and to stimulate hair growth on the head. No scientific studies aimed at testing these claims are known, but the oil treatment will definitely make hair shinier.

Castor oil is also used as a lubricant in engines. This makes use of another of its exceptional properties: its viscosity does not change much as the temperature is raised. Attempts to use castor oil in biodiesel were reported in Africa where the plant is cultivated for this purpose. As castor beans are not consumed directly by people, this practice does not cause food prices to rise.

Fig. 3.35 A common form of licorice candy is long and thin. (Copyright-free Wikipedia picture)

3.22 The Oldest Sugar-Free Candy: Licorice

Licorice candy (Choo Choo Bar, Crows, Good & Plenty, London drops, Vigroids, Pontefract cakes, Snaps, Turkish Pepper) may not be as popular as it used to be, but is still available in confectionery stores (Fig. 3.35). Sugarelly, a once popular British soft drink, was also made of licorice. The sweet taste is beyond doubt, but is it because of sugar?

Licorice (or liquorice in the UK) is the root of the legume *Glycyrrhiza glabra*, which is native to southern Europe and parts of Asia. The name of the plant is derived from the Greek γλυκύρριζα (glukurrhiza), meaning "sweet root", a compound noun containing γλυκύς (glukus), "sweet" and ρίζα (rhiza), "root". In German, the name of the candy is *Lakritz*, and the root is called *Süssholzwurzel* (sweet root). The long and thin, cable-like licorice candy has a more vulgar name: *Bärendreck* (Bär = bear, dreck = dirt, excrement).

Although licorice is now a candy, its first uses were medicinal. Assyrians, Egyptians, Chinese and Indians used it for various purposes millennia ago. Theophrastus referred to it as 'Scythian root' or 'sweet root' in his *Enquiry into Plants*, which proves the Asian origin, and recommended it for dry coughs. Today, the licorice plant grows mostly in Asia, Italy and Spain. Secondary roots growing from the stem of the root can reach 1–2 m in length. The roots are harvested, dried and chopped to prepare them for further processing.

Licorice still has medicinal uses because it is an expectorant, protects the mucus of the stomach and probably has some antiviral activity as well. Sometimes it is simply a sweetener in medical syrups. Its taste and health benefit are caused by the same family of compounds called triterpenoid saponins, mostly glycyrrhizinic acid, in which two molecules of glucuronic acid are attached to a triterpenoid base compound through glycosidic bonds. Glycyrrhizinic acid is mostly present in the plant

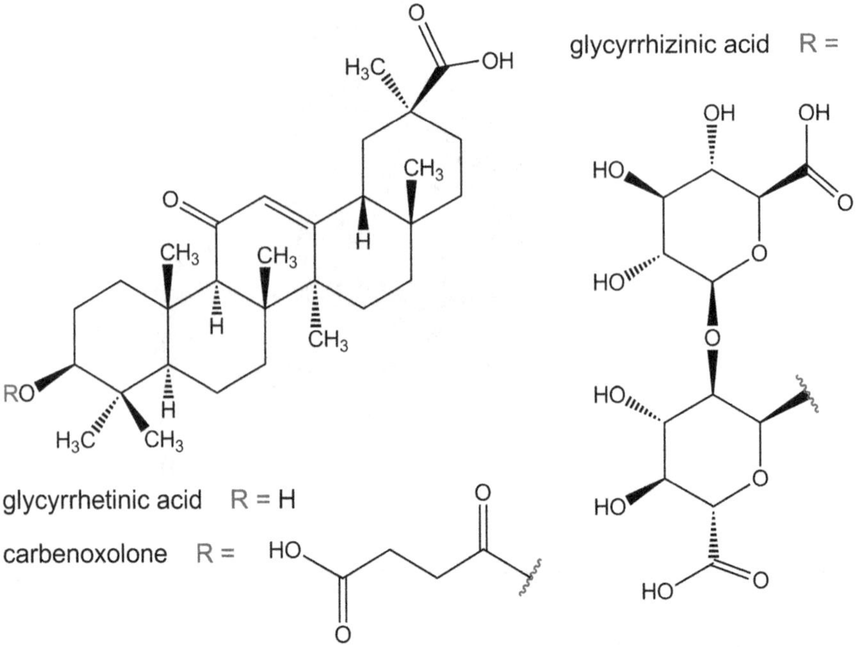

Fig. 3.36 Chemical structures of glycyrrhetinic acid (R=H), glycyrrhizinic acid, and carbenoxolone. (Authors' own work)

as sodium or potassium salt, collectively called glycyrrhizin. Glycyrrhizin is about 50 times sweeter than common sugar, or sucrose. The two glucuronic acid parts can be removed by acid hydrolysis leaving glycyrrhetinic acid behind. Strangely, neither glycyrrhetinic nor glucuronic acid is sweet. Apart from the taste, the medicinal potency of glycyrrhizinic acid (a glycoside) and glycyrrhetinic acid (an aglycone) is about the same.

The root usually contains 1–5 % glycyrrhizinic acid and a little bit more (4–8 %) sucrose (Fig. 3.36). As glycyrrhizinic acid is much sweeter, it dominates the taste. The root is seldom used for sweetening directly—usually its extract is preferred. The advantage of glycyrrhizinic acid extract is that it is heat stable, and survives a lot of industrial processing. Its degradation is only detectable in the presence of mineral acids or hydrolase enzymes. This reaction is sometimes intentionally started so that the glycyrrhetinic acid formed could be combined further with sodium succinate, yielding the drug carbenoxolone, which is effective against stomach ulcers.

Licorice candy must contain the highest possible levels of glycyrrhizinic acid from the root of the plant. The main ingredient is prepared by extracting the root with water at 130–150 °C under pressure. This extract is thickened: slow evaporation was preferred earlier, but vacuum evaporation is the method of choice today. The essence of the procedure has not changed much compared to the Dutch recipe followed in the eighteenth century: the thick, hot extract is pressed through a

Fig. 3.37 Licorice production in Italy at the beginning of the twentieth century. (Copyright-free picture from the book Magyary-Kossa Gy (1921): A hazai gyógynövények hatása és orvosi használata. Athenaeum Irodalmi és Nyomdai Rt., Budapest, pp. 133–134. (The effect and medical use of Hungarian medicinal plants—in Hungarian))

circular hole and the stick prepared in this way is cut to pieces or shaped (Fig. 3.37). Some records show that licorice candy was made as early as the thirteenth century, but the technology was different.

Beyond sweet root extract, other components have changed as time passed. Originally, the plant extract contributed much of the color and flavor. Today, the extract does not dominate the end product-added sucrose contributes to the sweet taste, too. The sucrose content of the marketed modern product occasionally reaches 60 %, while the licorice extract is about 1.5–4 %, and not all of this is glycyrrhizinic acid. To enhance the flavor, other sweeteners such as aspartame are sometimes used. Some organic acids (malic acid or citric acid), anise oil and caramel are used as additives, too. Caramel contributes to the dark color of the product that does not contain a high sweet root content any more. Vegetable carbon (food additive E153) is also used for this purpose. Even more surprisingly, some products contain as much as 8 % of ammonium chloride, which gives a salty taste that is quite popular in the Netherlands. In today's licorice candies, licorice plant is no longer dominant—flavored (strawberry, cherry, cinnamon) and ruby red colored varieties are also available. These do not resemble original licorice very much. In Italy, products close to the original can be found, but they have not gained much popularity throughout Europe.

The percentage of sweet root extract in licorice candy has decreased both for financial and health-related reasons. In addition to the high price of licorice extract, glycyrrhizinic acid may have undesired side effects (high blood pressure) when consumed in large quantities. The European Union proposed 0.1 g/day as the upper limit for ingestion of glycyrrhizinic acid in 2008. In most EU countries, 50 g of licorice candy is taken as a normal amount and the maximum allowable amount of glycyrrhizinic acid in licorice candy is 0.2 g in 100 g product. The actual extract used in the industry contains 10–20 % of glycyrrhizinic acid, so current production technologies ensure that consuming amounts larger than recommended is unlikely.

Although licorice candy often contains a great deal of sucrose, its specific sweetness is primarily caused by glycyrrhizinic acid. The chemical structures of these two compounds are quite different, but this is not surprising. Some other substances, such as amino acids, proteins, flavonoids and cumarins are also sweet despite being unrelated to sucrose. The sense of *sweet* taste is more complicated than *salty* or *sour*, where connections between taste and chemical structure are easier to establish.

3.23 Cannon Balls for Your Health: Antibiotics

Antibiotics are among the greatest of public enemies today, and some people even claim that their artificial origin is the origin of the harmful effects (→ **1.4**). Websites and magazine offer countless guaranteed natural "antibiotics" instead of "dangerous artificial drugs". Do the proponents of these views keep their confidence if they fall ill with pneumonia, TB (tuberculosis) or heart muscle inflammation? Are antibiotics truly dangerous? If so, is it simply because they are artificial?

Antiobiotics are chemotherapeutic agents. Microbiologists Paul Ehrlich (1854–1915) was the first to use the word chemotherapy in 1907, when he was searching for medicines against human sleeping sickness. This disease is caused by a single-cell pathogen, which is transmitted by tsetse flies. To cure the disease, a chemical was needed which kills the pathogen but not the host. In contrast to most other medicines, the desired effects are not upon human cells: any effect on the host is a potentially harmful side effect. Ehrlich did not find such a medicine for sleeping sickness, but similar research was also in progress for other diseases. The first breakthrough was the discovery of Salvarsan, which could be used to treat syphilis (→ **3.7**). The active ingredient was an arsenic compound, which is toxic toward humans, but much less so than toward the pathogen. The first chemotherapeutic agents had such severe side effects, they would not be approved as medicines today.

The first successes motivated scientists greatly, but development was painfully slow. If the cells of the pathogen are very different from human cells, it is not usually difficult to find a suitable chemotherapeutic agent, but the task gets more difficult as the differences get smaller. The idea was not conceived by human researchers: many plants (e.g. fungi) produce chemicals to poison other living organisms selectively. An accident led Alexander Fleming (1881–1955) to discover this phenomenon in 1928. He tried to grow bacteria, but his culture got infected with mold. Fleming noticed that bacteria had been killed near the mold and assumed that the mold produces a substance that has antibacterial effects. The Latin name of the mold species was *Penicillium notatum*: the active ingredient was duly identified and called penicillin (Fig. 3.38). Howard Walter Florey (1898–1968) took the next step by testing the antibacterial effect of penicillin in humans. Ernst Boris Chain (1906–1979) solved the problem of large scale production of the compound. Penicillin proved its benefit in World War II, by facilitating the prevention of bacterial infections among Allied troops. Before the 1940s, wound infections caused many more fatalities in wars than the wounds themselves. Fleming, Florey and Chain were awarded the 1945 Nobel Prize in physiology or medicine.

The discovery of penicillin caused a surge of research on fungi-produced antibiotics. These natural compounds were called antibiotics to distinguish them from chemotherapeutic agents. The use of both terms changed as time went on. Today chemotherapy primarily means cancer treatment with drugs, whereas antibiotics are compounds or substances used against bacterial infection, irrespective of their origin. Antibiotics today include semisynthetic compounds, made by modifying naturally occurring molecules to enhance their efficiency, and fully synthetic products as well.

Nowadays, antibiotics are primarily classified according to the mechanism of their action, with similarity of chemical structure as a secondary factor. Penicillin and its derivatives inhibit the formation of bacterial cell walls (Fig. 3.38). Cephalosporins have the same active mechanism. Other compounds are taken up into bacterial DNA to form unstable molecules (quinolones, metronidazole) or inhibit peptide synthesis (tetracyclines, aminoglycosides, macrolides). Some antibiotics (e.g. glycopeptides) exert a complex effect.

A fair number of antibiotics are natural or are prepared by minor modifications of natural compounds. This fact itself does not guarantee the lack of harm. For medicines, selectivity is critical, although in a few cases, even compounds with severe side effects are used in practice. Patients suffering from bronchial catarrh of bacterial origin are usually willing to face digestive disorders or diarrhea as a price of a speedy recovery. Interestingly, the primary source of these side effects is still not the action of drug on the human body: the antibiotics tend to kill useful intestinal bacteria, which aid digestion. A life threatening illness may even make bad side effects seem tolerable. Some antibiotics cause hearing loss or kidney damage, but the risk from these is still lower than from the lack of treatment. Carefully chosen doses usually minimize the risks of the treatment. If a substance is easily overdosed, its blood levels are continuously monitored and the doses are tailored to the specific needs.

The biggest single risk of antibiotics use today is generating resistant strains of bacteria. The rules of evolution work here as well, and mutations can produce pathogens that are no longer sensitive to a given antibiotic. If the treatment is insufficient, these resistant varieties can survive. One way to fight resistance would be to ensure that all the pathogens are killed by the treatment every time. This is easier said than done. When the doctor selects an antibiotic for treatment, usually nothing is known about the pathogen's resistance history. This can be tested in laboratory experiments, but those take time and the patient does not improve in the meantime. Usually, the doctor relies on her or his own experience with a particular type of disease in choosing a medicine. If this does not have an effect, a different one with a substantially different mechanism of action is selected again and again until an

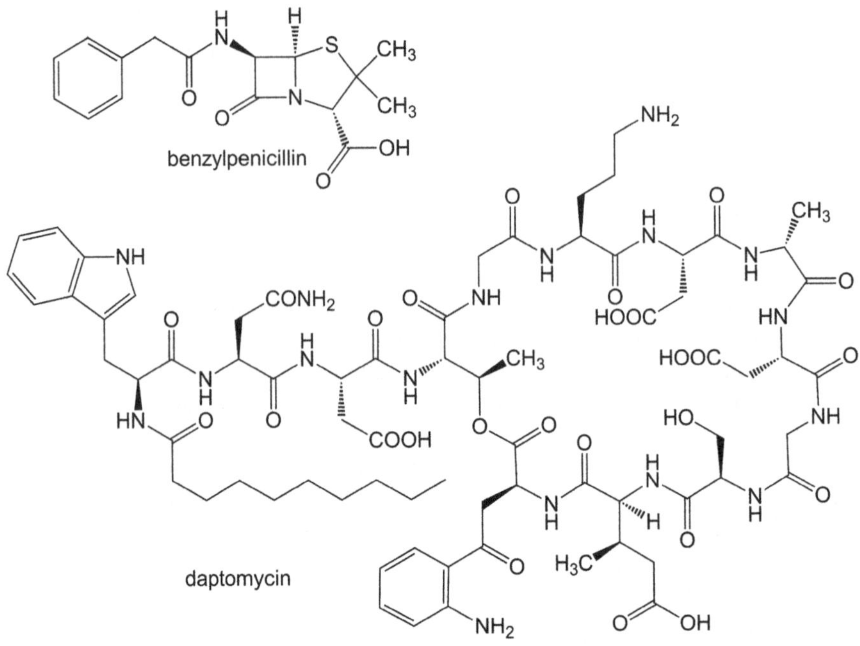

Fig. 3.38 Two natural antibiotics: the relatively simple benzylpenicillin, and daptomycin, which is complex both in its chemical structure and physiological effects. (Authors' own work)

effective drug is found. The cooperation of the patient is very important here, and has to survive an initial period without success. It is also very important that the patient should continue the full prescribed treatment even if he or she improves. If the antibiotic is under-dosed or the treatment is interrupted before all the bacteria are killed, resistant bacterial strains may evolve.

There are some antibiotics that are specifically reserved to treat highly resistant pathogens. These usually have multiple mechanisms of action against pathogens. They should only be used sparingly, as their frequent use would result in the development of strains that are resistant even to them.

Antibiotics were natural compounds at first, and although some of them are prepared by chemical methods today, natural origins are still vital. One of the latest antibiotics is a lipopeptide called daptomycin, which is produced by a fungus (Fig. 3.38). This compound is not the slightest bit less natural than substances promoted by the opponents of synthetic antibiotics such as garlic, the essential oil of tea plant, propolis, or grapefruit seed extract (→ **2.15**). The main difference is that active ingredients in medicines are very thoroughly tested and their efficiency is painstakingly proven. The maximum that can be said of natural alternatives is that some laboratory studies might have shown their bacterium-killing effects. If they met the stringent standards of pharmaceuticals, they would surely be mass produced by the pharmaceutical industry. It may well happen that some plant in the future will be shown to exert significant antibacterial effects, but this is highly unlikely for species that have already been studied.

The risks of antibiotic use are not at all connected to their origin. Semisynthetic and fully synthetic drugs may both be relatively harmless or may come with a number of side effects. Most of these side effects are unpleasant rather than dangerous to the health of the patient, and public concern about them is greatly exaggerated. Most of the time, side effects appear when antibiotics are not used properly. Without them, millions of people would die prematurely every year around the world.

3.24 Hearty Soybeans: Phytoestrogens

In recent times, there have been several noted sudden deaths among athletes (especially soccer players) as a result of heart failure (Welshman Robbie James, Marc-Vivien Foé from Cameroon, Hungarian Miklós Fehér, and Spanish Antonio Puerta). Typically, the exact cause of the death (cardiac arrest, hypertrophic cardiomyopathy, etc.) cannot be precisely determined but can only be suspected in these cases, which often gives breeding ground for rumors. One of the persistent rumors blames soybean and the estrogen-type compounds it contains. These claims are based on the theory of Hungarian professor István Csáky.

It is usually men who succumb to sudden heart failure, so women are likely to have some sort of protection. Among the many possible factors causing this protection, female sex hormones called estrogens have a distinguished place. 'Estrogen' is not a single hormone, but a group of compounds with similar chemical structures. The most significant members of this group are estrone, 17-β-estradiol, and estriol. The blood levels of these hormones are much higher in females than in men. Estrogens are known to protect the body from some heart diseases. Unfortunately, this effect is in addition to their normal hormone function, which makes it impossible to use estrogens as preventive heart medicines. A solution to this problem would be to find similar compounds with lower hormone activity. Some plant-derived compounds, called phytoestrogens, might serve this purpose well. It is here that soybeans enter the scene as they are among the most plentiful sources of phytoestrogens.

Prof. Csáky often repeats his highly counterintuitive theory in public. He exerts that plants with high levels a phytoestrogens do not only not decrease, but actually increase the risk of heart disease. He went as far as connecting the death of noted athletes to soy consumption. This claim is based on the results of American researchers, who gave soy-protein and milk-protein feeds to separate groups of mice. Soy-fed male mice showed some deterioration of heart function compared to the control group, but female mice did not show the same difference. Prof. Csáky explained the effect by assuming that phytoestrogens prevent the protective action of the small amounts of other estrogens in males. Taking an enormous leap from mice to humans, he pinpointed soy as one of the reasons of sudden heart failure in athletes. Predictably, the media had a field day with headlines such as 'Dangers of soybean', 'Soybeans responsible for tragedies', 'Soybean is the name of death'. Public understanding is not helped by the fact that legitimate science also knows something called 'Soybean Sudden Death Syndrome', but it is the plants that suddenly die of this disease...

Needless to say, articles based on this theory do not miss pointing out that soy is present in more than 20,000 food products including bread, cooking oil, cream, ice cream, and even meat products. Another "shocking" fact is that babies may consume an amount of hormones equivalent to 5–6 contraceptive pills every day with their soy-based formula. Apparently, one can imagine that all the troubles of humankind have been isolated to a single reason.

Before inspecting the evidence of the connection between so many sudden heart failures, a few sentences will be devoted to phytoestrogens since getting to know the enemy is usually a good idea. These compounds are quite diverse as far as the chemical structure goes, but there are some similarities: they all bind to estrogen receptors. The significance of these materials was recognized in the 1940s: diseases usually associated with high estrogen levels (infertility, miscarriage) were observed in sheep that ate a lot of clover (e.g. *Trifolium subterraneum*, subterranean clover or *Trifolium repens*, white clover). Phytoestrogens were detected in these plants, and this finding explained the earlier observations. Dozens of phytoestrogens have been identified since the initial discovery: isoflavones, lignans, stilbenes, and coumestans are the most common groups (Fig. 3.39).

Isoflavones are the most significant members of the phytoestrogen family, their level is the highest in food. Soybean, which is very important in the human diet, also contains isoflavones. Consequently, hundreds of scientific articles have been devoted to studying the main phytoestrogen components in soybean (genistein, daidzein, glycitein). In the human body, a compound named equol (Fig. 3.39) forms when these compounds are metabolized, and its estrogen effect is larger than that of any of the parent compounds.

Plant-derived lignans have no direct estrogen activity, but gut bacteria transform some of them to compounds with hormone-mimicking effects. Flax seeds and whole grain cereals contain significant amounts of such lignans.

The most important of the stilbenes is *trans*-resveratrol, relatively large amounts of which are present in red wine ($\rightarrow$ **2.10**). There is good reason to assume that this compound contributes to the beneficial effects of red wine consumption.

Coumestans are the least significant phytoestrogens from a human point of view. Coumestrol is the most noted member of this group that has hormone activity (Fig. 3.39), it is found in soy sprouts.

A common characteristic of phytoestrogens is that they compete with estrogen hormones for binding to estrogen receptors. As estrogen hormones bind very tightly, phytoestrogens only have marginal effects in women of reproductively active ages, who have high blood levels of endogenous estrogen. When the estrogen levels are low (post-menopause women, all men and children), the effect of phytoestrogens may become more noticeable. There are two kinds of estrogen receptors, they are labeled as α and β. Estrogen hormones bind to each of these, but phytoestrogens

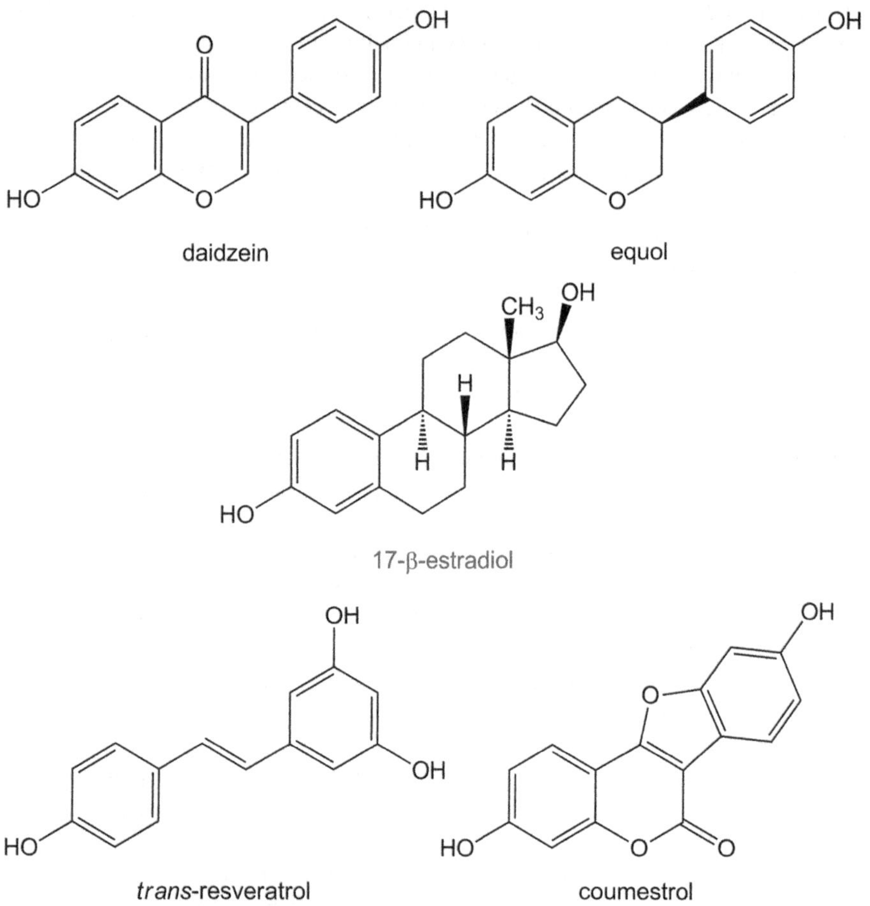

3.24 Hearty Soybeans: Phytoestrogens

Fig. 3.39 The chemical structures of some phytoestrogens and the main estrogen hormone, 17-β-estradiol. (Authors' own work)

prefer the β-receptor. It just so happens that most of the long-term benefits of estrogens are associated with this β-receptor.

The amount of phytoestrogens in the human diet depends very much on eating habits. In East Asia, soybean consumption is quite large so phytoestrogen exposure is more than ten times the European average (daily average of 100 mg in Asia, 5 mg for Europe). The long term effects must be much more pronounced in Asia. Observations made in the mid-twentieth century showed that the a few tumor types, osteoporosis and cardiovascular diseases are less common in Asia, and the symptoms connected to menopause are less severe than in Europe or North America. In addition to genetic factors, dietary habits were also identified as a possible reason. This assumption was supported by a further observation: Japanese people following North American habits do not seem to enjoy the original Asian benefits. A number of clinical studies have been conducted since then, most of which supported the positive effect of phytoestrogens on human health.

Insufficiently low levels estrogen seem to be connected to several symptoms of menopause, these can be treated by consuming plant estrogens. Osteoporosis speeds up in women with the onset of menopause as a result of falling estrogen levels, phytoestrogens can slow the process down, but scientific information in this question is quite ambiguous. Cardiovascular effects are similar: isoflavones seemed to lower cholesterol levels in month-long studies, but other investigations did not corroborate these results. Claims of the tumor-risk lowering effects for phytoestrogens are based on large-scale population studies. The true significance of the observations is unclear today. What is clear: any possible effect would only help in prevention, but not therapy.

As all substances with a physiological effect, phytoestrogens have side effects, too. These were uncovered when they were used for a prolonged time or in large doses. An overdose of dietary supplements or baby formula may present risks. These almost always contain soybean extracts, so the following discussion will be limited to this plant.

Soybean has been cultivated and consumed in China for millennia, it is a staple food there. Korea, Japan and other Asian countries imported the plant and the eating habits. There are a lot of processed forms, which are part of the diet even at a young age. Of these, tofu and mizo are not unknown in Europe, either. Although soybeans reached Europe as early as the eighteenth century, their true significance was only recognized in the second half of the twentieth century. In the US, cultivation began in the 1960s. The success was obvious: in a few decades, the US became the number one producer of the world (Japan is still a world leader in per-capita soy consumption). In North America and Europe, the plant is mostly a meat substitute in vegetarian diets.

Soybean is a very advantageous raw material in the food industry. About 40% of the bean is protein made up of essential amino acids. The oil of the plant is rich in unsaturated fatty acids (essential linoleic and α-linolenic acids, → **2.13**) and lecithin (about 20%). The seeds contain a lot of vitamin B and E, minerals and phytoestrogens.

Both the experience of several thousand years and the recent scientific evidence show that eating soy does not have any significant side effect. The only possible exception is caused by substances that inhibit digesting enzymes: the unpleasant effect may be diarrhea and flatulence. An increasing number of food products (e.g. sausages) also contain soy, but this cannot pose any problems for healthy adults. The reason for the popularity is that soy is less expensive and can be processed easier than meat in the food industry.

Soybean concentrates have been used increasingly since the discovery of the benefits from phytoestrogens. Most of the products are dietary supplements. Scientific evidence thus far has not uncovered any unwanted effects below the daily dose of 50–100 mg, which is equivalent to eating about 100 g soy a day, an average value in Japan. If the amount is significantly larger, side effects can be expected.

The soybean eating customs of the Far East must also be considered when the benefits are assessed, and not only when considering the risks. To alleviate symptoms of menopause, eating 20–80 mg of isoflavones as dietary supplements or as soybeans is sufficient. Other benefits (cancer prevention, for example) may not be accessible if soy consumption begins at an adult age. The evidence is unclear in this respect, but some of these effects of soy seem to be linked to puberty.

The rapidly increasing popularity of soy poses new challenges. This is especially true for applications which do not have precedence in East Asia. In the US, the early onset of puberty (sometimes as early as the age of three) is increasingly a problem. Typically, formula is used to feed American babies from the early age of 2 months. About one fourth of the formula sold is based on soy. The amount of isoflavones in this diet is about 8 mg/kg of body weight a day, which is much more than the typical intake of adults (3 mg/kg of body weight a day) in the Far East. Phytoestrogens were detected at micromole/liter levels in the blood of babies eating soy-based formula, this is high enough to exert estrogen activity. This phenomenon, although certainly reason for concern, is by no means unexpected. Babies on exclusive soy-based formula diet obviously consume more phytoestrogens than necessary, they would probably do better without this exposure. However, it is a distortion of facts to claim that this amount is equivalent to 5–6 contraceptive pills a day. Although this may be true as far as the amounts (weights) are concerned, contraceptive pills contain a lot more potent estrogens than phytoestrogens, so the two cases are not even comparable in terms of overall estrogen effect. As of today, the connection between early puberty and soy-based formula is not proven. Nevertheless, this is definitely a case in which the precautionary principle should be used: better safe than sorry. In Europe, soy-based formula is not as popular as in the US, so the question itself is irrelevant.

No data have been published in the scientific literature to connect soy consumption and sudden heart failures. In fact, some heart disease seems to be less common among those eating a lot of soybeans. The rumors mentioned in the beginning are based on a study on genetically modified mice, who suffered from artificially induced heart enlargement. The results were certainly interesting, but by no means transferable to humans. In addition, the study did not test sudden heart failure at all. Most experts did not give any credit to these claims. Very typically, Prof. Csáky tried to exploit the almost unanimous rejection of his colleagues. He began talking about the major financial interests behind hiding the 'truth'. In his opinion, the profit from the soy business is larger than from selling illegal drugs. Fans of conspiracy theories find this argument credible, and try to spread the information through websites or e-mails. This is just one of the examples where the results of a competent scientific study are misunderstood and used as a reference to blacklist an entire family of compounds. Phytoestrogens should be assessed based on facts to evaluate possible risks and benefits alike.

3.25 Willow Fever: Aspirin

Aspirin is one of the best known medicines around the world. It also has a long history, as it has been marketed in an essentially unchanged form for more than a century. Its application has widened over time. The development of Aspirin is one of the greatest success stories of plant-based pharmaceutical research. Willow bark, from which the drug was derived, is still used in medicine: surprisingly, its application seems to be on the rise and more versatile than that of Aspirin: websites and popular articles regularly promote its putative blood cleansing, mild anticoagulant, anti-fever, anti-inflammatory, catarrh drying, astringent, disinfecting, antifungal, pain killing, mild sedative and antispasmodic effects. Apart from the fact that blood cleansing ($\rightarrow$ **3.15**) and catarrh drying do not make any medical sense, this long list makes willow bark look close to a sort of wonder drug or panacea, which usually makes people with a healthy dose of common sense suspicious. If willow bark is so useful, why do medical professionals fail to use it directly, why do they take the trouble of making a drug similar to it?

The active ingredient in Aspirin is acetylsalicylic acid. It desires attention not only because of its rich past and present, but also because it gave rise to an entire family of drugs and its study led to the deeper understanding of inflammatory processes. NSAID is an abbreviation for non-steroidal anti-inflammatory drugs, about 30 known medicines belong to this family including Aspirin itself. Most of the family members are not related to acetylsalicylic acid chemically, their common property is their inhibiting effect on one of the enzymes important in inflammations.

The history of Aspirin began in eighteenth century England. Malaria was a common disease in Europe (!) at that time, as swamps along un-regulated rivers were hot breeding grounds for mosquitos. Malaria is an illness accompanied by high fever and the only known treatment at that time was the bark of a South American quina tree (*Cinchona* species). Today, it is clear that an alkaloid named quinine is the active ingredient in the bark: it kills the pathogen of malaria (parasitic *Plasmodium* protozoans). At the time, the effect observed was the reduction of fever, so the bark was used generally for this purpose as well (probably without much actual success). Quinine does not have real anti-fever effects, the destruction of the parasites comes with this "side effect".

The anti-inflammatory effects of willow bark were mentioned as early as 3500 years ago in Ebers Papyrus. Ancient Greek and Roman physicians used it for the same purpose, but then it was forgotten for centuries. Its re-discovery was an accident. Edward Stone (1702–1768), a vicar in Chipping Norton, Oxfordshire, wrote a letter addressed to the president of the Royal Society detailing 5 years of experiments and observations on the use of dried, powdered willow bark in curing fevers:

> There is a bark of an English tree, which I have found by experience to be a powerful astringent, and very efficacious in curing aguish and intermitting disorders. About six years ago, I accidentally tasted it, and was surprised at its extraordinary bitterness; which immediately raised me a suspicion of its having the properties of the Peruvian bark. (…) I have no other motives for publishing this valuable specific, than that it may have a fair and full trial in all it variety of circumstances and situations, and that the world may reap the benefits accruing from it.

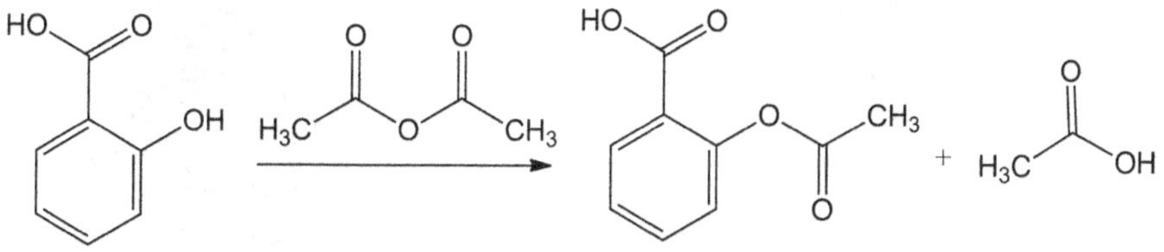

3.25 Willow Fever: Aspirin

Fig. 3.40 Chemical equation of the synthesis of acetylsalicylic acid. (Authors' own work)

Stone is credited with the (re-)discovery, but he misunderstood the effects of willow bark. It does not cure malaria at all, but it does reduce fever. Willow bark came to be used as an inexpensive alternative of quinine at the end of the eighteenth century in England. Under Napoleon's Continental Blockade, cinchona (or quina) first became very expensive then unavailable, which added further impetus to the widening use of willow bark. German pharmacist Friedrich Sertürner (1783–1841) isolated morphine in the early 1800s, which was a sort of revolution in plant chemistry and motivated a host of further studies on the active ingredients of plants. German pharmacologist Johann Andreas Buchner (1783–1852) isolated salicin, named after the Latin name for the white willow (*Salix alba*), in 1826. Twelve years after that, Italian Raffaele Piria (1814–1865) converted this substance into a sugar and a second component, which yielded salicylic acid upon oxidation. Physicians at that time recognized the effect of salicylic acid, but its medical application seemed hopeless as it strongly irritated the stomach. Further development was slow until 1897, when Felix Hoffmann (1868–1946), a chemist at the pharmaceutical company Friedrich Bayer & Co. reacted 100 parts of acetylsalicylic acid with 150 parts of acetic anhydride at reflux temperature for 3 h to obtain an acetylated product quantitatively (Fig. 3.40). The final sentence in his laboratory notebook ran:

> Durch ihre physikalischen Eigenschaften wie eine sauren Geschmack ohne jede Ätzwirkung unterscheidet sich die Acetylsalicylsäure vorteilhaft von der Salicylsäure und wird dieselbe in diesem Sinne auf ihre Verwendbarkeit geprüft.

This translates to:

> Due to its physical properties, such as an acid taste without any corrosive action, acetylsalicylic acid differs advantageously from salicylic acid and is being examined for its usefulness with just this in mind.

It appears that salicylic acid had already been prepared prior to 1897. Hoffman's achievement was to develop a very simple synthetic procedure and to recognize that it has effects similar to salicylic acid but with much weaker stomach irritation. The medical benefits were proven and the drug was in widespread medical use within 2 years. The name Aspirin was derived from "acetyl" (letter A in the beginning) and the botanical name for meadowsweet (*Spiraea ulmaria*), which also contains derivatives of salicylic acid: the old German name for salicylic acid was Spirsäure. The suffix '-in' is a common one in chemical names.

The drug's success surpassed all expectations. It is considered as a modern medicine even today, no other contemporary medical invention has the same stature.

Fig. 3.41 An advertisement for Aspirin in an American newspaper in the 1910s. (Copyright-free picture—authors' own work)

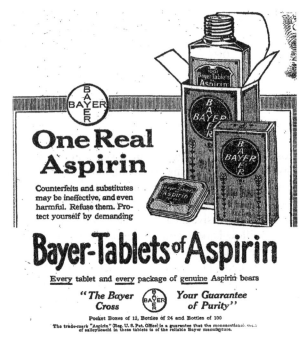

Now, acetylsalicylic acid (Fig. 3.41) is the active ingredient in many generic drugs (→ **3.1**), not only in Aspirin. Despite it widespread use, next to nothing was known about the mechanism of its action for decades. The breakthrough in this field was achieved by John Robert Vane (1927–2004), who discovered prostaglandins and leukotrienes. He was awarded the Nobel Prize in physiology or medicine in 1982. Acetylsalicylic acid inhibits an enzyme called cyclooxygenase, which catalyzes a key step in the formation of prostaglandins and leukotrienes. This mechanism is common to a lot of anti-fever drugs. Later, it was discovered that Aspirin also inhibits the formation of thromboxane A2, which has a role in blood platelet aggregation. As a consequence, acetylsalicylic acid, uniquely among anti-inflammatory drugs, slows down blood clot formation and may help prevent heart attacks and strokes. The daily dose in this use is much smaller than for the anti-inflammatory use.

So, how does willow bark compare to acetyl salicylic acid? The bark contains 1–2 % of salicyl derivatives, most of them as **glycosides**. The main ingredient is salicin (Fig. 3.42), which is also responsible for the bitter taste. In the intestine, salicin is solely hydrolyzed to give saligenin (salicyl alcohol), which turns into salicylic acid in the blood stream. The actual anti-inflammatory effect belongs to salicylic acid.

The advantages and drawbacks of willow bark over acetylsalycilic acid can be deduced from these facts. Salicin irritates the stomach less than salicylic acid or even Aspirin. Other properties are not so favorable, though: its effect takes a lot of time to develop (as salicin must be transformed to salicylic acid), and does not inhibit blood clot formation. Most importantly, its dose cannot be set with the required

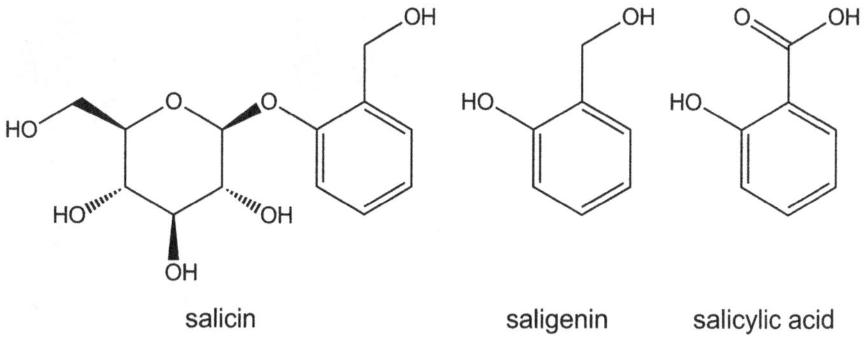

Fig. 3.42 Chemical structures of salicin, saligenin and salicylic acid. (Authors' own work)

precision. Willow barks from different sources may differ in salicin content, but this is not the only concern: the absorption, hydrolysis time and other properties of different glycosides are not the same. Therefore, the salicylic acid level in the blood is very difficult to predict. Willow bark typically contains 1.5% salicin. If all salicin is transformed to salicylic acid, some 10 g of willow bark must be consumed to be equivalent to a single pill of acetyl salicylic acid. This is a large amount: it cannot be eaten directly, preparing tea from such quantity is not that easy and would involve further uncertainty factors (temperature, time of extraction, amount of water).

Willow bark does not have all the beneficial effects of acetylsalicylic acid, it certainly has no effects that are unknown for Aspirin. Willow does not contain any other active ingredients that may be responsible for its supposed miraculous effects. None of the claimed "blood cleansing", "catarrh drying", astringent, mild sedative and antispasmodic effect have any support from scientific studies. Also, tea made from willow bark does not kill *Candida* fungi. These are sometimes well-intentioned inaccuracies, but other times, they are outright lies used in marketing.

Modern anti-inflammatory drugs are very safe and inexpensive. Willow bark is difficult to take in a controlled manner, and does not inhibit blood clot formation, so it has no importance as a modern pharmaceutical. This plant was very important in developing modern drugs, but its direct use would be a huge step backward.

3.26 Pill Secrets: The Efficiency of Drugs

The efficacy of medicines is often questioned. Even within a family, opinions may vary. Why do medicines have different effects on different people? Does the lack of effects on an individual render a medicine useless? Are there approved medicines that may not be effective at all?

On the surface, things are crystal clear. The basic requirement to register an active ingredient as a medicine is that it should have a desirable effect on humans. But when these effects are first tested, they are not tested on humans directly, but on cell cultures or receptors in laboratory studies. If these investigations yield positive

Fig. 3.43 Efficient but unsafe. Tens of thousands of babies were born with deformed limbs in the 1950s because their mothers took the medicine Contergan during pregnancy (one of the victims: film producer Niko von Glasow). (Copyright-free Wikipedia picture)

results, tests can be continued on animals, then on humans. There is a very high chance of failing these tests: one out of about ten thousand promising substances end up as actual medicines.

The last century has brought a sea of change in medicine development. The quality and quantity of evidence required has steadily increased, so has the effort necessary to be successful. This applies to physiological effects, safety and chemical purity alike. In the 1950s, women who used a medicine called Contergan (taken against nausea and to alleviate morning sickness in pregnant women) with the active ingredient thalidomide during pregnancy gave births to babies with deformed limbs (a condition called phocomelia, Fig. 3.43). This effect of the medicine was unknown before its approval as the safety tests required at that time were not as detailed as today. The memory of the Contergan scandal case was the driving force of introducing new and stricter legislation and requirements in drug approval.

Today, pharmaceutical companies sometimes pull certain medicines from the market even if authorities do not require them to do so. An example is the cholesterol lowering drug cerivastatin, which had side effects similar to other related drugs, but its manufacturer decided that these side effects were too common and stopped selling the medicine.

The efficacy is proved in human tests involving a sufficiently high number of patients taking quite a long time, so that there could be no doubt about the results. Not only is a placebo used as comparison (→ **3.6**), but other known similar medicines as well. These are called active comparator studies: one group of patients receives the already approved treatment, while the other group is on the tested medication. A lot of promising lead molecules are shelved because they are no better than drugs already in use.

The steady stream of new medicines developed makes human life better. However, drugs approved decades ago did not go through the same strict procedure and were not subjected to the same scrutiny. Dozens of medicines approved before 1940 are still sold. However, this apparent contradiction is not a cause for alarm. All of these early medicines were widely used in the last 70 years, and experience with thousands (often many more) patients show their effect and safety. This is much better than any current tests, where the time scales are understandably much shorter (Fig. 3.44).

However, some of the traditional drugs would probably not pass today's tests as new drugs. This is also true for medicines that were approved only 30–40 years ago. They do not get any less efficient as time goes on: the requirements increase. This never ending procedure guarantees constant progress in efficiency. A medicine might pass the test at a given time, but if a better one is discovered later, the old one will be gradually phased out. The pharmaceutical value of a given compound does not only depend on its effects, it also depends on the properties of competitor's products.

Efficacy is not the only game: safety (unwanted side effects and their consequences) and the nature of the treated disease are also major factors. Even mild side effects cause concern if the treated illness is not severe. On the other hand, serious side effects can still be acceptable in a medicine curing a life threatening disease. The tolerance limit for side effects is very different for cancer drugs and antacids.

The same substance might exert different effects on different people. This is often connected to the placebo or nocebo effect ($\rightarrow$ **3.6**), but again, reality is more complex than many people prefer to admit. Trust in a medicine is very important, but trust alone does not make a medicine effective. A drug might not have the expected effect even if the patient trusts it. This phenomenon is explained by **pharmacokinetic** and **pharmacodynamic** reasons.

The uptake (absorption), transport, degradation (metabolism) and excretion of active ingredients may show a lot of variability. These pharmacokinetic factors influence the efficacy of a drug individually, but their combined effect may be still stronger. The dose of a medicine is calculated for average conditions, which may not be representative for every use. Drug degradation is mostly carried out by the liver, the activity of enzymes shows genetically determined variability. To some extent, genes might render a particular medicine more or less effective for a particular group of people. Faster degradation usually means weaker effects, slower metabolism may lead to more severe side effects.

Pharmacodynamic factors are also important: the receptors, enzymes or other proteins targeted by the drug may show a lot of variability in distribution and amount, which also influences drug efficiency. Although personalized, genetics-based medicine is a dream of the distant future now, some observations for larger

Fig. 3.44 Belbarb, a drug no longer in use. (Copyright-free picture from http://www.decodog.com/inven/MD/md28496.jpg)

populations are already known. Some drugs against high blood pressure are less effective for African people because of genetic differences in receptors and genes.

Sometimes, it may even happen that drugs, otherwise very efficient under average conditions, fail to show improvement for certain people. This is not the fault of the medicine. The human body is such a complex chemical reactor that controlling all its parameters is next to impossible. Therefore, predicting the results of a treatment is somewhat uncertain.

3.27 In Quest of Active Ingredients: Herbs

Many people seem to agree that artificial synthetic medicines can be replaced by natural ones, which are more compatible with the human body. Herbs have a distinguished place among the natural healing methods: every illness seems to have a long-known remedy, which is less toxic and has fewer side effects. Was there ever a better example of wishful thinking?

There are a large number of highly toxic natural compounds (→ **1.3**). Synthetic and natural active ingredients have already been compared in a separate story in this book (→ **3.7**). This chapter examines if plant-derived compounds have weaker effects than synthetically prepared analogues.

There is no effect without side effects—this is a basic law in pharmacology. If plants contain effective compounds, they have side effects as well. Entire books have been written about the side effects of herbs—obviously without much of an effect on public opinions.

The medicinal history of the last century actually shows that a number of substances were phased out exactly because they had too strong effects. Several of them are alkaloids, which are plant-derived alkaline compounds. The chemical structures of alkaloids typically feature nitrogen-containing rings, which are connected to their strong effects on humans. A lot of neurotransmitter molecules in living organisms also contain nitrogen, so the relevant receptors can bind alkaloids as well, which may exert stronger effects than the original transmitter molecule.

Some alkaloids completely disappeared from medicine lists. Strychnine is not used any more as a stimulant since a few dozen milligrams of this substance is already lethal. Risk-cost-benefit analyses are very important in modern medicinal science, but this has not always been the case. In the case of strychnine, the risks are viewed today to be unacceptable compared to the expected benefits.

A survivor in the group of highly effective alkaloids is morphine (Fig. 3.45). It has been the most important painkiller for a long time after its discovery in the nineteenth century and it is still used in certain cases. The poppy plant (*Papaver somniferum*) contains it in the seeds, seed pods and straw. Despite its widely known narcotic effects, it can be used in certain cases to relieve pain without much risk from tolerance or dependence. Patients in the final stages of cancer or suffering from bad burns do not experience euphoria. In less serious illnesses, the side effects and the risk of dependence are of more concern and morphine is not usually used in such cases. Pharmaceutical companies have long aimed to develop analogs of morphine that have fewer side effects and lower risk of tolerance and dependence. Some results have been achieved, so compounds that have effects similar to morphine are already available to physicians. Their chemical structure, though, is often not similar to that of morphine.

Sometimes the effects of a synthetic compound may be marginally stronger than those of the natural alternative. D-Tubocurarine serves as an excellent example. South American indigenous people use a type of this substance as a special poison in arrows: they relax breathing muscles fast, so the animal is unable to move and

morphine   (1R, 2R)-tramadol

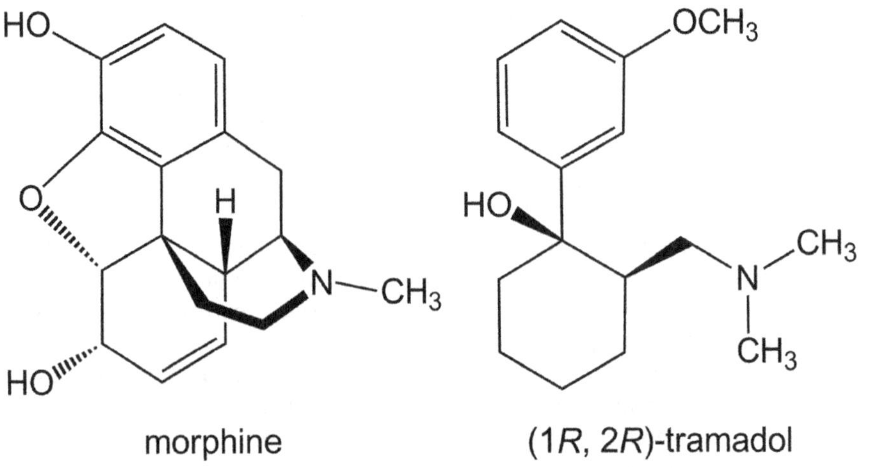

Fig. 3.45 The chemical structures of morphine and tramadol, a less potent painkiller developed from it. (Authors' own work)

dies under its effects (Fig. 3.46). The flesh of the killed animal can be eaten without risks as the poison is unable to reach the nervous system from the stomach or intestines. The name curare is used to refer to an entire group of such poisons. Tubocurare, a special type of curare, is usually stored in hollow bamboo tubes and was first studied by Harold King (1887–1956) in a museum specimen in the 1930s. He isolated the main active ingredient D-tubocurarine. The poison was also identified in its usual source, the bark of a South American tree (*Chondrodendron tomentosum*). The mechanism of the action was discovered by Claude Bernard (1813–1878) well before the chemical compound itself was isolated. This was an important achievement in science: it was the first case when the target receptor of an active compound was systematically studied. The chemical structure of tubocurarine was only discovered much later in the 1970s. One of the nitrogen atoms in the compound turned out to be a tertiary nitrogen, so it was connected to three carbon atoms. This was unexpected, earlier investigators assumed quaternary ammonium structure, in which a nitrogen is connected to four carbon atoms and bears a positive electric charge. This mistake proved to be a fruitful one: chemical analogs containing quaternary nitrogen atoms were synthesized and found to be more effective. D-Tubocurarine already has a reasonably strong effect, its **LD_{50}** value is lower than 1 **mg/kg of body weight** for intravenous injection. Some derivatives of this compound are used in surgeries as muscle relaxants.

3.27 In Quest of Active Ingredients: Herbs

Fig. 3.46 A South American indigenous man using blowgun dart dipped in curare. (Copyright-free Wikipedia picture)

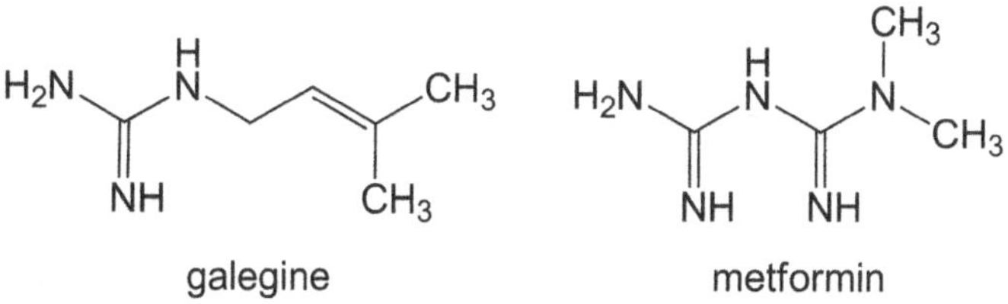

Fig. 3.47 Galegine and metformin: both of them lower blood sugar levels. (Authors' own work)

Sometimes a problem may arise from similar effects of an active medicinal ingredient and a herb. Goat's rue (*Galega officinalis*) is rapidly becoming a popular "underground" herb. This unusual phrase tries to call attention to the fact that no medical expert recommends it, yet it is apparently in wide use. There is no doubt about the effect: the plant contains guanidine derivatives (most significantly, galegine) that have been proven to lower blood sugar levels (Fig. 3.47). Goat's rue has been used as a medicinal herb for centuries, but its use has been gradually phased out since galegine and other derivatives (e.g. metformin) were introduced as medicines. This is more than reasonable. The galegine content of a plant must be known for medicinal use (domestic users have no idea about this), and goat's rue contains a number of other active ingredients, whose effects are less than desirable. Users seldom consider these possible problems. The equivalent antidiabetic drug metformin is sometimes not recommended for patients (e.g. pregnant women), the public trust in natural compounds may lead these people to drink the tisane of goat's rue instead. However, the effect of tea may be higher than that of the medicine. Some sellers even claim that goat's rue has no side effects in contrast with medicines based on guanidine molecules. Nothing can be farther from the truth. Marketing can be quite loud, and few people may be interested in the views of experts.

The examples listed above do not give a really balanced view in comparing natural and synthetic materials. They were only meant to show that the effects of natural herbs are by no means weaker than those of synthetic medicines. There are a lot of medicines on the market that have strong effects, but some herbs are also highly effective. Anticancer compounds (e.g. plant-derived paclitaxel; → **3.7**) usually have very strong effects, irrespective of their origins.

The real danger lies not in the strength of the effect, but in underestimating it. This is especially true if the amount of the active ingredient is not controlled sufficiently, which is often a problem when herbs are used.

3.28 Getting Rid of Acids: Alkaline Cures

The authors of this book are scientists, who have a moral obligation to tell people what they understand to be the truth. Therefore, the following sentences are quoted from a website with a great deal of reluctance:

> Foods are classified as acid-forming or alkalizing depending on the effect they have on the body. Acid-forming diets lead to acid indigestion that eventually creates a condition known as chronic acidosis which acidifies body tissue. This acidification process chokes off oxygen within cells and eventually compromises the immune system.
> This leads to all sorts of diseases such as cancer, heart disease, stroke, memory loss, nutrient deficiencies and poor athletic performance. All of these conditions are, for the most part, due to acid-forming diets that violate the acid/alkaline blood ratios of your body cells. One of the keys to cancer prevention is eating an anticancer diet, also referred to as the Acid Alkaline Diet. The A-A-A Diet™ is the only high alkaline diet that can alkalize and oxygenate body cells as you eat alkaline foods according to the acid alkaline food chart.

The alkaline diet is such a popular topic in websites and magazines today that trying to prove these claims to be scientific nonsense seems to be mission impossible against the marketing prowess of the advertisers. Yet, this is the challenge undertaken in this chapter. Let's begin where the story begins.

American psychic Edgar Cayce (1877–1945) was among the first to state that a healthy diet (called Cayce diet, is anyone surprised?) needs about 80 % alkaline-forming foods—mostly originating from plants. It is hard to do justice in the question of priority—several people seem to have come up with similar ideas at the same time. What is certain is that the idea surfaced in the United States in the first decades of the twentieth century. It was another American, Robert O. Young (1952-) who popularized the method immensely in a his *pH Miracle* series of books, hundreds of thousand copies of which were sold. This book was translated into more than 20 languages. Although Young is not a trained physician, he gives confident advice in all sorts of health issues. What is more, he developed a new theory of biology, which does not only expound how to cure illnesses, but explains the reasons as well.

Young improved and perfected the theories of alkaline diet first appearing in the early 1900s. This basically means he made up more complicated explanations.

3.28 Getting Rid of Acids: Alkaline Cures

He asserts that illnesses are caused by the acidification of blood, and the body tries to fight this process every way it can. During this fight, the body is exhausted and becomes susceptible to various illnesses (from the common cold to obesity, and even cancer). The acidic **pH** favors the growth of pathogens, so increases the risk of infectious diseases, including that caused by the dreaded *Candida* fungus. To neutralize harmful acids, the body removes ions from bones, this is the cause of osteoporosis. All these have a common reason in the background, so the alkaline diet prevents all the diseases. This is best done through a healthy diet: a lot of vegetables, fruit, fresh fish and nuts should be eaten, and acid-forming foods such as sugar, meat, eggs, dairy products, mushrooms, processed and refined grains, alcohol, coffee and chocolate should be avoided. Young even developed a diagnostic test: those concerned about getting acidified are advised based on a microscopic blood test.

Young's followers do not seem to care that their master does not cite any scientific results to support the value of his proposed diet for human health. The mentioned microscopic blood analysis is not considered a useful diagnostic method. Therefore, it cannot be used legally in the medicine. Today, everyone has the right to write an unscientific book of pure fiction, and with some luck, may even get rich as a result. The scientific literature, on the other hand, has a refined quality assurance process. The readers of Young's books are seldom aware of real scientific views on these topics. Many critics have already concluded that Young seems to lack basic chemical and physiological knowledge, and some of his views on microbiology were borrowed from books published in the nineteenth century. Those who try to explain the absurdity of Young's claims are often portrayed as members of a sort of scientific mafia that is bent on systematically harassing "geniuses" outside their usual circles. This image can be further strengthened by careful twisting of facts about legal cases brought against Young, most of which charged him with practicing medicine without a license. Unfortunately, a few actual doctors seem to have joined the followers of Young, most likely in an attempt to gain financial advantages.

The buzzword of alkaline diet is often accompanied by dietary advice that is contrary to the theory of the author of the "pH Miracle" book series. Eating vegetables and fruits is often recommended, but sometimes non-carbonated mineral water and lemon water is also claimed to be beneficial. Some negative lists include pasteurized milk and canned food in general. Trout is sometimes claimed to be less acid-forming, whereas others classify it as harmful among the fish. There is an infamous article where the advice on peanut butter is both to consume and avoid it. Careful mastication is important and breathing seems to be in agreement with the purposes of alkaline diets. For chemists, it is quite interesting to note that some of the recommended foods in the alkaline diet are actually quite a bit acidic...

The special products that can be purchased to aid an alkaline diet are quite numerous, there is no limit on the amount of money that could be spent on them. Those worried about acidification often think that minerals are removed from the body during the fight against low pH, so dietary supplements are popular among them. An especially useful effect is attributed to chlorophyll, concentrates of which are available who prefer not to eat green plants.

Young also emphasizes the quality and quantity of water consumed. Some of the followers assert that balancing the pH begins with superhydration, which means drinking exceptionally large amount of water (3–5 l/day). The quality of water consumed is also a central question. Water purifying and ionizing (whatever this could mean) devices are available for purchase. They are advertised to provide detoxified and ionized water, which has high antioxidant activity and is rich in oxygen. The following quotes from various websites—also included against the scientific instincts of the authors of this book—illustrate the total absence of basic scientific knowledge:

> Your body thrives in an alkaline environment since it is able to detoxify more efficiently than in an acidic environment. In an alkaline environment your tissues get rid of impurities more efficiently. When cancer patients come into my office to begin nutritional treatment, their bodies are almost always very acidic and toxic. My first task is to get their tissues alkalinized with alkaline water.....
>
> I have recommended alkaline, hexagonal water to even my youngest patients. In 2005 a ten-year old girl and her parents came to my office in South Carolina. The girl had crippling juvenile rheumatoid arthritis and weighed only fifty-two pounds. Her hands were swollen like mitts, and her knees were swollen as large as softballs. I put her on hexagonal, alkaline water, one to two quarts a day. A week and a half later she was pain free, and her swelling was significantly diminished. When she arrived she was wheelchair bound, but she was actually able to walk without pain after only a week and a half of drinking hexagonal, alkaline water. Her parents were ecstatic. We raised the pH of her tissues. After a month, her hands were almost normal size.
>
> There is significant evidence that Hexagonal Water moves within biological organisms with greater ease. It appears to enhance nutrient absorption and the removal of metabolic wastes. (Many who consume Hexagonal Water report a cleansing reaction within days or weeks.) This may be due to the smaller size of the molecular unit and its ability to enter and exit the cellular environment with greater ease. It may also be due to the specific organization of the hexagonal network which appears to enhance cellular communication and to support other structures within biological organisms.

Bottled alkaline water can be purchased in most developed countries. There are several brands on the market, some of which claim that their pH values are higher than 9. A lot of different medicinal effects are described, and sometimes surprising, and most probably misleading endorsements from various known organizations (such as the Red Cross) are quoted.

Some products advertised as alkaline are actually harmful. Miracle Mineral supplements were available from various websites in a number of countries, the claimed benefits included treatment of HIV, malaria, hepatitis and cancer. Marketing the brand was consequently banned in most countries. The chemical composition of the product explains the life threatening effects: it was sodium chlorite ($NaClO_2$), which had to be mixed with citric acid before use. Chlorous acid ($HClO_2$) is first formed in the induced reaction, and then chlorine dioxide (ClO_2) is produced. In the final product, chloride (Cl^-) and chlorite (ClO_2^-) ions are both present depending on the pH:

$$4\ H^+ + 5\ ClO_2^- \rightarrow 4\ ClO_2 + Cl^- + 2\ H_2O$$

$NaClO_2$ and ClO_2 are disinfectants, so they can be used for water purification, washing of foods or as an antibacterial treatment of surfaces if their concentration is sufficiently low. On the other hand, they are toxic when taken orally. It is also ironic (albeit not particularly significant) that the mixture of sodium chlorite and citric acid is clearly acidic.

Half-truths may be much more damaging than direct lies. Therefore, some of the distorted information about acidification that actually has factual origins is also worth mentioning. It is often claimed that tumor cells grow faster in an acidic environment, an anticancer agents are more effective in an alkaline medium. This is in fact true for cell cultures, but there is no evidence supporting a similar effect in living organisms.

The diet can only influence the pH of urine, but not of blood or of the liquids inside or outside cells. It is also known that some foods (e.g. meat) shift the originally somewhat acidic pH of urine into more acidic directions, whereas others (e.g. vegetables) make it more alkaline. However, there is no connection between the pH of urine and other body parts. Also, the diet does not influence the pH in other body fluids. This is quite fortunate, as pH must be finely regulated within the body, even seemingly minor deviations cause serious sickness or death. The pH of blood, for example, must be between 7.37 and 7.43. Modern lifestyle does not make the body acidic: the pH there has been constant for a few hundred million years. A shift of blood pH into the acidic direction (<7.35) already causes some symptoms, a larger change (<7.20) leads to blood circulation and disturbance of consciousness, and eventually death. If the blood turned more alkaline than normal, similar symptoms would follow. A blood pH below 6.8 or above 7.8 is an instant danger to human life. So the mechanisms keeping the pH constant in the body are actually quite sophisticated. Carbonic acid (H_2CO_3), which may shift the pH toward acidic value is dehydrated into carbon dioxide (CO_2) and leaves the body through the lungs as part of respiration. Similarly, "alkaline" ammonia (NH_3) is extracted into the urine in the kidneys. Buffer systems in the body do a very good job at maintaining a constant pH. Buffers in the body, such as the CO_2-H_2CO_3 system, some proteins and some phosphates, can neutralize both acids and alkalis, provided that their amounts are not too high. These buffer effects ensure that the pH in blood and other body fluids remains normal even if large amounts of acidic or alkaline food are consumed. The liquids inside cells have normal pH values between 6.9 and 7.4 depending on the type of the cell, and their tolerance toward changes is even worse than that of blood.

Alkaline diets also have some limited positive aspects as well. Increasing the amount of plant-derived foods in the diet at the expense of meat, refined grains and sugar is usually a healthy idea. However, taken to the extremes, these principles may cause serious shortages of nutrients necessary for health. Avoiding fats and oils may be a source of illness because of a lack of essential fatty acids, whereas cutting out dairy products would lead to problems in maintaining an adequate supply of calcium, vitamin D and proteins.

The scientific literature on the acid-base balance within the body is huge, only a small part is concerned about the advantages of alkaline foods, with limited positive indications, such as a slowdown in calcium loss. Of course, the sensational claims in the advertisements have no scientific background. Many of these advertised theories are intentionally misleading. Making the body alkaline by the diet is not possible, and any successful attempts would have life-threatening consequences.

3.29 "...upon the Face of the Waters": Mineral and Tap Water

The popularity of mineral water is on the rise. In Germany, Austria and Hungary annual average consumption is about 110 l/person, the same number is 150 l/person in France and 180 l/person in Italy, although the outstanding Italian popularity is not simply by choice: in Italy, tap water is not uniformly safe to drink. The rise of mineral water consumption is mostly at the expense of sweetened soft drinks, which seems favorable from a health perspective. Yet some questions arise. What is the difference between mineral water and tap water anyway? Are some varieties of mineral water healthier than others? Do all types of bottled water qualify as mineral water?

There is no consensus among experts on the exact recommended volume of daily fluid intake (this would have to include water in foods as well), but a minimal value of about 1.5 l is not an unreasonable guess. Based on this estimate, an adult needs at least 550 l of water in a year. There are no reliable statistics on the amount of tap water drunk, so all that can be stated with some certainty is that 20–30 % of the water required by the body is consumed as mineral water in Europe.

There is an additional notable drink: carbonated water with no minerals. This was first prepared by Hungarian inventor and priest Ányos Jedlik (1800–1895) in 1826 as a surprise to his fellow monks in the Benedictine order (Fig. 3.48). Although the original idea is credited to English chemist and clergyman Joseph Priestley (1733–1804), who found it to be a refreshing drink, it was clearly Jedlik who developed the industrial technology to produce carbonated water. The phrase 'soda water' was in common use in the United States before World War I to refer to carbonated water because carbonic acid (H_2CO_3) and soda (sodium carbonate, Na_2CO_3) were not clearly distinguished. Even today, soda means any sort of sweetened, carbonated drink in large parts of the US. Returning now briefly to the achievements of Jedlik: he had about 70 different inventions, the dynamo and a kind of electric engine may have been even more significant than carbonated water, but many Hungarians remember him because of the drink. Soda water was the very first Hungarian product to receive the label "Guaranteed Traditional and Special" from the European Union when Hungary joined it in 2004. In Hungary, there are still about 3000 families who earn their living by selling pure carbonated water and, in contrast to other European countries, most (about 65 %) of the mineral water consumed is carbonated.

Although it is not difficult to argue in favor of drinking mineral water, tap water is a clear winner if the price and the environmental footprint are considered. A liter of tap water seldom costs more than half a eurocent, the price of bottled mineral water is at least 100 times higher. Tap water needs no packaging material, and the

Fig. 3.48 Ányos Jedlik, Hungarian inventor, engineer, physicist, and Benedictine priest, the inventor of soda water. (Copyright-free Wikipedia picture)

environmental footprint of the non-existent advertising is also zero. Soda water is usually delivered in bottles that are re-usable dozens of times, and the production is often local, which translates to minimal transportation-related cost and environmental pollution. Mineral water, on the other hand, usually comes in bottles made of poly(ethylene terephthalate) (PET). Billions and billions of bottles get thrown away each year. Even environmentally conscious consumers, who recycle bottles, may not know that PET reprocessing is typically done in Asia, so an extensive amount of transportation is involved.

The question is complicated by a huge amount of easily available, but not quite reliable (mis)information. These are typically direct or indirect advertisements and often claim that healthy drinking water can only be obtained from tap water by water purifiers. (The reader will not have a hard time guessing whose interest it is to spread this particular opinion.) It is often pointed out that the tap water contains chlorine, chlorinated organic compounds, other chemicals that modify taste and odor, rust, scale and so on, which must be removed before consumption. It is sometimes added that drinking chlorinated water for 40 years is known to increase the risk of cancer by 70%. Even mineral water is debunked as expensive and causing (typically unspecified) health problems. This line of thought is designed to conclude that buying a water purifier is an absolute necessity. The simplest such purifiers use activated carbon and cost only dozens of euros, whereas more complicated, more expensive (and what a surprise, supposedly better) ones start selling at about one hundred euros. But is this in fact good business for the consumer?

The Latin question '*Cui prodest?*' (For whom it advances?) is very appropriate here. Concern about the quality of tap water and mineral water is entirely ill-conceived, it is exclusively motivated by financial interests. It may not be widely known, but tap water is probably the most often tested and most rigorously monitored product in developed countries. Regulations also typically require posting easily recognizable warnings on sources of water that is not safe to drink. The levels of impurities are regularly determined by scientific tests and action is taken if they get close to legal limits. Arsenic in tap water is often one of the largest concerns, misconceptions connected to this element are described in two separate stories ($\rightarrow$ **4.2, 4.6**).

A typical misconception about tap water is that it contains traces of various medicines. Taken literally, this is true. However, the detection of medicines in tap water is a sort of scientific exercise in finding miniscule amounts of a certain substance. Typical levels are on the picogram-nanogram/liter scale, whereas the usual single doses of medicine exceed 1 mg by far. So even drinking one million liters of tap water will not provide enough drug to exert any noticeable effect.

Others pinpoint chlorination as the reason why they refuse to drink tap water. Chlorination, which is a time-tested method of killing pathogens in water, certainly involves some very minor risks, but the absence of chlorination is infinitely more dangerous. Still, some generally credible civil organizations run a campaign against the use of chlorine ($\rightarrow$ **1.5**). This does not only ignore all scientific evidence but also endangers the lives of people in a number of developing countries. The formation of chloroform (or, using the systematic name, trichloromethane) is usually the main culprit in such concerns. Yet the amounts of chloroform detected in water are very low and even that minor amount is easily removed by letting water stand in open air for a few minutes because this organic substance evaporates rapidly. The sometimes recognizable aftertaste of chlorination can be removed by this simple treatment, too. Chloroacetic acids, chlorites and bromates may also form during chlorination, but again, the amounts are so low that it is a challenging task to find them even with the best instruments—the body simply does not notice such low quantities.

Tap water is generally safe to drink: most countries devote major financial resources to ensuring this. The use of water purifiers can only be advisable in certain exceptional cases: if tap water contains a lot of minerals, purification maybe a good idea before making coffee to prevent rapid formation of scale in the coffee machine.

Apparently, mineral water is a much more difficult target to attack. These waters are often regarded as beneficial because of their mineral content (Table 3.6). Yet some producers of water purifiers attempt to portray this fact as a disadvantage. There are a number of misleading experiments that supposedly show that water contains impurities. These are often similar to the one used in "detoxification devices" ($\rightarrow$ **3.15**) and rely on electrolysis, which precipitates "contaminants" into a brown residue. No residue forms if the water is free of minerals, which is supposed to demonstrate the burning need to buy a water purifier.

More reasonable concerns about mineral water have surfaced recently, mostly focusing on the possibility of antimony poisoning. Antimony trioxide is used as a catalyst during the manufacture of PET bottles, and the water in the bottle can dissolve some of this material. A series of experiments proved that several external factors (temperature, light, acidity of the water) influence how fast this dissolution occurs. However, the levels of antimony did not exceed safe values even under extreme conditions. Still, these studies serve as a warning: bottled mineral water cannot be stored indefinitely. It is usually a good idea if the storage place is reasonably cold and not exposed to much light.

The radioactivity of mineral waters has also attracted increasing attention, this might actually be responsible for a significant part of the radiation exposure in the population. The source of this radioactivity is mostly radon and radium from rocks. The World Health Organization (WHO) has published standards for ^{226}Ra and ^{222}Rn levels in mineral water. Meeting these sometimes requires actions from the manufacturers.

Drinking water from ion exchangers or reverse osmosis devices exclusively would lead to a shortage of trace elements and deficiency sickness after a while. Those who insist on avoiding tap water should occasionally drink mineral water. The high purity of waters coming from water purifiers is a myth that has been busted several times. Minerals and other contaminants can in fact be removed, but neglecting regular maintenance of the device may open the way for the growth of various pathogens. Observations also show that the owners of water purifiers often fail to follow the instructions of the manufacturer, so this mineral-free water poses a much greater overall pathogen risk than tap water does.

Because of chlorination, tap water does not contain pathogens, but the amount of minerals in it is significant, usually between 300 and 500 mg/l, which is close to the typical minimum value (500 mg/l) a mineral water should have. Hard tap water is usually considered disadvantageous. However, hardness is caused by the very same magnesium and calcium ions that are thought to be beneficial in mineral water! The mineral content in mineral waters typically changes between 500 and 4500 mg/l, but the exact legal definition varies depending on the country. In the US, the Food and Drug Administration (FDA) classifies mineral water as water containing at least 250 mg/l total dissolved solids, and originating from a geologically and physically protected underground source. No minerals may be added to this water. In the European Union, mineral water should be bottled at the source with no or minimal treatment. The removal of iron, manganese, sulfur and arsenic is permitted but no additions are allowed except for carbon dioxide. So this latter definition does not rely on measuring the amount of minerals. The label 'mineral water' can be used, independently of overall mineral content, if the concentration of a single biologically active component exceeds a minimum value. So there are oxymoronic mineral waters with very low (<50 mg/l) or low (<500 mg/l) mineral content, whereas high mineral content starts at 1500 mg/l.

Minerals and trace elements are necessary for normal body functions, but they can also be harmful in excessive quantities (Table 3.6). Healthy people can drink any mineral water in normal amounts, but people with certain illnesses should sometimes limit or entirely avoid consumption. It is usually not advisable to drink water with exceptionally high trace element content all the time. The Latin saying '*Varietas dēlectat*' (variety is the spice of life) is very wise as an advice when drinks are considered, and the varieties should not exclude tap water, either.

'*Non omne, quod nitet, aurum est*' (All that glitters isn't gold) —bottled water is not always mineral water. As already explained earlier, the legal definition of

Table 3.6 Components of mineral waters with their benefits and risks

Mineral component	Benefit	Risk
Sodium	Replacing salt (after sweating or exercise)	Heart disease, high blood pressure
Magnesium	Aiding healthy cardiovascular and muscle functions	Diarrhea
Calcium	Aiding bone formation, especially in pregnant or lactating women, or in elderly women	Kidney stone formation
Iron	Extra iron intake in the case of iron deficiency	–
Chloride	Helping digestion	–
Hydrogencarbonate	Relieving the symptoms caused by excessive amounts of gastric acid	Not recommended for people with low amounts of gastric acid
Sulfate	Laxative effect	Excessive amounts are not recommended
Fluoride	Reduction of tooth decay	Excessive amounts are harmful for the teeth of children
Iodide	Stimulates thyroid functions	Not recommended for people with highly active thyroid glands

mineral water is quite different in the US and the EU. Some mineral waters have confirmed health effects (e.g. laxative, or aiding digestion). Spring water does certainly have natural origins but is not necessarily mineral water, its quality standards are the same as for tap water, and mineral content is not a central issue. Bottled drinking water is also available: buying it does not make any sense in places with safe tap water. Tap water is sometimes also bottled with added minerals. In fact, the chemical composition of such water may be indistinguishable from that of mineral waters, only the origin is different.

Carbonated drinks contain carbonic acid, which is prepared by introducing high-pressure carbon dioxide into water. The higher the pressure, the higher the carbonic acid content. If the pressure is relieved (e.g. by opening a sealed bottle) carbonic acid is slowly transformed back to carbon dioxide in the form of bubbles. The chemical process of the formation and decomposition of carbonic acid is usually given as a single, reversible reaction:

$$CO_2 + H_2O \leftrightarrows H_2CO_3$$

High-quality carbonated water contains at least 7–8 g/l carbon dioxide, carbonated mineral waters typically feature somewhat lower values. Spritzer, a popular drink in Austria and Hungary, is made by mixing white wine with carbonated water. The use of mineral water for this purpose is not ideal: its sodium content may cause some saltiness in the final drink and, more importantly, it is against well-established traditions.

In addition to the official product categories described in the previous paragraphs, there is an enormous variety of oxygenated, deuterium-depleted, alkaline ($\rightarrow$ **3.28**) and pi waters available in supermarkets or specialized stores. An entire book could (and probably should) be written about these. Very few of them have any scientifically proven benefits. Nevertheless, the claims in the advertisements are often sensational and sound too good to be true even to a non-specialist.

In most developed countries, water supply companies and the state authorities spend a lot of money and effort to ensure the safety of tap water. It does not make much sense for consumers to spend more on buying water that does not bring any health benefits. The choice between mineral water, tap water, purified water or carbonated water should be a rational one or based on the taste rather than on irrational and unfounded claims. Yet regardless of the facts, making a few scientific-looking, but actually absurd claims on the miraculous effects of certain special types of water still seems to be a highly profitable business. *'O sancta simplicitas!'* (O holy simplicity!)

3.30 The Illusive Qualities of Plant Preparations

Most people seem to have strongly held opinions about plant-derived medicines: some unconditionally believe in them, whereas others think they are completely useless. In addition to issues of efficacy ($\rightarrow$ **3.7, 3.27**), there are a lot of misconceptions about the safety of products as well. Although a twenty-first century medicinal product should be of high quality according to the current level of scientific knowledge, in fact, this is often not the case. The primary reason is not the fact that a consistent quality is hard to guarantee for herbal preparations. In fact, quality control is not difficult at all. The legal background of marketing herbal products, on the other hand, does not always serve the best interests of the consumers.

Most herbal products belong to two major categories, drug-type and food-type products. There are several sub-groups in the former category (medicines, herbal medicines, traditional herbal medicines, medicinal products), but the quality of the products is not essentially different. For all kinds of pharmaceutical substances, regulations require guaranteeing the efficacy and harmlessness. No matter if a product is made of dried plant parts, extracts or pure active compounds, it is essential that the concentration of the active component should be kept constant. The key to achieving this constancy is the stable composition of the plant material. To achieve uniform quality, herbs are generally grown today rather than collected from nature.

Plant material collected in the wild (Fig. 3.49), with few exceptions, cannot be accepted in pharmacy. One of the reasons is that non-professional collectors do not necessarily collect the desired plants exclusively: they may also collect other similar-looking or even toxic plants. Some plants have several subspecies, which are very difficult to distinguish, but may differ significantly in composition. The chemical composition of plants is also influenced by environmental factors (soil, moisture, light conditions, time of harvest *etc.*) and even by the site and time of the

Fig. 3.49 Chamomile harvesting in the wild—similar-looking plants may get mixed. (Copyright-free picture from http://www.tiroler-kraeuterhof-naturkosmetik.com/informationen/kraeuterhoefe/pflegerhof-martha-mulser-kraeuter/)

harvest. Therefore, the use of plants grown under standardized conditions is almost exclusive in pharmacy. Pharmacopoeias specify the detailed quality requirements for the plant parts used in therapy. Processing (crushing, extraction, *etc.*) of the raw material follows strict quality standards so that the high quality of the raw materials can be guaranteed.

In spite of the strict rules, the steady quality of plants themselves cannot be guaranteed. There are time-tested methods to correct for the differences in chemical composition caused by environmental factors. Once the active ingredient of the plant is known, keeping the amount of this compound unchanged guarantees the constant quality and efficacy of the extracts. If the concentration is higher than the average in a plant, a slightly lower amount is used for the product and *vice versa*. This process is called standardization (Fig. 3.50). If the active compound is not known but the plant has a proven efficacy, the raw material (i.e. plant part) is quantified instead of standardization. In this case, the amount of one or more characteristic (but not necessarily effective) components are measured and kept constant.

If the effect of a herb is caused by several components instead of a single compound, certain benefits may arise. An example is milk thistle (*Silybum marianum*), whose liver-protecting effect is not associated with a single compound, but several structurally related substances. The effective mixture is called silymarin. Using a single compound of the silymarin complex would not be more advantageous than the application of silymarin itself. Therefore, milk thistle products always contain the characteristic mixture. No synthetic drug is available to replace silymarin in therapy (prevention and mitigation of liver damage). Silymarin products are standardized based on the active substance content.

To improve blood circulation, Ginkgo (*Ginkgo biloba*) extract is used, which is made of the leaves of the plant. The extract contains known quantities of two groups of active components (ginkgolids and bilobalide). It is unknown which of these two is responsible for the physiological effects. Therefore, Ginkgo extract is a quantified product.

Herbal products should also be free of pollutants such as other plant parts, heavy metals or pesticide residues. The quality is guaranteed not only by regular tests during production, but also by the professional supervision of drug licensing authorities.

3.30 The Illusive Qualities of Plant Preparations

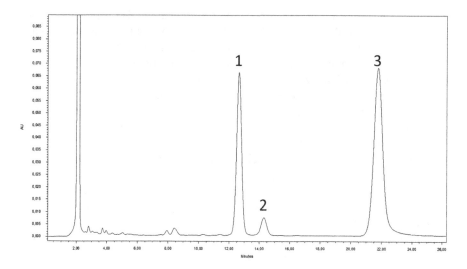

Fig. 3.50 HPLC chromatogram of a standardized soybean product. The amount of active substance is known and can be measured (isoflavones daidzein (1), glicitein (2) and genistein (3)). (Authors' own work)

Herbal products marketed as food are another cup of tea. Their quality is not guaranteed by the same strict regulations. Although herbs have long been marketed as food (*e.g.* tea, spice, or components of different products), the onset of the age of dietary supplements brought a sea of change. Dietary supplements are regulated in the EU by the Food Supplements Directive of 2002. Although a large group of dietary supplements contains minerals and vitamins only, herbal extracts are also present in many of them. There are thousands of products now, which pose new quality problems. The quality of medicinal plants used in food has never been regulated by strict pharmacopoeia standards. The requirements regarding strength were much looser, but safety was still a primary concern. Dietary supplements may even contain plants that have not been used in foods. Of course, the use of toxic and harmful substances is still forbidden.

Rules for marketing dietary supplements are much more liberal than for drugs or even foods. The composition and use of these products is not regulated by adequate laws. The manufacturer is supposed to be responsible for safety and effectiveness, but tests are never done in official laboratories to certify the claims of the producer. Notification or registration is the only legal requirement in several countries, which is by no means the same as obtaining an approval or authorization. The responsibility of the manufacturer is more moral than legal, which cannot guarantee the quality of all products. Most of the marketed supplements can be consumed safely, but even a few percent of low-quality products can add up to an unacceptably high risk to consumers.

The insufficient regulatory background and the bewilderingly wide range of products create a temptation for abuse. Dietary supplements are often marketed through foreign websites or agents—in this case, they are in fact illicit drugs. Even legally marketed products have problems. In the US, where the legal environment of dietary supplements is similar to that in the EU, a number of scientific papers reported that the composition of some of these products does not match the information found on the packaging. In the simplest such case, some expensive ingredients are omitted from the formulation, which causes lack of efficiency. More serious consequences may result if a component is overdosed, or if the product contains impurities not specified on the label, the source of which may be lack of care in production, or even deliberate action.

A very dangerous way of counterfeiting is complementing the supposedly plant-based product with a synthetic drug to enhance the effect. This practice is not only illegal, it is also highly unsafe. Laws forbid the direct medical uses of dietary supplements. As a result of counterfeiting, drugs are supplied to the consumers without their knowledge. Some of the compounds used in this kind of counterfeiting have strong effects, their unwitting use may even cause poisoning.

The first products that actually contained synthetic contaminants were advertised as traditional Chinese herbal preparations. In a report published in 1997, 2609 Chinese herbal preparations, collected from Taiwanese clinics, were tested. 23.7% of these preparations contained synthetic substances consistent with the intended effect of the product. However, counterfeiting is not an Asian problem, as the Chinese and Indian dietary supplements are also present in the European market, or anywhere in the global market. This was confirmed by a recently published article from French scientists, who tested 20 herbal weight loss pills and found that 16 contained synthetic drugs. These cases are known among experts, but only receive more publicity if the counterfeiting claims human lives, or leads to positive doping cases. There are many examples of both.

Scientific articles roughly identified the product groups most affected by counterfeiting and the materials most widely used in the process. The most common and most characteristic examples are stimulants and muscle growth pills that contain anabolic steroids, ephedrine or caffeine; weight loss pills that contain laxatives, metabolism-enhancing thyroid extract, fenfluramine or sibutramine (the latter two are now banned in the European Union); glucose-lowering tablets that contain drugs recommended as a treatment of diabetes. All the products listed above were marketed as herbal pills and none of them were labeled to contain a synthetic drug.

In recent years, not a few scientific articles have been published about the counterfeiting of herbal preparations and dietary supplements intended for the treatment of male sexual dysfunction (erectile dysfunction). The earliest reports appeared around the turn of the millennium, soon after the introduction of the first phosphodiesterase enzyme inhibitor drug called sildenafil (Viagra®) in 1998 (Fig. 3.51). Subsequently, an increasing number of articles detected the illegal use of sildenafil or its analogues in preparations of "herbal origin".

A series of tests carried out in Hungary analyzed 10 erection enhancement products, nine legal and one illegal. The analysis methods (thin-layer **chromatography**, gas chromatography and high-performance liquid chromatography) showed that six of the tested preparations contained synthetic phosphodiesterase inhibitors, one

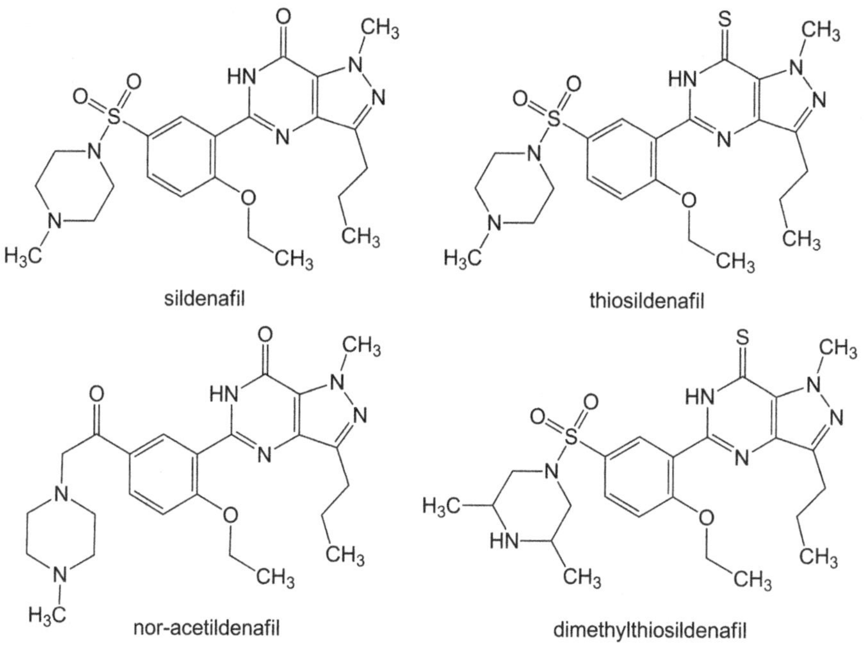

3.30 The Illusive Qualities of Plant Preparations

Fig. 3.51 The structural formula of sildenafil and three of its derivatives found in illegal "herbal" erection-enhancing drugs. (Authors' own work)

contained a large amount of caffeine, and another one contained yohimbine (an obsolete aphrodisiac). In many cases, the supposedly "herbal" products did not contain any plant extracts. Counterfeiting with phosphodiesterase inhibitors is a huge risk to consumers because these drugs may increase the risk of cardiovascular events. Medical tests must be carried out before taking erection enhancement products—it is no accident that legal Viagra is only available by prescription from a doctor. In the case of counterfeit products, this does not happen. Another major concern is that some of the counterfeit erection-enhancing products contain structurally modified forms of sildenafil that were never tested in humans.

There is no doubt about the safety and quality of herbal products marketed as medicines. The quality of these is guaranteed by the control and testing procedures of drug licensing authorities as well as by international standards, which even improved over the past decades. Dietary supplements are quite different: the manufacturer is solely responsible for the quality. Many of the producers are careful and even make efforts to follow the strict pharmaceutical standards. However, the recently reported anomalies demonstrate that current legislation does not provide the necessary protection for consumers. It is only a fraction of the products that have quality problems, and they can be easily filtered out with some caution. The most vulnerable product groups were described in the previous paragraphs. There is a general and simple advice: the consumer should be suspicious if the product promises an unrealistic, miraculous effect. Distributors of dubious legal status typically want to remain unknown and many of them do not even register their products. Lists of registered products are available from official websites, unlisted preparations should be avoided. However, the fact of being registered still does not guarantee quality. Chemical quality control would be a legitimate requirement in the twenty-first century, but laws are not strict enough in this sense. The reputation of bad quality products will be ruined sooner or later, but this might also lead to a general loss of trust in the herbs and dietary supplements. Even more importantly, the health of consumers is put to an unreasonable and avoidable risk.

3.31 The Antioxidant Story

The fear of serious diseases or the desire for healing easily attracts both healthy and sick people toward fast and easy solutions. The natural human instincts to maintain health or find healing have already made many people very rich. Antioxidants form a large group of seemingly universal, miraculous materials, at least according to advertisements. Their fortune rose sharply in the 1990s, but the story began much earlier, in 1957, when American researcher Denham Harman (1916-) published his theory about free radicals and their assumed effects on aging. His idea was that aging is caused by free radicals, which damage cells and tissues and are generated as a result of oxidative stress. The formation of free radicals is inhibited by antioxidants, so it seems logical to conclude that they slow down the aging process.

From this point, it is not much of a leap to conclude that any illness in connection with oxidative damage can be prevented by antioxidants. Moreover, some might even argue that these disorders are not only preventable, but also curable. This is an exaggeration that lacks scientific background, but sells well. The success of antioxidant products also requires the transfer of exciting new information to consumers. Newspapers broadcast mainly simple and supposedly striking news: they are willing to pick up somewhat sensationalist but scientific-looking reports from time to time.

Returning now to antioxidants: one of the first, very well-known representatives of the group was *trans*-resveratrol ($\rightarrow$ **2.10**), which is found in red wine. The compound was tested in several preclinical studies and its antioxidant effect was assumed to explain the benefits of moderate wine consumption. In the first decade of the twenty-first century, the compound made headlines because, as a journalist wrote, it prolongs life span. Unfortunately, the newspapers neglected to mention that this effect was proved only in yeast cultures. In addition, **flavonoids**, carotenoids and other antioxidants were in the news, too, without any human-related scientific background.

Today's stars are pomegranate with exceptionally high antioxidant activity ("Packed with powerful antioxidants and vitamins, this ruby-red fruit has been shown to be a cure-all for just about any ailment."), astaxanthin produced by some algae ("unique anti-oxidant and king of the carotenoids"), green tea ("rejuvenate

3.31 The Antioxidant Story

with a cup of Green Tea Super Antioxidant!" (→ **3.17**). Quite surprisingly, the new cure-all is chocolate ("chocolate and cocoa powder may be the next 'super foods' thanks to their high antioxidant content."). After scratching the attractive surface, the reality usually turns out to be murkier: these popular products are marketed by a powerful lobby of manufacturers, who invest a lot of money into research and development, marketing and media campaigns. Unfortunately, most of the claimed scientific evidence is not suitable to draw any conclusions for human effects. The tests are usually carried out under laboratory conditions, in flasks or on test animals. In fact, a thorough search in the scientific literature for one of these popular materials or plants will often reveal a number of publications that question the supposed beneficial effect.

What are these antioxidants that attract so many people and open so many wallets? Antioxidants form a collection of thousands of chemically diverse substances with one common feature: they are relatively easy to oxidize. Oxidation is an essential step in human metabolic processes and usually produces free radicals, which contain an unpaired electron and are relatively reactive as a consequence. If the quantity of these free radicals exceeds the normal level, they may damage some vital molecules such as proteins, fats, or nucleic acids. Although the body has extensive repair mechanisms against oxidative damage, neutralization (scavenging of free radicals) is not always possible in the long term.

Free radicals participate in a wide variety of chemical reactions. The most significant radicals in the human body are the superoxide radical ($O_2^{-\bullet}$) and the hydroxyl radical ($OH^\bullet$). These radicals cause the oxidative damage of biomolecules. Substances eliminating free radicals can be used to prevent these harmful processes. Free radicals can be decomposed enzymatically: superoxide dismutase, glutathione peroxidase and catalase enzymes are the major players in this game. Superoxide dismutase catalyzes the following reaction:

$$O_2^{-\bullet} + O_2^{-\bullet} + 2H^+ \rightarrow H_2O_2 + O_2$$

Glutathione peroxidase oxidizes glutathione by converting hydrogen peroxide to water:

$$H_2O_2 + \text{reduced glutathione} \rightarrow 2H_2O + \text{oxidized glutathione}$$

A cofactor (*i.e.* the key of enzyme activity) of the glutathione peroxidase enzyme is the element selenium. The enzyme contains cysteine residues in which sulfur is replaced by selenium. A deficiency of this element causes insufficient enzyme activity, so selenium is often considered as an antioxidant, although it has no such direct effect.

The reaction catalyzed by catalase is as follows:

$$2H_2O_2 \rightarrow 2H_2O + O_2$$

Table 3.7 Most significant antioxidants in the human body

Endogenous (produced by the body)	Exogenous (from the diet)
Superoxide dismutase	Vitamin C
Glutathione peroxidase	Vitamin E
Catalase	Polyphenols (flavonoids)
Ubiquinone (coenzyme Q_{10})	Carotenoids

There are no enzymes to scavenge hydroxyl free radicals, so they are decomposed in the body non-enzymatically. The same happens where the enzymatic reactions do not have enough capacity to decompose all the radicals formed in metabolic processes. In this case, the importance of antioxidants is crucial. They are easily oxidized, thus they can protect molecules and cell parts from oxidative damage. Glutathione, vitamins C and E play an especially important role in the human body. Vitamin C can reduce oxidized glutathione, thus enhancing the effect of glutathione, a crucial material in the oxidation-reduction balance of the body. Vitamin E protects cell membranes. The non-enzymatic decomposition of free radicals may form further free radicals. The main advantage of vitamins C and E is that the radicals formed from them are less reactive.

Antioxidants are needed to maintain balance in the body. They are present in the body if not consumed on purpose because some of them are produced endogenously. However, there are antioxidants which can only be introduced with food. If less of these compounds are consumed and, at the same time, the body is exposed to increased oxidative stress (*e.g.* smoking, alcohol consumption, X-rays, air pollution), in principle, oxidative damage becomes more intense. To prevent this, some suggest antioxidant compensation using concentrates (tablets containing purified materials). However, the situation is not simple at all: the antioxidant system of the human body is extremely complex. Increasing the ingested amount of a single material does not always reach the desired effect (Table 3.7).

The antioxidant theory is supported by results which show that some typical diseases of the elderly (Alzheimer's disease, cancer, diabetes, cardiovascular disorders) are often accompanied by the appearance of some markers of oxidative damage. However, the causal relationship is still unclear, so there is no way to tell cause from effect. The next paragraphs will summarize the main proven and assumed benefits of exogenous antioxidant intake.

Ascorbic acid is a vitamin, it is essential for the human body ($\rightarrow$ **3.13**). Vitamin C deficiency is connected to cardiovascular diseases, but there is no evidence that ascorbic acid doses above the physiologically required levels reduce the risk of any disease associated with oxidative damage.

Data about the preventive effect of vitamin E on cardiovascular diseases are contradictory. **Observational studies** showed that the consumption of vitamin E reduces the risk, but large **interventional studies** involving thousands of people could not verify this. The latter test types are more reliable in general. Similarly ambiguous is the relationship between vitamin E and cancer. There are numerous positive interventional studies, but an especially significant test showed that high

3.31 The Antioxidant Story

Fig. 3.52 Antioxidants in their most pleasant form. (Copyright-free picture from http://www.flickr.com/photos/flydime/384397661/)

doses of vitamin E increased cancer recurrence risk among patients who received radiation therapy. The reason is probably that the high antioxidant doses protect not only healthy, but also cancer cells. This is not a real threat if vitamin E is consumed in the normal diet (oilseeds, vegetable oils).

Flavonoids are a huge group of compounds containing more than 6000 individual chemicals. Their structures are diverse, so their biological effects are also very different. Flavonoids are polyphenolic compounds: they contain a large number of hydroxyl groups linked to aromatic rings. Their typical colors (ranging from yellow to red) are related to delocalized electron systems (Fig. 3.52), and their antioxidant effects can be explained by the presence of hydroxyl groups. By their chelate forming effect, they are able to bind metal ions that promote the formation of oxygen containing free radicals. It seems clear that people who consume a lot of flavonoid-rich food (such as fruit and vegetables) are less likely to suffer from cancer or cardiovascular diseases. It is quite possible that the positive effect is (also) caused by other substances present in vegetables and fruits: there is no evidence that the antioxidant activity of flavonoids is actually the primary positive effect. An example of a very thoroughly studied plant might help to illustrate this statement: regular consumers of green tea face a smaller risk of developing certain types of cancers. This positive effect is primarily attributed to a polyphenolic compound called epigallocatechin-3-*O*-gallate. Yet the primary mechanism of protection relies on other physiological

Table 3.8 Carotenoid sources in the diet

Carotene	Lutein/zeaxanthin	β-Cryptoxanthin	Lycopene
Carrot	Broccoli	Green pepper	Tomato
Savoy cabbage	Spinach	Squash	Melon
Squash	Brussels sprout	Papaya	Grapefruit (pink)

effects (increasing programmed cell death, inhibition of angiogenesis, *etc.*) and has nothing to do with antioxidants. The exact role of flavonoids is very difficult to assess as dozens of similar compounds are consumed daily, and very few studies have been carried out with a single pure flavonoid.

In addition to flavonoids, carotenoids also belong to the group of vegetable pigments. These orange-red substances are very lipophilic (soluble in fats). Although hundreds of representatives are known, only six of them (α-carotene and β-carotene, β-cryptoxanthin, lutein, lycopene, zeaxanthin) occur in significant quantities in the human diet (Table 3.8). Three of them (the α- and β-carotene and cryptoxanthin) are provitamins because they are converted to vitamin A in the body. Carotenoids show significant antioxidant activity in laboratory tests, but also behave as *prooxidants* in living organisms, which means that they can promote the oxidation of other substances. The significance of this effect is not detectable in observational studies about cardiovascular diseases because—as one would expect from an antioxidant—carotene consumption reduces the risk of disease overall. The opposite conclusions were drawn in several interventional studies of cancer prevention. The Alpha-Tocopherol Beta-Carotene (ATBC) study even had to be suspended because 20 mg/day doses of β-carotene significantly increased the incidence of lung cancer among male smokers. A similar conclusion was drawn as a result of the Carotene and Retinol Efficacy Trial (CARET) study. These two studies were enough to change the public opinion about carotenoids. Discussing the dangers rather than the benefits of carotenoids has become popular in the press.

However, the overall picture is by no means as dark. To demonstrate this clearly, the evidence should be interpreted correctly. It is a fact that a diet with high levels of carotene is advantageous (Table 3.8). But a very high-dose consumption of purified carotenoids (especially β-carotene and vitamin A) is associated with higher risk for humans exposed to increased oxidative stress. General statements about the effects of carotenoids are impossible. For example, several studies showed that lycopene (the red pigment in tomatoes) reduces the risk of prostate cancer, and observational studies of tomato consumers revealed no side effects.

A balanced diet rich in fruits and vegetables alone is sufficient to meet the antioxidant needs of the body. Other products containing antioxidants can also be used safely, but these are not the preferred forms. Antioxidants are present in fruits and vegetables as a mixture, which is better absorbed and utilized. Vitamins C and E—if the dosage limits are observed—are safe in concentrated forms. However, this is not necessarily true for newly developed antioxidant products marketed today.

It was already mentioned that antioxidants may behave as prooxidants under certain conditions. This is easy to explain: free radicals may decompose to produce

3.31 The Antioxidant Story

further radicals, which may be even more harmful than the originals. Before purchasing a new and widely advertised, super-powerful antioxidant, some thought should be given to the amount of risk involved. If even a well-known and apparently harmless material such as β-carotene can be dangerous under certain conditions, the same danger could certainly be present in a lesser-studied compound. The new compound, typically introduced without thorough scientific studies, may form new radicals in the body that are more dangerous than the old ones.

Allegations about the strength of antioxidants should be greeted with a healthy dose of skepticism. A number of completely different methods have been developed to measure the antioxidant capacity or free radical scavenging ability. Even if the same method is used for different compounds, this fact does not necessarily make the comparison of results informative as the body reacts quite differently to antioxidants than a mixture in a flask. A compound may seem a powerful antioxidant in a laboratory reaction, but much less effective in the human body.

There are several myths and urban legends about the absorption of antioxidants. As a rule, the absorption is better when consumed as part of the normal diet as the presence of other compounds in the consumed food promotes absorption. Of course, there are exceptions. The absorption of carotenoids is primarily affected by the lipophilicity (fat-liking character) of the medium. Capsules often contain oily solutions of carotenoids, which are absorbed much faster than from fresh tomato. However, if the tomatoes are baked in oil, the rates of utilization are similar. In the case of vitamin C, there is no significant absorption difference between capsules and natural sources. Vitamin C is also rapidly excreted from the body ($\rightarrow$ 3.13) independently of the source. In the case of vitamin E, there is some truth in what some manufacturers say about the better absorption of "natural" vitamins. The naturally occurring form of vitamin E is mainly α-tocopherol, but the synthetically obtained drug also contains biologically inactive and non-absorbable tocopherol derivatives.

Some studies show that "modern diets" may lack enough exogenous antioxidants. The easiest way to prevent the deficiency is to have a diverse diet that contains lots of fruits and vegetables. If this is not possible, tablets of vitamin C and E are acceptable as alternatives. Perhaps few people think of it, but a diet which is "too healthy" may also cause problems: the exclusion of oils and fats (for losing weight) leads to vitamin E deficiency. The consumption of high doses of purified antioxidants may have unexpected consequences, as has been demonstrated in the case of carotenoids and vitamin E. Chances of adverse or unseen consequences are even worse for a little-known (or often unknown) compound. In contrast to recent public opinion, the consumption of antioxidants does not necessarily protect the body against every disease associated with oxidative damage. It is also clear that these compounds do not cure any serious diseases. A well-balanced diet rich in antioxidants is healthy, but the numerous advertisements about antioxidants only benefit the financial health of the companies selling these antioxidant products.

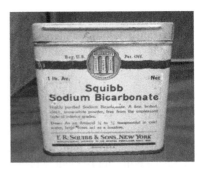

Fig. 3.53 Baking soda was used to cure not only stomach ache, but a variety of other conditions as well, including flu. (Authors' own work)

3.32 Stomach Ache: Baking Soda

Stomach ache is one of the most common digestion problems. Strong pain in the stomach can sometimes be a symptom of a serious disease (such as peptic ulcer), but it is more often caused by an unfavorable diet. Heavy meals or drinks (carbonated soft drinks, alcoholic drinks, sour fruits, caffeine, spicy or fatty food) are usually acidic or increase gastric acid formation. Although a number of modern antacids are available now, many people still stick to good old baking soda (Fig. 3.53). Older people usually rely on what they learned from their parents whereas the younger generation often prefers information from trusted web-based sources. Both of these advise the use of baking soda, the latter often mention stabilizing the **pH** as well. Is this actually true or just commercial pep talk?

Baking soda is a white crystalline substance. Chemists call it sodium hydrogen carbonate. It dissolves in water and gives a slightly basic solution. From a chemical point of view, it is suitable to neutralize gastric acid, which is mostly hydrochloric acid. The reaction is as follows:

$$NaHCO_3 + HCl \rightarrow NaCl + CO_2 + H_2O$$

Carbon dioxide formation accompanies reactions when baking soda neutralizes acids. This is also important for the food industry as gas formation makes dough lighter ($\rightarrow$ **2.18**). Gas formation is exactly why baking soda is less than ideal for medicinal applications. Although the neutralization reaction occurs very rapidly in water and the stomach ache is alleviated, another symptom arises because of the gas formation. Prolonged use of baking soda in large doses may have more serious consequences. Hydrogen carbonate ions may enter the blood stream and shift the pH into basic direction. This is not just a hypothetical problem, a condition called metabolic alkalosis is often diagnosed as a very real medical problem. The other product of the neutralization reaction, sodium chloride, does not pose any direct danger. In the long term, though, there are concerns even about the innocent-looking common salt. The World Health Organization (WHO) recommends limiting sodium chloride intake to no more than 5 g a day, which is about a teaspoonful. A larger amount will increase the risk of high blood pressure. Routine stoichiometric calculations show that 5 g of sodium chloride forms in the reaction of 8 g of baking soda. Consuming two teaspoonful of baking soda produces more common salt than the recommended daily maximum amount, so overconsumption might lead to high blood pressure.

Regardless of the side effects, baking soda is preferred by many because its effect is almost instantaneous, whereas other antacids need about half an hour. Everything has a price, however: the rapid and efficacious effect is also over in a short time. Another dose of baking soda may be needed within 30 min. The human body is dynamic, introducing base into the stomach does not shift the pH easily, nor does it change in the long term. Digestive enzymes in the stomach work best at low pH, and the body often tries to restore the optimal conditions by forming more acid if it is consumed by adding a base.

Because of these disadvantages, baking soda is disappearing from medicines. It is still used in effervescent tablets combined with some solid organic acid (citric acid or tartaric acid), which form carbon dioxide bubbles with baking soda.

Using calcium carbonate as an antacid has the same disadvantages as baking soda, except for the formation of sodium chloride. An additional danger for calcium carbonate, especially in combination with sodium hydrogen carbonate, is called the milk-alkali syndrome, which is caused by high levels of calcium in the blood and may result in kidney damage.

Despite what has been said, sodium hydrogen carbonate and calcium carbonate are by no means dangerous substances that should be avoided. The only conclusion here is that stomach ache can be treated with modern medicines better. Gastrointestinal cramps should be relieved with spasmolytics, whereas in case of excessive acidity, drugs that reduce acid production of the stomach (these belong to the groups of proton-pumps inhibitors and histamine2 receptor antagonists) are more effective than acid neutralizing compounds. The two traditional substances should only be used occasionally because of the risks of long term use. The human body is a complex system, in which even a simple chemical reaction might have unintended consequences.

3.33 Does Hot Pepper Cause Ulcer?

Patients suffering from stomach ulcer know well that the symptoms get worse if they eat spicy food. Hot pepper has quite a distinctive taste and patients tend to avoid it. However, recent scientific research does not support this apparently common sense opinion about the danger of hot pepper.

Most spices contain high levels of essential oils, which are known to increase the flow of blood to the stomach and enhance the production of gastric acid. The acidity does not help the healing process of an existing ulcer. Hot pepper also has these physiological effects yet there is increasing evidence that it does not make ulcers worse. Regular consumption may even help prevent ulcers. To understand the reasons, it is worth having a closer look at the plant and its ingredients.

Chili pepper was originally native to tropical America, but it has been grown world-wide since Christopher Columbus brought it to the Old World. Its path to Central and Eastern Europe went through Hungary, the word 'paprika' in English is one of very few having Hungarian origins (another one is 'coach' meaning the vehicle). Hungarian linguists, on the other hand, know quite well that even in Hungarian, the word was borrowed from Slavic *piperka*, which means pepper.

In culinary and medicinal uses, three species dominate: *Capsicum annuum*, *Capsicum chinense,* and *Capsicum frutescens*. The peppers (technically, these are berries) of all three plants taste hot, but *Capsicum annuum* has milder varieties. The other two are also known as chili pepper and all of the varieties (e.g. Habañero, Serrano, Tabasco, and Cayenne) are hot. Pepperoncini is chili pepper that is not fully ripe.

One of the main ingredients in pepper is L-ascorbic acid. Hungarian Nobel laureate Albert Szent-Györgyi (1893–1986) was the first to isolate this compound from pepper, this achievement was one of the reasons why he received the Nobel prize. The hot taste of pepper is caused by capsaicinoid compounds, which are chemically related to vanillyl amine. Capsaicin is the most significant substance in this family. Pepper contains capsaicinoids at levels of 0.1–1 %. These are colorless, crystalline compounds that do not dissolve well in water but are soluble in oils. This solubility issue is also the reason why their hot taste cannot be efficiently removed by rinsing with water. The pungency of pepper is measured on the Scoville scale, which is an empirical measure of capsaicinoid content (Fig. 3.54). The scale basically gives the highest dilution at which the hot taste is still detectable. Some sweet red peppers score 0 on this scale (no hot taste felt), whereas the hottest common chili peppers are around 200,000. Pure capsaicin scores 15,000,000 whereas the current record holder pepper itself has a value of 1,300,000. There are better scientific methods today to characterize pungency, but the Scoville scale has a public appeal that is hard to beat.

When ingested, capsaicin enhances saliva and gastric acid formation and also facilitates stomach movements. In principle, these are not favorable for someone suffering from ulcer, but there are additional effects as well. Capsaicin was the first compound that was discovered to bind to a family of receptors called vanilloid receptors in the body. The inhibitors of these receptors are intensely studied today as they have extremely high painkilling potential. Hungary is still a major player in the scientific research of pepper, including capsaicin studies.

Vanilloid receptors are named so because they are specific to vanilloyl derivatives. These receptors are actually selective channels through which calcium and magnesium ions can move through the membranes of neurons. They may also open as a result of pH change. When the channel is open for a long time, neurons are desensitized. The human body also produces substances that bind to this receptor. The most notable of these is anandamide, but its exact physiological role is unknown.

The protecting effect of hot pepper and capsaicin on stomach mucus was proven in animal tests, and it is also known that vanilloid receptors are activated in the process. The protecting effect is pronounced at low doses, much higher amounts are harmful. At low levels, capsaicin causes the formation of a substance beneficial for

3.33 Does Hot Pepper Cause Ulcer?

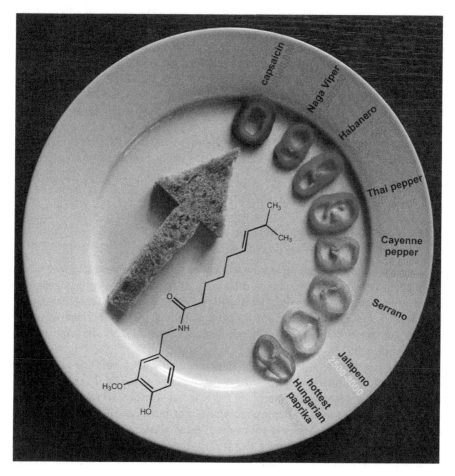

Fig. 3.54 The pungency of peppers on the Scoville scale and the structure of capsaicin, the "hot" component of pepper. (Authors' own work)

the mucus of the stomach, which improves blood flow, but desensitized neurons destroy the effect. Capsaicin also inhibits the growth of *Helicobacter pylori* bacteria, which often have a role in ulcer formation.

To make matters more confusing, human tests show that capsaicin increases gastric acid formation and has an irritating (even damaging) effect on the stomach mucus. However, the amounts used in these studies were unreasonably high, which would be too much even for pepper-loving Koreans (they are the record holders in average consumption, which is 9 g/day/person). Average capsaicin concentration in hot Korean food is 0.3 **mmol/l**, the same figure for Indian food is 0.1 mmol/l. The hot taste is felt from 0.003 mmol/l.

Epidemiological test in the Far East showed that stomach ulcer is less common among those who eat a lot of pepper than less hot-minded people. In another study, hot pepper was shown to decrease mucus damage caused by acetylsalicylic acid, the active ingredient of Aspirin (incidentally, Aspirin and many other painkillers may cause stomach ulcer, → **3.25**).

The vanilloid receptor is also known to be strongly affected by a capsaicin analog compound called resiniferatoxin, which was first isolated from a cactus-like plant (*Euphorbia resinifera*). Lafutidine, an antacid, is more efficient against ulcers than other similar medicines, which may also be the consequence of an interaction with the vanilloid receptor.

Although there is no scientific evidence to prove that hot pepper cures ulcers, the lack of effect has not been proved, either. It seems likely that moderate consumption lowers the risk of ulcer formation.

3.34 Cocaine—Drug, Narcotic, or What?

Cocaine has a deserved bad reputation, which mostly comes from movies or media news about Mexican or Colombian drug traffickers. It is more or less generally believed that cocaine is an illegal narcotic. In a strict medical sense, this is not true. What is true is that the use of cocaine is prohibited by most national laws and international treaties together with classic narcotics. The rules are the same, but the effect is different.

In a medical sense, a narcotic is a substance that causes narcosis (*nomen est omen*). Narcosis primarily means a *decrease in activity*, maybe decrease of pain, hallucinations, a decrease of interest in the external world, a sort of introversion. Prohibited narcotics include heroin, morphine, marijuana, and LSD (lysergic acid diethylamide). Cocaine does not have any of these classic narcotic effects. Medically, it is a *stimulant*, a substance that increases activity. This has dangers of its own, so prohibiting the use of cocaine without doubt serves the interests of society in general.

Cocaine is found in a shrub called coca (Latin name: *Erythroxylon coca*), so it is a natural substance (Fig. 3.55). The plant is cultivated in the Northeastern part of South America, primarily in Colombia, Peru and Bolivia, and on the Indonesian island of Java. Its stimulating effect has been well known for a long time. Among the Incas, the nobility was privileged enough to enjoy its effects. When the Spanish conquistadors occupied South America, chewing coca leaves was a very common habit there. This could not be stopped even by brute force, so it had to finally be included in daily schedules later on. Coca-Cola (the readers can probably guess the origin of this name!) contained significant amounts of cocaine until 1903. Incidentally, American authorities wanted to force its manufacturer to get rid of the caffeine content as well, but a jury stopped this proposed regulation. Extracts of coca leaves are still used in the production of the soft drink, but cocaine itself is carefully removed.

Why do many people believe that cocaine is a narcotic? In purely medical terms, this view is incorrect, but it is not untrue when legal terms are considered. The same

3.34 Cocaine—Drug, Narcotic, or What?

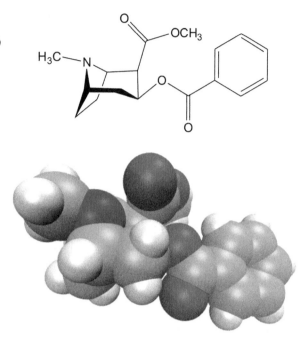

Fig. 3.55 The chemical structure and 3D model of cocaine. (Authors' own work)

is true for a few other prohibited stimulants such as amphetamine or metamphetamine. The use and sale of these substances is prohibited because they pose a danger to their users, and through the usual behavior of users, they also pose a serious danger to the entire society. This is especially true for cocaine. When considering legal prohibitions, it would not make good common sense to create several lists of banned substances just because of their different ways of acting. So the only such list is associated with narcotics, although some of the substances on it are actually not narcotics in a medical sense. This fact does not make the law less reasonable: many experts rank cocaine as the second most dangerous drug for society. Heroin is number one (Fig. 3.56).

Cocaine is often used to fight fatigue or to excite someone's mood or senses. A serious overdose may cause hallucinations, but this is not typical. It does not cause physiological dependence, in deviation from most narcotics. This means that cocaine addicts can be cured without using medicines, by simply denying them the substance. But a cocaine addict usually becomes very deeply depressed upon withdrawal, and they often seek out the drug again, even at the expense of committing crimes. This is called psychological dependence. In the case of cocaine, this is very strong. So cocaine is usually classified as highly addictive, although this is not physiological. Tolerance is also a well know phenomenon for narcotics. This means that a regular user requires gradually increasing amounts to achieve the same effect. Tolerance is unknown for cocaine.

Cocaine is also listed on the prohibited substances of the World Anti-Doping Agency. It is not in the category of narcotics, it is among the stimulants. Interestingly

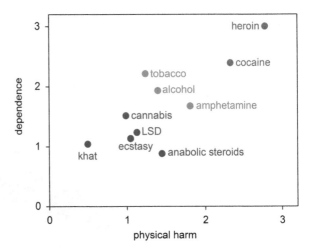

Fig. 3.56 The health damage and dependence potential of some drugs. (Authors' own work)

enough, cannabinoids, which are typical narcotics, are not listed in the category of narcotics, either. They have a category on their own. The prohibited stimulants in sports must not be used during competitions. Technically, out-of competition use is not banned, but in many countries, legislation prohibits them all the time for everyone. Cocaine use in sports is typical for athletes who are over their prime, and feel that they are not as sharp as they used to be. Cocaine helps to re-create the feeling of being the best. Swiss tennis player Martina Hingis was caught using cocaine in 2007 during the Wimbledon tournament. She denied the doping charges, but the evidence was strong: she was disqualified for 2 years. This was only a matter of principle at that time, as she already announced her retirement because of unrelated health reasons.

Chapter 4
Catastrophes, Poisons, Chemicals

4.1 Is the Use of Cyanogen Bromide Forbidden in Hospitals?

In February 2008, a box labeled cyanogen bromide was found in Markhot Ferenc Hospital in Eger, a town in northern Hungary (Fig. 4.1). The box, maybe not surprisingly, in fact contained cyanogen bromide. The story was complicated by the fact that privatization of the hospital was in progress at the time and inflamed emotions. As it is becoming routine for issues involving "chemicals" (→ **1.1**), desperate soul-searching began. One of the most influential evening news on TV said that the substance "…was last used in large quantities as a weapon in World War I and it is very difficult to find today" (February 15, 2008). For a self-respecting chemist, misinformation like this is enough to reach for the keyboard of a computer.

Nomen est omen, this triatomic molecule (Fig. 4.2) seems destined to induce fears because of the word 'cyan' in its name. It is or was also mentioned under names such as bromine cyanide, bromocyan, bromocyanide, bromocyanogen, cyanobromane, and cyanobromide, none of which sound less ominous than cyanogen bromide.

The word cyan originates from Ancient Greek κυανος (kýanos, meaning dark blue) and refers to the color of the dye called Prussian blue. It seems that the spelling *kyan-*, which would be in better agreement with Greek etymology, is only used in chemical names in the Czech and Slovak languages in Europe. Another (English) remnant of this etymology appears in the name of blue-colored silicate mineral *kyanite*.

Cyanogen bromide is a volatile solid, which dissolves readily in water, diethyl ether, and alcohol. It is a highly toxic substance through inhalation, ingestion and skin contact. It is usually considered about 2–3 times less toxic than hydrogen cyanide, but the exact lethal dose is unknown. In the developed countries, the permissible workplace exposure limits in the air are 5 mg/m^3 (as cyanide, time-averaged concentration in an 8-h shift), and 20 mg/m^3 (as cyanide, peak concentration).

Cyanogen bromide is prepared by reacting metal cyanides with bromine in water or carbon tetrachloride. The properties of cyanogen bromide resemble those of interhalogens (a compound formed by several halogens). When it reacts with

Fig. 4.1 Markhot Ferenc Hospital in Eger. (Authors' own work)

Fig. 4.2 Space filling model of cyanogen bromide. (Authors' own work)

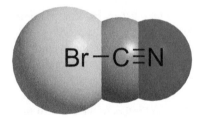

water, both hydrogen bromide and hydrogen cyanide can be formed. In basic medium, it decomposes to give a mixture cyanate, bromide, cyanide and hypobromite ions. (The dominant process is $BrCN + 2\ NaOH = NaOCN + NaBr + H_2O$ with a minor reaction of $BrCN + 2\ NaOH = NaCN + NaOBr + H_2O$). Analogous cyanogen iodide forms cyanide and hypoiodite ions under the same conditions ($ICN + 2\ NaOH = NaOI + NaCN + H_2O$). These reactions are employed during decontamination: a dilute solution of cyanogen bromide is reacted with a mixture of sodium hydroxide and bleach (sodium hypochlorite) in a molar ratio of 1:1:2. The cyanide ion formed in the process is further oxidized by hypochlorite ion into much less dangerous cyanate ion (OCN^-), and the final decomposition products are ammonia and carbon dioxide. The reaction can be violent if not controlled sufficiently, so care must be taken during decontamination.

Cyanogen bromide reacts with hydroxylamine in alcohol or ether solvent explosively, the products are hydrogen cyanide and hydrogen bromide. With organic substrates, the electropositive cyanide part is the usual site of reaction, such processes are used in the preparation of highly valuable intermediates (brominated derivatives, cyanamides, dicyanamides, cyanates, thiocyanates, secondary amines, guanidines, hydroxyguanidines, nitriles, carboxylic anhydrides, ureas, thio- and selenoureas). Purchasing and working with cyanogen bromide is not usually considered a particularly onerous task for trained experts but a number of routine precautions must be observed. As assumed by chemists from the moment the story broke, the box of cyanogen bromide found in the hospital was there for analytical purposes.

Cyanogen chloride, which is analogous to cyanogen bromide, was in fact used as a chemical weapon in World War I. Cyanogen bromide was deemed inconvenient to use as a weapon as it is a solid at room temperature and its melting and boiling points are unusually close (52 and 62 °C), which causes sublimation even in freezer storage. Both cyanogen bromide and cyanogen chloride are used in fumigants (gases used in pest control) as marker substances, their intense smell indicates the presence of other, more toxic, but odorless chemicals. Cyanogen chloride is quite poisonous, in a high enough dose, it causes death within 6–8 min. It is listed in schedule 3A of the Chemical Weapons Convention. This means that it may be produced in large commercial quantities for purposes not prohibited under the Convention, but all production must be reported to the Organization for the Prohibition of Chemical Weapons. A derivative of cyanogen chloride called trichloroisocyanuric acid is used in the preparation of a bactericides, disinfectants, textile bleaching agents, dyes, pesticides and melamine ($\rightarrow$ **2.6**). Annual production of trichloroisocyanuric acid was about 80,000 t in 1987, the dominant application was swimming pool disinfection. Cyanogen bromide is not listed by the Chemical Weapons Convention. Cyanogen fluoride and cyanogen iodide have very little significance. However, hydrogen cyanide (another schedule 3A substance) is produced on a scale exceeding 1 million t a year (!). About 41 % of this amount is used to manufacture nylon, 28 % for acryl plastics.

4.2 Poison in Groundwater: Arsenic

Arsenic may easily earn one of the top spots in any list of poisons. It has been a popular tool of murder ever since the age of the Ancient Rome. History teaches that poison master Locusta of Gaul (unknown birth date—69 AD) got a secret summons from the Empress Julia Augusta Agrippina (15/16–59 AD), the fourth wife of Emperor Claudius (10 BC—54 AD), to supply poison for the murder of her own spouse. Nero (37–68 AD), the son of Agrippina, rescued Locusta from execution and in return called upon her to supply poison to murder Britannicus (41–55 AD), the son of Claudius and his third wife Valeria Messalina (17/20–48 AD). Arsenic poisoning is still considered a possible cause of the death of French emperor Napoleon Bonaparte (1769–1821), although current scientific facts render this story to be no more than a persistent urban legend (→ **4.6**). Other rulers, notable persons or even famous animals fell victim to accidental or deliberate arsenic poisoning (e.g. Francesco I de' Medici, George III of Great Britain, American explorer Charles Francis Hall, China's penultimate emperor Guangxu, King Faisal I of Iraq, Australian racehorse Phar Lap, → **4.6**).

Mass arsenic poisoning cases include the infamous Tiszazug murders from Hungary: more than 40 (some estimates are over 300) people, mainly men, were murdered by their female relatives in a period between 1914 and 1929. The case and the following court trial received a lot of attention from the media. Several movies and the book *The Angel Makers* by Jessica Gregson were based on this true story, which shed light on the great moral crisis of the traditional peasant culture. The traditions included neglecting the sick, the elderly and the disabled, which proved conducive to violent solutions of family problems. Additional factors leading to the murders involve changes in Hungarian family structures and inheritance patterns, and the impact of political events such as the end of mass emigration, the break-up of the Austro-Hungarian Empire, the autarchic policies of the "successor states" and the failure of land reform.

Arsenic is not a particularly abundant element in the crust of the Earth, although a few minerals contain it as a major component (realgar, As_4S_4; orpiment, As_2S_3; arsenolite, As_2O_3). Today, the element is not prepared from these ores, as more than enough arsenic is produced as a byproduct (about 50,000 t per year measured as arsenic oxide) in various metalworking processes. Arsenic is used as an additive in metalworking, glassmaking or semiconductor production. Gallium arsenide (GaAs) is often the essence of light emitting diodes (LEDs). Arsenic is also found in air in the form of volatile compounds (arsine, AsH_3 and other similar derivatives). Volcanic activity releases about 3000 t of arsenic, whereas bacterial life processes and burning of fossil fuels are responsible for 20,000 and 80,000 t of emission each year, respectively.

Elemental arsenic and its insoluble compounds are non-toxic, whereas soluble inorganic and organic forms of the element may be poisonous. Needless to say, the amount (or dose) matters a lot. Trivalent arsenic is often toxic, the most common form is arsenite (AsO_3^{3-}) salts. Pentavalent arsenates (AsO_4^{3-}) are less of a concern, many of them are insoluble. This is also true in general: the properties of a certain element (including toxicity) very much depend on the form (most commonly the oxidation state) found in the compound. Determining which of the possible forms occur in a sample is called speciation.

Arsenic as a poison usually means white arsenic, which goes by the chemical name arsenic(III) oxide and formula As_2O_3. In relatively large doses, it is highly poisonous. Its excretion with feces or urine is slow. Gradual accumulation causes a characteristic set of symptoms, which resemble the progress of other diseases with slow disintegration of body functions (Addison's disease, tumors). Skin abnormalities may be the most noticeable (symmetric hyperkeratosis on palms or soles, pigmentation on spots exposed to a lot of light), and mucous membranes may become dehydrated. Several different findings seem to indicate that arsenic is carcinogenic, but this effect depends on the solubility of the particular arsenic compound. An

4.2 Poison in Groundwater: Arsenic

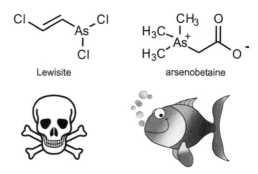

Fig. 4.3 Toxic and non-toxic: the structural formulas of chemical weapon Lewisite, and non-toxic arsenobetaine (occurring in fish). (Authors' own work)

increase in cancer risk is only observed above the concentration of 150 **ppb**, this is 15 times the value specified in the recommendations of the World Health Organization (WHO). The body can develop tolerance toward small amounts of arsenic, which may also have stimulating effects. Eating arsenic to this end was not uncommon in the nineteenth century: herders and hunters in the Austrian states Tyrol and Styria often used this method to improve their climbing skills. Horse traders also used arsenic to make the hair of their horses look shiny.

Salvarsan (606, Arsphenamine) is probably the most well-known organic compound of arsenic. It was first prepared by Sahachiro Hata and Paul Ehrlich in 1909, and was used as a cure of syphilis and sleeping sickness until penicillin was discovered. An infamous organic substance containing arsenic is the chemical weapon Lewisite, which is very toxic and is readily absorbed through skin (Fig. 4.3).

The daily human intake of arsenic may reach even 1 mg as a few types of foods (mussels, shrimp and edible mushrooms) contain quite significant amounts. Another possible source of arsenic exposure is drinking water, primarily because of its high daily volume of consumption. The European Union and the WHO set the maximum value of arsenic in drinking water as 10 µg/l (10 **ppb**). This value had earlier been set at 50 ppb in several countries. In Europe, 66 Hungarian villages do not meet this limit (140,000 inhabitants): the arsenic content of drinking water is higher than 30 ppb. This level is between 10 and 30 ppb in 335 more Hungarian settlements (1,200,000 inhabitants). The final deadline for Hungary to comply with the above EU limit is June 30, 2015.

Where does this magic number, 10 µg/l come from? The devil, as always, is in the details. The EU standard is based on the dietary customs of average Europeans. In most of Western Europe, eating sea fish contributes about 80 µg a day to the arsenic load of the body, and it is known that a healthy human body can tolerate 100 µg without any noticeable effects. The difference between the two values is 20 µg, which leads to a limit of 10 µg/l if the average daily consumption of water is estimated to be 2 l. Too bad for more than a million of Hungarians. Or is it? Hungary is a landlocked country. Sea fish consumption there is a tiny fraction of the European average, a daily exposure of 80 µg is a highly unrealistic overestimate. Taking Hungarian dietary customs into account, a legal limit of 30 µg/l for drinking water would provide the same protection as 10 µg/l does in other parts of Europe. In fact,

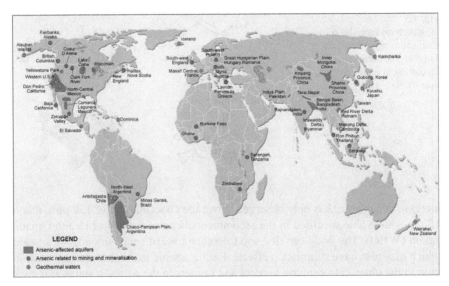

Fig. 4.4 Worldwide levels of arsenic contamination. (http://www.bgs.ac.uk/research/groundwater/images/arsenic_map_big.jpg, permission obtained from copyright-owner)

even with the relatively high arsenic concentration of water in some areas, Hungarians face one of the lowest arsenic exposures throughout Europe. The legal history of arsenic content in sea fish is also interesting: the limit was abolished in 2003 as evidence showed that the arsenic is present in non-toxic forms (such as arsenobetaine, Fig. 4.3). Yet the clearly disproved logic behind the limit for drinking water has remained in effect. It seems that civil or environmental organizations do not believe in scientific advice when they argue that regulations are unreasonably strict…

In Hungary, enforcing the legal limit of 10 µg/l for drinking water will clearly not result in any benefits for society. Yet, the one-size-fits-all approach of the legislation still requires action. Most of the areas with drinking water of high arsenic content are in Békés County. A possible solution would be to borrow water from the neighboring Romanian wells, which give water with low arsenic content. The legal limit can be met by mixing the local and borrowed water. To that end, a cross-border cooperation began to unfold: the water supply companies of the two regions formed a joint venture, each providing about half of the start-up investment of 165,000 € with the primary objective of obtaining all necessary legal permits and building the infrastructure needed. Fortunately, only about 30 % of the available water resources in the Romanian county of Oradea are used. The cross-border agreement would increase this number to 60 %. An alternative solution, which might hold some promise for developing countries as well, relies on a filter that removes about 98–99 % of arsenic from drinking water using a cereal by-product, crushed straw and chaff.

Imagined or actual problems of arsenic in the European water supply are dwarfed by the severity of similar troubles in developing countries (Fig. 4.4). At least 42 significant cases of poisoning are known where the origin of arsenic was in the groundwater used for drinking. Throughout the world, about 57 million people drink water that is heavily polluted with arsenic. India and Bangladesh are affected the most. In the latter country alone, 28–35 million people only have access to water with very high arsenic content, a few 100 times larger than the 10 µg/l limit recommended by WHO. The origin of the problem should not be left unexplained here. In these tropical countries, the main sources of water were dug wells or open-air containers which remained exposed for a long time (the monsoon provides more than enough rain). The pathogens of infectious diseases such cholera, typhus, dysentery and hepatitis often thrived in this water. To fight these diseases, WHO recommended constructing tube wells. This was at first effective against the spread of infections, but arsenic eventually appeared in this water, mostly as a consequence of the presence of bacteria in deeper regions. An attempt to solve a problem created another one, which is arguably more serious.

Most researchers agree that arsenic is weathered into the groundwater through natural processes in these areas, although the exact sources are unknown: they may be in the overlying soils of an aquifer or deeper down, within the aquifer itself. The original source of the arsenic in the Bay of Bengal is almost certainly rocks containing sulfide minerals in the rapidly eroding Himalayas. The breakdown products of these primary minerals include iron (oxyhydr)oxides, which can transport arsenic incorporated in their structures or adsorbed on their surfaces. Great rivers such as the Ganges transport this material, which is then deposited on the floodplains and contributes to the sediments hosting the aquifers that are now accessed by the shallow tube wells that provide the drinking water for millions. The microbial reduction of iron(III) and arsenic(V) requires organic matter, thus the carbon loading in an aquifer plays an important role in determining its arsenic level. The resulting arsenic(III) is more soluble than arsenic(V) and is thus also more available. Human activities that provide organic matter may seriously contribute to the increasing arsenic content of groundwater. Rice is particularly susceptible to arsenic contamination when irrigated with arsenic-tainted groundwater and grown in paddy fields under reducing conditions: it can take up ten times more arsenic than rice grown without flooding.

How can the problem of arsenic contamination be mitigated? There are several possible ways of decreasing the levels of arsenic in drinking water (oxidation-sedimentation, coagulation-sedimentation-filtration, sorptive filtration, membrane techniques). However, the choice of the appropriate method, especially in the poor countries where this problem is most striking, depends on several factors (efficiency, economy of proposed solution, maintenance, monitoring of arsenic level, social acceptance of introduced technology, etc.). Piped water and deep wells are certainly promising solutions that could replace tube wells, as deep aquifers are much less exposed to arsenic contamination.

Arsenic is naturally present in foods, it is unlikely to be completely eliminated from them. Its control, however, presents a big challenge all over the world in the years coming. Yet this control and efforts to remove arsenic should be based on rational arguments and a detailed understanding of what risks the different forms arsenic actually pose. Otherwise, even major efforts may result in no human health benefit at all.

4.3 Don't Touch the Spilled Mercury!

In his *Book of explanation*, prominent Islamic polymath Abu Mūsā Jābir ibn Hayyān (722–815, name Latinized as "Geber") writes:

> ...the metals are all, in essence, composed of mercury combined and coagulated with sulfur [that has risen to it in earthy, smoke-like vapors]. They differ from one another only because of the difference of their accidental qualities, and this difference is due to the difference of their sulfur, which again is caused by a variation in the soils and in their positions with respect to the heat of the sun...

Jābir did not use the words sulfur and mercury here to refer to the actual elements. In alchemy, sulfur was used to symbolize flammability, odor, oily nature, volatility, and bristliness, whereas mercury stood for metallic nature, inflammability, fusibility, and ductility. The sulfur-mercury theory, which became known in Europe through the works of Islamic alchemist, is a variation of the ancient Chinese *yin* and *yang* philosophy, which points out that seemingly opposite forces are actually interconnected and interdependent in the nature. The concept of affinity, the forces holding substances together, also appears in the sulfur-mercury theory.

The vapors of mercury are toxic, the typical legal limits for workplace exposure are 0.08 mg/m^3 (8-h average) and 0.32 mg/m^3 (maximum). There are incentives to limit the uses of mercury worldwide. The European Union banned all exports of metallic mercury and certain mercury compounds and mixtures from 15 March 2011 (Regulation (EC) No. 1102/2008 of the European Parliament and of the Council of 22 October 2008). Mercury is seldom found in everyday life. The typical case, a broken mercury thermometer, is becoming less common as the liquid metal is not used any more in measuring devices. So, what should be done if an old mercury thermometer breaks? The general advice given everywhere is that good ventilation must be maintained and the mercury should be neutralized by sulfur powder. Ventilation is without doubt a good advice, but sulfur powder is another issue. This is a common misconception even among experts, reinforced by numerous textbooks and websites. And even if it were an efficient remedy, sulfur powder is seldom found outside chemical laboratories....

In fact, all sulfur powder can do on spilled mercury is to decrease its surface tension a little bit. This may help control the otherwise quite mobile drops of mercury, but does not transform mercury chemically. The liquid metal does in fact react with sulfur to form mercuric sulfide under suitable conditions according to the chemical equation $Hg + S = HgS$. Mercuric sulfide has two forms: red cinnabar and black metacinnabarite. The latter slowly converts to the former if left alone. The reaction of mercury and sulfur is a favorable one in terms of energy, but it does not occur at room temperature.

To treat a mercury spill, a better solution is to use some absorbent such as Mercon™ (composition not fully disclosed), HgX® (a mixture of sodium thiosulfate and ethylenediaminetetraacetic acid), Mercsorb® (zinc, citric acid, sulfur, and silica gel) or Mercurisorb™ (silver nitrate on silica gel). The last one is the most expensive and the most efficient at the same time. This substance is a silver-containing silicate,

4.3 Don't Touch the Spilled Mercury!

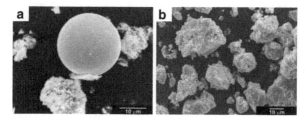

Fig. 4.5 The reaction of mercury and sulfur in a ball mill. In the image taken by an electron microscope, sulfur powder particles are attached to mercury drops after 15 min (**a**), whereas mercuric sulfide particles are covered by excess sulfur powder after 60 min (**b**). (Permission obtained from copyright-owner)

in which adsorption, salt and amalgam formation contribute to the process. The weight of mercury that it can take up is about the same as the weight of the absorbent material. Mercury containing waste collected by these absorbers is still hazardous waste, so care must be taken to find a proper hazardous waste collection site.

What are the conditions necessary for mercury and sulfur to react? Spanish researchers reported a special method, in which mechanical energy is used to induce a chemical process (which is, incidentally, called a mechanochemical reaction). If a mixture of sulfur and mercury is placed into a ball mill operated at high intensity (400 revolutions per minute), black metacinnabarite particles appear in 15 min and there is little unreacted sulfur and mercury remaining after an hour (Fig. 4.5). This process transforms elemental mercury into a mercury salt, which is much safer to handle. The conditions are extreme compared to just pouring sulfur onto mercury balls: mercury drops are continuously made smaller and smaller, and friction also enhances the temperature.

The unusual physical properties of mercury make its handling quite burdensome. It is about 13 times denser than water, its surface tension is 7 times higher than the already high value for water. Surface tension causes a tendency in a liquid to assume a shape with the smallest surface-to-volume ratio (which is spherical, by the way) in the absence of external forces. As a result of surface tension, certain objects or small animals do not sink in water despite their larger density (Fig. 4.6).

The spherical size of a drop of liquid is distorted depending on the magnitude of the forces acting between the liquid and the surface. The stronger the interactions, the more spread out the drop will be. Wetting properties are often given by the contact angle of a drop of liquid on a surface (Fig. 4.7). This contact angle is close to 0 for water on glass, but 140° for mercury on glass. Adding the huge density of mercury, it is understandable why mercury drops move on glass (and on most other) surfaces even upon the slightest influence.

Wettability only makes sense for a pair of substances. Water drops on the surfaces of plants behave the way they do because the wettability is low. The phenomenon was first studied on lotus plants. Therefore, it is often called the lotus effect. The lotus effect is caused by a special surface structure: the lotus leaf features 5–10 μm high warts that are 10–15 μm away from each other, which form a protective layer

Fig. 4.6 The strange image of a dandelion floating on the surface of water is a consequence of the bent interface between the water and the dandelion, which acts as an optical lens. (Authors' own work)

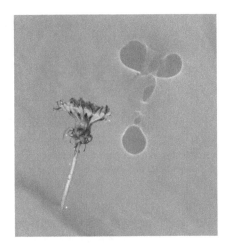

Fig. 4.7 Contact angles with a few materials (**a–c**) and a surface (*S*). Water drops on a blade of grass (**d**). (Authors' own work and copyright-free Wikipedia picture)

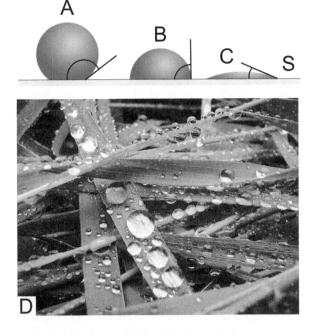

on the surface. The contact angle for water increases dramatically because of this special structure, it can be as high as 160°. Only about 2–3 % of the surface of the drop is in actual contact with the surface of the plant. The small dirt particles on the leaf do not adhere to the plant strongly, and are easily washed away by the water drops. This effect is already used in technology to prepare self-cleaning surfaces.

4.4 Was DDT of More Harm Than Use?

In October 1943, the German occupying army left Naples in ruins: the water and sewage system was blown up. Washing became a problematic task in the cold winter weather: poor hygiene made the environment inviting to body lice, especially on people living in caves, and an epidemic of louse-born typhus followed. By January 1944, the number of infections rose to 700. Allied forces had to act fast and their coordinated efforts were successful. With financing from the Rockefeller Foundation, 1.3 million people were treated against body lice with 3.2 million doses of newly discovered DDT, and local people were vaccinated with an extract of chicken embryos. The typhus epidemic was stopped, an achievement never seen before in history.

DDT is an abbreviation of the old name of the compound 4,4'-dichlorodiphenyl-trichloroethane. It was first prepared by Austrian chemist Othmar Zeidler (1859?–1911) in 1874, its insecticide potential was only discovered 60 years later in the systematic studies of Swiss chemist Paul Hermann Müller (1899–1965), who was awarded the 1948 Nobel Prize in Physiology or Medicine for his insights.

Commercial DDT contains several active ingredients: about 77% of the 4,4'-isomer (hence, it is often called DDT, which might cause confusion as the mixture is also marketed under the same name), 15% of the 2,4'-isomer, and the remaining 2% is DDE and DDD (Fig. 4.8). DDT affects neurons: it disrupts the normal functions of sodium ion channels at the receptors of neurotransmitters. The channels cannot close properly, so neurons remain in an excited state for a prolonged time, which seriously disturbs motor functions and senses. Insects are more sensitive to DDT than mammals, and because mammals also detoxify their bodies more efficiently, the initial observations did not show any adverse side effects.

DDT was successfully used after World War II against lice, fleas, malaria, yellow fever and mosquitos spreading typhus. It is estimated that the use of DDT saved the lives of about 50 million people and prevented 1 billion cases of infectious disease. According to the World Health Organization (WHO), 1 billion human malaria cases were prevented by the use of DDT until 1971. More than 18 million t of DDT have been used since 1940, about 80% of this in agriculture. In addition to the notable success in fighting epidemics, DDT was used for other purposes as well: in a dose of 1 kg/ha, it proved effective against the caterpillar larvae of Bordered White (*Bupalus piniaria*) without observing any side effects. The same amount is also sufficient against mosquitoes carrying malaria, but a 25-times higher quantity was used against bark beetles. Initially, DDT seemed to be completely harmless for humans.

The first concerns began surfacing when birds and fish started to die because of DDT derivatives accumulating in insects and insect larvae. Robins (*Erithacus rubecula*) became extinct in several areas. Grebes living near Clear Lake (USA) also died in mass in 1954. DDT is a persistent substance: it degrades very slowly in the environment. It takes 5–25 years to reach 95% degradation in soil. At the same time, DDT accumulates in body fat: the sediment of Lake Michigan (USA) only contained 0.02 **ppm** of DDT but gulls had levels as high as 98 ppm. Peregrine

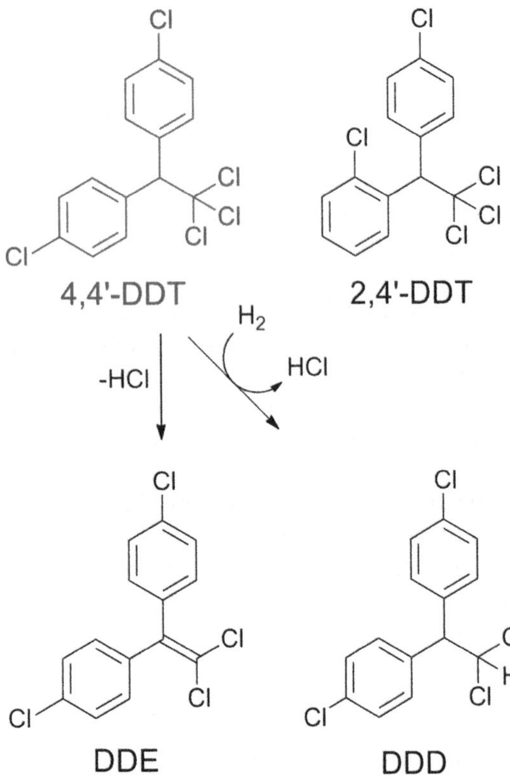

Fig. 4.8 Structures of 4,4'-DDT, 2,4'-DDT, DDE, and DDD, the constituents of 'DDT' (the last two are also formed from 4,4'-DDT in living organisms). (Authors' own work)

falcons, common kestrels, and bald eagles were not killed, but DDT caused the shells of eggs to have very low calcium levels and they often broke under the females. The sensitivities of different birds are quite different: quails, pheasants and chickens, often used in toxicology studies, tolerate DDT quite well, whereas birds of prey and fish-eating birds tend to be very sensitive, with the possible exception of the California condor. Noted entomologist J. Gordon Edwards (1919–2004), made an impassioned plea in a posthumously published article, which actually supported the use of DDT: "The ban on DDT, founded on erroneous or fraudulent reports and imposed by one powerful bureaucrat, has caused millions of deaths, while sapping the strength and productivity of countless human beings in underdeveloped countries. It is time for an honest appraisal and for immediate deployment of the best currently available means to control insect-borne diseases. This means DDT."

In 1962, at the height of DDT overuse, Rachel Carson (1907–1964) published a book titled *Silent Spring*, which is generally considered as the most important single event in the formation of the modern environmental movements. Carson shared a vision about a world without bird singing as a consequence of the blatant overuse of pesticides. The book became very influential, it facilitated the ban of DDT, beginning with Hungary in 1968 and the US following suit in 1973.

4.4 Was DDT of More Harm Than Use?

Another serious concern about pesticide residues is their carcinogenic effects. Initial worries have proved to be exaggerated. American biochemist Bruce N. Ames (born in 1928) developed a method to test carcinogenic and mutagenic properties of materials, now called the Ames test. It was Ames himself who, together with his co-worker Lois Gold, absolved pesticides by writing: "Neither epidemiology nor toxicology supports the idea that synthetic industrial chemicals are important for human cancer. [...] Although some epidemiologic studies find an association between cancer and low levels of industrial pollutants, the associations are usually weak, the results are usually conflicting, and the studies do not correct for potentially large confounding factors like diet. Moreover, the exposure to synthetic pollutants is tiny and rarely seem plausible as a causal factor when compared to the background of natural chemicals that are rodent carcinogens."

The debate about carcinogenicity barely faded away when the first reports appeared decrying the damaging effects of pesticides on the human immune system. Many experts urged the increased use of immunotoxicological testing in pesticide approval processes.

Another problem was presented by endocrine disrupting effects. Some artificial chemicals including pesticides caused developmental and reproduction problems in various animals and humans. These compounds are now called endocrine disrupting chemicals (EDC) and exert their effects primarily not on the exposed individuals, but their fetuses, and the negative effects only develop later, in mature offspring. Depending on the method of their action, these substances mimic estrogen (female) or androgen (male) hormones, or inhibit these hormones. One of the best studied examples is the alligator population in Lake Apopka (USA). These alligators had reproductive problems, the testicles of males showed deformations, testosterone levels were lower and penises were smaller. Disorders were diagnosed in the sex organs of female alligators as well, and their estrogen levels were abnormally high. The suspected reason was the presence of pesticide DDE in the lake (Fig. 4.8). J. Gordon Edwards again pointed out that the problem is complex and DDE cannot be singled out as a culprit.

Organic phosphates used to replace DDT have caused hundreds of fatalities in accidents or murders. Pyrethroid pesticides are also used for the same purpose, but resistance toward this agent caused an increase in malaria rates. In the South African province KwaZulu-Natal, 4000 cases of malaria were reported in 1995, the number jumped to 27,000 by 1999 as a result of banning DDT. At a 2000 meeting in Johannesburg, representatives of World Health Organization (WHO), Greenpeace and World Wide Fund for Nature (WWF) agreed that DDT should be allowed again in malaria-infected areas, primarily for indoors use.

The history of DDT is not a simple one, but it has been further complicated by the rather extreme human reaction that began to turn against its use (Fig. 4.9). The silent spring envisioned by Carson has never arrived, the frightening conclusions of the book were indeed unfounded (critics say outright erroneous). Carson and her followers also share some responsibility for helping to develop public chemophobia ($\rightarrow$ **1.1**) and encourage a hysterical approach toward chemicals.

Fig. 4.9 The cover of the book about the political and scientific history of DDT, published in 2010. (Permission obtained from copyright-owner)

Harmful properties of substances used in the environment cannot be predicted in each case, their use must be continuously monitored to detect possible negative effects. Last, but not least: a suspected or even a real negative environmental effect does not automatically void the benefits of the substance. Advantages and risks, costs and benefits must be assessed objectively (→ **1.2**). The history of DDT reconfirms the fact that risks cannot be excluded from human life, they can only be exchanged. In the meantime, dioxins (→ **4.5**) have taken the position of DDT as the number one public concern….

4.5 Could Dioxin be the Most Toxic Substance?

The US Army used about 50,000 t of defoliants (chemicals that cause leaves to fall off) in the jungle battles of the Vietnam War between 1962 and 1969. *Agent Orange*, the 1:1 mixture of 2,4-dichlorophenoxyacetic acid (2,4-D) and 2,4,5-trichlorophenoxyacetic acid (2,4,5-T), gained a particularly high dose of infamy. The use of this agent was responsible for discharging from 20 to 100 kg of dioxin into the environment.

4.5 Could Dioxin be the Most Toxic Substance?

Fig. 4.10 Ukrainian president Viktor Yushchenko was poisoned by dioxin in 2004. In his blood, dioxin levels were 6000 times the normal values and the long-lasting effects of chloracne are still visible on his face in this picture from 2006. (Copyright-free Wikipedia picture)

After the war, veterans suffering from a range of health problems (baldness, impotence) sued the manufacturers of Agent Orange, who agreed to pay them $ 180 million in an out-of-court settlement. The dioxin levels in the blood of some veterans were 5 **ppt** (parts per trillion), which is on average a slightly higher than the values for the general population, varying between 2 and 20 ppt. Troops directly involved in Agent Orange transport had higher levels. About a quarter of a million veterans were checked, 27,000 of whom were hospitalized because of suspected dioxin-related health problems. Five soldiers were diagnosed with a disease called chloracne, which became widely known after Ukrainian president Viktor Yushchenko was poisoned with dioxin in 2004 (Fig. 4.10). The investigation on the health of veterans was concluded with a lengthy 1993 review which analyzed results published in 6400 individual studies. The analysis linked only five illnesses with the use of the defoliant: soft tissue sarcoma, Hodgkin's disease, non-Hodgkin's lymphoma, chloracne, and *porphyria cutanea tarda* (another disease with blistering of the skin). No statistically significant connections were revealed between Agent Orange and the incidence of the following diseases: leukemia, bone cancer, birth defects, gastrointestinal cancer, skin cancer, urinary bladder cancer, or brain tumor. Even when a connection was established by the statistical methods, it was unclear whether Agent Orange or dioxin is responsible.

The history of manufacture and application of dioxins is plagued with industrial accidents and environmental problems (Nitro, USA: 1949; Ludwigshafen, Germany: 1953; Duphar, Netherlands: 1963; Coalite Chemicals, UK: 1968; Seveso, Italy: 1976; Love Canal, USA: 1978; Times Beach, USA: 1983; Belgium: 1999; Europe: 2007, 2011, → **2.21**). The site of the first industrial accident in 1949 was a chemical plant of Monsanto in Nitro, West Virginia. The town was named after explosive nitrocellulose, which was manufactured there during World Wars I and II. 2,4,5-T was produced in this plant, and 122 workers were diagnosed with chloracne. Their health history was monitored by the Department of Environmental Health of the University of Cincinnati for 30 years. Surprisingly, exposed workers lived longer and experienced fewer health problems than their peers.

The accident of Seveso is much better known. At half past seven in the morning of July 10, 1976 a reactor used for manufacturing chlorophenol pesticides in a chemical plant exploded in the Italian town of Seveso, about 13 km from Milan. About 16 kg of dioxin was released into the environment, 3 kg of which was dispersed

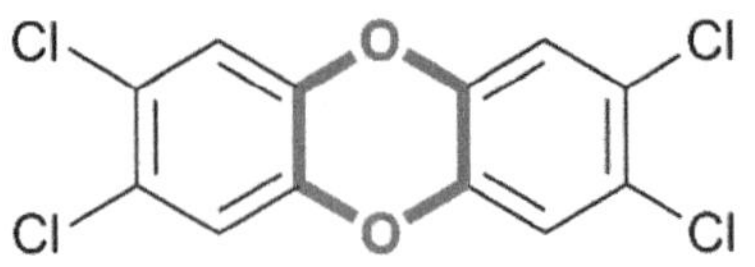

Fig. 4.11 The structure of 2,3,7,8-tetrachlorodibenzo[b,e][1,4]dioxin (TCDD), the most common compound of the family of dioxins (the dioxin motif is highlighted in *bold* and *red*). 2,3,4,7,8-Pentachlorodibenzo[b,d]furan (PCDF) is the most common substance in the similar group of chlorinated dibenzofurans. (Authors' own work)

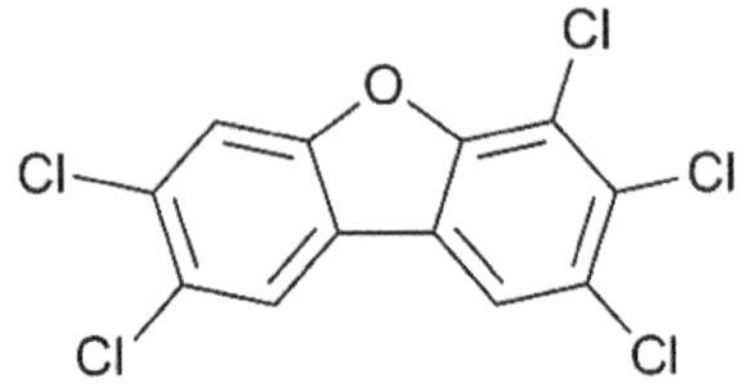

in the small town of 17,000 residents. Close to 2000 animals (sheep, rabbits, pets) were killed, 250 people were evacuated, 180 of whom developed chloracne. One hundred and fifty pregnant women were also exposed, thirty of whom opted for abortion. 29 of the aborted fetuses were entirely healthy, one showed symptoms of Down syndrome. This latter disease is a genetic disorder, which cannot be explained by exposure to dioxin. Similar results were published in an analysis of 1500 related births. A study published in 1988 analyzed 15,291 births, and the rate of disorders found (5%) was not any higher than the statistical average. The residents of Seveso were monitored for a decade by the Department of Occupational and Environmental Health of the University of Milano, no increase in cancer rates were seen, not even in the most polluted neighborhoods. Pier Alberto Bertazzi of the same institute published an even more detailed analysis in 1993: the health records of 37,000 local residents were compared with 182,000 others who were not exposed to dioxin. Among Seveso residents, a few rare cancer types showed a slight increase in rate (fewer than ten cases overall). The group living in the most polluted area (724 people) did not contribute to this increase at all, the extra cases were diagnosed among people living in areas of medium or low pollution levels. Despite these facts, dioxin inherited the public role of DDT ($\rightarrow$ **4.4**) as one of the most fearsome substances.

The material called dioxin is not a single substance, it is a whole family of related compounds. 'Dioxin' often means the most common chemical in this group, which is 2,3,7,8-tetrachlorodibenzo[b,e][1,4]dioxin (TCDD) (Fig. 4.11). There are 75 possible varieties of dibenzodioxins that contain 1–8 chlorine atoms. Seven of these compounds are toxic, TCDD is the most common of these. To make matters more complicated, dibenzo[*b,d*]furans have similar structures and properties. There are 135 such derivatives with 1–8 chlorine atoms, 10 of which are toxic. As a shortcut,

4.5 Could Dioxin be the Most Toxic Substance?

Table 4.1 Toxicity data for TCDD and other substances

Toxicity of TCDD for various species		Toxicities of other substances	
Animal	LD_{50} (mg/ttkg)	Substance (animal/human, way of intake)	LD_{50} (mg/kg of body weight)
Guinea pig (*Cavia* species)	0.0006–0.002	botulinum toxin (human, estimated)	0.000000001
Rhesus monkey (*Macaca mulatta*)	<0.070	tetrodotoxin (mouse, ingested)	0.334
Rat (*Rattus* species)	0.022–0.045	sodium cyanide (rat, ingested)	6.4
Dog (*Canis familiaris*)	0.100–0.200	nicotine (rat, ingested)	50
		nicotine (mouse, ingested)	3
		nicotine (human, ingested)	0.5–1.0
Hamster (*Cricetus* species)	1.1–5.1	alcohol (rat, ingested)	7060

experts refer to chlorinated dibenzodioxins and dibenzofurans collectively as 'dioxins'.

TCDD is the most toxic in this dioxin group, some individual **LD_{50}** values are listed in Table 4.1.

Table 4.1 shows that the toxicity data vary in a huge range for different animal species, which mandates extreme caution in drawing conclusions. Human toxicity data are not known for dioxins, but some observations imply that people are more resistant to poisoning than hamsters. The right side of the table shows toxicity data for other substances. It is clear that TCDD is not among the most poisonous substances.

The notable human tolerance toward dioxin probably has its roots in history. Dioxin was found in 8000-year old sediments in Japan and dioxin levels of 30 ppt were detected in 170-year old soil samples from Rothamsted, England (90 ppt was the measured level in 1995). Industrial development obviously increased the amounts of dioxin, but its origin in the early or pre-historical samples is a much more interesting question. In 1977, evidence was found that dioxins are formed regularly in combustion processes. Wood contains about 0.2% of chlorine in the form of chloride salts, some of which is transformed into dioxins during burning. About 5 billion t of wood are burnt annually on Earth, which may form about 55 t of dioxin. The human body has learnt to tolerate dioxins in small amounts.

How small is this amount? A human body contains about 5000 **picograms** (pg) of dioxin at any given time, the daily intake is estimated to be 1–3 pg. The **half-life** of TCDD is about 7 years. The World Health Organization (WHO) and national agencies have very diverse views on the tolerable daily intake: 0.0062 pg (USA), 1 pg (Germany), 2 pg (UK), 10 pg (WHO, Canada). This disagreement has not been resolved yet. The participants of an international conference in 1990 agreed that the lowest value in this list is way too low, whereas the highest is too high. The US Environmental Protection Agency recommended relative toxicity data and risk assessment for 28 different dioxins in 2011. The toxicity of 2,3,4,7,8-pentachlorodibenzo[*b,d*]furan (PCDF) was estimated to be 0.3 relative to TCDD. These recommendations are still being analyzed, but their methodology has also been criticized by several experts.

4.6 Was Napoleon Murdered with Arsenic?

Napoleon Bonaparte died on May 5, 1821 on the island of Saint Helena. His physician diagnosed stomach ulcer or stomach cancer as the cause of death. The former emperor of France was still relatively young, only 52 at the time, and was not generally known to be in bad health. So suspicions of deliberate poisoning naturally arose shortly after his death, but were quickly laid at rest by credible medical reports. In the twentieth century, Sten Forshufvud (1903–1985), a Swedish dentist published a book in his native language entitled, *Napoleon Murdered?* In 1983, he co-authored an English book titled *Assassination at St. Helena.* A laboratory study in 1965 found abnormally high amounts of arsenic in the emperor's hairs. Some people, including even experts, rushed to the conclusion that Napoleon was in fact murdered with arsenic. In response, a few skeptical historians pointed out that building a murder case usually requires identifying a possible murderer and a motif as well, both of which were missing. The results of the hair analysis itself were anything but definitive: they compared the arsenic content of Napoleon's hairs with samples from the mid-twentieth century. Even today, the high arsenic content in drinking water is a major scientific problem in several parts of the world, for instance in Bangladesh ($\rightarrow$ **4.2**).

The word arsenic in the popular literature usually means white arsenic, which goes by the chemical name arsenic(III) oxide and formula As_2O_3. This is a white powder, which dissolves in water readily. Its solution has no taste, smell or color. Given in one dose, one fourth of a gram can cause the death of an adult. It was the weapon of choice for Julia Agrippina, probably the most infamous poison master in the history of Ancient Rome. First she murdered her own husband with arsenic, then Emperor Claudius's third wife in order to become empress by marrying Claudius. The next in line was Britannicus, son of Claudius. Then the emperor in effect signed his own death warrant by declaring Nero, the son of Agrippina, to be his heir. Nero followed in the footsteps of his mother but did not learn much of her sophistication as he simply used a knife to get rid of her. In the 1500s, Italian Cesare and Lucrezia Borgia also used arsenic to pave the way to the papacy for their father, who was elected to be pope and chose the name Alexander VI. Or at least this is how some powerful historic legends tell the story. Much less doubtful is the fact that Phar Lap, one of the most famous Australian race horses ever, fell victim to arsenic poisoning. Unambiguous scientific evidence was found in 2010, when some of the preserved hair of the unlucky creature was analyzed.

Napoleon had a major influence on history. Such personalities are often surrounded by various legends. To prove this point, it is enough to read a few lines from Leo Tolstoy's epic novel, *War and Peace*:

> The French alphabet, written out with the same numerical values as the Hebrew, in which the first nine letters denote units and the others tens, will have the following significance:

4.6 Was Napoleon Murdered with Arsenic?

a	b	c	d	e	f	g	h	i	k	l	m	n	o	p
1	2	3	4	5	6	7	8	9	10	20	30	40	50	60
q	r	s	t	u	v	w	x	y	z					
70	80	90	100	110	120	130	140	150	160					

Writing the words L'Empereur Napoleon in numbers, it appears that the sum of them is 666, and that Napoleon therefore the beast foretold in the Apocalypse.

The often quoted line of thought has a hidden problem: the sum of letter numbers in '*l'empereur Napoléon*' is only 661. It takes a mistake in French grammar to bring it up to 666: '*le empereur Napoléon*'. Pierre Bezukhov, the main character in the novel, even manages to turn his own name into '*l'russse Besuhof*', which also gives 666 as the sum of letter numbers, through some creative transcription form Cyrillic to Latin letters and (quite appropriately) another mistake in the use of the French definite article. However, instead of rejecting number mysticism, Pierre chooses to conclude that he himself is the nemesis of the French emperor.

Returning to Napoleon's hairs: Italian physicists made a concerted effort to study them in 2008. They used a scientific method called neutron activation analysis: the sample was irradiated with neutrons for some time (a nuclear reactor is usually needed for this), which caused the formation of short-lived radioactive atoms, the amounts of which could be easily measured without actually destroying the object being studied. An advantage of this method is that very small amounts of elements can be detected. The particular hairs were carefully selected in this study. Contemporary human hairs were chosen randomly. From museums around Europe, the scientists obtained hairs both from King of Rome Napoleon II (1811–1832) and empress Josephine (1763–1814), the son and first wife of Napoleon Bonaparte, and of course from Napoleon himself, not only from the island of Saint Helena, but also form his childhood (age of 1) in Corsica and from the island of Elba, the scene of his first exile. The arsenic content of these samples is shown in Table 4.2.

The interpretation of the results from extensive studies requires a lot of care, which includes something that is called statistical analysis. Top-quality scientific data always give an estimate of their error, which means the precision of the results is also measured. It is quite impossible to measure anything with 100% precision in science. There is always a source of error, which introduces uncertainty into measured final results. This uncertainty is expressed by the numbers after the ± signs in Table 4.2. The measured value of 8.3 ± 0.9 **ppm** for the hair of one-year old Napoleon in Corsica implies that the actual content of arsenic is somewhere between $8.3 - 0.9 = 7.4$ ppm and $8.3 + 0.9 = 9.2$ ppm. At first sight, this may seem a lot of uncertainty for a scientific result. However, this is only a question of what the data are needed for.

Table 4.2 shows quite nicely that the concentration of arsenic in people's hair is about a hundredfold lower today than in the hairs of Napoleon or his son. Empress Josephine's hair had somewhat smaller arsenic concentrations, but the values would still be outstandingly high in the twenty-first century. The high values only reflect the fact that arsenic was much more common in paints, medicines and preservatives at earlier times, so its amount in the human body was also understandably higher.

Table 4.2 Concentrations of arsenic in hair samples form the eighteenth, nineteenth and twenty-first centuries

Contemporary hairs	As (ppm)
1	0.086±0.019
2	0.056±0.018
3	0.110±0.035
4	0.124±0.039
5	0.024±0.011
6	0.071±0.028
7	0.054±0.017
8	0.040±0.014
9	0.043±0.031
10	0.048±0.020
Napoleon II	*As (ppm)*
1812, 1	9.4±1.0
1812, 2	6.1±0.6
1816, 1	12.6±1.3
1816, 2	9.9±1.0
1821, 1	9.9±1.1
1821, 2	11.2±1.3
1826, 1	7.6±0.8
1826, 2	8.5±0.9
Napoleon	*As (ppm)*
Corsica 1770, 1.	8.3±0.9
Corsica 1770, 2.	6.3±0.7
Elba 1814, 1.	4.4±0.5
Elba 1814, 2.	3.5±0.4
Saint Helena, 5 May 1821, 1	13.1±1.3
Saint Helena, 5 May 1821, 2	16.7±1.7
Saint Helena, 5 May 1821, 3	14.2±1.4
Saint Helena, 5 May 1821, 4	17.0±1.7
Saint Helena, 5 May 1821, 5	15.4±2.3
Saint Helena, 5 May 1821, 6	18.9±2.2
Saint Helena, 6 May 1821, 1	15.2±2.0
Saint Helena, 6 May 1821, 2	9.7±1.6
Empress Josephine	*As (ppm)*
1814, 1	0.8±0.4
1814, 2	1.2±0.5

These findings do not imply poisoning by any means. There is no dramatic difference between the arsenic contents of the hairs of Napoleon and his son. Historical records show that Napoleon II died of tuberculosis, and there was no reason to suspect poisoning, let alone murder with arsenic. Even more revealing is the fact that hairs from Napoleon himself at the age of 1 already contained high amounts of arsenic compared to today's average values. Napoleon could not have been poisoned on Saint Helena—at least not with arsenic. Had this actually happened, hairs form Saint Helena would show a much higher arsenic content than earlier ones.

Despite all the efforts of scientists, legends will continue to surround the death of Napoleon. Strong survival skills make legends what they are.

4.7 The Resin Wars: Formaldehyde

The Consumer Product Safety Commission (CPSC) of the US asked the Government Accountability Office (GAO) to investigate the effect of formaldehyde resins, commonly found in clothing, on the health of consumers. The report, published in August 2010, stated that the formaldehyde content of textiles began to fall in the 1980s, but it might still cause allergy in sensitive people (about 2% of the population despite the low levels detected (below 20 **ppm**).

The use of formaldehyde resins to reduce wrinkling of materials in the textile industry was introduced in the 1920s. Formaldehyde levels in textile products are regulated in 13 countries, which resulted in a distinct decrease of the concentrations from their values of about 100 ppm 25 years ago. In 1984, 67% of the measurements showed formaldehyde levels exceeding 100 ppm, the same number was less than 2% in 2003. This was not entirely the effect of legal limits, industry took some voluntary action, too, including developing novel, more resistant resins. Earlier, formaldehyde-urea and formaldehyde-melamine resins were widely used, and they slowly released some formaldehyde. The quoted levels may still seem high as the US Occupational Safety and Health Administration (OSHA) allows 0.5 ppm of formaldehyde in workplace air on average for an 8-h shift. However, the quoted concentrations are referenced to the pure resin and not the whole textile product. The small amount of formaldehyde released from the resin reacts promptly on the surface of the skin: no connection was found between formaldehyde levels in air and textiles. Experience in the last decades shows that some skin symptoms might arise if the formaldehyde levels in the resin are higher than 300 ppm (Fig. 4.12).

The most widely used formaldehyde resin is Bakelite, which is prepared from formaldehyde and phenol and was developed by Belgian chemist Leo Hendrik Baekeland (1863–1944) between 1907 and 1909. Powdered wood was used as a filling material in this plastic, it was heatproof and a good insulator. Bakelite was the first plastic manufactured on an industrial scale, but—contrary to some beliefs—LP discs have never been made of it.

What are the health risks of formaldehyde? Its uses are quite diverse, it is found in textiles, wood products (especially recently made furniture), photographic materials, shaving foams, and air fresheners. Tobacco smoke, exhaust gases and the air in plastic factories is usually polluted with this substance. James A. Swenberg's 1980 study found that inhalation of formaldehyde induces squamous cell carcinomas (a form of cancer) in the noses of rats. Since then, the connection between diseases and formaldehyde inhalation has been studied exhaustively in animal tests. Human studies found an increased risk of larynx and pancreas cancer as a result of formaldehyde and coal powder exposure. Formaldehyde also caused crosslinking of DNA in the nose tissues of rats and monkeys, which damages genes and inhibit the reproduction of DNA. Earlier suspicions linking formaldehyde to leukemia and asthma have been disproved.

The aqueous solution of formaldehyde (formalin) is a well-known disinfectant, it is indispensable in anatomical preparations. Thomas Mann, German writer in the

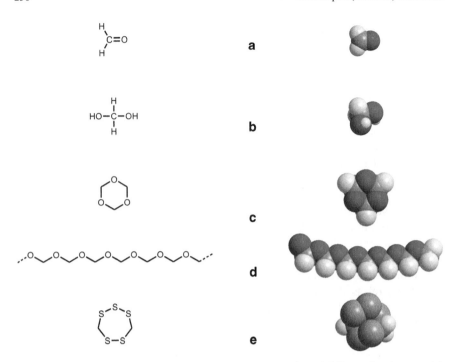

Fig. 4.12 The 2D and 3D structures of formaldehyde in a number of different disguises: gas (**a**), aqueous solution (formaldehyde hydrate, **b**), solid trimer (trioxane, **c**), solid polymer (paraformaldehyde, **d**) and the odor substance of shiitake mushrooms (lenthionine, **e**). (Authors' own work)

twentieth century described the disinfectant use in his novel *The Magic Mountain* (*Der Zauberberg*):

> "Fumigated it, eh? That's ripping," he said loquaciously and rather absurdly, as he washed and dried his hands. "Methyl aldehyde; yes, that's too much for the bacteria, no matter how strong they are. H_2CO. But it's a powerful stench. Of course, perfect sanitation is absolutely essential." (Translation: H. T. Lowe-Porter)

Formaldehyde is also present at significant levels in a number of foods such as onions, and especially those that are preserved by smoking. Shiitake (*Lentinus edodes*) is probably the mushroom that has the longest history of cultivation, current world production is about 600,000 t. The formaldehyde concentration in it is very high, usually between 100 and 500 ppm, but cooking removes most of it, so this is not a significant health risk. The source of formaldehyde in shiitake is a substance called lenthionine, which has been shown to inhibit the aggregation of platelets in blood and is being tested as a drug preventing thrombosis.

Why is carcinogenic formaldehyde so abundant in nature? Formaldehyde (the term methyl aldehyde in Thomas Mann's book means the same but is not much used any more) is one of the simplest organic substances (Fig. 4.12a). It is common in interstellar space and it plays an important role in adding and removing methyl (CH_3) groups in biological systems. Such enzyme-promoted methylation reactions have been long known to be significant both in forming and preventing cancer.

4.7 The Resin Wars: Formaldehyde

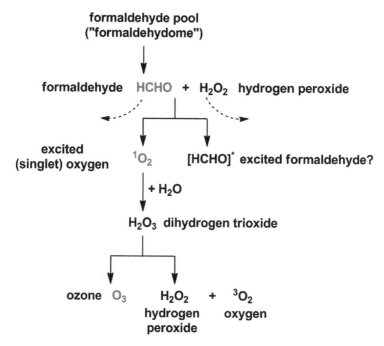

Fig. 4.13 The role of formaldehyde in the formation of singlet oxygen and endogenous ozone. (Authors' own work)

Some methylated compounds (e.g. *N*-nitrosodimethylamine) cause tumorous cell proliferation, whereas others (such as *N*-methylascorbinogen, a methylated variety of vitamin C) show antitumor activity. The mechanisms through which methylation and demethylation reaction affect cancer formation are not known yet.

Formaldehyde is mostly present bonded to a number of different molecules (glutathione, L-arginine) in living organisms. Deciduous trees were observed to have a dramatic increase of formaldehyde methylated compound levels in the spring. This phenomenon is connected to the onset of the growing season. There is a similar jump in formaldehyde level in the fall as well, but the level of methylated compounds drops sharply at that time, which causes the leaves to fall. Demethylation processes dominate under these conditions. Experimental evidence shows that there is a fast and primary formaldehyde cycle in biological system, and the derivatives of the amino acid L-methionine play an important role in it.

Formaldehyde and related compounds form continuously both inside and outside cells. Interaction with hydrogen peroxide can drastically increase the physiological effects. Formaldehyde and hydrogen peroxide react in various cell parts to give singlet oxygen ($^1O_2$) and excited formaldehyde, which is a more reactive form of the compound, in which electrons have higher-than-normal energies (Fig. 4.13). The most recent studies show that highly reactive biological oxidants (such as dihydrogen trioxide, H_2O_3; ozone, O_3), formed in the reaction of singlet oxygen and water, are the actual species that are responsible for inhibiting or killing antibodies

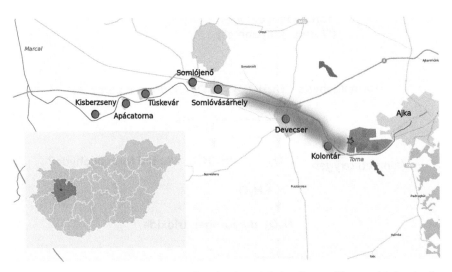

Fig. 4.14 The area around the town of Ajka after the red sludge disaster. The *asterisk* denotes the location of dam failure. (Copyright-free Wikipedia picture)

or antigens. Singlet oxygen seems to be a major player in the tolerance to certain diseases, therefore formaldehyde may also be significant.

Formaldehyde has three sources in the human body: (a) under controlled conditions, as part of normal biological processes (endogenous formaldehyde cycle); (b) as a result of random processes (demethylase, peroxidase, semicarbazide-sensitive amino oxidase enzymes); (c) exogenously, from external sources (air, water, food, environmental pollution). Formaldehyde is by no means just a harmful byproduct in biological systems, it has essential functions, which are not yet known in sufficient depth. The question is not whether formaldehyde occurs in living organism. The question is its amount and the role it plays in the complex and dynamic cycle of methylation and demethylation.

4.8 The Great Hungarian Red Mud Deluge

One of the worst environmental catastrophes of the last decade in Europe occurred on October 4, 2010 (Monday) in Hungary. Between 12:10 and 12:25, the northwestern dam of red sludge reservoir No. X, which belonged to the alumina plant of MAL Hungarian Aluminum Production and Trade Company Ltd., broke in a village named Kolontár. A huge amount of red mud flowed out (1,876,622 metric t, as it was confirmed later). It contained 7–8 % highly alkaline red sludge. Between 97 and 98 % of the thick red sludge actually remained in the reservoir. The alkaline sludge flooded the surrounding villages of Devecser, Kolontár, Somlóvásárhely, Tüskevár, Apácatorna and Kisberzseny, and about 2500 acres of mainly agricultural areas (Fig. 4.14). The flood of dilute sludge killed 10 people, significantly eroded

4.8 The Great Hungarian Red Mud Deluge

Fig. 4.15 The area around Kolontár closed by police—this was the place hardest hit by the mud wave. The *arrow* shows the height of the sludge level. (Picture by Ákos Stiller, permission obtained from copyright-owner)

buildings, and caused considerable damage to the natural environment. About 400 homes were flooded but this was only in the badly hit village of Devecser, and dozens of people had to be evacuated from other villages (Fig. 4.15). The news spread all over the world press. This was the most serious accident in the 120 years old history of industrial aluminum production.

"It is not the red mud itself, but the huge amounts of muddy, alkaline water from the red sludge reservoir in Kolontár that caused the disaster on Monday. The victims did not die due to poisoning or burns, but because of the flood. The red sludge itself is not considered hazardous waste, it can only cause damage to health when it dries out and spreads through the air. The base content of red sludge reservoirs is much more dangerous as it can cause burns, kill plants and aquatic organisms." These were the words of Hungarian journalist Sándor Joób (index.hu) on the very next day.

The accident had far-reaching environmental, economic, political and social consequences. As expected, a lot of correct and erroneous decisions were made, and a lot of information—both reliable and unreliable—was published. After the disaster, a number of politicians and some activists of Greenpeace thoroughly discredited themselves by issuing inaccurate and hysteria-prone reports. In the ensuing pandemonium, the events were declared to be an industrial disaster before revealing the real reasons. This unfortunate statement automatically excluded Hungary from the financial support mechanisms of the European Union, which can only be used in the case of natural disasters. After the catastrophe, the civil protection authorities made huge efforts to build entirely unnecessary barriers between reservoir No. X and Kolontár, although it was clear that the rest of the mud in reservoir No. X. was not going to move anywhere, and the new dams would be unable to stop the possible flow of mud from the neighboring reservoir No. IX. The defense laws were amended on the initiative of Defense Minister Csaba Hende to include a passage that authorizes the government to place any entity under state supervision at any time. This "Einstand" law did not specify what events should trigger this action. MAL, which was worth about 100 million € and held by Hungarian private investors, was put under state supervision on October 11, 2010.

Under public pressure, responsible persons were named before the end of the investigation, which was highly inappropriate. MAL CEO Zoltán Bakonyi, who was responsible for the operation of the red sludge reservoir, made the truly outrageous and cynical statement that "in the mud, there are no hazardous free radicals; it is a harmless substance which can be removed by a strong splash of water". Despite the lack of credible communication, it is quite possible that no regulations were violated during the operation of the establishment, which was built in the 1980s. Mr. Bakonyi was arrested on October 11, but a court released him two days later because no sufficient grounds were found for detention. It was easy to blame the irresponsible attitude of MAL for the disaster, but the shortcomings of economic and environmental regulations and their enforcement were equally important. The state secretary of Ministry of Rural Development responsible for environmental affairs, Mr. Zoltán Illés—strengthening the misconception that red mud, in contrast to earlier Hungarian regulations, is not considered a hazardous waste in the EU—hinted that the previous socialist government changed the environmental classification of red mud "because of self-interest" (*i.e.*, because the owners had actively supported that government). A few days later, the very same man took a 180-degree turn and announced an initiative that red mud should be regarded as hazardous waste "in the whole world". It did not matter at all that EU regulations were quite logical: red sludge does not qualify as hazardous waste only if it does not contain hazardous materials. It is the authorities that should have decided whether the sludge in the reservoir is "other wastes containing dangerous substances from physical and chemical processing of metalliferous minerals", or "red mud from alumina production other than the wastes mentioned in the above." The environmental permit of the operators classified the waste into the latter, harmless category. In early 2002, when Mr. Illés was chairman of the Hungarian Parliament's Committee on Environmental Protection and a professor at the Central European University (CEU), he visited the plant with his students, and found everything in order—as also reported in the weekly newsletter of MAL. Before the tragedy occurred at the end of September in 2010, the representatives of the head of the environmental authority responsible for issuing permits, Ms. Andrea Zay (appointed directly by Zoltán Illés) visited the plant as a preparation for extending the permit, set to expire in 2011, for another 5 years. They did not have any objections. Investigative journalists of two Hungarian journals (*HVG* and *Magyar Narancs*) have done a lot to find all the people responsible for the event, but this mission seems doomed because the case involves too many people with important roles in the government. In September, 2011, the responsible environmental authority issued an enormous fine (450 million €) against the company MAL. MAL appealed the decision as this sum is about four times the value of the company. Some 15 employees of the company MAL have been sued in a trial that started on September 24, 2012. The reconstruction works at the site of the disaster, covered by the Hungarian state, cost 127 million €. 7.5 million €, collected by civil organizations and individuals Europe-wide have been distributed in the area among the citizens and villages affected.

The clarification of the root causes of the dam failure is still in progress. At present, the most likely explanation is that the dam was built in the wrong place: there was a creek bed at the north wall. Moreover, leakage was observed both by the northern and the much thicker western dam, and the dams were heightened several times. The reservoir was probably overloaded.

Alumina is the most important raw material for the production of aluminum, and red mud is a by-product of alumina production. During the operation, 1.5–2 t of red sludge is produced for every ton of alumina. The sole raw material in the process is bauxite, a significant quantity of which occurs in Hungary. To produce 1 t of alumina, 1.9–3.6 t of bauxite is needed depending on the exact composition. Aluminum oxide is dissolved from bauxite by an alkali, sodium hydroxide (NaOH) treatment at high temperature and pressure to form aluminate base, $Na[Al(OH)_4]$. At present, 90% of the all alumina in the world is produced by this method, which was patented by Austrian chemist Carl Josef Bayer (1847–1904) in 1892. From the aluminate solutions, solid aluminum hydroxide $[Al(OH)_3]$ is crystallized, which is then separated from the sodium hydroxide solution by filtration. Sodium hydroxide is then recycled. Calcination of aluminum hydroxide (alumina hydrate) gives alumina (Al_2O_3). Aluminum is obtained by electrolysis (electrical current) from alumina dissolved in molten cryolite (Na_3AlF_6) at high temperature, about 1000 °C. Electrolysis usually needs a lot of electricity and is quite expensive, but the technology is so refined that the price of aluminum is only high compared to iron, but not to other metals. Even in comparison with iron, the higher price is often compensated by the favorable properties of aluminum, such as low density and good mechanical properties.

The residue of the sodium hydroxide treatment, called red mud, is insoluble in strongly alkaline medium. Strictly speaking, this phrase should only be used for these insoluble residues of the Bayer technology. However, the term red mud is also used in a much broader context today: it also means the highly alkaline slurry that forms in the Bayer technology and contains about 30% solids. To adapt to this change, specialists increasingly use the term bauxite residue instead of the original red mud. The highest possible amount of sodium hydroxide is washed out from red mud for recycling. The insoluble residue is still highly alkaline: the pH is generally between 10 and 11, sometimes can even reach 12. Values of pH higher than 12 clearly indicate that red mud was not washed properly for sodium hydroxide recovery. Usual ways of handling red mud include dumping into sea water, wet storage, dry stacking and dry storage. The method used in Hungary until 2011 was the cheapest available: wet storage. Dry storage was introduced from 28 February 2011. Red sludge is still only stored, and not processed in Hungary. The environmental risks of maintaining the huge reservoirs are very large, so strict rules must be observed during storage. The huge amount of red mud is an unavoidable by-product of the Bayer alumina production. Over the last decades, a total of 55 million t of red mud had accumulated in the reservoirs across the country. Worldwide, the deposited mass of red mud was 2.7 billion t in 2007, and the amount increases by 119 million t per year. Red sludge in Hungary is the biggest source of potentially hazardous waste because of its alkali content.

Red mud may be used to produce titanium, germanium, vanadium, a red dye, as a coagulant in water treatment, as an additive for brick manufacturing (maximum amount of 15% because its radioactivity is somewhat higher than the usual background), and it can also be used for road construction. The high sodium content is usually quite problematic when red mud is used as a raw material for iron production. The patent of Hungarian inventors János Palla and István Röczei might provide a solution, but the process is currently used only on small scales. Obviously, the real breakthrough would be economical recycling of red mud instead of its simple accumulation. This is not only a technological problem, but also a matter of economic regulations. As long as storage in different forms is cheaper than the re-use, the producers will hardly change their attitudes.

During the red mud disaster, the public interest was highest in the alkali and heavy metal content as well as the radioactivity. The dust of the completely dried sludge was also an issue.

Red mud itself is not toxic, but the residual sodium hydroxide content makes it a hazardous substance. The main cause of the environmental damage was the high pH (around 13) of the red mud slurry, i.e., the liquid part was strongly alkaline, about ten times more basic than household bleach. The red mud stored in the region typically contains 5–8% sodium, but a substantial part of this is present and bonded in sodium aluminum hydrosilicates. Sodium hydroxide is widely used in large quantities in the paper, textile, soap and detergent industries and, of course, in bauxite processing. When the dried red mud powder enters respiratory tracts, it may cause irritation as dust usually does, but the high pH makes this effect even stronger.

The main component of red mud in the region was iron oxide (40–45%), which caused the red color. Iron is a heavy metal, but not a toxic one. It is even essential to life in small quantities ($\rightarrow$ **1.12**). In addition, red mud contains some rare metals, too, but their overall concentration is below 0.3%. Furthermore, it also contains the elements vanadium (0.1%) and titanium (2.7%). Many of these constituents are chemically bonded, which prevents their dissolution under usual circumstances. Therefore, the dissolution of heavy metals (chromium, lead, mercury, thallium, cadmium) from red mud—in sharp contrast with the initial claims of Greenpeace—poses no real risk. On the contrary, testing of soils contaminated by wastes from mines and heavy industry showed that the addition of red mud can significantly reduce the availability of heavy metals because it stabilizes these substances. For example, the water soluble lead content of red sludge is lower than the water soluble lead content of normal soils. The alkaline liquid part of red mud dissolves heavy metals less readily than pure water does. The metals present in red mud (except arsenic) have very low mobility. Under strongly alkaline conditions, aluminum can be dissolved, but solubility decreases very steeply with a decrease in pH. The toxic metals from red sludge typically enter the upper 10 cm layer of the soil, but the risk to the environment is minimal because of the poor solubility. Belgian researchers have recently grown barley on soil to which 5% of red mud from Hungary was added, and found that the shoot yield of barley seedlings was affected by 25%. Control experiments using sodium hydroxide led to the conclusion that the decrease is entirely due to the high sodium content of the soil (a phenomenon called salinization).

Naturally occurring radioactive materials in bauxite are transferred almost entirely into the mud during processing. The degree of enrichment relative to bauxite is a factor of 2 or 3. The radioactivity of the red mud in the disaster was about ten times higher than the average radioactivity of the surrounding soil, but it is still substantially below the legal limits, so red mud is not officially considered a radioactive material. Similar substances are known in the literature as NORM materials (*Naturally-Occurring Radioactive Materials*). Another source of radioactivity was radon gas in the air. Radioactivity values measured on site were around 20% of the latest, recently confirmed recommendations of the World Health Organization (WHO) for homes. It is very likely that the main source of this radon gas was not the red mud that caused the catastrophe: red sludge in the intact reservoirs, or the relatively radium-rich nearby waste-piles are both known to give off radon.

When red sludge dries, the increased amount of dust formed should also be considered. To estimate the radioactive dosage from this source, information about the size range of red mud dust particles is needed. The samples from different locations mainly (47–95%) contained particles of about 20 μm, 35–68% were about 10 μm, and 20% were smaller 5 μm. For inhalation, particles smaller than 1 μm cause the most concern. The measured values would make 0.2% of the annual average inhaled radioactivity assuming a constant inhaled amount of dust. All of these facts clearly show that the red sludge catastrophe did not increase public exposure to radioactivity at all. Dust levels did not exceed the dangerous levels in the disaster area. Human dust protection is only necessary if unfavorable weather conditions (drought, strong winds) increase the concentration of dust to high values. In this case, the public exposure may be similar to that of a dusty work site, which may pose hazards to the health even if the powder is inert.

Viewed from some distance, every disaster brings some benefits, too. Some processes have to be re-evaluated, and the way things are conducted in the future will be different from the previous practices. Technologies should be reviewed and more stringent safety measures must be taken.

4.9 Death in a Single Grain

Barium sulfate seems to be a notorious substance, at least according to the popular press. On December 4, 2003 a traffic accident in Hungary involved a truck that was transporting barium sulfate. Some of the media reported the accident and noted that a single grain of the substances could cause death.

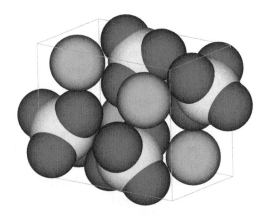

Fig. 4.16 Crystal structure of barium sulfate. (Copyright-free Wikipedia picture)

Everyone is entitled to his own opinion—but not to his own facts. In this instance, readers and viewers were misled big time. Barium sulfate is benign. What is more, doctors often ask people to drink its suspension in water. It is not listed as a hazardous substance in the European Union (67/548/EEC). In the US, standard NFPA 704 classifies it at 0 (without any known danger) for health, fire safety and reactivity. It is difficult to imagine where the incorrect information originated as a minimal effort in a web browser reveals the properties of the substance. It is true that water-soluble salts of barium are often poisonous, although the poisonous dose is much higher than a single grain for all of them. Barium sulfate, on the other hand, is benign because it does not dissolve in any liquid: this is the primary property most chemists can recall about it.

Barium sulfate is chemically the same as the natural mineral called barite (Fig. 4.16). These are unusually heavy crystals (its density is high), which are relatively easy to identify even for those less experienced in mineralogy. The main use of barium sulfate is in oil drilling. It also has a very attractive, bright white color and high reflection, so it is used in photographic papers on which expensive art books are published. Its medical use is typical for X-ray tests of the intestine and bowels. It is called a contrast agent: it absorbs X-rays much better than any human tissues, so it helps form a much more informative picture (Fig. 4.17). The barium sulfate goes through the body unchanged, it does not interact with anything.

A literary mystery is also connected to barium sulfate. In Sir Arthur Conan Doyle's crime story titled *A Case of Identity*, a conversation between Sherlock Holmes and Dr. Watson goes like this.

> 'Well, have you solved it?' I asked as I entered.
> 'Yes. It was the bisulphate of baryta.'
> 'No, no, the mystery!' I cried.
> 'Oh, that. I thought of the salt that I have been working upon.'

There is a huge mystery here. The name bisulphate of baryta would correspond to the chemical formula $Ba(HSO_4)_2$. A substance with this formula is unknown in modern science. What could have been the salt Sherlock Holmes was working upon?

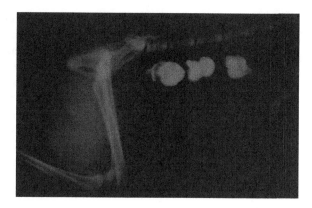

Fig. 4.17 The gastrointestinal tract of a cat in an X-ray image taken using barium sulfate as a contrast agent. (Authors' own work)

4.10 A Closer Look at Acids

For most people, all acids are liquids. This is most probably caused by the fact that the common acids in everyday life, such as hydrochloric acid or battery acid are indeed liquids, and nitric acid or phosphoric acid, which one may remember from school, are also liquids.

However, a careful look in a supermarket will reveal, perhaps to the customer's surprise, white powders labeled 'citric acid' and 'tartaric acid'. In pharmacies, ascorbic acid is often sold as the pure active ingredient of vitamin C, and it is also a white powder. Can't chemists make up their minds?

Well, actually, the case is much worse. Experienced chemists ordinarily mean different things when they use the word 'acid'. The concept of acid has three different, but commonly used definitions in science. After all this, it may not be too surprising that a fourth one will be discussed first.

Originally, acid meant something sour. Similarly, bases were bitter things. These original definitions are no longer used in science. It would be far from practicable or healthy to taste every substance to decide whether it is a base or acid. However, this was not uncommon in the eighteenth century. Carl Wilhelm Scheele (1742–1786) was the discoverer of 'blue acid', which is an aqueous solution of hydrogen cyanide. In his notes, he also described the taste of blue acid. He survived this particular experiment, but his practice of tasting of chemicals may have contributed to his early death.

The first scientific definition in this new field was described by a Nobel laureate, Swedish chemist Svante Arrhenius (1859–1927). His theory, which is now called the Arrhenius acid-base concept, states that acids are substances that produce hydrogen ions in an aqueous solution. Hydrogen ions (H^+) do not exist in solutions because they are always attached to at least one water molecule- often written as hydroxonium ion (H_3O^+). More detailed studies show that H_3O^+ ions are not very common, either, and hydrogen ions are often attached to more than one water molecules. But this is only a matter of notation and different chemists use H^+, H_3O^+, $H_5O_2^+$ or $H_7O_3^+$ to mean essentially the same thing.

The Arrhenius model is still widely used, and substances that are called 'acid' are usually acids in the Arrhenius sense. Since citric acid, for example, produces hydrogen ions in solution, it is therefore an acid. Incidentally, it also tastes sour and is edible at the same time. The same is true for tartaric acid. The most commonly consumed form of tartaric acid is wine, but it is mixed with alcohol there, which may have a significant health effect. In the Arrhenius concept, a substance can be called acid or base without further specifications (bases decrease the concentration of hydrogen ions). In the remaining two theories, strictly speaking, a substance cannot be called acid or base. All that can be said is that a certain substance behaves as an acid (or base) in a certain chemical reaction with another reactant.

The second recognizable theory is named after Danish Johannes Nicolaus Brønsted (1879–1947) and English Thomas Martin Lowry (1874–1936). This theory interprets acids as substances which donate hydrogen ions to another substance. The substance accepting the hydrogen ion is the base. It may seem peculiar, but there are a few known reactions in which sulfuric acid (H_2SO_4), despite its name, plays the role of a base:

$$HSO_3F + H_2SO_4 \rightarrow SO_3F^- + H_3SO_4^+$$
acid   base

In the Brønsted–Lowry theory, water no longer plays a special role, and the chemical reaction can take place in any solvent.

This classification was still not quite satisfactory for American chemist Gilbert Lewis (1875–1946)—or at least one could imagine this had he not published his thoughts in the same year as Brønsted and Lowry (1923). Lewis' notion was that hydrogen is present in many substances, but not all of them, whereas electrons are present in every substance. So, in the Lewis acid-base theory, electrons (or electron pairs, to be more precise) play central roles. In this theory a molecule that can accept an electron pair plays the role of an acid. Conversely, a base can donate an electron pair:

$$(CH_3)_3 B + :NH_3 \rightarrow (CH_3)_3 B:NH_3$$
Lewis acid Lewis base

After four diverse definitions, it must be noted that none of them makes reference to the physical state of the substance. Accordingly, acids can be gases, liquids or solids. To introduce an additional complication, the most common acids in chemistry are not pure substances but aqueous solutions. Concentrated hydrochloric acid, for example is a 36% aqueous solution of hydrogen chloride molecules. It would be a little more precise to call it a solution of hydrochloric acid, but chemists are a lazy bunch and they tend to economize on the use of words. Despite the differing theories from these chemists, and despite the rigor that scientists adhere to, none of these theories satisfactorily encompasses the great range of chemical substances and our perception and usage of them.

Fig. 4.18 The model of the ozone molecule. (Copyright-free Wikipedia picture)

4.11 O, the Fresh Smell of Ozone

It is commonly thought that ozone is an agreeable feature of the atmosphere. If this were not true, then why would the lack of ozone (ozone hole) pose such a problem? Few people realize that the hole in the ozone layer is high above us ($\rightarrow$ **1.9**), so ozone must primarily dwell in the clouds. In fact, ozone is quite poisonous, so it would be more than unpleasant in the air we breathe.

Ozone is a variety (chemists use the word allotrope) of the element oxygen (Fig. 4.18). The most common variety is the oxygen molecule, which is composed of two chemically bonded oxygen atoms and makes up about one fifth of the air. In contrast, ozone consists of molecules in which three oxygen atoms are bonded together: one of them is in the middle and is connected to both of the other atoms—the other two are only connected to the center. This difference between the allotropes of oxygen is critical as far as chemical and physical properties go. Diatomic oxygen molecules are not very reactive at room temperature and this is the reason why there are plenty of them in the air. Ozone, on the other hand, is very reactive and unstable, and more importantly, it breaks down practically every organic material including human tissue. If there are no other possible partners, an ozone molecule will even react with another ozone molecule to form diatomic oxygen molecules. Diatomic oxygen has no odor, and if for no other reason, there is so much of it around that humans get used to it. Ozone, however, has a particularly unpleasant stink, which may be familiar to those who have noticed something out of the ordinary next to an operating copier. This stink is so distinctive that it gave birth to the name 'ozone': Christian Friedrich Schönbein (1799–1868), a Swiss chemist, coined this word in 1840 based on the ancient Greek 'ozein' (όζειν), which means to smell. Ozone forms in air during electrical discharges. This is even mentioned in the novel of Thomas Mann, *The Magic Mountain* (*Der Zauberberg*):

> ..."But what is it I smell?"
> "Oxygen," said the Hofrat. "What you notice in the air is oxygen. Atmospheric product of our little private thunderstorm, you know. Eyes open!" he commanded. "The magicking is about to begin." Hans Castorp hastened to obey. (translation: H. T. Lowe-Porter)

Well, the great German writer was not wrong about the peculiar smell from the private thunderstorm (electrical discharge). But his mistake was in identifying the cause. Oxygen, as already mentioned, has no smell and is continuously present in the air. It is ozone that is responsible for the unpleasantries. Incidentally, too much oxygen in the air would not be without adverse effects, either. Combustion would be much more intense and some biological reactions would be accelerated. French writer Jules Verne wrote a novel about the possible effects of two much oxygen in the air. The title is *Dr. Ox's Experiment*.

Even at low levels, ozone is highly toxic to most living orgasms as the processes that occur with diatomic oxygen slowly and in a controlled manner are usually very rapid with ozone. Oxidizing smog (named after Los Angeles—the reader can guess the reason) has ozone as one of its most harmful components. In megacities, ozone levels are often reported in the media on a daily basis. The toxic properties of ozone become useful when microorganisms must be killed—it is an excellent disinfectant. In some water treatment plants, ozone is generated on site, kills bacteria quickly, and then the excess decomposes rapidly. An even more positive property about ozone is the protective layer it forms in the upper atmosphere that shields the surface of Earth from harmful ultraviolet radiation ($\rightarrow$ **1.9**).

4.12 The Diamond Thermometer

Cold as a diamond. The older generation may still remember Chris Norman's song titled *Some Hearts are Diamonds* from 1986. In esoteric texts, cold diamonds are often contrasted with fiery rubies.

But are diamonds really cold? Few people are privileged enough to decide this question experimentally with a suitably sized piece of diamond. A few witnesses with reliable senses may even swear that the queen of precious stones does in fact have a cold touch. Experimental observations, though, are not limited to jewelry stores. One can try touching things in the kitchen as well, although not something taken out of the fridge or from the oven. The temperature of a wooden spoon or the plastic handle of a knife feels normal. The metal blade of a knife, on the other hand, is definitely cold. Or is it? Temperature measurements with a thermometer will reveal something unexpected—all objects in the kitchen have exactly the same temperature, and in addition, this is the same temperature as the air in the kitchen.

There are very well established natural laws, called the laws of thermodynamics, working behind the scenes. Along with a few other features, these laws state that objects in contact with each other, absent any external heating or cooling, will reach a state called thermal equilibrium, in which their temperatures will be equal. Under normal kitchen conditions, half an hour is more than enough to reach this thermal equilibrium, so all objects in contact with air will have identical temperatures. Why then, do they feel so different?

There are two key scientific facts here. First, the human body heats itself, and usually maintains a temperature around 36 °C. So a human hand is usually warmer than the air in kitchen. If you touch something, you begin to heat it at the same time. A piece of ice melts in human hands much faster than ice left alone. However, this does not itself explain why you feel one object cold and another warm. A second important phenomenon is heat conduction. Heat is the disordered movement of

4.12 The Diamond Thermometer

Table 4.3 Thermal conductivities of selected substances

Substance	Thermal conductivity (W m^{-1} K^{-1})
Diamond	1000
Silver	430
Copper	380
Gold	320
Aluminum	240
Table salt	6.3
Ice	2.0
Concrete	1.7
Sand	1.3
Glass	0.8
Water	0.6
Wood	0.04–0.4 (depending on type)
Air	0.025

the invisibly small particles of a material, and it radiates to every object. What is more, it radiates in different objects at different rates. If a material is a good thermal conductor, heat will spread within it rapidly, and the temperatures of its different parts quickly become equal. In poor thermal conductors, the same process is much slower. If you touch a good thermal conductor, it feels cold as the relatively small amount of heat that can be transferred from your hand is led away to other parts of the object so quickly your hand cannot heat up the thin layer that is in direct contact. If the object has low thermal conductivity, a very thin layer is heated up, but the temperature change does not radiate efficiently, so even though your fingers feel warm, the whole object does not heat up significantly.

Thermal conduction is very important in certain areas of everyday life. For example, a dish intended for cooking must be made of a good heat conductor because this is the only way to ensure that the food in it will also be heated. Metals are usually good conductors of heat, but not equally so. The best are silver and copper. Silver cooking dishes would probably be too expensive even for a king, so copper is the material of choice for top quality cooking dishes—these are not inexpensive, either. The thermal conductivities of a few materials are listed in Table 4.3. Among the listed substances, it is not a metal, but a gem that has the highest value. This does not necessarily mean that diamonds are the best heat conductors of all known materials, but it is certainly close to the top. At the bottom of the list are substances with low thermal conductivities which are called good insulators. An even better thermal insulator is air, so doubly glazed windows are good insulators because of the air trapped between the two layers of glass. This fact has paramount importance in the kitchen as only a few centimeters from a hot oven, the air temperature is already tolerable to humans.

Returning to diamonds, the high thermal conductivity of diamonds is certainly mysterious. In physics, a rule called the Wiedemann–Franz law states that the electric conductivity of a substance is proportional to the thermal conductivity. Briefly put, this law says that good thermal conductors are good electrical conductors at the same time. Strictly speaking, this is only true for metals. The best counterexample happens to be the diamond mineral—it possesses higher thermal conductivity than any metals, but it is also an excellent electric insulator (Table 4.3).

4.13 Everlasting Love: Diamonds

Thanks to the decades-long and expertly nuanced advertising campaign of diamond sellers, everybody seems to believe that diamonds are the most beautiful, most spectacular and most valuable of the precious stones, and men have no better way to express their appreciation for the beauty of women than buying one (at least in America). Furthermore, based on similar disinformation, everyone is convinced that diamonds are forever. If nothing else, the title of a 1971 James Bond movie *Diamonds are Forever* would convince all skeptics.

This slogan appears to have been created by US novelist and screenwriter Anita Loos in her 1925 book titled *Gentlemen Prefer Blondes*, the film adaptation of which starred Marylin Monroe in 1953. Prompted by the advertising agency Ayers, the leading diamond seller in the world, De Beers began using this slogan in 1948 and has continued to make profit on it ever since. In a March 1999 special issue, the journal *Advertising Age* named this catchphrase the most successful advertising slogan in the twentieth century. Additionally, a website is dedicated to this phrase alone. All this hype does not ensure that the statement is scientifically correct.

It is an amazing chemical fact that diamonds are anything but eternal, especially compared to other precious stones. Diamonds are a variety of the element carbon. Another variety of carbon is the well-known and not even particularly expensive graphite. Coals used for heating or electricity production contain carbon in the form of graphite, mixed with some other substances. A scientific study of carbon leaves very little room for doubt: at room temperature and atmospheric pressure, graphite is more stable than diamond, because diamonds slowly—extremely slowly!—convert to graphite under normal conditions. To prevent a major panic among diamond owners, another scientific fact should be stated: this conversion cannot be detected even with the best instrumentation available today, even after a thousand years. A diamond ring can improve the marriage prospects of the great-great-great-great-grandson of the original buyer.

How are diamonds formed then? The answer is not very complicated. As emphasized above, graphite is more stable than diamond at room temperature and atmospheric pressure. Increasing the pressure and the temperature changes things, and at a certain point, diamond becomes the favored form. Information such as this is often displayed on graphs called phase diagrams. The phase diagram of the carbon element is shown in Fig. 4.19. This shows which conditions favor diamond or graphite. In addition, there is also information on conditions at which the inter-conversion of the two forms is very slow. The less stable form is called metastable under conditions like these. There are conditions where graphite is stable and diamond is metastable (for example, room temperature and atmospheric pressure, which is approximately 1 bar), whereas change may result in the opposite case, whereupon diamond is stable and graphite is metastable (for example 100,000 bar and 1000 °C). What is

4.13 Everlasting Love: Diamonds

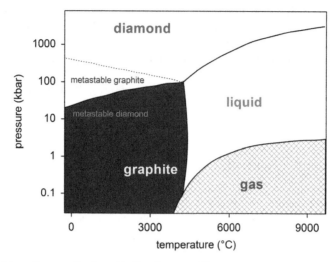

Fig. 4.19 Phase diagram of carbon. (Authors' own work)

more, there are even conditions where graphite converts to diamond quite rapidly. To achieve this, a pressure higher than 600,000 bar is needed. Such high pressures are common 100–150 km deep underground where diamonds can form. If it reaches the surface somehow, a miner is usually very happy to collect the financial benefit of finding it.

Diamonds are in fact an exceptional precious stone because of their extreme hardness and light refraction, the second of which makes it sparkle beautifully. A common, but not completely accurate view is that diamonds are the hardest of all known substances. This statement is not easy to prove or disprove as most crystalline substances are anisotropic, which means that their properties (hardness included) depend on the orientation in which they are studied. Crystals of rhenium diboride (ReB_2) are harder in certain orientations than diamonds are in any orientation. The same is true for another substance called ultra-hard fullerene, which—no small irony involved—is another, very rare form of carbon, and can be metastable under any conditions.

Returning to the robust persuasion powers of the diamond industry, in the James Bond film mentioned earlier, the evil protagonist, Ernst Stavro Blofeld, needs numerous diamonds to build a laser that can shoot down spaceships or satellites. The writer, Ian Fleming was probably reflecting about glowing diamonds when he wrote the story. However, there is a scientifically weird, but valid point that among precious stones, the ruby is best suited for constructing a laser. To put it more bluntly—ruby lasers are well-known, whereas diamond lasers are completely unknown.

To add one final item to the list of misconceptions about diamonds, let us quote a scientific blunder of Lieutenant Columbo. In the episode titled *Ashes to Ashes*, the legendary detective mentions to a mortician that nothing happens to the stone of a diamond ring during cremation. This would be true for most other precious stones, but not for diamonds. As it is made of carbon, it can burn and produces carbon dioxide at the high temperatures of a crematorium. This misconception is not significant for solving the murder case though, because the bit of material that survives cremation and serves as decisive evidence is actually a piece of metal shrapnel... This is possible provided that the melting point of the metal is high enough, which is not unlikely for ammunition.

So, diamonds are certainly hard, but do not last forever. Both fire and (very) long age are mortal enemies of these precious stones.

4.14 Is pH 5.5 Neutral?

Cosmetics with a **pH** of 5.5 are well known from advertisements or drugstore shelves. Most of these products are shower gels, moisturizers or other creamy stuff that can be rubbed directly into the skin, and it is implied that a pH of 5.5 is an important and skin-friendly property of these products. Unfortunately, this leads people to conclude that a friendly pH is a neutral pH. In first year university chemistry, a sizeable part of the students selects 5.5 as neutral pH despite the fact that they were supposed to absorb something else.

The concept of pH is not among the simplest ones in chemistry. However, as these advertisements prove, pH sometimes finds its way even into mass media. The approximate meaning of pH is the negative, ten-based logarithm of the concentration of hydrogen ion measured in units of **mol/dm³**. This may sound complicated enough, but to prevent an otherwise foreseeable attack of physical chemists on the authors of this book, it needs to be mentioned that the concept of pH has a more rigorous scientific definition, which focuses upon molality-based activities. To understand this more rigorous definition, reading a few chapters of a textbook on thermodynamics is indispensable. For our more simplified purpose, that is probably not necessary, but some explanation is still beneficial.

Pure water does not simply contain molecules since ions are also present, as the following reaction also shows:

$$H_2O = H^+ + OH^-$$

This process bears the respectable name of '*self-dissociation of water*' and is presumed to be an equilibrium reaction, because it concludes when the species indicated on the left and right side of this chemical equation (reactants and products) are present at the same time. This means the order of the reactants and products in the above chemical equation could even be reversed. The reaction itself is also called reversible. In pure water, the same amounts of hydrogen (H^+) ions and hydroxide ions (OH^-) form. The notation used for hydrogen ion is still a matter of debate in today's science. Some experts stubbornly stick to the notation H_3O^+, which is called oxonium or hydroxonium ion (→ **4.10**). It is difficult to formulate justice here, as reality itself is much more complex than any single chemical formula could express. But as far as notations go, using the simpler one seems... well, simpler.

4.14 Is pH 5.5 Neutral?

Table 4.4 The pH of a neutral solution at different temperatures

T (°C)	Neutral pH
0	7.47
10	7.26
20	7.08
25	7.00
30	6.92
37	6.81
40	6.77
50	6.64
60	6.52
70	6.41
80	6.31
90	6.21
100	6.13

This disagreement on notation does not really change anything important—a solution is neutral if the amount of hydrogen and hydroxide ions in it are equal. In chemistry, the quantity of a substance in a set volume is preferred and is called concentration. The usual unit of this is mol/dm^3. The concentration of a substance in a solution is 1 mol/dm^3 or 1 M if 1 l (which is the same as 1 dm^3) of the solution contains 1 mol of the substance. It is also useful to remember that every substance in a solution has a concentration of its own much like a large storeroom of fruits needs to keep track of the amounts of pears, apples and other fruits individually. During the self-dissociation of pure water at 25 °C, an equilibrium state is reached in which the concentrations of both hydrogen and hydroxide ions are 1×10^{-7} mol/dm^3. If this number seems very small, it is useful to imagine a very small droplet of water (approximately 0.00001 l): it would still contain 6×10^{11}, i.e. 600 billion, hydrogen ions. A further consequence of this self-dissociation is that whenever the concentration of hydrogen ion increases because of the addition of another substance, the concentration of hydroxide ions must decrease, and vice versa. If the solution contains more hydrogen ions than hydroxide ions, it is called acidic. If the opposite is the case, then the solution is basic. If the concentration of hydrogen ions increases to 0.1 mol/dm^3, the solution will be acidic, and its pH will be 1 (because this is the negative ten-based logarithm of 0.1).

With this information, one can conclude that a pH of 5.5 is lower than neutral. When the concentration of hydrogen ion is larger than the concentration of hydroxide ion, the solution is acidic. Still, it must be remembered that the surface temperature of a person's skin is not exactly 25 °C. Table 4.4 shows the dependence of neutral pH on temperature. At 37 °C, the neutral value is 6.81. Therefore, 5.5 must still be acidic.

Anyone interested in why pH values gained such prominence in the scientific world should also note that many of the properties of a solution do not depend on its actual constituents—pH is all that matters. A solution of pH 14 is highly harmful for the eyes and the skin, and it is not important what causes this basic nature. In Hungary in 2010, there was a major catastrophe of 'red mud' In this disastrous waste spill, the highly-basic pH of the 'mud' was the primary hazard, and not the metal content of the mud, as the media reported at the time ($\rightarrow$ **4.8**).

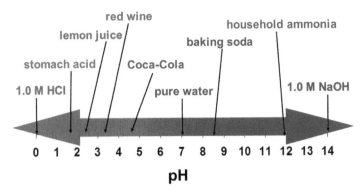

Fig. 4.20 The pH of some solutions in everyday life. (Authors' own work)

However, it would be misleading to say that the human body is only compatible with neutral solutions. The usual pH scale of aqueous solutions ranges from 0 to 14 and the pH of a few common solutions is shown in Fig. 4.20. Human blood has a slightly basic pH value (between 7.35 and 7.45) in healthy people. If someone's blood is neutral (pH of 6.8 considering the temperature), it is already of sign a serious illness. Pure rainwater is slightly acidic as air contains some carbon dioxide, which forms carbonic acid when dissolved in water. Coca-Cola contains a little more carbonic acid than rainwater does, and since it also contains phosphoric acid, it is more acidic than rainwater. In the stomach, conditions are highly acidic to help digestion. The acidity is not a problem there, but the esophagus and the mouth do not tolerate these conditions easily, and this is partly why it is very unpleasant if the stomach contents decide to see daylight through the way they entered.

Finally, the authors, as chemical professionals, should defend the manufacturers of cosmetics with pH values of 5.5. There is absolutely no evidence that these companies ever made a claim for the 'neutrality' of their product. The most they do is declare that it is skin neutral. On a few websites, it is even explained that pH 5.5 is actually acidic, and skin neutral refers to a pH that is the same as the pH of the skin surface. Strictly speaking, pH only makes sense in aqueous solutions. Still, the term skin-neutral is not unacceptable. Such cosmetics do not change the acid-base properties of skin, and even though no one has proved that this is beneficial for the skin, it is certainly not harmful, either (Fig. 4.20).

4.15 Sorrel Soup Scare: Oxalic Acid

After a scientific lecture about a newly developed UV-meter during a chemical conference in 1995, a heated debate evolved about the toxicity of oxalic acid. For some witnesses, this debate was quite memorable for two reasons. First, neither of the sides in the debate was familiar with basic toxicology and second, the question

4.15 Sorrel Soup Scare: Oxalic Acid

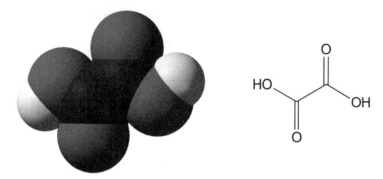

Fig. 4.21 Space-filling model and the chemical structure of oxalic acid. (Copyright-free Wikipedia picture)

itself was irrelevant because the UV-meter only contained oxalic acid in its chemically bound form, which is by no means the same as oxalic acid itself. Oxalic acid occurs in many edible plants, so the question is not entirely academic.

In everyday life, a substance such as white arsenic, or even oxalic acid, is often deemed poisonous, or to use a more highbrow word, toxic. This statement by itself does not mean much, as the devil is in details, more specifically, in the amount of the substance. This idea has been stated by quite a number of people in a variety of ways. One of the first examples comes from Paracelsus, a scientist in the sixteenth century:

> All things are poison, and nothing is without poison; only the dose permits something not to be poisonous.

This can be worded in a shorter version as well: "The dose makes the poison."

Thomas Mann, German writer in the twentieth century had his own take on this in his novel *The Magic Mountain* (*Der Zauberberg*):

> The truth was…, in the world of matter, that all substances were the vehicle of both life and death, all of them were medicinal and all poisonous, in fact therapeutics and toxicology were one and the same, man could be cured by poison, and substances known to be the bearer of life could kill at a thrust, in a single second of time. (Translation: H. T. Lowe-Porter)

Although neither Paracelsus nor Thomas Mann is accepted as an authority of scientific questions today, they both made an excellent point in this particular instance—every poisonous substance has an amount that is too low to cause any detectable effect in living organisms. To put it in a different way, all substances that are necessary for a healthy life can be harmful if the amount is excessive. Shakespeare's aversion against too much of a good thing does not lack scientific justification.

Before jumping into our analysis of the poisonous amount of oxalic acid, it may be useful to become familiar with some of its properties. Oxalic acid is a white crystalline solid (most acids are not liquids! → **4.10**), which dissolves quite readily in water (Fig. 4.21). It was first isolated from wood-sorrel by a French scientist in 1688, and subsequently, in 1776, noted Swedish chemist Carl Wilhelm Scheele

Table 4.5 The oxalic acid content of a few vegetable plants

Plant	Content of oxalic acid in weight percent
Spinach	0.97
Rhubarb	0.50
Garden sorrel	0.30
Beetroot	0.30
Cauliflower	0.15
Cabbage	0.10
Pea	0.05
Tomato	0.05

artificially created it by reacting sugar with nitric acid, also proving the prepared substance to be identical to its natural form. For science historians, there is a vital anecdote regarding this substance: German chemist, Friedrich Wöhler prepared oxalic acid from an inorganic substance called cyanogen, which contributed significantly to disproving the *vis vitalis* theory (i.e. the view that organic substances can only be formed in living organisms).

The name oxalic acid refers to the Latin name of wood-sorrel, which is *Oxalis*. As far as its chemical structure goes, it is the simplest dicarboxylic acid, composed of two carbon, four oxygen, and two hydrogen atoms. It occurs naturally in a number of plants (see Table 4.5). However, a word of warning should accompany the data given in this table: the reader will find very different numbers in other sources. There are several reasons for this, one of the most important is the fact that the oxalic acid content depends quite a bit on the growing conditions even if an identical plant is studied. Another reason is that the percentage stated occasionally refers to the entire plant as 100%, with only the green parts in other instances, and still another practice is to dry the plant carefully before measuring. With this word of caution, the table actually reveals that even edible plants contain quite noticeable amounts of oxalic acid, for example 100 g of spinach may contain 1 g of the acid. Had it occurred to anyone to ask Popeye if he felt all right after a hearty meal?

There are actually some more facts which prove that oxalic acid is not so toxic after all. The most common kidney stone is composed of calcium oxalate, which is the calcium salt of oxalic acid. This substance is not exactly the same as oxalic acid itself, but it can only be formed from oxalic acid in the body. Patients with kidney stones seldom feel well, but oxalic acid is not always responsible for this medical condition. Some people have kidney stones composed of uric acid, and they feel just as bad as those whose kidney stones are composed of have calcium oxalate.

The value **LD$_{50}$** is often used to show how toxic a given substance is. Most scientific sources put the LD$_{50}$ of oxalic acid somewhere around 400 mg/kg of body weight. What this means is that 25 g of oxalic acid may cause the death of an adult weighing 65 kg with 50% probability. This value is quite high compared to other well-known poisons such as potassium cyanide. Nevertheless, there is a documented example of a fatality from oxalic acid poisoning. A 53-year old man, who had diabetes, was an alcoholic and smoked quite heavily died of oxalic acid poisoning after consuming about half a liter of sorrel soup, which contained 6–8 g of oxalic acid. In the 1950s, the tragic story of a child was recorded who ate a fatal amount of rhubarb. Physicians know exactly why oxalic acid is harmful—it lowers the level of calcium in blood.

Finally, it should not be left unnoted that in the UV-meter mentioned in the first paragraph, there was no pure oxalic acid at all. It contained a chemically bonded form of oxalic acid, which—from a toxicity viewpoint—is quite different from oxalic acid itself. This is by no means surprising, because common salt, which is added to most food, forms from a chemical reaction between a metal (sodium), which reacts vigorously with water, and a gas (chlorine), which was used as a toxic chemical weapon in World War I. The substance produced in the reaction, NaCl, has properties that are unrelated to the properties of the predecessors. This process of 'toxic' parents giving birth to 'non-toxic' children is a cornerstone of modern chemical science.

4.16 The Mysteries of Magnetism

Magnetic phenomena have always attracted a sort of mystic attention. After all, magnets can cause movement without any visible cause. Many people seem to think that magnets attract all metals. Nothing can be farther from the truth, and this can simply be demonstrated by a refrigerator magnet. This type of magnet remains on the refrigerator door or is suitable to collect dropped needles. However, gold or silver jewelry, objects made of lead or aluminum foil do not seem to mind the presence of a magnet. Not all metals are magnetic, after all. What's more, magnets sometimes attract objects that are not made of metal.

Physicists classify substances into three groups based on their magnetic properties: ferromagnetic, paramagnetic, and diamagnetic. Ferromagnetic materials are those that can be magnetized in an everyday sense. These behave as magnets on their own when they are placed into a magnetic field, and the magnetic properties often persist even after the external magnetic field is switched off. Permanent magnets lose their magnetic properties if heated above a characteristic temperature called Curie temperature, which is different for every material. Of the common metals, iron, cobalt and nickel are ferromagnetic, and in alloys containing these metals, there is a good chance of ferromagnetism. The Curie temperature for iron is 770 °C, that of cobalt is 1130 °C, whereas the value for nickel is 358 °C. Among the elements, no other is ferromagnetic. The closest is the rare earth metal gadolinium, which has a Curie temperature of 19 °C, which would probably be considered a permanent magnet in the coldest parts of the world, although it would not be used much because it is too expensive. Some nonmetallic substances may also be ferromagnetic. These are compounds of metals with nonmetallic elements with the most noted example being magnetite (Fe_3O_4), which even takes its name from this property. The same is true for pirrhotite ($Fe_{(1-x)}S$ where x = 0 to 0.2) or chromium dioxide, CrO_2), which may still be familiar for those who remember the higher quality cassette tapes.

The other two groups of materials, paramagnetic and diamagnetic, are not influenced notably by magnetic fields and the difference between them cannot be created in this way. For chemists, though, the difference is quite significant: paramagnetic substances typically contain unpaired electrons, whereas diamagnetic materials do not. This information is sometimes important when the substance is categorized or its application possibilities are assessed.

Returning to ferromagnetic materials: extremely high magnetic fields are needed in certain scientific or medicinal laboratories. Since these strong fields these cannot be generated by permanent magnets, electromagnets are needed in which the magnetic properties of strong electric currents are used. Very strong currents can only be maintained in superconductors, which can only be maintained at very low temperatures, close to absolute zero. Large magnets are used for NMR (nuclear magnetic resonance) in science, and MRI (magnetic resonance imaging) in medicine. Operating such instruments requires very special attention—the use of ferromagnetic objects in such laboratories must be avoided completely. Therefore, there are always warnings on the doors of such laboratories such as the one shown in Fig. 4.22. A careless visitor in such a room might easily find his or her keys flying toward the large magnet, or in less fortunate cases, credit cards or some magnetic data recording devices may become unusable. Everyone with a pacemaker should stay a good distance away from an NMR or MRI instrument to avoid health problems. Once upon a time, there was a chemistry student who had a steel needle in her heart because of a childhood accident. Physicians did not remove the needle because the surgery would have been too risky. This colleague of ours could never enter an NMR laboratory.

Recently, there have been reports of creating very strong permanents magnets, which are suitable for use in NMR instruments without electronics or ultra-cooling. These are made of a special alloy of samarium and cobalt and can be made very small. However, the same extreme care must be taken in handling this magnet as for the large superconductor magnets.

Experienced experts can classify the magnetic properties of most materials simply by looking at them. However, for an object with a metallic shine about which nothing further is known, there is no obvious way to decide whether it is ferromagnetic, paramagnetic or diamagnetic. Identifying ferromagnetic substances with a single glance is special ability only given to a precious few. These lucky few also share the property of being fictional. For example, James Bond instantly recognizes the replacement teeth of Jaws as ferromagnetic in the movie *The Spy who Loved Me*, and, of course, uses this information at the critical moment of their struggle. Agent 007 also unzips a woman's dress using a magnet at the beginning of the movie *Live and Let die*. All attempts to reproduce this scene in real life have been miserable failures, as none of the real-life zippers were made of a ferromagnetic material. The third movie of the Terminator series (*The Rise of the Machines*) features John Connor, who recognizes that the sometimes-liquid material of the TX terminator of Skynet is ferromagnetic. This saves him some valuable time when the chase approaches a large particle accelerator, in which strong currents generate strong magnetic fields—which is, incidentally, true.

Fig. 4.22 Warnings on the door of an NMR laboratory. (Authors' own work)

4.17 Water: The Extravagant Liquid

Water is by far the most commonplace liquid in everyday life. Some 70% of the surface of Earth is covered by oceans and about two thirds of the body mass of a human is water. The word water occurs in the second or third sentence of the Bible ("And the Spirit of God moved upon the face of the waters."), depending on the actual translation. Aristotle, the most influential Greek philosopher listed water as an element in addition to air, fire and earth. One might think that such a liquid must also be an ordinary liquid. Nothing can be farther away from the truth!

In fact, water is an exceptional substance from both a chemical and physical point of view. A website dedicated to water (www.lsbu.ac.uk/water/) lists no fewer than 69 of its extraordinary properties, although some of these 69 are directly connected by causal relationships. What is more, it is exactly these unusual properties that make water indispensable.

The molecules of water are made up of one oxygen and two hydrogen atoms. Hydrogen is the most abundant element in the Universe, oxygen is the third most abundant. (For those who feel this information incomplete: helium is the second, but it does not form any chemical compounds.) This abundance does not mean that the Earth must necessarily have a lot of water as the composition of Earth is quite different from that of the Universe. Instead of diving into more philosophical lines of thought, it is probably best to note only the empirical fact that water is a very common substance on the surface of Earth, and also a major player in chemically bonded forms of rocks.

H_2O is the smallest molecule that is not a gas at the temperature and pressure common on the surface of Earth. (For a few exacting chemists: LiH and BeH_2 are solids with molar weights lower than that of water, but they are not formed of molecules.) It is often said that water has an abnormally high melting point (0 °C) and boiling point (100 °C). Normal would mean here the values expected based on the properties of similar molecules: the boiling point of methane (CH_4) is −162 °C, ammonia (NH_3) −33 °C, hydrogen sulfide (H_2S) −60 °C and hydrogen fluoride (HF) 20 °C. The reason for this phenomenon has been known for quite some time: water features a delicate network of hydrogen bonds, in which each water molecule is bonded to four neighboring molecules. This connectivity means that it is difficult to separate water molecules without high temperatures. It should not be forgotten that ammonia and hydrogen fluoride molecules also form hydrogen bonds, but they do not exist in such an extensive network.

Another famously irregular property of water is that it expands (its volume increases) upon freezing. This might have caused some trouble for those who wanted to rapidly cool a bottled drink in the freezer but forgot about it. This expansion upon freezing can break the bottle and produce a notable sound effect as well, often resulting in a human reaction expressed by the question 'What has exploded in the freezer?' If water expands upon freezing, then ice must have a smaller density than liquid water does. Indeed, ice cubes float on water, although only a small part is above the liquid level. This is true for ice cubes as large as icebergs, as the tragedy of Titanic reminds everyone. This abnormal property is even more important than a few naval catastrophes may imply. The lower density of ice causes natural surface waters to freeze beginning from the top. Deep lakes and rivers do not freeze entirely even in a very cold winter, thus leaving some liquid water at the bottom and allowing aquatic life to survive until spring comes.

Another anomalous property of water was nicely described by the extremely productive French science fiction writer Jules Verne (1828–1905) in his novel titled *Hector Servadac*:

> Before the evening of this day closed in, a most important change was effected in the condition of the Gallian Sea by the intervention of human agency. Notwithstanding the

increasing cold, the sea, unruffled as it was by a breath of wind, still retained its liquid state. It is an established fact that water, under this condition of absolute stillness, will remain uncongealed at a temperature several degrees below zero, whilst experiment, at the same time, shows that a very slight shock will often be sufficient to convert it into solid ice.

It had occurred to Servadac that if some communication could be opened with Gourbi Island, there would be a fine scope for hunting expeditions. Having this ultimate object in view, he assembled his little colony upon a projecting rock at the extremity of the promontory, and having called Nina and Pablo out to him in front, he said:

"Now, Nina, do you think you could throw something into the sea?"

"I think I could," replied the child, "but I am sure that Pablo would throw it a great deal further than I can,"

"Never mind, you shall try first."

Putting a fragment of ice into Nina's hand, he addressed himself to Pablo:

"Look out, Pablo; you shall see what a nice little fairy Nina is! Throw, Nina, throw, as hard as you can."

Nina balanced the piece of ice two or three times in her hand, and threw it forward with all her strength. A sudden thrill seemed to vibrate across the motionless waters to the distant horizon, and the Gallian Sea had become a solid sheet of ice.

These paragraphs describe a well-known, although still exceptional property of water, or to be more precise, super-cooled water. If water is cooled without any movement, its temperature can actually be brought 10–15 °C lower than its melting point—it does not freeze. In liquid water, there is quite a bit of disorderliness but during very careful cooling, water molecules are not able to find their positions in the ordered structure of ice. Therefore, water remains a liquid even below its melting point until the first tiny movement suddenly provides the initial push for the formation of the highly ordered ice crystal. Verne correctly describes the phenomenon: the entire body of water suddenly freezes at this time.

Super-cooling is not an exclusive characteristic of water, it can be done with other liquids as well, but it is surprisingly easy for water because of the special hydrogen bonding network. In fact, under very special conditions at high pressure, water can even be super-cooled to −130 °C. This extremely unstable state can freeze upon heating!

Finally, the extremely high specific heat of water should also be noted. This means that—compared to other substances—a great deal of heat is needed to raise or lower the temperature of water. Conversely, water is an excellent medium for cooling or heating. Those living close to sea benefit from this fact enormously—the proximity of a huge body of water has a cooling effect in the summer, and a warming effect in the winter.

4.18 Fire in Space

The title might seem natural at first glance to anyone who watches movies. For example, in the 1998 movie *Armageddon*, starring Bruce Willis, there is an extensive fire aboard a Russian space station and the viewers have no reason to doubt the authenticity of the scene. In addition, real-life events may also ring a bell to readers familiar with the history of space exploration: on January 27, 1967, there was a fire on board spaceship Apollo 1, which killed astronauts Roger B. Chaffee, Virgil I. Grissom and Edward H. White II. What sort of doubt could remain after these?

Well, Hollywood movies seldom care about scientific accuracy, although there are respectable exceptions. The misfortune of Apollo 1, however weird this might be, was not in space, but during training on the surface of Earth. In fact, the spacecraft was not even named Apollo 1 at the time of the accident because only spaceships actually making it to space are numbered. Apollo 1 is an exception to commemorate the three astronauts who perished on board.

A tragic accident like this is always followed by a long and painstaking investigation. The gas filling the cabin of Apollo 1 was not air, which contains about 80% nitrogen and 20% oxygen, but pure oxygen, whose pressure was about one fifth of the usual value on the surface of Earth. For human breathing, this is essentially the same as air except that nitrogen was absent, which is unnecessary for life anyway. An unforeseen effect of this change in the gas composition was that a fire started by an electric shortcircuit spread with dramatic speed in the cabin. NASA later tried to test this effect in unmanned automatic spacecraft in Earth orbit. However, it seemed impossible to start a fire under those conditions. Later, crews aboard space shuttles made similar observations.

Apparent contradictions between observations can often be explained by carefully noting the differences between the experimental conditions. The obvious difference between the experiment on the surface of Earth and in space is the absence or lack of gravity, and everything can be understood following this clue. There are three main conditions of starting a fire: a substance that burns, a medium that provides oxygen for burning, and a suitably high temperature for ignition. In spacecraft, or even in an apartment on the surface of earth, there are many substances that can burn (such as wood or plastic), and the medium providing oxygen is ambient air. If one heats up a small part of the combustible material, burning will start. In this process, the gases formed (most often carbon dioxide) cannot provide the oxygen for burning to continue. However, the gas formed from combustion is hot, and its density is lower than the density of ambient air. Therefore, this hot gas rises making way for more cold air to approach the burning material. This difference in temperature ensures a continuous supply of oxygen for combustion. Oxygen is usually the key player—if someone wants to speed up a camp fire or a grill, blowing on it is a very effective way to do so. This causes more oxygen to reach the burning material faster.

On board a spacecraft, however, the temperature difference does not cause the burning product to rise. To put it in a simplified manner, there is no 'up' direction in space. The burning product cannot escape, and the oxygen necessary for further combustion cannot be replenished. The fire will extinguish itself. When fire fighters do their job on the surface of Earth, their primary concern is almost invariably preventing more air to reach the fire, not cooling it down, as might be suggested by the quantity of water they sometimes use. In a spacecraft, astronauts would have to maintain effective ventilation to keep up a prolonged fire. A similar sequence of thought about burning also explains why rockets are very heavy. They must carry the medium necessary for combustion in addition to the fuel itself.

This has been a wonderful bit of musing thus far, but the history books of space exploration give reason for some more explanations. At 10.35 PM (Moscow time) on February 23, 1997, there was a major fire aboard the Russian Mir space station. All of the six-membered American-German-Russian crew had to take part in fighting the fire, which took about 90 s—at least according to the official records. It took about a day to get rid of the last bits of smoke formed in this fire, the astronauts had to wear gas masks for hours. Is there a mistake somewhere in the previous line of thinking?

Well, the answer is a definite yes and no at the same time. The medium that supports a fire can actually be different from air. On board the Mir, the fire was caused by a chemical called lithium perchlorate. This can provide the oxygen for burning even in the absence of air if it is in contact with some combustible material. In the accident, all the astronauts reacted very rapidly and professionally to the events, thanks to their excellent training, and the damage caused was minimal. However, all of them thought that the fire was burning for a much longer time than the official records note—this is actually a known psychological effect of fires burning in confined spaces. In addition, it was also discovered that some plastics may burn in a way that also ensures some mixing of nearby gases and thus maintains combustion. Zero-g fire safety is an active area of research nowadays.

Aboard the International Space Station, the fire alarm has sounded twice, erroneously on both occasions. On September 18, 2006, a bit of solid material found its way into the oxygen supply system and caused the false alarm. On June 14, 2007, the eventual cause was a computer error.

Although there are fire extinguishers on a space station, these do not use water because that would add too much unnecessary weight and would be too expensive to put on Earth orbit. A funny-sounding remark in the history of the Soviet space station Salyut 6 nicely illustrates the code of water use in space: on January 5–6, the crew tested a portable shower cabin for the first time, which was later used *once a month* by the permanent crew members. One might even wonder if the training program of cosmonauts included some preparation for the frequency of shower use in space.

4.19 Is Natural Gas Toxic?

In a common scene from films, but unfortunately sometimes in real life, people try to commit suicide by opening the gas valve. If someone attempts this, it is not only evidence of a spectacular chemical misconception, but also endangers the lives of others. Nowadays, gas in the pipeline is not poisonous, but above a certain amount, it forms an explosive mixture with air. The spark triggering the explosion is usually not initiated by the person attempting the suicide, but an unfortunate neighbor or delivery person.

The natural gas delivered through the pipelines is mostly methane (CH_4) mixed with smaller amounts of ethane (C_2H_6). Both of these compounds belong to the family of hydrocarbons, which are quite well known in chemistry for their refusal to engage in chemical reactions. Poisoning usually requires chemical reactions within the human body, so a nonreactive substance cannot be toxic. In addition, the density of natural gas is lower than that of air, so the choking effect from carbon dioxide formed in wine cellars ($\rightarrow$ **4.20**) is not significant, either. One of the very few reactions of methane and ethane is combustion, and high temperature is required to initiate it. If the level of natural gas reaches high levels in a closed room, and enough combustible material is available, a spark is enough to start the combustion.

For the reasons given above, the main risk during gas leaks is that of an explosion and not poisoning. The very low reactivity of ethane and methane also causes these gases to be odorless. Gas leaks, however, must be recognized very early on. To assist recognition, gas supply companies add foul-smelling substances such as ethyl mercaptan (C_2H_5SH) or *tert*-butyl mercaptan (C_4H_9SH) to the pipelines in minor quantities. These compounds are very odorous, and the human nose can recognize them at very low, and certainly harmless levels.

A drawback to the use of methane and ethane is that they cannot be economically transported in a gas cylinder. Two other gases, propane and butane, are suitable for this sort of transportation. Their mixture is called LPG gas (liquefied petroleum gas), and it is extensively used in remote or underdeveloped locations where gas cannot be obtained from the pipelines. The main components of LPG are odorless as well, and foul-smelling additives are used in the cylinders for safety.

Given all these facts, one may actually wonder why there is such a widespread belief in the toxicity of natural gas. This belief has its roots in history. Until the middle of the twentieth century, gas in the pipelines actually contained a highly toxic component called carbon monoxide (CO). This coal gas was by no means the same as natural gas: it was obtained by heating coal to a controlled temperature, and the carbon monoxide content could reach 8%. Carbon monoxide is called a silent killer because it is colorless and odorless, and its presence is undetectable by human senses. Its toxicity is caused by its ability to bind strongly to hemoglobin in blood, so after some time, it completely prevents oxygen from reaching the cells. Coal gas was replaced by natural gas because of the toxicity of carbon monoxide. Unfortunately, this was not enough to prevent fatal accidents due to carbon monoxide poisoning, as carbon monoxide is also formed during the imperfect combustion of substances containing carbon, for example coal or natural gas.

In the United States alone, inadequately serviced gas-burning devices are responsible for about 10,000 hospitalizations and 250 fatalities each year. Car engines also produce some carbon monoxide, and this is why it is dangerous to run the engine indoors, although the operation of car engines have become much more efficient and much less dangerous in recent decades thanks to the development of catalytic converters. For this same reason, the ventilation of tunnels must be planned and maintained very carefully. The most deadly carbon monoxide poisoning in history occurred in a tunnel on 2 March 1944 in the Italian town of Balvano. A crowded train with a steam engine got stuck in the Armi tunnel. The number of fatalities was above 500.

The most famous victim of carbon monoxide poisoning was most probably Émile Zola (1840–1902). The chimney in his apartment was totally blocked on September 28, 1902, so carbon monoxide formed during combustion could not leave and caused the death of the famous French writer. Decades later, a roof repair specialist admitted on his deathbed to intentionally blocking this chimney. His motive was that he was fiercely opposed to Zola's political views.

Modern gas from pipelines does not cause any problems like this, but carbon monoxide formed during imperfect combustion is still a concern. It is a good idea to have gas appliances tested regularly by a specialist. Investing in a carbon monoxide detector is not very expensive, and may make good common sense especially in rooms where babies sleep.

4.20 Respiratory Affairs: Carbon Dioxide

In wine-producing regions, accidents sometimes occur late in the fall when careless visitors to wine cellars suffocate. The culprit is carbon dioxide gas.

It seems logical that a gas causing fatal accidents must be poisonous, but actually, the situation is a little more complicated. Carbon dioxide causes suffocation, which means it blocks an adequate supply of oxygen to the body. Although water can also cause suffocation, it is not considered toxic. However, there is a further twist to this story—carbon dioxide can actually cause suffocation even in the presence of a normal amount of oxygen. It is difficult to do justice to this topic here, but experts usually do not classify carbon dioxide as a toxic substance. For example, there is significant carbon dioxide in soft drinks, yet the effect is refreshing rather than harmful.

Carbon dioxide is a molecule in which a carbon atom is bonded to two oxygen atoms (Fig. 4.23). It is heavier than air (for the more scientific minds: its density is higher than the density of air), so it can be poured like an invisible liquid to some extent. A common demonstration of this property is when carbon dioxide is poured onto a burning candle, which is immediately extinguished. Carbon dioxide should definitely not be confused with carbon monoxide (not even under pressure from ill-informed journalists), which is the other common oxide of carbon, highly toxic, and has the same density as air ($\rightarrow$ **4.19**).

Carbon dioxide may accumulate in wine cellars when wine is produced from must, the technical precursor of wine. The reason for this chemical reaction is that some yeast-like microbes convert sugar in the must into alcohol, and subsequently, release carbon dioxide. Grape sugar ($C_6H_{12}O_6$)—not surprisingly—is the most common sugar in grapes. Its fermentation can be chemically represented by the following equation:

$$C_6H_{12}O_6 \rightarrow 2\ C_2H_5OH + 2\ CO_2$$

Fig. 4.23 The space filling model of carbon dioxide. (Copyright-free Wikipedia picture)

Since carbon dioxide is heavier than air, it accumulates in the lower part of cellars and begins to force out the breathable air. When wine fermentation is at full throttle, a cellar may become an unhealthy place. Combustion, like breathing, needs air, so a burning candle can be used to test if the air in the cellar is healthy enough. The candle must, of course, be held low, as carbon dioxide accumulation starts there. Carbon dioxide also collects in a number of caves, and spelunkers need to take special care in such places.

Accumulation of carbon dioxide was one of the major problems during the Apollo 13 mission failure. In spacecraft, there are special devices to replenish used air. During the accident, the three members of the Apollo 13 crew spent a long time in the cabin that was designed for lunar landing. The CO_2 scrubber in this module was designed for two people so the carbon dioxide level increased to a dangerously high value during the mission. To solve this problem, the crew had to adapt a carbon-dioxide filter into an incompatible apparatus. This is a very interesting part of the famous Hollywood movie about Apollo 13 mission.

As already stated, if a large amount of carbon dioxide prevents oxygen from reaching the cells in the body, suffocation results. This is easy to understand when the gas entirely replaces air. However, even a 10% replacement of carbon dioxide in air can be fatal despite the fact that there is enough oxygen around. An oversimplified reason is that carbon dioxide cannot leave the blood in the lungs when the outside concentration is high. Therefore, there is no space in the gas-carrying molecules in the blood to absorb oxygen from the air. Carbon dioxide concentrations as low as 1% can cause dizziness, 5% causes panic in humans, and 8% causes unconsciousness and death within a few minutes. As already explained, in these cases, the essential oxygen does not reach the brain or other cells. Interestingly, the brain regulates breathing not by monitoring the oxygen in the blood, but it is carbon dioxide that plays a decisive role. A manifestation of this fact is the physiological phenomenon called hyperventilation, which occurs when someone breathes very fast. An abnormally low concentration of carbon dioxide in the air causes a shift in blood pH towards the alkaline range, which slows down the blood's oxygen release and results in even faster breathing. Many people know that a simple cure for this state is to breathe into a bag. In this case, exhaled carbon dioxide re-enters the lungs, and returns the carbon dioxide level of blood to normal.

4.20 Respiratory Affairs: Carbon Dioxide

Carbon dioxide was responsible for two horrible accidents in the Central African Republic of Cameroon. Along a road by Lake Monoun, authorities found 37 dead bodies on August 16 in 1984. An investigation revealed that a small earthquake, which was barely detectable by instruments, caused the volcanic lake to release quantities of carbon dioxide, which displaced the air on the shore. The killer lake struck again in 1986, killing 1746 humans, in addition to countless domestic animals. A less dangerous form of volcanic carbon dioxide release occurs from mofettes, cracks in the planet's crust associated with volcanoes, which are known in many parts of the world. In addition, murder by carbon dioxide suffocation is the central theme in an episode of the popular crime show CSI Las Vegas (Season 5, Episode 23 '*Iced*')

A state similar to carbon dioxide suffocation can manifest even without the presence of the gas. Altitude sickness, which occurs in the mountains, means a general lack of oxygen in the body. At high altitudes, the overall amount of oxygen in air is much lower than under normal conditions, so the oxygen level in the blood also drops. The human body can adapt to these conditions to some extent. In time, the body can respond by forming more red blood cells and can tolerate a lower oxygen concentration. The ultimate limit of human adaptation is approximately the highest mountain on Earth. The top of Mt. Everest can be reached without oxygen masks for a sufficiently fit hiker, but it seems to be rather unwise to risk one's life for this end. Airplanes are, of course, hermetically closed. The oxygen level is kept high artificially in them, in other words, they are pressurized, to a level which is about equivalent to conditions at 2400 m above sea level, which is comfortable to healthy people. Unfortunately, the history of aviation records a case when pilots forgot to switch on the pressurizing instrument in the plane. Helios flight 522 left Larnaca in Cyprus on August 14, 2005 bound for Athens. The pilots did not recognize their mistake after take-off, and soon they fell into a deep coma. The automatic pilot flew the plane to Athens and began circling until the fuel supply was exhausted. There were no survivors.

News about carbon dioxide regularly reaches the popular media nowadays. Quite sound, but still not totally unequivocal evidence shows that this gas is the main culprit for the increasing greenhouse effect and global warming. In 2013, the carbon dioxide content of air exceeded 0.040%, which is a monumental increase over the pre-industrial level of 0.021%. This is a problem because carbon dioxide molecules allow sunshine to pass, but block a large part of the infrared radiation emitted by Earth. The energy remains in the atmosphere and eventually causes excessive warming.

Global warming is undeniable ($\rightarrow$ **1.12**). Many experts also agree that it is primarily caused by human activities, but this is a more complicated question. Gasoline, natural gas and coal produce carbon dioxide when they are burnt, so there is substantial financial interest in denying the connection between carbon dioxide and global warming. Scientific experts without obvious financial interests occasionally also belong to the small but rather vocal group of skeptics in this regard. The most famous of these might be Danish scientist Bjørn Lomborg, who wrote the book *The*

Skeptical Environmentalist in 1998, which stirred a major upheaval. In 2007, he published another book titled *Cool It. The Skeptical Environmentalist's Guide to Global Warming*. Apparently, Lomborg is willing to follow the scientific method, which involves revising opinions in the light of fresh evidence: in his 2010 and 2011 interviews, he no longer denied the connection between human activities and global warming. However, his view remained that global warming is just one of the challenges the humankind must face, and it is by no means the number one on the list.

4.21 The True Death Toll of Chernobyl

Most people want simple answers to simple questions. The simplicity of an answer is often more important than the truth, especially if the producer of a television show devotes 10 s to it.

When and where will the next major airplane accident occur? The question is not very complicated, yet nobody knows the answer. If someone knew, the accident could be prevented. Similarly, there are simple questions to which no scientifically correct answers can be given—at least not short ones. Kurt Gödel (1906–1978), a mathematician from Austria once even proved that whatever previous information we might have, it is always possible to ask a yes or no question that cannot be answered.

How many people died in the Chernobyl accident? This is not a very complicated question, but cannot be answered with a single number. Civil nuclear energy has a sizeable group of both proponents and opponents. The latter group tends to oversimplify this question and give frighteningly high number in hundreds of thousands, sometimes even millions. This has nothing to do with reality, but is due to ignorance or business interests.

The answer is very difficult to give because the main health risk of the Chernobyl accident was the released radioactive material, which caused people, mostly in today's Belarus, to suffer extra radiation exposure. The level of radiation was not high enough to cause radiation sickness directly—but it increased the risk of certain types of cancer. However, cancer also occurs among people not exposed to radiation, and the reason for cancer cannot always be clearly identified. So the only solution for our question about the number of victims is statistical risk assessment.

The most competent assessment was made in 2005 by the international group called Chernobyl Forum. Among the members were the International Atomic Energy Agency, six specialized organizations of the United Nations (WHO, UNDP, FAO, UNEP, UN-OCHA, UNSCEAR), the World Bank, as well as the governments of Ukraine, Russia and Belarus. The report goes into excessive detail when it analyzes the possible number of victims. During the accident, two workers died because of injuries that were unrelated to radiation. Among those who were working on decontamination, 134 developed radiation sickness, 28 of them died in 1986, and a further death occurred in that year because of cardiovascular failure. Among those who survived the first year, 19 died between 1987 and 2004. However, the

4.21 The True Death Toll of Chernobyl

relationship between the accident and the death is rather doubtful in most of these cases. Altogether, no more than 50 direct victims can be linked to the events.

As already stated, the increased risk of cancer was probably more significant. As a consequence of the accident, approximately 600,000 people suffered radiation exposure that was significantly larger than the natural background level. Everyone dies sooner or later, and on a purely statistical basis, in a group of this size, cancer will be the eventual cause of death in about 100,000 cases. The statistical analysis showed that the extra radiation exposure increased this number by about 4000. This is a highly uncertain number—it could be 1000 or 2000. An additional 5 million people were exposed to extra radiation somewhat above the natural background after the event, but any health effect of this is likely to be undetectable. So this assessment showed that about 4000 deaths are likely to be caused by the accident (some of them have not yet occurred), but less than 50 of these victims can actually be identified by name. For the 2011 Fukushima accident ($\rightarrow$ **4.22**), more time will be needed to complete a risk assessment. Not a single fatality is known yet. There were seven deaths among emergency workers in the 2 years following the events, but none of these could be linked to radiation exposure. In February 2013, the WHO issued a report with some preliminary data, which seemed to indicate that radioactive material released during the Fukushima event does not increase the cancer risk for people living outside the immediate vicinity of the facility. People living very close face lifetime risks of some cancer types 4–7 % higher than normal, with the numbers somewhat higher for emergency workers. It should be recognized that the normal risks are quite small, so a 4–7 % increase still makes a very small risk. Needless to say, Greenpeace condemned the report for downplaying the risks with the 20-km evaluation zone, but did not issue any scientific analysis or data.

In Eastern Europe, stories linking the deaths of truck drivers who were in the area during the accident, or shortly thereafter, sometimes appear in the media. The personal tragedies in the background are often very moving, but the implied cause and effect relationship does not have any scientific basis. One must think of the millions who have lived continuously in the area. Any negative health effects must appear in that group first.

Another common view in the press is that the nuclear industry is trying to paint a much more positive picture of the event than actually exists. This rings true as trivializing the events is definitely within the financial interest of the industry. This sort of thinking is called an '*ad hominem*' argument because it primarily deals with the person making the claims and not their scientific precision. There is a catch to this sort of argument independently of objective observations—how can we gain knowledge about a question if we declare everyone partial just because of some expertise in the area? And as long as personal arguments are concerned, maybe one should think of the large number of engineers and scientist who work in nuclear power plants. These are highly trained and knowledgeable experts, for whom finding a different job would not be very much of a problem. Still, these people work in the nuclear industry and their families live close to nuclear facilities. Can they all be so unethical as to knowledgeably and continuously put their own children into danger just to live a little better?

4.22 Were There Nuclear Explosions in Chernobyl or Fukushima?

On April 26, 1986, in a country what was called the Soviet Union at the time (today it is Ukraine), the most serious civil nuclear accident occurred in human history. Most people on Earth have heard about this event. It is a great pity that society along with the press was mostly content with a very shallow version of the story, on the intellectual level of a scary movie, despite the fact that well written and scientifically accurate accounts of the events for non-experts are available from multiple sources. Nuclear scientists could have experienced a frightening feeling of *Déjà vu* in March 2011, when serious explosions occurred in four of the six reactors of the Japanese nuclear power plant Fukushima Daiichi. In the Japanese case, the accident was not caused by human irresponsibility, but by a natural catastrophe. The relatively slowly progressing events continued to make headlines for weeks on, but the news was sometimes so incorrect that one might suspect intentional distortion behind the mistakes.

A frequently repeated claim about the two accidents was that a *nuclear explosion* occurred (for example, a Youtube video of the Fukushima even is explicitly described as a "Reactor 3 Nuclear Explosion"). What else could be a serious accident in a nuclear power plant? The truth, however, is that it is physically impossible for a nuclear explosion to occur in a nuclear reactor very much like it is impossible to separate a glass of warm water into half a glass of cold and half a glass of hot water. During the Chernobyl accident, there were two serious explosions. The first one was a *thermal explosion* in the morning of 26 April, 1986 at 1:23:49 AM. Eleven seconds later, a second, *chemical explosion* followed. Four reactors in Fukushima were left without cooling because of an earthquake and the consequent tsunami, and a *chemical explosion* followed in all four of them. The difference between a chemical explosion and a nuclear explosion is exactly the same as the difference between a conventional World War II bomb and a nuclear bomb. Photos are revealing: the Chernobyl explosion destroyed a solid concrete building, but not the neighboring one (Fig. 4.24). The devastation of the atomic bomb dropped on Hiroshima, which was actually very small in its own class, left no wall intact in an area hundreds of meters wide.

This may be unbelievable at first glance, but making an atomic bomb is a much more difficult task than building a nuclear reactor. For those not feeling comfortable with technical explanations, checking a history book might serve as evidence. The first nuclear reactor began operating on 2 December 1942 by Enrico Fermi and Leo Szilard in the basement of a sports stadium in Chicago. The first atomic explosion only occurred on 16 July 1945. This was a weapons test near Alamogordo, New Mexico in the United States. In the meantime, World War II raged and the most talented scientists in the world were working day and night on the weapon development with practically unlimited financial resources and with the urgent knowledge that the enemy was making similar efforts. Yet, two and a half years were needed

4.22 Were There Nuclear Explosions in Chernobyl or Fukushima? 291

Chernobyl　　　　　　　　Hiroshima

Fig. 4.24 A picture of the Chernobyl reactor after explosion (*left*) and the city of Hiroshima after the devastation of the atomic bomb (*right*). (Copyright-free Wikipedia pictures)

to overcome the problems. By that time, the war was not yet over but it had already been won.

There are two kinds of nuclear explosions: the first is based on nuclear fission, the second on nuclear fusion. In everyday chemical processes, the nuclei of atoms never change. In contrast, nuclear reactions involve transformations of atomic nuclei. The energy released during nuclear fission is immeasurably larger than the energy gained from chemical processes. In a power plant, a single kilogram of natural uranium is as valuable as 10 t of highly purified coal (this a rude estimate, but it gives the general picture). Nuclear fusion can produce even more energy.

Nuclear fission can only occur for very few nuclei. Of these, only the 235 isotope of uranium occurs on Earth. An example of a nuclear fission process is given as:

$$^{235}U + n \rightarrow {}^{142}Ce + {}^{91}Zr + 3n$$

The symbols represent nuclei here, the number before the symbol is the mass number, whereas n means a neutron. In this process, an incoming neutron causes the nucleus of a uranium-235 atom to split into two smaller parts. Three neutrons are produced in the meantime, which can split three more uranium nuclei, which will produce nine more neutrons, which will split nine new nuclei, which will produce twenty-seven neutrons, which will split twenty-seven nuclei, which will produce eighty-one neutrons and so on. A process like this is called a chain reaction and can produce a huge amount of energy in a very short time. Needless to say, the described process is an idealized case. In reality, a sizable fraction of neutrons produced escapes into the environment without causing subsequent nuclear fission, so maintaining a chain reaction is not always an easy business. The first-generation nuclear weapons were based on a chain reaction of nuclear fission. More advanced and more powerful nuclear weapons also involve nuclear fusion reactions, but this is irrelevant for this chapter.

A thermal explosion is entirely different. This can be modeled by boiling water in a closed container (WARNING: Do not try this experiment, it may lead to serious injury.) After enough heat was absorbed by the container, formation of steam results in such a high pressure inside that cannot be borne by the walls. To avoid this, all closed systems contain safety valves that open at a pre-set pressure limit that is deemed safe and release excess steam to the environment and prevent the explosion from happening. The first explosion in Chernobyl was of this type because of the cooling water present in the reactor (the water circulation was intentionally stopped because of an experiment!).

A chemical explosion is in essence a reaction that is complete in a very short time and produces a lot of energy. In Chernobyl and Fukushima, this was the well know hydrogen-oxygen explosion, which also caused the tragedy of the Hindenburg dirigible on 6 May 1937:

$$2H_2 + O_2 \rightarrow 2H_2O$$

Under normal working conditions, neither hydrogen nor oxygen is present in a nuclear reactor. In these accidents, hydrogen was formed because the temperature was much higher than the safe maximum, and the cooling water began to react with the zirconium present in the reactor wall. In Chernobyl, oxygen entered the reactor because the previous explosion damaged the reactor wall: the second explosion could not have happened without the first. In Fukushima, hydrogen escaped to the environment from more or less intact reactors and the explosion occurred outside the reactor in the building housing it.

Finally, it is important to emphasize that the explosions themselves were not the major concerns in either of these accidents. It was the radioactive isotopes that were released.

4.23 What Was the Gulf War Syndrome?

The recent military history of the USA supplied many things to friends and foes alike. The First Gulf War (1990–1991) was fought to liberate Kuwait after an Iraqi invasion. American soldiers in and after this war began showing symptoms doctors could not explain for quite a long time. Approximately 250,000 of the 697,000 veterans who served in the war, mainly in combat, were afflicted with the illness that is now called the Gulf War Syndrome. Veterans often complained of fatigue, chronic headache, memory and concentration disorders, gastro-intestinal problems, skin conditions, muscle and joint pain, and a number of other, less unambiguously described symptoms. These symptoms differed significantly from those of post-

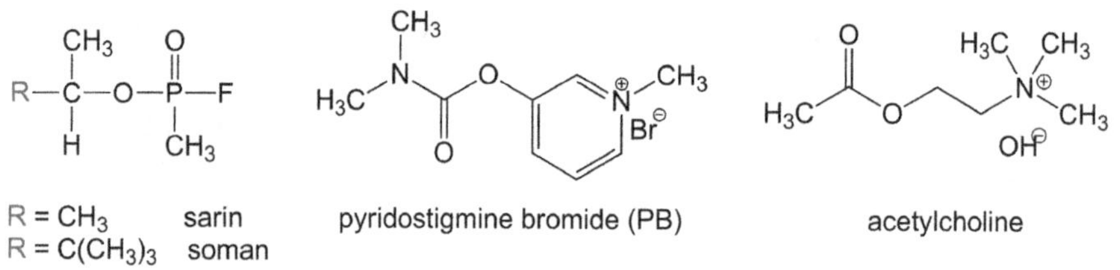

4.23 What Was the Gulf War Syndrome?

Fig. 4.25 Structures of nerve agents sarin and soman, developed by German scientists during World War II, and their antidotes, pyridostigmine bromide and acetylcholine. (Authors' own work)

traumatic stress disorder (PTSD), which was observed for the veterans of other wars. Few of the soldiers suffering from the Gulf War Syndrome recovered fully or even saw improvement in their health as years went by. The proposed causes included a chemically inert paint (chemical agent resistant coating, CARC), solvents and other chemicals, tainted food and drinking water, depleted uranium shells, fuels and their combustion products, smoke from burning oil wells, pesticides, psychological stressors, a chemical named pyridostigmine bromide, sand particles or particles from smoke, anthrax and other vaccines.

Seventeen years after the war, the Research Advisory Committee on Gulf War Veterans' Illnesses (RAC-GWVI) prepared a 465-page report on the findings and conclusions. The symptoms were analyzed in detail. The anthrax vaccine, combat stress, and oil-well fires were ruled out as possible causes. The following possible contributing factors were identified:

4.23.1 Nerve Agent Antidote Pyridostigmine Bromide (PB)

PB is used in the US Army as an antidote against the nerve gases sarin and soman (Fig. 4.25). PB is not a vaccine. As a 90-mg dose taken before a nerve gas attack, it increases the efficiency of other antidotes. The use of PB was approved by the FDA for this purpose in 2003 (12 years after the First Gulf War). Another medical use of PB used since 1955 is in treating a disease named *myasthenia gravis*, where the necessary daily dose is 1500 mg (!). PB inhibits an enzyme called acetylcholine esterase, so it slows down the decay of the neurotransmitter acetylcholine. Acetylcholine is a physiological antidote of adrenaline, and is formed at sites where there are parasympathetic nerve impulses. Acetylcholine dilates peripheral blood vessels, lowers blood pressure, slows down heartbeat and intensifies gastrointestinal motility. Its effect is short-lived as the enzyme acetylcholine esterase hydrolyzes it to choline and acetic acid.

About half of the veterans of the First Gulf War received PB tablets before combat. An epidemiological study of *myasthenia gravis* and the Gulf War Syndrome showed that PB alone is not likely to have caused the syndrome, but may be a contributing factor in the presence of other risk factors.

4.23.2 Organophosphate Type Pesticides and Insect Repellents

These substances are inhibitors of the enzyme acetylcholine esterase. The US Army used them to fight insects and pests during the war. The report of RAC-GWVI found a dose-dependent causal connection between the use of these substances and the Gulf War Syndrome. High concentration levels of organophosphate pesticides damage the cognitive functions of the nervous system and disrupt the normal operation of endocrine glands.

4.23.3 Nerve Agent Sarin (Fig. 4.25)

Veterans with Gulf War Syndrome often showed some symptoms of mustard agent and nerve gas poisoning. Coalition forces often encountered low, non-lethal levels of chemical and biological weapon materials. Chemical alarms went off 18,000 times. The Iraqi army did not use chemical warfare agent directly in attacks, but these compounds were released during the destruction of stockpiles or in coalition raids of Iraqi military facilities. Around 125,000 American and 9000 British troops were exposed to mustard and nerve agents during the demolition of rockets with chemical warheads at Khamisiyah. Nerve gases are strong inhibitors of acetylcholine esterase.

4.23.4 Depleted Uranium Shells

About 320 t of depleted uranium shells were used during the war in tanks and cannons. Depleted uranium is slightly radioactive and gives off toxic fumes after impact. It interferes with kidney, brain, liver and heart functions, and is known to be nephrotoxic (causing kidney damage), teratogenic (harmful for fetuses), immunotoxic (immune system poisons), neurotoxic (nerve poisons) and potentially carcinogenic. Depleted uranium was also used during conflicts in the Middle East, but nothing similar to the Gulf War Syndrome was uncovered there.

The RAC-GWVI report concluded that the combined effect of the above mentioned factors, but especially acetylcholine esterase inhibitors such as PB, organophosphate pesticides and sarin nerve gas, may be the cause behind the syndrome, which was corroborated by animal tests. However, the Advisory Committee did not declare the case closed and advised further studies with an annual funding of at least $ 60 million from the US government. This story illustrates quite clearly how much effort and money is needed to understand complex medical-chemical phenomena and reveal the cause-effect relationships in the background.

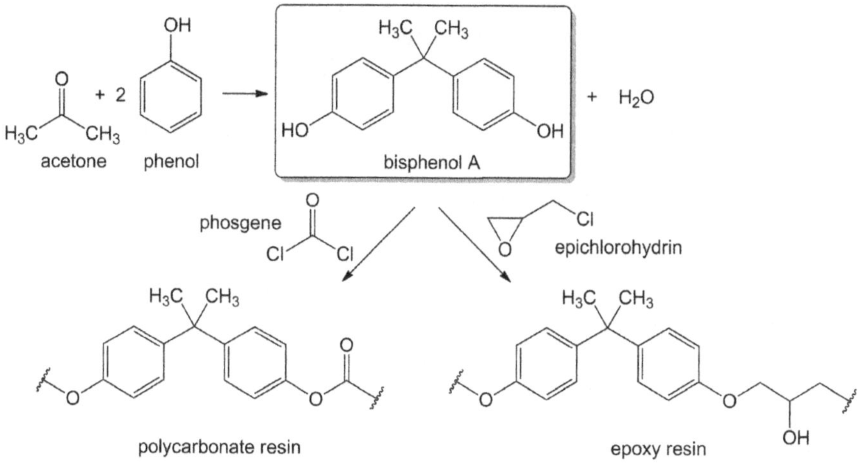

Fig. 4.26 Structure and synthesis of bisphenol A (BPA) and the most significant plastics prepared from it. (Authors' own work)

4.24 The Strange Case of Bisphenol A

"Can shopping receipts make you impotent?" This was the title of a letter published on July 13, 2010 in *The Daily Telegraph*. The letter was written by Professor Richard M. Sharpe, who wanted to comment on an article titled "Sex and shopping—how retail therapy really is bad for men's health and fertility" by science journalist Richard Alleyne in the 30 June issue. The journalist quoted German urologist Frank Sommer, who found that bisphenol A (BPA) in the heat-responsive paper used for printing receipts may lower testosterone levels in men and increase estrogen production. Estrogens are female hormones, and their heightened level decreases male sexual activity.

What is BPA and what is it used for? BPA is a precursor of resistant polycarbonate and epoxy plastics. It was first synthesized in 1888 (Fig. 4.26). Commercial use of BPA began when epoxy resins were developed in the 1950s. Epoxy resins are used as protective covering layers on metals (food cans, pipelines), dental adhesives, DVDs, cell phones, contact lenses, car parts, and sports equipment. In shopping receipts, BPA is a slightly acidic activator that causes color change when pressure or heat is applied. Because of its widespread use, BPA is produced on a vast scale, 5.4 million t a year.

Is fear of BPA reasonable? As usual, the devil is in the details. At present, experimental information does not show that bisphenol A (BPA) causes impotence, especially not in the minor amount used in receipts. In the scientific literature, there is a single publication that studied this possibility, and it found no causal connection between sexual activity and the level of exposure to BPA in Chinese men. Animal experiments did not link BPA and impotence, either.

The amount of BPA in shopping receipts was tested in a recently published scientific study. Stores in Boston gave customers 8 cm wide and 30 cm long pieces of paper as receipts on average. After alcohol extraction at 35 °C, the BPA content of these receipts was found to vary between 3 and 19 mg. Only a tiny fraction of this contributes to human exposure as a result of handling receipts. In the regulations issued by the European Food Safety Authority (EFSA) and the US Food and Drug Administration, the **tolerable daily intake** (TDI) of BPA is 0.05 **mg/kg of body weight**, which is 3.75 mg for a 75-kg adult.

These TDI values for BPA have been criticized during the past years, and have been reviewed three times recently. The authorities concluded each time that there is no reason to set lower limits. What is more, an earlier limit of 0.01 mg/day/kg of body weight was increased fivefold. The debates have remained heated, though. Opponents of BPA use claim that the long term effects of BPA are harmful even at lower levels, and they recommend a 5000-fold reduction (!) of TDI to 0.01 µg/day/kg of body weight. Their arguments run counter to common sense as they imply that BPA is more dangerous in smaller doses. Rochelle W. Till and coworkers of the Research Triangle Institute in North Carolina carried out in vivo experiments on rats using low levels of BPA in 2002. No harmful effects were found. Frederick S. vom Saal and John Peterson Myers criticized Till's work because rats, they said, were less sensitive to BPA exposure. Till repeated his experiments with mice in 2008: the results were the same. This time, the criticism targeted the mice food used, which might have contained phytoestrogens (substances in plants that are similar to estrogen-type female hormones) and could have made the effect of BPA undetectable. In response, Till remarked that arguments against BPA are not based on experimental data or known facts, but are primarily generated by irrational fears.

BPA is a known endocrine disrupting chemical (EDC), but this effect is slight and has been known since the 1930s. This fear of EDC activity led to a total ban of BPA in polycarbonate plastic bottles intended for babies. The chemical industry was not really shaken by this ban as this use was only a tiny fraction of the market. On the other hand, finding substances that rival the favorable properties BPA is not easy, and more significantly, not cheap. Isosorbide, which can be produced by artificial sweetener sorbitol, is a possible replacement. A promising alternative of polycarbonate plastics is named Tritan co-polyester, and is made from dimethyl terephthalate, 1,4-cyclohexanedimethanol (CHDM), and 2,2,4,4-tetramethyl-1,3-cyclobutanediol (CBDO). However, replacing BPA in all current applications would be difficult, very expensive, and—most importantly—unlikely to save anyone from any harm.

Donald R. Kelsey called attention to the fact the usual human diet exposes people to phytoestrogens to much higher levels than any contact with BPA, whose endocrine disrupting effects are weak to begin with (Fig. 4.27). This fact still seems to be unnoticed by the media, most probably because the popular belief in the harmlessness of natural substances is so solid that any warning to the contrary is widely ignored (→ **1.1, 1.3, 1.4**). Phytoestrogens in beans, olive oil, fruits, vegetables, grains, tea, coffee, wine, beer and chocolate cannot be meaningfully evaluated to-

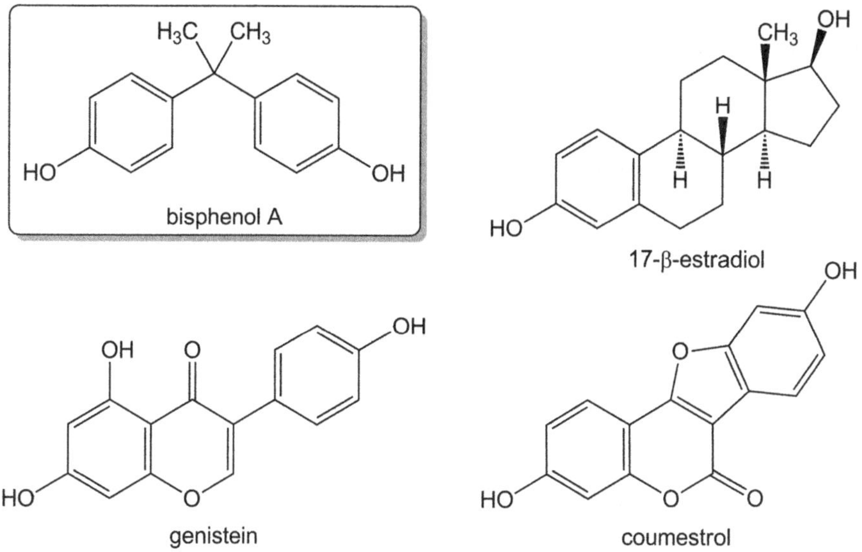

4.24 The Strange Case of Bisphenol A

Fig. 4.27 Chemical structures of BPA, estradiol and a few phytoestrogens. (Authors' own work)

gether as they have other physiological effects as well, which range from beneficial to harmful.

Opponents of BPA use often refer to the precautionary principle (*Vorsorgeprinzip*). This principle states that it is the responsibility of the user to prove a substance or technology harmless if the earlier scientific data are inconclusive or insufficient. This sounds logical to some as any new material or method can have risks that are unknown in the beginning. In other words, better safe than sorry. However, a more nuanced approach to the problem must also recognize that the overreliance on the precautionary principle can lead to absurdity. Very few of the great inventions in human history (beginning with the use of fire) would have been achieved by overly cautious people. The precautionary principle also has its group of critics, who argue that it does more harm by discouraging innovation than the expected benefit from avoiding some problems.

The safety of BPA was investigated in no fewer than 5000 studies! German experts, including the president of the German Society of Toxicology, published a review article in April 2011, which concluded that (a) toxicological data in multigeneration rodent studies are reliable, (b) the **half-life** of BPA in the human body is less than 2 h as it is excreted as sugar and sulfate derivatives in urine, (c) the daily exposure is well below the tolerable daily intake, and (d) taking every known factor into account (method of consumption, exposure of babies and children, possible effects in addition to the estrogen, different legislations in different countries), available evidence indicates that BPA exposure represents no noteworthy risk to the health of the human population, including newborns and babies.

Is this the end of the BPA debate? Toxicologically speaking, it should be. Sociologically speaking, it is highly unlikely.

4.25 Death in the Danube: Cyanide Spills

The last year of the previous millennium was less than kind to the Szamos, Tisza and Danube rivers in Hungary, Romania and Serbia. At the end of January, huge quantities of cyanide entered these waterways and played havoc on the aquatic life. Needless to say, the Hungarian media went into a frenzy, and international journals also took notice. The chemical magazine *Chemistry & Industry* called the accident 'the worst disaster since Chernobyl' in its February 21, 2000 issue. American *Newsweek* magazine ran a short story titled 'Death in the Danube' accompanied by the main photo attraction of the accident, which were masses of dead fish. The picture shown there must have actually been taken along the Tisza River, because fish deaths did not occur on such a massive scale in the Danube, where the pollution arrived in a much diluted form. A leading scientific journal, *Science,* ran a long story about the accident, but it was also not free of seriously flawed information. On an accompanying map, the Danube River was erroneously drawn. The map ended the big river in of its small tributaries, the Ipoly river, rather following the Danube's actual route through Austria and Germany.

Among chemists, major debates were inflamed by a story on a popular evening news in Hungary. The journalists could not be blamed in this instance as they actually asked someone who looked like a real expert.

In the story, a laboratory experiment was shown. A sample was prepared which contained cyanide in the same concentration as the Tisza River did at its worst. Fish soon died in this water. When iron(II) sulfate was added to another sample, fish survived. Anyone could have thought this was quite convincing and no error could have been made. Still, a small, but important mistake was uncovered later. The demonstrating expert did not take the trouble of getting enough information about the actual composition of the Tisza during the cyanide pollution wave. The lab sample had indeed the same concentration of cyanide that Tisza had, but it was quite different in other respects. The experiment was repeated using an actual polluted sample. Unfortunately, the evening news was no longer interested. Fish in these experiments died even after the addition of iron(II) sulfate. This apparent discrepancy called for an explanation, which was found without much of an effort.

As already stated, the water used in the television experiment was only similar to the Tisza water in its concentration of cyanide. The pollution, in fact, came from a gold mine. In a gold mine, metals that are above gold in the periodic table (silver and, mainly, copper) are also present. So the pollution that was called a 'cyanide spill' in the press actually contained a number of additional components. The presence of copper itself would have not been a problem, but its interaction with cyanide ions fundamentally altered the chemical properties. Even among chemists, heated debates followed among those who were not aware of all important facts.

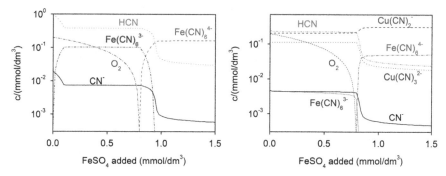

Fig. 4.28 Change of the concentration of species present as a function of the amount of added iron(II) sulfate under the typical conditions of the 2000 cyanide spill neglecting (*left*) and considering (*right*) the presence of copper. (Authors' own work)

The method presented in the TV interview is based on a complex formation reaction, in which iron(II) bounds to cyanide ion and forms $Fe(CN)_6^{4-}$ ions (named hexacyanoferrate(II) ions), which are much less toxic than cyanide ions (actually, the sodium, potassium and calcium salts of this anion are used as chemical additives known in the EU as E535, E536 and E538, respectively, as anticaking agents for table salt). Complex formation reactions may be very complicated even for chemists, who use species distribution diagrams to aid understanding. Such a species distribution diagram is shown in Fig. 4.28. Experts can draw multiple conclusions from a graph like this. The y axis displays concentrations on a logarithmic scale (i.e. moving from a lower to an upper tick represents a tenfold increase). On the x axis, the amount of added iron(II) sulfate is shown: the zero point here is the polluted water without anything added to it, i.e. the dead fish experiments from the TV interview. The chemical species with the highest concentration is HCN, that is hydrogen cyanide, which is well known to be quite poisonous ($\rightarrow$ **4.26**). There is also dissolved oxygen (O_2) and cyanide ions (CN^-) present. By adding a sufficient amount of iron(II) sulfate, the concentrations of both HCN and CN^- can be greatly decreased, because cyanide ions are mostly bound in the form of $Fe(CN)_6^{4-}$. The concentration of toxic species is low in a solution like this, which is not that dangerous to fish. Dissolved oxygen disappears from the solution, but it is easily replenished from the air in the laboratory experiment. In a river-sized reactor, whose surface area is much smaller compared to the total volume, this would be much more of a problem than in the laboratory experiment.

As already foretold, the presence of copper causes fundamental changes. First, it explains why HCN did not slowly evaporate from the river water. Although this process was also slowed down by the cold winter weather, it would still have been much faster if copper had not been present. Copper also forms cyano complexes very much like iron does, but they are more stable than the iron(II) complexes, and, more importantly, they are as toxic as the cyanide ion itself. A species distribution diagram which includes the presence of copper is shown in Fig. 4.28, right panel. The y axis here represents the actual state of the river at the worst time of the pollution wave. Here, the extremely toxic cyano complex $Cu(CN)_2^-$ is the major player. If iron(II) sulfate is added to this solution, very little $Fe(CN)_6^{4-}$ is formed. Similarly to the previous case, dissolved oxygen disappears form the solution, and a closer look at the graph reveals that the concentration of $Cu(CN)_2^-$ actually increases a little bit upon the addition of iron(II) sulfate. As a result, the solution remains as toxic as it was.

Cyanide use in gold mines is widespread in the world, so those who tried to blame the Australian owner of the mine for using outdated technology were off the mark. Cyanide spills also occur every now and then. A memorable accident occurred in Kyrgyzstan along the Barskaun river on May 20, 1998. A truck transporting sodium cyanide fell into the river from a bridge. After the accident, an expert report tried to identify the possible methods of chemical intervention for cyanide spills. The final conclusion was that no chemical method offered advantages that outweighed the risks of the intervention. This is true even if no copper is present in the cyanide spill.

4.26 Sparkling Cyanide: The Ultimate Poison?

Cyan is a popular poison in detective stories, and readers may often have the impression that cyan is the most poisonous material known. Potassium cyanide was listed as number 4 on an Internet list of the 10 deadliest poisons. A closer look on this list will, however, reveal to anyone that the 10 deadliest poisons claimed are not the most poisonous substances known.

First of all, some definitions about the names may be useful. Chemists never use the word 'cyan' to refer to a particular substance. It has a general meaning indicating the presence of a carbon and nitrogen atom bonded to each other through a triple covalent bond. In everyday life, the word cyan most often means gaseous hydrogen cyanide (HCN, cyan gas) or white powder potassium cyanide (KCN). These are indeed dangerous poisons. However, not all materials that contain the word 'cyan' or 'cyano' in their names are poisonous. Those who doubt this should look up vitamin B_{12} in a good dictionary. The chemical name of this essential molecule is cyanocobalamine, for a good reason.

Cyanides are popular in crime stories because their action is fast and spectacular. If someone is murdered by potassium cyanide, death is rapid and the symptoms leave no room for doubt about the fact that a crime has occurred. The reason for this is the way cyanide ions exert their effect—they stop an enzyme called cytochrome oxidase in mitochondria from working properly. This enzyme is essential in every living cell as it has a vital role in the reaction of glucose with oxygen, which produces energy. If the enzyme is paralyzed by cyanide ions, the entire energy-producing sequence stops. The symptoms and the speed of their development resemble suffocation, which is correct, because the cells cannot absorb oxygen.

Potassium cyanide and hydrogen cyanide have very similar effects on humans as stomach acid changes potassium cyanide into hydrogen cyanide. This is also true

4.26 Sparkling Cyanide: The Ultimate Poison?

Table 4.6 Various poisons and their lethal doses for adults

Poison	Lethal dose (mg)
Botulinum toxin	0.0002
Polonium-210	0.001
Ricin	0.1
Sarin	1
Strychnine	50
Potassium cyanide	50
Arsenic(III) oxide	250

for a number of other toxic cyanides. Some other cyanides, for example iron and gold cyanides, do not react with acid. Potassium hexacyanoferrate(II) is used as an additive during winemaking (in Europe known as the food additive E536).

The lethal dose of a poison is not always easy to define. This depends on the weight, age and health of humans and a number of other factors that cannot be considered simultaneously. Despite this, it is quite customary to characterize substances by both an LD_{50} value and a lethal dose. Such data are given in Table 4.6. One of the most toxic substances known is botulinum toxin (also known as botox), whose lethal dose is less than 1 µg ($\rightarrow$ 1.3). Nevertheless, as injections, botox can be safely used in cosmetics to remove wrinkles from the face. The ingestion of only 1 µg of polonium-210 is enough to cause a lethal radiation sickness, as tragically proved by the case of former Russian spy Alexander Litvinenko. A barely visible grain of ricin is more than enough to kill someone. This fact was recognized by the former Bulgarian secret service and the poison was used in 1978 to murder Georgi Markov in London (the event is recalled as the "Umbrella Murder"). One of the most toxic nerve gases, sarin, has a lethal dose about 1 mg. This substance was used in a terrorist attack of the sect Aum Shinri Kyo on April 19, 1995, in the Tokyo subway system.

The table shows that potassium cyanide has about the same lethal dose as another infamous poison, strychnine. Strychnine is also popular in detective stories, which may be caused by the fact that the action, although quite different from that of cyanide, is also very rapid and easy to recognize. Another important point in these stories is that strychnine is one of the most bitter substances known, which is important for the stories as it creates an opportunity for the writer to dwell on the question of how this characteristic taste was hidden from the victim.

Cyanide poisoning in real life usually occurs in accidents. A bitter irony of reality is that a method to effectively treat the poisoning is actually well known. The problem is that the victim is usually dead by the time the first paramedic or physician reaches the scene. This unfortunate fact is true even where the antidotes are kept at hand because of a recognized risk of cyanide poisoning.

Hydrogen cyanide is often described in books as having a bitter almond odor. Unfortunately, a sizable minority of humanity is genetically unable to recognize this odor. In contrast, everyone can smell the malodorous stink of hydrogen sulfide (H_2S). In fact, hydrogen sulfide is more toxic than hydrogen cyanide, but its characteristic stink makes it easy to recognize so that fatal poisoning can only occur if someone is unable to get away from a closed room where the gas is accumulating.

Fig. 4.29 Symbolic structures of chain polymers (sculptures by Béla Vízi, brass and wood). Note the relative position of repeating units in each row. (Permission obtained from copyright-owner)

Unfortunately, medical books on toxicology sometimes seem to exaggerate the danger of hydrogen sulfide. But the warning 'extremely dangerous' applied to hydrogen cyanide is not without a good reason. Although the lethal dose is smaller for hydrogen sulfide, the risk of poisoning is much greater for hydrogen cyanide.

Finally, let us recall an actual crime story where hydrogen cyanide was used as the weapon. In Isaac Asimov's *The Death Dealer* (later republished under the title *A Whiff of Death*), one chemist is murdered by another chemist who puts sodium cyanide into a jar labeled 'sodium acetate'. It is a basic work-safety rule that chemicals must not be tasted, so, this is not what the murderer has planned. The killer knows that the victim often pours acid onto sodium acetate, but with the switch of sodium cyanide instead of sodium acetate, hydrogen cyanide gas is liberated and it is already too late by the time the victim realizes that something is amiss. It is not an accident that Asimov created this method of murder—before becoming a professional writer, he worked as a chemistry professor.

4.27 Poly… What? Plastics vs. Natural Materials

The use of macromolecules or polymers has accompanied the entire human history. The most common polymers known today are called plastics and they are truly indispensable in everyday life. Their world production was about 230 million t in 2009, with virtually all of them made from natural gas or petroleum. The structures of polymers resemble an extremely long pearl rope. This is a somewhat simplified picture as the pearls may occasionally be quite different and the forces holding them together might also vary (Fig. 4.29).

The first semisynthetic polymer, celluloid, was prepared by Alexander Parkes in 1855. Adolph Spitteler and W. Kirsch prepared plastic from milk protein (casein) and formaldehyde in 1899. Buttons, handles, pens and piano keys were made from the new material and it was patented under the name Galalith (aka Erinoid in the United Kingdom). Fully synthetic Bakelite was fist formulated by Leo Hendrik Baekeland (1863–1944) in 1907, and the age of plastics began with the discovery and large-scale industrial production of vulcanized rubber (1910), PVC (1926), polystyrene (1931), synthetic rubber (1931–1935), polyethylene (1933), nylon

Table 4.7 Annual world production and recycling codes of the most important plastics

Origin	Plastic (abbreviation)	Estimated world production (million t, 2008)	Recycling code(s)
Synthetic	Poly(ethylene terephthalate) (PET)	49	1
	Polyethylene (high density: HDPE, low density: LDPE)	70	2, 4
	Poly(vinyl chloride) (PVC)	26	3
	Polypropylene (PP)	29	5
	Polystyrene (PS)	24	6
	Acryl polymers	3	7
Semi-synthetic	Vulcanized rubber	9	
	Viscose	4	
	Polylactic acid	0.5	7

(1935), polyacrylonitrile (1941), silicones (1943), PTFE (1945, the best known brand name is Teflon by DuPont Co.), and polypropylene (1954) to name only the most important.

What makes artificial plastics so attractive compared with long-used natural polymers such as wood, paper, cotton, wool, silk, horn, or natural rubber (caoutchouc)? Synthesized plastics can be easily formed into almost any shape, they are resistant to environmental effects, heat, chemicals, and they are inexpensive (these properties, of course, differ depending on the type of plastics). Natural polymers have some advantages, too, primarily that they are typically more biodegradable than synthetic materials. As environmental pollution worsens, this property is becoming increasingly important. Humankind must use resources efficiently and must try to prevent unnecessary problems in the environment.

'Back to Nature' is the solution then? Looking at a few numbers in Table 4.7 may prevent hasty judgments regarding this question. About 50 million t of plastics are produced in Europe each year from petroleum. The production of natural polymers is about 0.1% percent of this total virtually all of which is polylactic acid. These numbers are not promising: it would be lunacy to expect that polylactic acid (Fig. 4.30) or any other semi-synthetic and biodegradable polymer could replace petroleum-based plastics in any foreseeable future.

Polylactic acid is indeed attractive for certain purposes, because the energy needed in its production from fossil fuels is about 30% less than for other plastics. It has disadvantages, though. It does not tolerate heat as well and is about 30% more expensive that common plastics that can be used for the same purpose. The price issue has legal aspects as well—no environmental taxes must be paid for products made of polylactic acid.

Biodegradability of materials such as natural polymers or polylactic acid seems to be a big issue today. The reader should not fail to notice how paradoxical this issue is. Chemists are widely expected to create substances that degrade fast! For centuries, the opposite was true. Materials such as metals, wood, and paper had to be made more durable and more resistant to external influences. Who would want food packaging that degrades before the food is consumed? Biodegradability actually

Fig. 4.30 Product made of polylactic acid degrades reasonably fast in nature to give water and carbon dioxide. (Copyright-free picture)

day 0　　　　day 12　　　　day 33　　　　day 45

contradicts another modern requirement: sustainability. Sustainability requires that the smallest possible amount of waste should be produced. Nothing should be simply thrown away and materials should be recycled. The tacit premise behind biodegradable substances is that they are intended to be thrown away. On the balance of sustainability and biodegradability, natural polymers lose their advantage.

Recycling of plastics, because of their large scale use, is an everyday practice in the developed world. Plastics are classified into seven groups depending on their origin, with recycling codes indicating which group (Table 4.7). Europe produces about 25 million t of plastic waste each year, most of which can be recycled either in its original form (nylon, PET) or after chemical transformation. The 27 EU member states, Switzerland and Norway recycled 54 % of their plastic waste in 2009. In the most developed nations, the recycling rate is as high as 84 %. Experience shows that the success rate of the collection of plastic waste for recycling depends primarily on the product type, and not on the chemical composition.

Can petroleum-based plastics be transformed into bio-plastics? Two cases will serve as examples here.

Polyethylene is produced from ethylene gas, which is produced from ethanol. Brazil is a world leader in the sugar cane-based production of ethanol ($\rightarrow$ **2.23**). The first bio-polyethylene plant (with a capacity of 200 thousand tons a year) was opened in 2009 in the state of Rio Grande do Sul. If the price of petroleum, the usual feedstock, is above $ 50 per barrel, the technology is profitable. The price of petroleum has been higher than $ 100 a barrel since 2010. Vinyl chloride and PVC can be produced from bioethylene readily, and such a process is already used in India. It may sound odd, but as long as there is salt in the oceans, chlorine, hydrochloric acid and PVC are renewable resources…

In addition to find alternative feedstock for manufacturing known polymers, there is continuous research aimed at developing new materials, especially from biomass, which is also renewable and is produced in vast amounts (200 billion t a year). Only about 3 % of this is used today, but it is unclear how much this rate can be increased in the future without causing damage. An obvious idea would be using carbohydrates, a major part of the biomass, more extensively. However, the price of petroleum is still not high enough to do this economically. Only one carbohydrate-based product, furan, is already economically competitive with petroleum-based materials. Furans can be made by an acidic treatment of agricultural waste products. One of the best known furans is 5-hydroxymethylfurfural (HMF), which can be converted into a material very similar to PET, the material used in plastic bottles.

4.28 The Difference between a Rose and a Cornflower

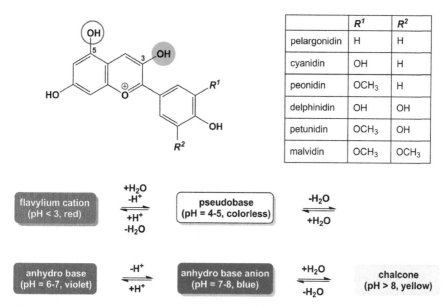

Fig. 4.31 The colors and structures of different anthocyanidins depending on the pH of the medium (the base decomposition of chalcone is not shown). The dyes in roses are derivatives of cyanidin, peonidin and pelargonidin in which the carbohydrate units are connected to positions 3 and 5. (Authors' own work)

4.28 The Difference between a Rose and a Cornflower

Nature has provided us with all colors of rainbow. This can also be observed in food of animal and plant origin. In another chapter ($\rightarrow$ **2.8**), we have already treated the peculiar red color of meat, now let's turn our attention to the colors of plants. The most important color pigments in plants are chlorophylls (green color), carotenoids (from yellow to red), and **flavonoids** including anthocyanin dyes (red, blue and violet). Anthocyanin dyes occur in the vacuoles, the membrane-bound organelles of plant cells which are basically enclosed compartments. A common chemical motif in these dyes is the anthocyanidin core-bearing hydroxyl groups, which often contains carbohydrate units attached in position 3 (or sometimes even in 5), and sometimes are esterified with organic acids (Fig. 4.31). The carbohydrate patterns of the anthocyanins in blue grapes depend characteristically on the grape's origin (European or American). In European grapes, carbohydrate units are only present in position 3 (green circle), whereas both positions 3 and 5 (blue circle) are occupied in grapes from the New World. Wines from these two continents can be chemically distinguished based on this observation.

The color of anthocyanin dyes depends very much on the **pH** of the medium, and the presence of at least five different species must be considered (Fig. 4.31). In a strongly acidic medium, anthocyanins are red. A colorless compound named pseudobase is formed by the addition of water between pH 4 and 5. With increasing pH

(6–7), the elimination of water gives a violet anhydrobase, and a further increase in basicity gives the blue anion. These transformations are reversible in the pH range between 2 and 8. In a more basic medium (pH > 8) the ring in the middle opens in an irreversible process (not shown in Fig. 4.31).

This phenomenon was observed quite early in science. Nobel laureate Richard Willstätter (1872–1942), who contributed to the study of chlorophylls, introduced a hypothesis in 1913: different colors might be caused by the same anthocyanin dyes depending on the characteristic pH of the plant. The colors of cornflower (*Centaurea cyanus*) and French rose (*Rosa gallica*) are both caused by cyanine (a D-glucose derivative of cyanidin). Japanese botanist Keita Shibata and his organic chemist brother Yuji Shibata questioned Willstätter's hypothesis as early as 1919 because such extreme pH changes were unreasonable in living organisms. The Shibata brothers measured pH in the petals of flowers and found it moderately acidic or neutral (4.6–5.5 in the vacuoles of rose and 4.6 for cornflower). So differences in pH alone could not explain the different colors. Despite these unambiguous results, many chemistry textbooks still mention the flawed explanation.

The actual explanation was given as a result of many years of research. The pH of 4–6 in the vacuoles of plants would be consistent with the colorless pseudobase form of anthocyanins or a little bit of the violet anhydrobase. This pseudobase, on the other hand, is understood to be rare in Nature. In addition to pH, the interactions of anthocyanins with each other, other polyphenol type pigments (such as **flavonoids** or flavons), and metal ions such as aluminum, gallium and iron(III) collectively determine the color. Desert bluebells (*Phacelia campanularia*) are true to their name as their flowers are blue, which is caused by aggregation between phacelianin molecules and their interaction with aluminum(III) and iron(III) ions.

Kosaku Tadeka and his coworkers achieved major new results in exploring the structure of the cornflower dye (called protocyanin complex) in 2005: four metal ions (one iron, one magnesium, and two calcium ions) were shown to form a complex with 6–6 molecules of other anthocyanin and flavone pigments. In the vacuoles of rose petals, no metals were found.

The flowering plant morning glory (*Ipomoea tricolor*), an herbaceous annual or perennial twining liana, seems to prove Willstätters hypothesis. Heavenly blue anthocyanin (HBA), a complicated derivative of peonidin, is present both in the purple bud and in the sky blue flower. The pH in the vacuoles changes from 6.6 to 7.7 during flowering as a result of the exchange of hydrogen, sodium and potassium ions. Even in this case though, other intermolecular interactions are important for the formation of the intense blue color in addition to the shift of pH.

Another interesting example is hortensia (*Hydrangea macrophylla*). The dye in this flower is the D-glucose derivative of delphinidin. Gardeners have long known that the color of the hortensia flower depends on the pH and alum content of the soil. Low pH and high aluminum levels result in pink flowers. Figure 4.32 shows the structure of the blue dye: it is a complex of delphinidin glucoside with quinic acid pigments and aluminum ions. The pH in the vacuoles is between 4.1 (blue flower) and 3.3 (pink flower).

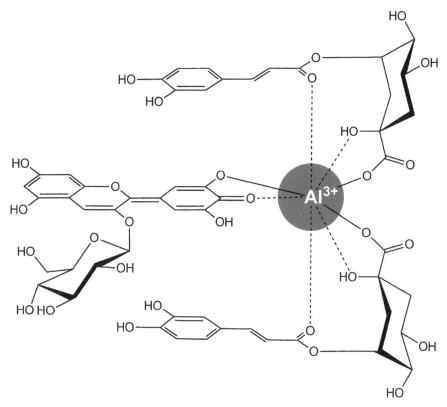

Fig. 4.32 The structure of the dye in *blue* hortensia. (Authors' own work)

4.29 The Erin Brockovich Mystery: Chromium Salts

The true value of a cup of water can only be understood in the desert. The average annual rainfall in the Mojave desert in California is below 150 mm. Water is a treasure there. It was a very nasty surprise to the members of the community of Hinkley, located on the periphery of the Mojave desert, to find that their water supply was polluted with toxic chromium salts. Mysterious diseases leading to cancer appeared and lawyer Ed Masry and his assistant, Erin Brockovich, sued the US West Coast energy corporation Pacific Gas & Electric Company (PG&E), which was thought to be responsible, on behalf of the 634 community members. After the legal battle, PG&E was ordered to pay a settlement amount of $ 333 million to be distributed among the plaintiffs in 1996. PG&E paid $ 295 million more to about 1100 other plaintiffs in California during 2006 and finally an additional $ 20 million to the residents of Hinkley. The plot of the 2000 movie entitled *Erin Brockovich* (Fig. 4.33) was based upon this story. Lead actress Julia Roberts won the 2001 Best Actress Academy Award for her performance.

Fig. 4.33 Erin Brockovich (Oscar winner Julia Roberts) in the movie titled Erin Brockovich. (Copyright-free Wikipedia picture)

Before going into the particular details of the Hinkley case, it is time to become familiar with chromium compounds. Chemical elements occur in various compounds in nature, which often have totally different effects on humans depending on the route they enter the body. Chromium(III) compounds containing *trivalent* chromium (positively charged Cr^{3+} ion) are essential for the human body as they play an important role in the metabolism of glucose in the form of the Glucose Tolerance Factor that increases the availability or the effect of insulin, a hormone that regulates sugar levels in the blood. Insufficiently low amounts (less than 6 µg/day) of chromium(III) in the diet may cause chromium deficiency, a condition causing in increased risk of cardiovascular diseases. The recommended daily intake is 35 µg for men and 25 µg for women. Larger amounts of chromium(III) are secreted, but this could still lead to liver or kidney problems. Large amounts of chromium(III) salts may irritate the skin upon direct contact. At any given time, a healthy human body stores 6–12 **milligrams** (mg) of chromium in the form of chromium(III) salts.

However, another important type of chromium compounds contains *hexavalent* chromium as negatively charged chromate (CrO_4^{2-}) or dichromate ($Cr_2O_7^{2-}$) ions. Water-soluble chromates are taken up by the body and mainly transformed into chromium(III) salts in the liver. These hexavalent chromates have an irritating and corrosive effect on the skin, eyes and lungs upon contact. Insoluble chromates cannot enter the body. When inhaled, their small particles can remain in the lungs for a long time and may cause lung cancer after an exposure of at least six months. The tolerable levels of soluble chromates are surprisingly high: 25 **ppm** of chromates did not prove toxic in long-term experiments on rats. These phenomena are quite normal: the chemical (oxidation state) and physical (solubility) properties of an element and the method of entry (inhalation, digestions) influence its biological effects greatly.

PG&E operated compressors in Hinkley to supply natural gas. These compressors had cooling towers which were covered with rust-preventing paint. The paint used between 1952 and 1966 contained hexavalent chromium and this was a source of chromium salts in the groundwater. The average chromate levels were 1.2–3.1 **ppb** in Hinkley, but much higher (7.8 ppb) close to the compressor. At peak time, chromate levels reached a maximum of 31.8 ppb. The current legal limit for chromium in groundwater is 0.06 ppb in California, but this was only introduced in 2009. Before that, the limit was 50 ppb. So, the measured values were lower than the limits at the time but much higher than the limits introduced later.

The *Erin Brockovich* movie and the general antipathy against dominant corporations (called the Goliath effect) contributed greatly to a panic about chromium(VI) salts. However, experts in environmental toxicology, for example Ana Gonzalez, Kuria N'Dungu and Russell Flegal (University of California, Santa Cruz) very much doubt that ingested chromates could cause the diseases in Hinkley. They found that chromium in soil mostly occurs as chromate salts but their distribution is not uniform, so they are unlikely to have originated from a single source. Their measured concentrations were between 6 and 36 ppb, not very far from the legal limit before 2009. In 2005, these scientists hypothesized that chromates are formed in *natural processes* from chromium(III) salts as a result of the oxidizing effect of manganese dioxide, a common mineral. Some additional evidence of this process was found in 2010, and the hypothesis was also strengthened by observations in the Brazilian municipality of Urânia.

The California Cancer Registry shows that cancer rates in Hinkley were essentially the same from 1988 to 2008. There were 196 registered cases from 1996 to 2008, which is actually fewer than expected based on demographical statistics and regional medical history. However, the evaluation of the data is altered by highly incomplete records as the investigations only began two decades after the use of the chromate-based paint was abandoned, and people moving out of the region or dying in this period could not be tracked.

In the fall of 2010, high levels of chromates were found in Hinkley despite the extended length of time that had passed since PG&E stopped using the paint and implemented an upgrade to its water treatment technology. High chromate levels have been found in other US communities as well, with the top value (12.9 ppb) in Norman, Oklahoma. Erin Brockovich returned to Hinkley and was ready for anoth-

Fig. 4.34 Erin Brockovich in Hinkley, California in December, 2010. (Picture of San Bernardino Sun, permission obtained from copyright-owner)

er fight (Fig. 4.34). Will she be the one to solve the mystery of chromium(VI) salts? Unlikely. The scientific quest promises to be a long and difficult one. Although it is logical to assume that chromium(VI) in the environment is the result of industrial activities, the facts actually imply natural processes are the main culprit. The new legal limit in California (0.06 ppb) was proposed by an organization called Environmental Working Group. This new limit does not make much sense for experts, but probably keeps the society unnecessarily alert about chromium.

Glossary of Terms

Acceptable Daily Intake or Allowed Daily Intake (ADI) → *Dose-Response Relationship/Curve*

Allergen
The *allergen* is a material which triggers an allergic reaction.

Allopathy
The term *allopathy* was created by Christian Friedrich Samuel Hahnemann (1755–1843) (from the Greek prefix άλλος, állos, "other", "different" and the suffix πάθος, páthos, "suffering") in order to distinguish his technique (homeopathy) from the traditional medicine of his age. Today, *allopathy* means a medicine based on the principles of modern pharmacology.

Anaphylactic shock
Anaphylaxis (or an anaphylactic shock) is a whole-body, rapidly developing allergic reaction, which may lead to lethal respiratory and circulatory failure.

Antibody
Antibodies are proteins produced by the immune system to neutralize exogenous (external) substances.

Chromatography, chromatogram
Chromatography is the common name of different techniques used to separate mixtures of compounds. HPLC stands for high-performance liquid chromatography. A *chromatogram* is the pattern of separated substances obtained by *chromatography*.

Colloidal sol
A *colloidal sol* is a suspension of very small solid particles in a continuous liquid medium. *Colloidal sols* are quite stable and show the Tyndall effect (light scattering by particles in a colloid). They can be quite stable. Examples include blood, pigmented ink, and paint. *Colloidal sols* can change their viscosity quickly if they

are *thixotropic*. Examples include quicksand and paint, both of which become more fluid under pressure.

Concentrations: parts per notations

In British/American practice, the *parts-per notation* is a set of pseudo-units to describe concentrations smaller than thousandths:

1 ppm (parts per million, 10^{-6} parts)	One out of 1 million, e.g. mg/kg or one drop of water in 50 l or approximately 30 s in 1 year
1 ppb (parts per billion, 10^{-9} parts)	One out of 1 billion, e.g. μg/kg or one drop of water in 50 m^3 or approximately 3 s in a century
1 ppt (parts per trillion, 10^{-12} parts)	One out of 1 trillion, e.g. ng/kg or one drop of water in 50,000 m^3 (which is the volume of water in 20 Olympic swimming-pools) or approximately 3 s in 100,000 years

Contact dermatitis

Contact dermatitis is an allergic disease causing dry skin, redness and itching, which develops as a result of delayed-type hypersensitivity reaction.

Dose-response relationship/curve

The physiological effects of various materials depend on the amount (concentration) applied in a specific way (known as *dose-response curve*). At low doses, usually there is no measurable effect. Then, the effect begins to grow more and more steeply with increasing concentration, finally reaching a saturation value at maximum impact (curve **A**). The different sections of these so-called sigmoid curves contain important information about the different stages. The maximum effect of a drug is difficult to measure accurately because of the saturation characteristic, so the comparisons are mainly made based on the amount required for 50% of efficiency that is the *median effective dose* (ED_{50}). For direct comparison, the dosages in each case are calculated to a unit of body weight, for example, in milligrams/kilogram of body weight (mg/kg of body weight).

The healing effect is always associated with the toxic effects, but the latter (hopefully) belongs to significantly higher concentrations (curve **B**). The toxic effect is also characterized by a sigmoid curve, the amount required for 50% of the effect is called the *median lethal dose* (LD_{50}), which is the amount of drug that will kill half of the sample population of a specific test animal within 24 h. The LD_{50} of a substance strongly depends on the species of the test animal and the route of entry (swallowed, dermal, inhalation etc). For humans, such data are usually unavailable for obvious reasons. Available human LD_{50} data are often gained from the identifiable death of a famous person. Further human values can be estimated from animal tests. However, these values are rather uncertain because of the extrapolation processes used. Relative classification of poisons varies from time to time, materials with an LD_{50} less than 25 mg/kg of body weight (in rats, by ingestion) are usually considered very strong poisons.

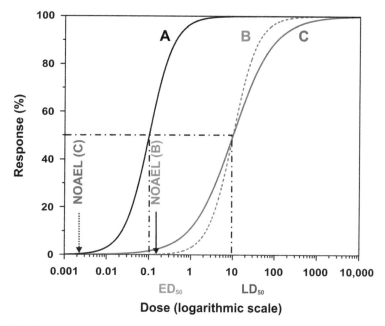

Fig. 1 The dose-response curve

For toxic gases, the product of the *median lethal concentration* (LC_{50}) and the *exposure time* (t) is given as LCt_{50} (expressed in mg × min/m³ units) instead of LD_{50}.

The ratio of the median lethal dose (LD_{50}) and the median effective dose (ED_{50}) gives the *therapeutic index* ($TI = LD_{50}/ED_{50}$), which expresses how safe it is to use the drug the higher the *therapeutic index* of a material is, the safer it is. In the case of diazepam (a common sedative), this value is around 100, while for digoxin (a cardiac glycoside), the TI is only 2–3, so the latter must be used very carefully. The actress Judy Garland died in 1969, at the age of 47, because she took too much Seconal, a sedative with small therapeutic index.

The *slope of the dose-response curves* is also an important characteristic, because it shows the development of the impact as a function of the increase of the dose: the steeper the curve, the faster the therapeutic or toxic effects are reached. Another feature of the toxic effect is the foot point of the curve, the highest concentration where there was no toxic effect observed yet: this is the level of the *non-observable adverse effect level* (or *NOAEL: no adverse effect level*). The *NOAEL* values are directly influenced by the *slopes* of the curves: materials with the same LD_{50} values can yield very different *NOAEL* values (see the *NOAEL* values belonging to curves in **B** and **C**). Sometimes, the so-called *non-observable effect level* (or *NOEL: no effect level*) value is given, at which absolutely no impact can be measured. This is obviously smaller than the *NOAEL* value. The determination of the *NOEL* is usually very inaccurate.

The *NOAEL* value is especially important for non-medicinal applications (e.g. food additives, household materials, workplace safety values) as an upper limit which must be adopted when exposure limits are determined. Even moderately toxic materials may be used for certain purposes if a large safety factor is applied. This safety factor is usually close to a hundred (sometimes even a thousand), i.e. the *acceptable* or *allowed* or *tolerable daily intake* (*ADI* or *TDI*) is calculated as follows: *ADI* = *NOAEL*/safety factor.

Long-term occupational exposure is determined in a similar process: these have various names in the various countries (US: *threshold limit value, TLV*; Great Britain: *workplace exposure limit, WEL*; Germany: *Arbeitsplatzgrenzwert, AGW*, formerly *Maximale Arbeitsplatz-Konzentration, MAK*; France: *valeur limite d'exposition professionnelle (VLEP)*; Spain: *límite de exposición profesional*; Italy: *valore limite di soglia*; Holland: *Maximaal Aanvaarde Concentratie, MAC-waarde*; European Union: *indicative occupational exposure limit value, IOELV*; Russia: *преде́льно допусти́мая концентра́ция*, ПДК) and define the same maximum allowable air pollution value of a hazardous substance in the workplace, expressed as a concentration.

In some cases, e.g. for carcinogenic materials, trace elements and vitamins, the *dose-response curves* are more complex.

Double blind study

A double blind study is the most be suitable method to test the efficiency of a medicine objectively. Randomization, i.e. the random classification into groups excludes the distortion of results caused by inhomogeneities. A double blind design ensures that neither the physician nor the patient knows whether the patient receives the test drug or a placebo. For this reason, expectations on the treatment do not interfere with the results.

Drug

Drug is a word of Germanic origin. Its meaning was 'dry, dried'. In the Middle Ages, the word *drug* was also used to refer to medicine (and poison), which can be explained by the fact that most medicines were made of dried herbs in those times. From the nineteenth century, *drug* also means narcotic.

Epidemiological study

An *epidemiological study* is the study of health status in a given population and the analysis of factors influencing it.

Ergotism

Ergotism is the effect of long-term ergot poisoning, traditionally due to the consumption of the alkaloids produced by the *Claviceps purpurea* fungus that infects rye and other cereals. It is also known as ergotoxicosis, ergot poisoning or Saint Anthony's Fire or Holy Fire. Ergot poisoning is a proposed explanation of bewitchment. The symptoms of ergot poisoning include convulsive symptoms (painful seizures and spasms, diarrhea, paresthesias, itching, mental effects including mania or psychosis, headaches, nausea and vomiting) and gangrenous symptoms of the fingers and toes (desquamation or peeling, weak peripheral pulses, loss of peripheral sensation, edema and ultimately the death and loss of affected tissues).

Esters
Esters are the products of condensation reactions between acids and alcohols. The other product of such reactions is water. Fruit *esters* are volatile compounds found in fruits. One class of lipids (triglycerides) are *esters* formed in the reaction between glycerol and various fatty acids.

Flavonoids
Flavonoids are plant pigments belonging to the class of polyphenolic compounds. Chemically, they are 2-phenyl-1,4-benzopyrone derivatives.

Glycemic index
The *glycemic index* characterizes how much a carbohydrate-containing food increases the blood glucose concentration. It shows how certain foods increase the blood glucose levels after eating, using glucose as a reference. Its value is influenced by numerous factors, such as the fiber content of the given food, the food preparation procedures, and the consistency of the food.

Glycosides
Glycosides are organic compounds in which a carbohydrate is attached to some other compound, usually a small organic molecule (called aglycone). The bond formation happens between the so-called glycosidic hydroxyl group of the carbohydrate part and the hydroxyl (OH), amino (NH_2) or thiol (SH) groups of the aglycone forming the corresponding *O*-, *N*- or *S*-glycoside by water elimination. The glycosyl units are simple or complex carbohydrates. In plants, *glycosides* are often used to store different active substances, which can be activated by an enzyme-induced cleavage of the sugar moiety. In animals and humans, *glycosides* are also responsible for the natural detoxification of the body since toxins mainly leave the body as water-soluble *glycosides*.

Gram (g) → Very small and very big quantities

Half-life
The *half-life* is equal to the length of time during which a continuous, monotonically decreasing value is halved. A typical example is the decay of radioactive nuclei. The amount of these nuclei shows an exponential decrease. During $t =$ *half-life* time, one half of the starting amount is decayed; during $t = 2 \times$ *half-life*, one half of the previous 50% decays again, so, only one fourth of the starting amount remains undecayed; after $t = 3 \times$ *half-life*, the same happens, so, only one eight of the starting quantity remains and so on. After $t = 10 \times$ *half-life*, less than one thousandth of the starting amount remains un-decayed. A radioactive isotope of carbon (carbon-14) has a *half-life* of 5730 years, and the mathematical relationship between the time (t)

and the remaining amount of radioactive nuclei described in the previous sentences is the theoretical background of radiocarbon dating. Other radioactive nuclei decay much more slowly or more quickly, e.g. the uranium-238 isotope has a *half-life* of 4.468 billion years and iodine-123, used for the treatment of thyroid, has a *half-life* of only 13 h.

The transformation/elimination of the substances in the body can also be characterized by their *half-lives*, but these processes usually show a more complex pattern than a single exponential. The *half-life* of water in the human body is 7–14 days, for the anticancer agent cisplatin 30–100 h, and for the hormone epinephrine (adrenaline), the *half-life* is only 2 min.

HDL → High density lipoprotein

High density lipoprotein (HDL)
High density lipoproteins transport cholesterol from the atherosclerotic deposits, thereby reducing the incidence and severity of cardiovascular diseases.

HPLC → Chromatography, chromatogram

Interventional study
In an *interventional study*, the investigators give the research subjects a particular medicine or other intervention. Usually, they compare the treated subjects to subjects who receive no treatment or standard treatment. Then the researchers measure how the subjects' health changes. *Cf.* **Observational study**.

Ischemic heart disease
Ischemic heart disease is associated with poor oxygenation of the heart muscles.

LDL → Low-density lipoprotein

Ligand
A *ligand* is a substance that specifically binds to the active site of a molecule or receptor.

Low density lipoprotein (LDL)
Low density lipoproteins are compounds which play a role in the development of atherosclerosis and in cholesterol transport.

Maximum air pollution concentrations → Dose-response relationship/curve

Median effective dose (ED_{50}) → Dose-response relationship/curve

Median lethal dose (LD_{50}) → Dose-response relationshipcurve

Median lethal concentration (LC_{50}) → Dose-response relationship/curve

Mediterranean diet
The *Mediterranean diet* is typical on the northern shore of the Mediterranean Sea. It is rich in vegetables and fruit. Instead of red meat, fish consumption is widespread. All this is combined with low fat and high vegetable oil consumption, together with some wine drinking during meals. Beneficial health effects of the *Mediterranean diet* are proved by many studies.

Meta-analysis
Meta-analysis is a detailed, cumulative analysis of data from several similar studies, which is usually based on scientific publications. One of its many advantages is that the assessment or testing of any hypothesis can be carried out on a much larger sample set than in any single tests. It is widely used both in clinical and **epidemiological studies**.

mg/kg of body weight
Milligrams/kilogram of body weight, a unit used in **Dose-response relationship/curve**

Microgram (µg) → Very large and very small quantities

Milligram (mg) → Very large and very small quantities

mmol/l or mmol/liter → Molar concentration

Molar concentration
Molar concentration (c) is defined as the amount of a constituent (n) (usually measured in moles—hence the name) divided by the volume of the mixture (V), thus c = n/V. The usual unit of molar concentration is **mol**/dm^3 or **mol**/liter.

mol
mol is the abbreviation of mole. Mole is a unit of measurement used in chemistry to express amounts of a chemical substance, defined as the amount of any substance that contains as many elementary entities (e.g., atoms, molecules, ions, electrons) as there are atoms in 12 g of pure carbon-12 (^{12}C), the isotope of carbon with relative atomic mass 12. This corresponds to the Avogadro constant, which has a value of $6.02214129(27) \times 10^{23}$ elementary entities of the substance. It is one of the seven base units in the International System of Units. One thousandth of mole is millimole (abbreviated mmol), often used in concentration unit, e.g. **mmol/liter** or **mmol/l**.

mol/dm^3 → Molar concentration

Mortality
Mortality means the rate of death.

Nanogram (ng) → Very large and very small quantities

No adverse effect level (NOAEL) → **Dose-response relationship/curve**

NOAEL → **Dose-response relationship/curve**

No effect level (NOEL) → **Dose-response relationship/curve**

NOEL → **Dose-response relationship/curve**

Observational study
Observational studies are studies in which the natural evolution of some processes are monitored and analyzed without interference. *Cf.* **Interventional study**.

Ozonolysis
Ozonolysis is the treatment of a chemical compound with ozone (O_3), a highly reactive allotropic form of oxygen. The process is mainly used in the chemical industry for the conversion of alkenes, and in the wastewater treatment to destroy organic compounds. In the latter case, the reaction by-products must be removed by a charcoal treatment.

pH
The pH of a solution is a number which gives its acidity or basicity. The pH is the ten-based negative logarithm of the hydrogen ion concentration given in **mol/dm^3** unit. The pH generally ranges from 0 to 14. A value of less than 7 indicates an acidic, a value of greater than 7 is an alkaline, and a pH = 7 value indicates a neutral solution at room temperature.

Pharmacokinetics
Pharmacokinetics examines the temporal fate of a drug (or substances) administered externally to a living organism.

Pharmacodynamics
Pharmacodynamics is the study of the biochemical and physiological effects of drugs on the body and the mechanisms of drug action.

Pharmacological study
A *pharmacological study* tests the mechanism of action for a drug.

Phytotherapy
Phytotherapy is the proper use of scientifically tested herbs and their products in the prevention and treatment of illnesses.

Picogram (pg) → **Very large and very small quantities**

ppb → **Concentrations: parts per notations**

ppm → **Concentrations: parts per notations**

ppt → **Concentrations: parts per notations**

Relative risk
Relative risk is the ratio of the probability of an event occurring (for example, developing a disease) in an exposed group to the probability of the event occurring in a comparison, non-exposed group. A relative risk greater than 1 indicates an increased risk.

Reverse osmosis
Reverse osmosis is the inverse of natural osmosis under pressure. Osmosis is the solvent flow from a more dilute solution toward a more concentrated one through a semi-permeable membrane. The process is driven by the osmotic pressure of the concentrated solution, which depends on the chemical identities of the solvent and the dissolved substance, as well as on their ratio. If the external pressure acting on the more concentrated solution is higher than its osmotic pressure, *reverse osmosis*, i.e. a solvent flow toward the more diluted solution happens through the membrane. *Reverse osmosis* is most commonly known for its use in drinking water purification from seawater, removing the salt and other effluent materials from the water molecules.

Sickle-cell anemia
Sickle-cell anemia (SCA) or sickle-cell disease (SCD) or drepanocytosis, is a hereditary blood disorder, characterized by red blood cells that assume an abnormal, rigid, sickle shape. The process of 'sickling' decreases the cells' flexibility and results in a risk of various life-threatening complications. This sickling occurs because of a mutation in the haemoglobin gene. Individuals with one copy of the defunct gene display both normal and abnormal haemoglobin. The average life expectancy of patients with this condition was estimated earlier to be 42–48 years but today, thanks to better management of the disease, patients can live into their 70s or beyond.

Sickling → **Sickle-cell anemia**

Tannins or tannic acids
Tannins are water-soluble compounds containing more than one phenolic hydroxyl groups, which precipitate (tan) the proteins. In a broader sense, substances which convert the raw skins of animals to leather (substances such as alum, aluminum, chromium, titanium, and zirconium salts, aldehydes, etc.) are called *tanning agents*.

TDI → **Dose-response relationship/curve**

Therapeutic index (TI) → **Dose-response relationship/curve**

TI → **Dose-response relationship/curve**

Thixotropy/thixotropic

Thixotropy is the reduced viscosity of a material caused by a shear stress such as mixing. However, when the stirring is terminated (resting), the viscosity more or less returns to the original, higher value after a time. For example gelatin, honey, butter, paint or wet beach sand show *thixotropic* behavior.

Dilatancy is a similar phenomenon in which the force applied causes a temporary increase in viscosity and the viscosity rests when the force ceases again. Starch gum (aka oobleck, ooze, goop, gloop, glurch) shows such behavior.

Tolerable daily intake (TDI) → Dose-response relationship/curve

Very large and very small quantities

Quantities can be well perceived only if they are close to the human scale. *Quantities* which are significantly larger or smaller are usually difficult to handle for us. Here are some practical examples of *mass* units:

5.976 trillion (5.976×10^{21}) t	The mass of the Earth
200 billion (2×10^{11}) t	The estimated mass of biomass annually formed on the Earth
1 ton (1 t, 1000 kg)	The mass of a small car
1 kilogram (1 kg)	The mass of 1 l of water
1 gram (1 g)	One thousandth of a kilogram, approximately the mass of a peanut eye
1 milligram (1 mg, 0.001 g)	One thousandth of one gram, the mass of 2-3 poppy seeds
1 microgram (1 µg, 0.000001 g, 10^{-6} g)	One millionth of a gram or one thousandth of a milligram
1 nanogram (1 ng, 0.000000001 g, 10^{-9} g)	One billionth of a gram or one millionth of a milligram
1 picogram (1 pg, 0.000000000001 g, 10^{-12} g)	One trillionth of a gram or one billionth of a milligram

Vitalism, vis vitalis theory

One of the main stages of the development of organic chemistry was the life-force (*vis vitalis*) theory. Until the early 1800s, scientists at the time, most notably the Swedish chemist Jöns Jacob Berzelius (1779–1848) believed that organic compounds cannot be prepared in simple chemical reactions, since they may be formed only in living organisms by the action of a "life-force". The *vis vitalis* theory did not live long, because Friedrich Wöhler (1800–1882) synthesized oxalic acid in 1824, and urea in 1828, both in simple chemical reactions. With the failure of *vitalism*, the development of organic chemistry began.

Sources and Further Reading

General Reading

H.-D. Belitz, W. Grosch, P. Schieberle (2009): Food chemistry. 4th ed. Springer, Berlin.
F. Braudel (1982): Civilization and capitalism, 15th–18th Century. Volume 1. The structures of everyday life. HarperCollins Publishers, New York.
G. F. Combs (2008): The vitamins—fundamental aspects in nutrition and health. Elsevier Academic Press, Burlington.
T. P. Coultate (2002): Food. The chemistry of its components. 4th ed. Royal Society of Chemistry, Cambridge.
R. M. Eddy (2000): Chemophobia in the college classroom: Extent, sources, and student characteristics. *J. Chem. Educ.*, **77**, 514–517.
J. Emsley (1994): The consumer's good chemical guide, W. H. Freeman-Spektrum, Oxford.
J. Emsley (1998): Molecules at an exhibition: portraits of intriguing materials in everyday life. Oxford University Press, Oxford.
J. Emsley (2003): Nature's building blocks: An A–Z guide to the elements. Oxford University Press, Oxford.
J. Emsley (2008): Molecules of murder. Criminal molecules and classic cases. Royal Society of Chemistry, Cambridge.
J. Emsley (2010): A healthy, wealthy, sustainable world. Royal Society of Chemistry, Cambridge.
P. Freedman (ed.) (2007): Food. The history of taste. Thames and Hudson, London.
B. Goldacre (2009): Bad science. Fourth Estate, London.
N. N. Greenwood, A. Earnshaw (1984): Chemistry of the elements. Pergamon Press, Oxford.
J. Gregory, S. Miller (1998): Science in public. Communication, culture, and credibility. Basic Books, Cambridge.
P. Le Couteur, J. Burreson (2004): Napoleon's buttons: how 17 molecules changed history. Jeremy P. Tarcher, New York.
T. Mann (1927): The magic mountain (translated by H. T. Lowe-Porter). Secker and Warburg.
J. Schwarcz (2009): Science, sense and nonsense. Doubleday Canada, Toronto.
H. This, M.-O. Monchicourt (2011): Building a meal: from molecular gastronomy to culinary constructivism (translated by M. B. Debevoise). Columbia University Press, New York.
J. Timbrell (2002): Introduction to toxicology. 3rd ed. Taylor and Francis, London.
J. Timbrell (2005): The poison paradox. Oxford University Press, Oxford.
M. Toussaint-Samat (2009): A history of food. 2nd ed., Wiley-Blackwell, Chichester.
R. L. Wolke, M. Parrish (2002): What Einstein told his cook. Kitchen science explained. W. W. Norton & Company.

Special Reading Related to the Individual Essays

H. Ibsen (1936): Peer Gynt: a dramatic poem (translated by R. Farquharson Sharp), J. B. Lippincott, Philadelphia, Act Five, Scene V, p. 218.

1.1 Fear of the Unknown: Chemicals

Anon. (2006): Making sense of chemical stories. Sense about science. London. http://www.senseaboutscience.org/resources.php/5/making-sense-of-chemical-stories (last access: February 25, 2014).
R. M. Eddy (2000): cited work.
http://www.snopes.com/science/dhmo.asp (last access: February 25, 2014).
http://www.stuff.co.nz/national/politics/38005 (last access: February 25, 2014).
http://www.dhmo.org/linking/ (last access: February 25, 2014).
http://en.wikipedia.org/wiki/DHMO (last access: February 25, 2014).
http://en.wikipedia.org/wiki/Chemophobia (last access: February 25, 2014).

1.2 Life as a Risky Business

J. Emsley (1994): cited work, pp. 141–145.
J. H. Fremlin (1985): Power production: what are the risks? Adam Hilger, Bristol.
M. Henderson (1987): Living with risk. The British Medical Association guide. John Wiley and Sons, Chichester, pp. 67–73, 87–95, 147–155.
S. Lichtenstein, P. Slovic, B. Fischhof, M. Layman, B. Combs (1978): Judged frequency of lethal events. *J. Exp. Psychol. Human Learning Mem.*, **4**, 551–578.
M. G. Morgan, Engineering and Public Policy/Carnegie-Mellon University Graduate Research Methods Class (1983): On judging the frequency of lethal events: a replication. *Risk Anal.*, **3**, 11–16.
http://en.wikipedia.org/wiki/Risk_society (last access: February 25, 2014).
http://en.wikipedia.org/wiki/Risk_assessment (last access: February 25, 2014).
https://www.toxicology.org/ai/fa/riskassess.pdf (last access: February 25, 2014).

1.3 Natural Products: A Delusion of Safety

B. N. Ames (1991): Natural carcinogens and dioxin. Sci. Total Environ., 104, 159–166.
S. Everts (2011): Vegetative warfare. Chem. Eng. News, 89(5), 53–55.
J. D. Mann (2004): How to poison your spouse the natural way. A kiwi guide to safer food. JDM and Associates, Somerfield, New Zealand, 3rd edn.
http://en.wikipedia.org/wiki/Vitalism (last access: February 25, 2014).
http://en.wikipedia.org/wiki/Median_lethal_dose (last access: February 25, 2014).
http://en.wikipedia.org/wiki/Botulinum_toxin (last access: February 25, 2014).
http://en.wikipedia.org/wiki/Ames_test (last access: February 25, 2014).

1.4 Man-Made Commodities and Safety Issues

J. Lovelock (2009): The vanishing face of Gaia. A final warning. Basic Books, New York.
M. Shermer (2008): Wheatgrass juice and folk medicine. Why subjective anecdotes often trump objective data. *Sci. Am.*, **299**(2), 42. http://www.scientificamerican.com/article.cfm?id=how-anecdotal-evidence-can-undermine-scientific-results (last access: February 25, 2014).
http://www.snopes.com/medical/toxins/toxins.asp (last access: February 25, 2014).
http://en.wikipedia.org/wiki/E_number (last access: February 25, 2014).
http://www.nestle.ch/de/nutrition/news/themes/Pages/topics_320.aspx (website no more available, last access: November 10, 2010).
http://www.senseaboutscience.org.uk/pdf/MakingSenseofChemicalStories.pdf (last access: February 25, 2014).

1.5 The Cholera Pandemonium: The Blind Leading the Blind

W. McDonough, M. Braungart (2002): Cradle to cradle: remaking the way we make things. North Point Press, New York.
http://en.wikipedia.org/wiki/E_number (last access: February 25, 2014).
http://www.senseaboutscience.org.uk/pdf/MakingSenseofChemicalStories.pdf (last access: February 25, 2014).
http://www.greenpeace.org/international/Global/international/planet-2/report/1998/3/cholera-and-chlorine.pdf (last access: February 25, 2014).
http://www.colorado.edu/geography/gcraft/warmup/cholera/cholera_f.html (last access: February 25, 2014).
http://www.who.int/csr/don/1998_02_25/en/index.html (last access: February 25, 2014).
http://www.scoop.co.nz/stories/WO1105/S00700/rising-cholera-cases-in-haitis-ouest-province.htm (last access: February 25, 2014).
http://chlorine.americanchemistry.com/Chlorine-Benefits/Safe-Water/Chlorine-Misinformation.html (last access: February 25, 2014).

1.6 Regulating Chemicals: Maybe or Maybe Not

C. H. Arnaud (2010): Transforming toxicology. *Chem. Eng. News*, **88**(51), 37–39.
B. Borrell (2010): America pushes to overhaul chemical safety law. *Nature*, **463**, 599.
M. English (2010): European Commission responds to chemical testing story. *Nature*, **465**, 420.
N. Gilbert (2009): Chemical safety costs uncertain. *Nature*, **460**, 1065.
N. Gilbert (2010): Crucial data on REACH not disclosed. *Nature*, **464**, 1116–1117.
T. Hartung, C. Rovida (2009): Chemical regulators have overreached. *Nature*, **460**, 1080–1081.
Regulation (EC) No 1907/2006 of the European Parliament and of the Council of 18 December 2006 concerning the Registration, Evaluation, Authorisation and Restriction of Chemicals (REACH), establishing a European Chemicals Agency.
J.-F. Tremblay (2011): The dark side of Indian drug-making. *Chem. Eng. News*, **89**(1), 13–14.
http://en.wikipedia.org/wiki/Registration,_Evaluation,_Authorisation_and_Restriction_of_Chemicals (last access: February 25, 2014).
http://en.wikipedia.org/wiki/Responsible_Care (last access: February 25, 2014).

1.7 Biowaste: Biotechnology in Perspective

S. K. Ritter (2008): Calling all chemists. *Chem. Eng. News*, **86** (33), 59–68.
R. A. Sheldon (2007): The E factor: fifteen years on. *Green Chem.* **9**, 1273–1283.
http://en.wikipedia.org/wiki/Virtual_water (last access: February 25, 2014).

1.8 Ice Skating and the Science of Sliding

C. A. Liberko (2004): Using Science Fiction to Teach Thermodynamics: Vonnegut, Ice-nine, and Global Warming. *J. Chem. Educ.*, **81** (4), 509.
R. Rosenberg (2005): Why is ice slippery? *Physics Today*, December Issue, 50–55. http://lptms.u-psud.fr/membres/trizac/Ens/L3FIP/Ice.pdf (last access: February 25, 2014).
http://globalsyntheticice.com/default.aspx (last access: February 25, 2014).

1.9 So What's Happening to the Ozone Hole?

R. A. Kerr (2011): First detection of ozone hole recovery claimed. *Science*, **332**, 160.
M. Salby, E. Titova, L. Deschamps (2011): Rebound of Antarctic ozone. *Geophys. Res. Lett.*, 38, L09702.
I. Williams (2001): Environmental chemistry. A modular approach. Wiley, Chichester.
http://ozone.unep.org/Ratification_status/ (last access: February 25, 2014).

1.10 Predicting the Elements: Success and Failure

E. R. Scerri (2007): The Periodic table. Its story and its significance. Oxford University Press, New York, pp. 123–157.
J. W. van Spronsen (1981): Mendeleev as a speculator. *J. Chem. Educ.*, **58** (10), 790–791.
J. W. van Spronsen (1969): The priority conflict between Mendeleev and Meyer. *J. Chem. Educ.*, **46** (3), 136–139.

1.11 Vedic Wisdom: Lead and Ayurveda

180 perc, MR1 Kossuth Rádió, March 17, 2008. http://www.mr1-kossuth.hu/m3u/00327032_2609916.m3u (last access: May 5, 2011).
S. H. Ali (2010): Treasures of the Earth. Need, greed, and a sustainable future. Yale University Press, New Haven, London.
J. Emsley (2003): cited work, pp. 226–233.
M. László F. (2007): Az úgy hagyott Magyarország 5.—A Csonkabihar: Megköszönni a morzsákat is (Abandoned Hungary, part 5. Thanking for the chips—in Hungarian) *Magyar Narancs*, July 26, 2007. http://mancs.hu/index.php?gcPage=/public/hirek/hir.php&id=15028 (last access: February 25, 2014).

J. Timbrell (2005): cited work, pp. 136–142.
http://en.wikipedia.org/wiki/Lead%E2%80%93acid_battery (last access: February 25, 2014).
http://en.wikipedia.org/wiki/Ayurveda (last access: February 25, 2014).
http://en.wikipedia.org/wiki/Gout (last access: February 25, 2014).

1.12 Manipulating Weather: Ocean Fertilization

R. Black (2009): Setback for climate technical fix. *BBC News*, 23rd March 2009. http://news.bbc.co.uk/2/hi/7959570.stm (last access: February 25, 2014).
P. W. Boyd, T. Jickells, C. S. Law, S. Blain, E. A. Boyle, K. O. Buesseler, K. H. Coale, J. J. Cullen, H. J. W. de Baar, M. Follows, M. Harvey, C. Lancelot, M. Levasseur, N. P. J. Owens, R. Pollard, R. B. Rivkin, J. Sarmiento, V. Schoemann, V. Smetacek, S. Takeda, A. Tsuda, S. Turner, A. J. Watson (2007): Mesoscale iron enrichment experiments 1993–2005: synthesis and future directions. *Science*, **315** (5812), 612–617.
J. Emsley (2003): cited work, pp. 205–211.
T. D. Jickells, Z. S. An, K. K. Andersen, A. R. Baker, G. Bergametti, N. Brooks, J. J. Cao, P. W. Boyd, R. A. Duce, K. A. Hunter, H. Kawahata, N. Kubilay, J. laRoche, P. S. Liss, N. Mahowald, J. M. Prospero, A. J. Ridgwell, I. Tegen, R. Torres (2005): Global iron connections between desert dust, ocean biogeochemistry, and climate. *Science*, **308** (5718), 67–71.
H.-G. Körber (1982): Alfred Wegener. BSB B. G. Teubner Verlagsgesellschaft, Leipzig. 2., erweiterte Auflage.
P. C. Soares (2010): Warming power of CO_2 and H_2O: correlations with temperature changes. *Int. J. Geosci.*, **1**, 102–112.
http://canov.jergym.cz/vyhledav/varian19/ironword.pdf (last access: February 25, 2014).
http://www.shabdkosh.com/hi/translate?e=iron&l=hi (last access: February 25, 2014).
http://en.wikipedia.org/wiki/Iron_fertilization (last access: February 25, 2014).
http://www.whoi.edu/oceanus/viewArticle.do?id=34167 (last access: February 25, 2014).
http://www.climos.com/pubs/2009/Climos_Why_OIF-2009-03-12.pdf (last access: February 25, 2014).
http://en.wikipedia.org/wiki/Plankton (last access: February 25, 2014).
http://carbonsequestration.blogspot.com/ (last access: February 25, 2014).
http://www.uri.edu/news/releases/?id=5708 (last access: February 25, 2014).
http://isisconsortium.org/ (last access: February 25, 2014).
http://www.nature.com/news/2009/090401/full/458562f.html (last access: February 25, 2014).
http://www.awi.de/de/aktuelles_und_presse/selected_news/2009/lohafex/ (last access: February 25, 2014).
http://en.wikipedia.org/wiki/Global_warming (last access: February 25, 2014).

2.1 Test Your Cranberry Pie: Vitamin C and Benzoates

H.-D. Belitz, W. Grosch, P. Schieberle (2009): cited work, pp. 405, 417–420, 449–450.
M. Hickman (2007): Caution: Some soft drinks may seriously harm your health. *The Independent*, May 27th, 2007. 27, http://www.independent.co.uk/life-style/health-and-families/health-news/caution-some-soft-drinks-may-seriously-harm-your-health-6262638.html (last access: February 25, 2014).
Indications of the possible formation of benzene from benzoic acid in foods. BfR Expert Opinion No. 013/2006, 1 December 2005. http://www.bfr.bund.de/cm/245/indications_of_the_possible_formation_of_benzene_from_benzoic_acid_in_foods.pdf (last access: February 25, 2014).

P. W. Piper (1999): Yeast superoxide dismutase mutants reveal a pro-oxidant action of weak organic acid food preservatives. *Free Radical Biol. Med.*, **27**, 1219–1227.

Questions and Answers on the occurrence of benzene in soft drinks and other beverages.

J. Timbrell (2005): cited work, p. 266.

http://www.fda.gov/Food/FoodborneIllnessContaminants/ChemicalContaminants/ucm055131.htm (last access: February 25, 2014).

http://www.senseaboutscience.org/for_the_record.php/40/sodium-benzoate (last access: February 28, 2014).

2.2 Food Dyes: The Good, the Bad and the Ugly

JECFA Expert evaluation of the permitted dye Sunset Yellow FCF: http://www.inchem.org/documents/jecfa/jecmono/v17je29.htm (last access: February 25, 2014).

JECFA Expert evaluation of the phased out dye Citrus Red 2: http://www.inchem.org/documents/jecfa/jecmono/v46aje10.htm (last access: February 25, 2014).

D. McCann, A. Barrett, A. Cooper, D. Crumpler, L. Dalen, K. Grimshaw, E. Kitchin, K. Lok, L. Porteous, E. Prince, E. Sonuga-Barke, J. O. Warner, J. Stevenson (2007): Food additives and hyperactive behaviour in 3-year-old and 8/9-year-old children in the community: a randomised, double-blinded, placebo-controlled trial. *Lancet*, **370**, 1560.

Reply of EFSA (European Food Safety Authority) to the McCann report:http://www.efsa.europa.eu/en/efsajournal/pub/660.htm (last access: February 25, 2014).

http://www.southampton.ac.uk/mediacentre/news/2007/sep/07_99.shtml (last access: February 25, 2014).

http://www.southampton.ac.uk/mediacentre/news/2008/apr/08_65.shtml (last access: February 25, 2014).

2.3 The Organic Vegetable Hype

A. D. Dangour, S. K. Dodhia, A. Hayter, E. Allen, K. Lock, R. Uauy (2009): Nutritional quality of organic foods: a systematic review, *Am. J. Clin. Nutr*, **90**, 680–685 and references cited.

P. Santamaria (2006): Review—Nitrate in vegetables: toxicity, content, intake and EC regulation. *J. Sci. Food Agric.*, 86, 10–17.

2.4 "You're the One for Me, Fatty": Concocted Fats

http://en.wikipedia.org/wiki/E_number (last access: February 25, 2014).

2.5 Fat Matters: Margarine vs. Butter

J. Beare-Rogers, A. Dieffenbacher, J. V. Holm (2001): Lexicon of lipid nutrition (IUPAC Technical Report). *Pure Appl. Chem.*, **73**, 685–744.

H.-D. Belitz, W. Grosch, P. Schieberle (2009): cited work, pp. 656–661.

S. C. Bevan, S. J. Gregg, A. Rosseinsky (1976): Concise etymological dictionary of chemistry. Applied Science Publishers, London, p. 83.

M. M. Chrysan (2005): Margarines and spreads. In: *Bailey's industrial oil and fat products*. 6th ed., Vol. 4. (Ed: F. Shahidi) John Wiley and Sons, Hoboken, pp. 33–73.
T. P. Coultate (2002): cited work, pp. 86–90, 109–112.
B. Erickson (2013): Good-bye, trans fat. *Chem. Eng. News.*, **91**(46), 8.
P. Freedman (ed.) (2007): cited work, pp. 242–244.
F. D. Gunstone (2008): Food applications of lipids. Chap. 27. In: *Food lipids. Chemistry, nutrition, and biotechnology*. 3rd ed. (Eds: C. C. Akoh, D. B. Min) CRC Press, Hoboken, pp. 683–703.
D. Mozaffarian, A. Aro, W. C. Willett (2009): Health effects of *trans*-fatty acids: experimental and observational evidence. *Eur. J. Clin. Nutr.*, **63**, S5–S21.
J. Schwarcz (2009): cited work, pp. 6–9.
M. J. Sadler (2005): *trans*-Fatty acids. In: *Encyclopedia of human nutrition*. 2nd ed. Vol. 2. (Eds: B. Caballero, L. Allen, A. Prentice), Elsevier, Amsterdam, pp. 230–237.
M. Toussaint-Samat (2009): cited work, pp. 199–200.
http://www.imace.org (last access: February 25, 2014).
http://www.snopes.com/food/warnings/butter.asp (last access: February 25, 2014).
http://en.wikipedia.org/wiki/Conjugated_linoleic_acid (last access: February 25, 2014).
http://en.wikipedia.org/wiki/Michel_Eug%C3%A8ne_Chevreul (last access: February 25, 2014).
http://en.wikipedia.org/wiki/Margarine (last access: February 25, 2014).
http://en.wikipedia.org/wiki/Ghee (last access: February 25, 2014).

2.6 Fake Food and Kidney Stones

J. Emsley (2010): cited work, pp. 26–27.
B. Erickson (2008): Food Safety. Food and Drug Administration sets melamine safety level in food. *Chem. Eng. News*, **86**(41), 9.
F. Grases, A. Costa-Bauzá, I. Gomila, S. Serra-Trespalle, F. Alonso-Sainz, J. M. del Valle (2009): Melamine urinary bladder stone. *Urology*, **73**, 1262–1263.
H. Ogasawara, K. Imaida1, H. Ishiwata, K. Toyoda, T. Kawanishi, C. Uneyama, S. Hayashi, M. Takahashi, Y Hayashi (1995): Urinary bladder carcinogenesis induced by melamine in F344 male rats: correlation between carcinogenicity and urolith formation. Carcinogenesis, 16, 2773–2777. http://carcin.oxfordjournals.org/content/16/11/2773.abstract (last access: February 25, 2014).
H. Ogasawara, K. Imaida, H Ishiwata, K. Toyoda, T. Kawanishi, C. Uneyama, S. Hayashi, M. Takahashi, Y. Hayashi (2009): Urinary bladder carcinogenesis induced by melamine in F344 male rats: correlation between carcinogenicity and urolith formation. *Carcinogenesis*, **16**, 2773–2777.
Pediatric Environmental Health Specialty Units (2009): Melamine: information for pediatric health professionals. http://www.coeh.uci.edu/PEHSU/factsheets/Melamine_Factsheet.pdf (last access: February 25, 2014).
S. L. Rovner (2009): Silver lining in melamine crisis. Deadly adulteration of Chinese milk drives development of analytical methods to detect contaminant in food products. *Chem. Eng. News*, **87**(21), 36–38.
Q. Sun, Y. Shen, N. Sun, G. J. Zhang, Z. Chen, J. F. Fan, L. Q. Jia, H. Z. Xiao, X. R. Li, B. Pus (2010): Diagnosis, treatment and follow-up of 25 patients with melamine-induced kidney stones complicated by acute obstructive renal failure in Beijing Children's Hospital. *Eur. J. Pediatr.*, **169**, 483–489.
X.-H. Sun, L.-M. Shen, X.-M. Cong, H.-J. Zhu, J.-L. Lv, L. He (2011): Infrared spectroscopic analysis of urinary stones (including stones induced by melamine-contaminated milk powder) in 189 Chinese children. *J. Pediatric Surgery*, **46**, 723–728.
World Health Organization (2008): Melamine and cyanuric acid: toxicity, preliminary risk assessment and guidance on levels in food. 25 September 2008—Updated 30 October 2008. http://www.who.int/foodsafety/fs_management/Melamine.pdf (last access: February 25, 2014).

C.-F. Wu, C.-C. Liu, B.-H. Chen, S.-P. Huang, H.-H. Lee, Y.-H. Chou, W.-J. Wu, M.-T. Wu (2010): Urinary melamine and adult urolithiasis in Taiwan. *Clin. Chim. Acta*, **411**, 184–189.

X. Zhang, T. Chen, Q. Chen, L. Wang, L. J. Wan (2009): Self-assembly and aggregation of melamine and melamine-uric/cyanuric acid investigated by STM and AFM on solid surfaces. *Phys. Chem. Chem. Phys.*, **11**, 7708–7712.

X.-B. Zhang, J.-L. Bai, P.-C. Ma, J.-H. Ma, J.-H. Wan, B. Jiang (2010): Melamine-induced infant urinary calculi: a report on 24 cases and a 1-year follow-up. *Urol. Res.*, **38**, 391–395.

http://en.wikipedia.org/wiki/Lactose_intolerance (last access: February 25, 2014).
http://en.wikipedia.org/wiki/Kjeldahl_method (last access: February 25, 2014).
http://en.wikipedia.org/wiki/2008_Chinese_milk_scandal (last access: February 25, 2014).
http://en.wikipedia.org/wiki/2007_pet_food_recalls (last access: February 25, 2014).
http://en.wikipedia.org/wiki/Melamine (last access: February 25, 2014).

2.7 The Coming Shortage: Vanilla and Menthol

M. J. W. Dignum, J. Kerler, R. Verpoorte (2001): Vanilla production: technological, chemical, and biosynthetic aspects. *Food Rev. Internatl.*, **17**, 199–219.

M. B. Hocking (1997): Vanillin: synthetic flavoring from spent sulfite liquor. *J. Chem. Educ.*, **74**, 1055–1059. http://jchemed.chem.wisc.edu/Journal/Issues/1997/Sep/abs1055.html (last access: February 25, 2014).

Lawrence, B. M. (2007): Mint. The genus *Mentha*. Medicinal and aromatic plants—Industrial profiles. CRC Press, Boca Raton.

http://en.wikipedia.org/wiki/Ig_Nobel_Prize (last access: February 25, 2014).
http://improbable.com/ig/ig-pastwinners.html (last access: February 25, 2014).
http://www.japanprobe.com/?p=2875 (last access: February 25, 2014).
http://www.terradaily.com/reports/Japanese_Researchers_Extract_Vanilla_From_Cow_Dung.html (last access: February 25, 2014).
http://en.wikipedia.org/wiki/Vanillin#Chemical_synthesis (last access: February 25, 2014).
http://en.wikipedia.org/wiki/Vanilla (last access: February 25, 2014).

2.8 Red Alert: Meat Colors

H.-D. Belitz, W. Grosch, P. Schieberle (2009): cited work, pp. 563, 573–577, 775, 810, 907.

R. C. Claudia, J. C. Francisco (2010): Effect of an argon-containing packaging atmosphere on the quality of fresh pork sausages during refrigerated storage. *Food Control*, **21**, 1331–1337.

J. Itten (1970): The elements of color (translated by Ernst van Haagen). John Wiley and Sons, New York, p. 83.

J. Y. Jeong, J. R. Claus (2011): Color stability of ground beef packaged in a low carbon monoxide atmosphere or vacuum. *Meat Sci.*, **87**, 1–6.

O. Sorheim, H. Nissen, T. Nesbakken (1999): The storage life of beef and pork packaged in an atmosphere with low carbon monoxide and high carbon dioxide. *Meat Sci.*, **52**, 157–164.

D. Whitford (2005): Proteins. Structure and function. John Wiley and Sons, Chichester. pp. 40, 67–70.

http://en.wikipedia.org/wiki/Nitrite#Nitrite_in_biochemistry (last access: February 25, 2014).
http://en.wikipedia.org/wiki/E249 (last access: February 25, 2014).
http://en.wikipedia.org/wiki/Sodium_nitrite (last access: February 25, 2014).

2.9 Moldy Business: Whole-Grain Cereals

M. Beyer, M. B. Klix, H. Klink, J.-A. Verreet (2006): Quantifying the effects of previous crop, tillage, cultivar and triazole fungicides on the deoxynivalenol content of wheat grain—a review. *J. Plant Dis. Protect.*, **113**, 241–246.

Food Standards Agency (2007): The UK code of good agricultural practice to reduce fusarium mycotoxins in cereals. http://www.food.gov.uk/multimedia/pdfs/fusariumcop.pdf (last access: February 25, 2014).

RASFF: Annual report on the functioning of the RASFF 2004. http://ec.europa.eu/food/food/rapidalert/report2004_en.pdf (last access: February 25, 2014).

RASFF: Annual Report 2005. http://ec.europa.eu/food/food/rapidalert/report2005_en.pdf (last access: February 25, 2014).

RASFF: Annual Report 2006. http://ec.europa.eu/food/food/rapidalert/report2006_en.pdf (last access: February 25, 2014).

RASFF: Annual Report 2007. http://ec.europa.eu/food/food/rapidalert/report2007_en.pdf (last access: February 25, 2014).

RASFF: Annual Report 2008. http://ec.europa.eu/food/food/rapidalert/report2008_en.pdf (last access: February 25, 2014).

RASFF: Annual Report 2009. http://ec.europa.eu/food/food/rapidalert/report2009_en.pdf (last access: February 25, 2014).

RASFF: Annual Report 2010. http://ec.europa.eu/food/food/rapidalert/docs/rasff_annual_report_2010_en.pdf (last access: February 25, 2014).

RASFF: 2011 Annual Report. http://ec.europa.eu/food/food/rapidalert/docs/rasff_annual_report_2011_en.pdf (last access: February 25, 2014).

RASFF: 2012 Annual Report. http://ec.europa.eu/food/food/rapidalert/docs/rasff_annual_report_2012_en.pdf (last access: February 25, 2014).

The Rapid Alert System for Food and Feed (RASFF): Annual report on the functioning of the RASFF 2003. http://ec.europa.eu/food/food/rapidalert/report2003_en.pdf (last access: February 25, 2014).

2.10 Red Wine and the Vineyard Blues

V. Cheynier (2005): Polyphenols in foods are more complex than often thought. *Am. J. Clin. Nutr.*, **81**, 223S–229S.

A. C. Cordova, L. S. Jackson, D. W. Berke-Schlessel, B. E. Sumpio (2005): The cardiovascular protective effect of red wine. *J. Am. Coll. Surg.*, **200**, 428–439.

S. R. Coimbra, S. H. Lage, L. Brandizzi, V. Yoshida, P. L. da Luz (2005): The action of red wine and purple grape juice on vascular reactivity is independent of plasma lipids in hypercholesterolemic patients. *Braz. J. Med. Biol. Res.*, **38**, 1339–1347.

Csupor D., Szendrei K. (2006a): A szőlő, a bor és az alkohol I. rész (Grape, wine, alcohol. Part 1—in Hungarian). *Családorvosi Fórum*, **7**, 66–68.

Csupor D., Szendrei K. (2006): A szőlő, a bor és az alkohol II. rész (Grape, wine, alcohol. Part 2—in Hungarian). *Családorvosi Fórum*, **7**, 54–57.

P. L. da Luz, S. R. Coimbra (2004): Wine, alcohol and atherosclerosis: clinical evidences and mechanisms. *Braz. J. Med. Biol. Res.*, **37**, 1275–1295.

C. Drahl (2009): Revisiting resveratrol. *Chem. Eng. News*, **87**(50), 36–37.

J. E. Freedman, C. Parker 3rd, L. Li, J. A. Perlman, B. Frei, V. Ivanov, L. R. Deak, M. D. Iafrati, J. D. Folts (2001): Select flavonoids and whole juice from purple grapes inhibit platelet function and enhance nitric oxide release. *Circulation*, **103**, 2792–2798.

M. Ganguli, J. Vander Bilt, J. A. Saxton, C. Shen, H. H. Dodge (2005): Alcohol consumption and cognitive function in late life: a longitudinal community study. *Neurology*, **65**, 1210–1217.

W. Greenrod, C. S. Stockley, P. Burcham, M. Abbey, M. Fenech (2005): Moderate acute intake of de-alcoholised red wine, but not alcohol, is protective against radiation-induced DNA damage ex vivo—Results of a comparative in vivo intervention study in younger men. *Mutation Research/Fundamental and Molecular Mechanisms of Mutagenesis*, **591**, 290–301.

M. Gronbaek (2004): Epidemiologic evidence for the cardioprotective effects associated with consumption of alcoholic beverages. *Pathophysiology*, **10**, 83–92.

S. Honsho, A. Sugiyama, A. Takahara, Y. Satoh, Y. Nakamura, K. Hashimoto (2005): A red wine vinegar beverage can inhibit the renin-angiotensin system: experimental evidence in vivo. *Biol. Pharm. Bull.*, **28**, 1208–1210.

R. S. Jackson (2008): Wine science. Principles and applications. Elsevier, Amsterdam, 3rd edn., pp. 686–706.

A. Jamroz, J. Beltowski (2001): Antioxidant capacity of selected wines. *Med. Sci. Monit.*, **7**, 1198–1202.

J. G. Keevil, H. E. Osman, J. D. Reed, J. D. Folts (2000): Grape juice, but not orange juice or grapefruit juice, inhibits human platelet aggregation. *J. Nutr.*, **130**, 53–56.

L. B. Mann, J. D. Folts (2004): Effects of ethanol and other constituents of alcoholic beverages on coronary heart disease: a review. *Pathophysiology*, **10**, 105–112.

M. V. Moreno-Arribas, M. Carmen Polo (2009): Wine chemistry and biochemistry. Springer, New York, pp. 571–591.

K. J. Mukamal, M. Cushman, M. A. Mittleman, R. P. Tracy, D. S. Siscovick (2004): Alcohol consumption and inflammatory markers in older adults: the Cardiovascular Health Study. *Atherosclerosis*, **173**, 79–87.

P. Pignatelli, A. Ghiselli, B. Buchetti, R. Carnevale, F. Natella, G. Germano, F. Fimognari, S. Di Santo, L. Lenti, F. Violi (2006): Polyphenols synergistically inhibit oxidative stress in subjects given red and white wine. *Atherosclerosis*, **188**, 77–83.

J. Schwarcz (2005): Let them eat flax. ECW Press, Toronto, pp. 59–62.

D. Shanmuganayagam, M. R. Beahm, H. E. Osman, C. G. Krueger, J. D. Reed, J. D. Folts (2002): Grape seed and grape skin extracts elicit a greater antiplatelet effect when used in combination than when used individually in dogs and humans. *J. Nutr.*, **132**, 3592–3598.

A. Takahara, A. Sugiyama, S. Honsho, Y. Sakaguchi, Y. Akie, Y. Nakamura, K. Hashimoto (2005): The endothelium-dependent vasodilator action of a new beverage made of red wine vinegar and grape juice. *Biol. Pharm. Bull.*, **28**, 754–756.

WHO Technical Report Series (2003): Diet, nutrition and the prevention of chronic diseases. WHO Geneva.

2.11 Popeye's Spinach and the Search for Iron

Anon. (1935): Spinach over-rated as source of iron. *The Science News-Letter*, **28**(749), Aug. 17, p. 110.

T. J. Hamblin (1981): Fake! *Br. Med. J.*, **283**, 1671–1674.

G. Kohler, C. Elvehjem, E. Hart, E. (1936): Modifications of the bipyridine method for available iron. *J. Biol. Chem.*, **113**, 49–53.

M. Sutton (2010): Spinach, iron and Popeye: Ironic lessons from biochemistry and history on the importance of healthy eating, healthy scepticism and adequate citation. *Internet J. Criminol.* http://www.internetjournalofcriminology.com/ijcprimaryresearch.html (last access: February 25, 2014).

USDA National Nutrient Database for Standard Reference, Release 26, 2013.

Sources and Further Reading 331

http://www.cracked.com/article_18517_the-7-most-disastrous-typos-all-time.html (last access: February 25, 2014).
http://www.de-fact-o.com/fact_read.php?id=2 (last access: February 25, 2014).

2.12 The Perplexing Spice: Curry

R. Hänsel, O. Sticher (2010) Pharmakognosie—Phytopharmazie. Springer Medizin Verlag, Heidelberg, 986–991.
A. A. Oyagbemi, A. B. Saba, A. O. Ibraheem (2009): Curcumin: from food spice to cancer prevention. *Asian Pacific Journal of Cancer Prevention*, **10**, 963–967.
J. M. Ringman, S. A. Frautschy, G. M. Cole, D. L. Masterman, J. L. Cummings (2005): A potenti6M. Toussaint-Samat (2009): cited work, pp. 449–451.
W. Wongcharoen, A. Phrommintikul (2009): The protective role of curcumin in cardiovascular diseases. *Int. J. Cardiol.*, **133**, 145–151.
http://en.wikipedia.org/wiki/Curry_Tree (last access: February 25, 2014).
http://www.suite101.com/content/curry-spices-of-life-a98586 (last access: February 25, 2014).
http://www.surreycc.gov.uk/sccwebsite/sccwspages.nsf/LookupWebPagesByTITLE_RTF/History+of+Curry?opendocument (last access: February 25, 2014).
http://www.menumagazine.co.uk/book/curryhistory.html (last access: February 25, 2014).
http://www.indepthinfo.com/curry/ (last access: February 25, 2014).
http://www.hub-uk.com/interesting/curry-history.htm (last access: February 25, 2014).
http://www.ehow.com/list_7618790_components-curry-powder.html#ixzz1IolSc3XW (last access: February 25, 2014).
http://en.wikipedia.org/wiki/Curry (last access: February 25, 2014).

2.13 Bigger the Better? The Alpha and Omega of Fatty Acids

M. N. D. Di Minno, E. Tremoli, A. Tufano, A. Russolillo, R. Lupoli, G. Di Minno (2010): Exploring newer cardioprotective strategies: omega-3 fatty acids in perspective. *Thromb. Haemost.*, **104**, 664–680.
C. Klemens, D. Berman, E. Mozurkewich (2009): A meta-analysis of perinatal omega-3 fatty acid supplementation on inflammatory markers, allergy, atopy, and asthma in infancy and childhood. *Am. J. Obstet. Gynecol.*, **201**, S228.
Omega-3 Fatty Acid Content in Fish. http://fn.cfs.purdue.edu/fish4health/HealthBenefits/omega3.pdf (last access: February 24, 2014).
G. L. Russo (2009): Dietary n-6 and n-3 polyunsaturated fatty acids: From biochemistry to clinical implications in cardiovascular prevention. *Biochem. Pharmacol.*, **77**, 937–946.
A. P. Simopoulos (2006): Evolutionary aspects of diet, the omega-6/omega-3 ratio and genetic variation: nutritional implications for chronic diseases. *Biomed. Pharmacother.*, **60**, 502–507.
A. P. Simopoulos (2008a): The importance of the omega-6/omega-3 fatty acid ratio in cardiovascular disease and other chronic diseases. *Exp. Biol. Med.*, **233**, 674–688.
A. P. Simopoulos (2008b): The omega-6/omega-3 fatty acid ratio, genetic variation, and cardiovascular disease. *Asia Pac. J. Clin. Nutr.*, **17**, 131–134.
A. P. Simopoulos, D. X. Tan, L. C. Manchester, R. J. Reiter (2005): Purslane: a plant source of omega-3 fatty acids and melatonin. *J. Pineal Res.*, **39**, 331–332.
http://en.wikipedia.org/wiki/Omega-9_fatty_acid (last access: February 25, 2014).
http://en.wikipedia.org/wiki/Essential_fatty_acid (last access: February 25, 2014).

2.14 Sweet as Birch: Xylitol

H.-D. Belitz, W. Grosch, P. Schieberle (2009): cited work, pp. 432–443, 862–891.
F. Braudel (1981): cited work.
E. K. Dunayer: New findings on the effects of xylitol ingestion in dogs. *Veterinary Medicine*, 2006, **101** (12), 791–797.
J. Emsley (1995), cited work, pp. 31–59.
P. Freedman (2007): cited work, pp. 48–49, 157–159, 218–220.
F. Gare: The sweet miracle of xylitol. Basic Health Publication, Laguna Beach, 2003.
H. Schiweck, A. Bär, R. Vogel, E. Schwarz, M. Kunz (2005): Sugar alcohols. In: Ullmann's Encyclopedia of Industrial Chemistry, 7th edn., Wiley-VCH Verlag GmbH & Co. KGaA, Weinheim, Vol. 25, pp. 413–440.
http://en.wikipedia.org/wiki/Xylitol (last access: February 24, 2014).
http://en.wikipedia.org/wiki/Sugar_alcohol (last access: February 24, 2014).
http://en.wikipedia.org/wiki/Sucrose (last access: February 24, 2014).
http://en.wikipedia.org/wiki/Glycemic_index (last access: February 24, 2014).

2.15 Bittersweet Effects: Grapefruit Seeds

J. J. Cerda, F. L. Robbins, C. W. Burgin, T. G. Baumgartner, R. W. Rice (1988): The effects of grapefruit pectin on patients at risk for coronary heart disease without altering diet or lifestyle. *Clin. Cardiol.* **11**, 589–594.
S. Gorinstein, A. Caspi, I. Libman, H. T. Lerner, D. Huang, H. Leontowicz, M. Leontowicz, Z. Tashma, E. Katrich, S. Feng, S. Trakhtenberg (2006): Red grapefruit positively influences serum triglyceride level in patients suffering from coronary atherosclerosis: studies in vitro and in humans. *J. Agric Food Chem.* **54**, 1887–1892.
K. He, K. R. Iyer, R. N. Hayes, M. W. Sinz, T. F. Woolf, P. F. Hollenberg (1998): Inactivation of cytochrome P450 3A4 by bergamottin, a component of grapefruit juice. *Chem. Res. Toxicol.* **11**, 252–259.
K. K. Mandadi, G. K. Jayaprakasha, N. G. Bhat, B. S. Patil (2007): Red Mexican grapefruit: a novel source for bioactive limonoids and their antioxidant activity. *Z. Naturforsch. C* **62**, 179–88.
S. L. Milind (2007): Citrus Fruit Biology, Technology and Evaluation. Academic Press, pp 125–190.
P. S. Negi, G. K. Jayaprakasha (2001): Antibacterial activity of grapefruit (Citrus paradisi) peel extracts. *Eur. Food Res. Technol.* **213**, 484–487.
O. A. Oyelami, E. A. Agbakwuru, L. A. Adeyemi, G. B. Adedeji (2005): The effectiveness of grapefruit (Citrus paradisi) seeds in treating urinary tract infections. *J. Altern. Complement. Med.* **11**, 369–371.
L. Reagor, J. Gusman, L. McCoy, E. Carino, J. P. Heggers (2002): The effectiveness of processed grapefruit-seed extract as an antibacterial agent: I. An in vitro agar assay. *J. Altern. Complement. Med.* **8**, 325–332.
A. M. Rincón, A. M. Vásquez, F. C. Padilla (2005): Chemical composition and bioactive compounds of flour of orange (*Citrus sinensis*), tangerine (*Citrus reticulata*) and grapefruit (*Citrus paradisi*) peels cultivated in Venezuela. *Arch. Latinoam. Nutr.* **55**, 305–310.
G. Takeoka, L. Dao, R. Y. Wong, R. Lundin, N. Mahoney (2001): Identification of benzethonium chloride in commercial grapefruit seed extracts. *J. Agric Food Chem.* **49**, 3316–3320.
A. Vikram, P. R. Jesudhasan, G. K. Jayaprakasha, B. S. Pillai, B. S. Patil (2010): Grapefruit bioactive limonoids modulate E. coli O157:H7 TTSS and biofilm. *Int. J. Food Microbiol.* **140**, 109–116.

T. von Woedtke, B. Schlüter, P. Pflegel, U. Lindequist, W. D. Jülich (1998): Aspects of the antimicrobial efficacy of grapefruit seed extract and its relation to preservative substances contained. *Pharmazie* **54**, 452–456.

http://www.thecandidadiet.com/grapefruitseed.htm. (last access: September 15, 2013).

http://www.globalhealingcenter.com/benefits-of/grapefruit-seed-extract. (last access: September 15, 2013).

2.16 Sweet Dreams Without Sugar: Artificial Sweeteners

European Food Safety Authority's FAQ on aspartame: http://www.efsa.europa.eu/en/faqs/faqaspartame.htm?wtrl=01#21 (last access: February 24, 2014).

Official Journal of the European Union, **54**, 12 November, L295 (2011) (http://eur-lex.europa.eu/LexUriServ/LexUriServ.do?uri=OJ:L:2011:295:FULL:EN:PDF) (last access: February 24, 2014).

Regulation (EC) No. 110/2008 of the European Parliament (http://www.wipo.int/wipolex/en/details.jsp?id=5422 (last access: February 24, 2014).

Schink, B., Zeikus, J. G.: Microbial methanol production. A major end product of pectin metabolism, *Current Microbiol.*, **4**, 387–389 (1980).

2.17 An Ointment for your Throat: The Secrets of Olive Oil

G. K. Beauchamp, R. S. Keast, D. Morel, J. Lin, J. Pika, Q. Han, C. H. Lee, A. B. Smith, P. A. Breslin (2005): Phytochemistry: ibuprofen-like activity in extra-virgin olive oil. *Nature*, **437**, 45–46.

M. de Lorgeril, P. Salen, F. Paillard, F. Laporte, F. Boucher, J. de Leiris (2002): Mediterranean diet and the French paradox: two distinct biogeographic concepts for one consolidated scientific theory on the role of nutrition in coronary heart disease. *Cardiovasc. Res.*, **54**, 503–515.

Local Food-Nutraceuticals Consortium (2005): Understanding local Mediterranean diets: a multidisciplinary pharmacological and ethnobotanical approach. *Pharmacol. Res.*, **52**, 353–366.

M. Mancini, J. Stamler (2004): Diet for preventing cardiovascular diseases: light from Ancel Keys, distinguished centenarian scientist. *Nutr. Metab. Cardiovasc. Dis.*, **14**, 52–57.

M. Servili, R. Selvaggini, A. Taticchi, S. Esposto, G. Montedoro (2003): Volatile compounds and phenolic composition of virgin olive oil: optimization of temperature and time of exposure of olive pastes to air contact during the mechanical extraction process. *J. Agric. Food Chem.*, **51**, 7980–7988.

A. Trichopoulou, C. Bamia, D. Trichopoulos (2005): Mediterranean diet and survival among patients with coronary heart disease in Greece. *Arch. Intern. Med.*, **165**, 929–935.

K. P. Townsend, D. Pratico (2005): Novel therapeutic opportunities for Alzheimer's disease: focus on nonsteroidal anti-inflammatory drugs. *FASEB J.*, **19**, 1592–1601.

WHO Technical Report Series (2003): Diet, nutrition and the prevention of chronic diseases. WHO Geneva.

Y. Zhou, Y. Su, B. Li, F. Liu, J. W. Ryder, X. Wu, P. A. Gonzalez-DeWhitt, V. Gelfanova, J. E. Hale, P. C. May, S. M. Paul, B. Ni (2003): Nonsteroidal anti-inflammatory drugs can lower amyloidogenic Abeta42 by inhibiting Rho. *Science*, **302**, 1215–1217.

http://www.scientificpsychic.com/fitness/fattyacids1.html (last access: February 25, 2014).

http://en.wikipedia.org/wiki/Olive_oil (last access: February 25, 2014).

2.18 To Add or Not to Add? Food Additives

Natural glutamate content of different food: Monosodium Glutamate (MSG): A safety assessment. Food Standards Australia New Zealand, Technical Report Series No. 20., June 2003 http://www.foodstandards.gov.au/scienceandeducation/publications/technicalreportserie1338.cfm (last access: February 25, 2014).

2.19 There's Salt, and then There's Salt

R. L. Wolke, M. Parrish (2010): cited work, pp. 63–70.
http://www.heart.org/HEARTORG/GettingHealthy/NutritionCenter/HealthyDietGoals/Sea-Salt-Vs-Table-Salt_UCM_430992_Article.jsp (last access: March 23, 2014)
http://www.mayoclinic.org/healthy-living/nutrition-and-healthy-eating/expert-answers/sea-salt/faq-20058512 (last access: March 23, 2014)
http://science.howstuffworks.com/innovation/edible-innovations/salt2.htm (last access: March 23, 2014)

2.20 Sugar for Your Tea? White or Brown?

R. L. Wolke, M. Parrish (2010): cited work, pp. 22–25.
http://www.nyu.edu/pages/mathmol/library/sugars/ (last access: February 25, 2014).

2.21 The Thickening Stuff: Guar Gum Gumbo

J. Emsley (1994): cited work, p. 203.
J. Timbrell (2005): cited work, pp. 257–258, 311.
J. T. Tuomisto, J. Tuomisto, M. Tainio M. Niittynen, P. Verkasalo, T. Vartiainen, H. Kiviranta, J. Pekkanen (2004): Risk-benefit analysis of eating farmed salmon. *Science* **305** (5683), 476–477. http://www.sciencemag.org/content/305/5683/476 (last access: February 25, 2014).
http://en.wikipedia.org/wiki/Carat_%28mass%29 (last access: February 21, 2014).
http://en.wikipedia.org/wiki/Guar_gum (last access: February 25, 2014).
http://vernon.tamu.edu/center_programs/agronomy/guar/guar_bean.php (last access: February 25, 2014).
http://www.hort.purdue.edu/newcrop/afcm/guar.html (last access: February 25, 2014).
http://www.lsbu.ac.uk/water/hygua.html (last access: February 25, 2014).
http://en.wikipedia.org/wiki/Polychlorinated_dibenzodioxin#Dioxin_exposure_incidents (last access: February 25, 2014).

2.22 Is Caffeine Free of Risk?

H.-D. Belitz, W. Grosch, P. Schieberle (2009): cited work, pp. 938–970.
S. Berger, D. Sicker (2009): Classics in spectroscopy. Isolation and structure elucidation of natural products. Wiley VCH, Weinheim. pp. 25–52.

F. Braudel (1985): cited work, pp. 252–265.
T. P. Coultate (2002): cited work, pp. 327–328.
J. Emsley (1998): cited work, pp. 9–14.
J. Emsley (2010): cited work, pp. 158–159.
P. Freedman (2007): cited work, pp. 214, 217, 246–247, 330–331.
P. Le Couteur, J. Burreson (2004): cited work, pp. 260–269.
J. B. Leikin, F. P. Paloucek (2008): Poisoning and toxicology handbook. 4th ed. Informa Healthcare, New York. pp. 162–163.
B. A. Weinberg, B. K. Bealer (2002): World of caffeine. The science and culture of the world's most popular drug. Routledge, New York.
L. K. Wolf (2014): Caffeine boosts memory in humans. *Chem. Eng. News*, January 20, **92** (3), 34.
http://en.wikipedia.org/wiki/Sylvia_Gerasch#Caffeine (last access: February 25, 2014).
http://en.wikipedia.org/wiki/Caffeine(last access: February 25, 2014).
http://en.wikipedia.org/wiki/Theobroma_cacao (last access: February 25, 2014).
http://www.wada-ama.org/Documents/World_Anti-Doping_Program/WADP-Prohibited-list/2014/WADA-prohibited-list-2014-EN.pdf (last access: March 23, 2014).
http://chineseculture.about.com/library/symbol/np/nc_tea.htm (last access: February 25, 2014).

2.23 The Biofuel Dilemma

C. Drahl (2011): Retooling a bacterial biofuel factory. *Chem. Eng. News.*, **89**(15), 36–37.
J. Emsley (2010): cited work, pp. 78–101.
S. Hahlbrock (2011): Grüner fliegen (Flying greener). *Lufthansa Magasin*, 7, 57–60.
I. T. Horváth, H. Mehdi, V. Fábos, L. Boda, L. T. Mika (2008): gamma-Valerolactone—a sustainable liquid for energy and carbon-based chemicals. *Green Chem.*, **10**, 238–242.
M. Mascal, E. B. Nikitin (2008): Direct, high-yield conversion of cellulose into biofuel. *Angew. Chem. Internatl. Ed.*, **47**, 7924–7926.
G. Olah, A. Goeppert, G. K. S. Prakash (2009): Beyond Oil and Gas: The Methanol Economy, 2nd, Updated and Enlarged Edition. John Wiley and Sons, Chichester.
S. K. Ritter (2007): Biofuel bonanza. *Chem. Eng. News.*, **85** (25), 15–24.
S. K. Ritter (2008): Calling all chemists. *Chem. Eng. News*, **86** (33), 59–68.
http://www.chemistryviews.org/details/ezine/1024397/Jet_Fuel_from_Waste_Cellulose.html (last access: February 25, 2014).

2.24 Perfect Timing: Egg Cooking

P. Barham, L. H. Skibsted, W. L. P. Bredie, M. B. Frost, P. Moller, J. Risbo, P. Snitkjaer, L. M. Mortensen (2010): Molecular gastronomy: a new emerging scientific discipline. *Chem. Rev.*, **110**, 2313–2365.
H.-D. Belitz, W. Grosch, P. Schieberle (2009): cited work, pp. 546–562.
H. This (2006): Molecular gastronomy. Exploring the science of flavor. Columbia University Press, New York, pp. 29–31.
H. This (2009): Molecular gastronomy, a scientific look at cooking. *Acc. Chem. Res.*, **42**, 575–583.
H. This, M.-O. Monchicourt (2010): cited work, pp. 27–43.
C. Vega, R. Mercadé-Prieto (2011): Culinary biophysics: on the nature of the 6X °C egg. *Food Biophys.*, **6**, 152–159.
R. L. Wolke, M. Parrish (2010): cited work
http://blog.khymos.org/2009/04/09/towards-the-perfect-soft-boiled-egg/ (last access: February 25, 2014).

http://blog.khymos.org/2011/04/18/perfect-egg-yolks/ (last access: February 25, 2014).
http://blog.khymos.org/2011/04/23/perfect-egg-yolks-part-2/ (last access: February 25, 2014).
http://en.wikipedia.org/wiki/Sodium_silicate (last access: February 21, 2014).
http://en.wikipedia.org/wiki/Molecular_gastronomy (last access: February 25, 2014).

2.25 Food Fraud: Then and Now

F. Braudel (1985): cited work, p. 129.
H.-D. Belitz, W. Grosch, P. Schieberle (2009): cited work, pp. 467–497, 857–859.
J. Emsley (2010): cited work, p. 25.
P. Freedman (ed.) (2007): cited work, pp. 333–357.
A. H. Hassall (1855): Food and its adulterations. Longman, Brown, Green, and Longmans, London.
C. Petrini (2011): Slow Food: Collected Thoughts on Taste, Tradition and the Honest Pleasures of Food. Chelsea Green Publishing.
http://press.princeton.edu/video/wilson/index.html#top (last access: February 25, 2014).
http://www.rsc.org/Education/EiC/issues/2005Mar/Thefightagainstfoodadulteration.asp (last access: February 25, 2014).

3.1 Are Generic Medicines the Same as the Original?

A. S. Al-Jaizizi, S. Bihareth, I. S. Eqtefan, S. A. Al-Suwayeh (2008): Brand and generic medications: Are they interchangeable? *Ann. Saudi Med.*, **28**, 33–41.
G. Borgherini (2003): The bioequivalence and therapeutic efficacy of generic versus brand-name psychoactive drugs. *Clin. Ther.*, **25**, 1578–1592.
J. A. Guimaraes Moirais, M. do Rosários Lobato (2010): The New European Medicines Agency guideline on the investigation of bioequivalence. *Basic Clin. Pharmacol. Toxicol.*, **106**, 221–225.
http://en.wikipedia.org/wiki/Bioavailability (last access: February 25, 2014).
http://en.wikipedia.org/wiki/Generic_drug (last access: February 25, 2014).
http://ec.europa.eu/competition/sectors/pharmaceuticals/inquiry/communication_hu.pdf (last access: February 25, 2014).

3.2 The Vitamin that Never was: B_{17}

S. R. Adewusi, O. L. Oke (1985): On the metabolism of amygdalin. 1. The LD_{50} and biochemical changes in rats. *Can. J. Physiol. Pharmacol.*, **63**, 1080–1083.
J. H. Carter, M. A. McLafferty, P. Goldman (1980): Role of the gastrointestinal microflora in amygdalin (laetrile)-induced cyanide toxicity. *Biochem. Pharmacol.*, **29**, 301–304.
C. A. Kousparou, A. A. Epenetos, M. P. Deonarain (2002): Antibody-guided enzyme therapy of cancer producing cyanide results in necrosis of targeted cells. *Int. J. Cancer*, **99**, 138–148.
I. J. Lerner (1981): Laetrile: a lesson in cancer quackery. *CA Cancer J. Clin.*, **31**, 91–95. http://caonline.amcancersoc.org/cgi/reprint/31/2/91 (last access: February 25, 2014).
S. Milazzo, E. Ernst, S. Lejune, K. Schmidt (2006): Laetrile treatment for cancer. *Cochrane Database System. Rev.*, **15**, CD005476.

S. Milazzo, S. Lejeune, E. Ernst (2007): Laetrile for cancer: a systematic review of the clinical evidence. *Support. Care Cancer*, **15**, 583–595.
B. Wilson (1988): The rise and fall of laetrile. *Nutrition Forum*, **5**, 33–40, http://www.quackwatch.org/01QuackeryRelatedTopics/Cancer/laetrile (last access: February 25, 2014).
http://www.quackwatch.org/04ConsumerEducation/News/apricotseeds.html (last access: February 25, 2014).
http://en.wikipedia.org/wiki/Ernst_T._Krebs (last access: February 25, 2014).
http://en.wikipedia.org/wiki/Amygdalin (last access: February 25, 2014).
http://tomtit.blog.hu/2008/05/24/manioka_1 (last access: February 25, 2014).
http://www.cancernet.nci.nih.gov/cancertopics/pdq/cam/laetrile (last access: February 25, 2014).

3.3 Joint Efforts: Glucosamine and Chondroitin

Glucosamine/Chondroitin Arthritis Intervention Trial (GAIT). Primary and ancillary study results. http://nccam.nih.gov/research/results/gait/ (last access: February 25, 2014).
D. Scott, A. Kowalczyk (2007): Osteoarthritis of the knee. http://clinicalevidence.bmj.com/ceweb/conditions/msd/1121/1121_G2.jsp (last access: February 25, 2014).
J. Schwarcz (2009): cited work, pp. 48–54,70–71.
T. E. Towheed, L. Maxwell, T. P. Anastassiades, B. Shea, J. Houpt, V. Robinson, M. C. Hochberg, G. Wells (2008): Glucosamine therapy for treating osteoarthritis. *Cochrane Database of Systematic Reviews*, Issue 2, 1–55. DOI: 10.1002/14651858.CD002946.pub2.
S. Wandel, P. Jüni, B. Tendal, E. Nüesch, P. M. Villiger, N. J. Welton, S. Reichenbach, S. Trelle (2010): Effects of glucosamine, chondroitin, or placebo in patients with osteoarthritis of hip or knee: network meta-analysis. *British Med. J.*, **341**, c4675. http://www.bmj.com/content/341/bmj.c4675 (last access: February 25, 2014).
http://en.wikipedia.org/wiki/Methylsulfonylmethane (last access: February 25, 2014).
http://en.wikipedia.org/wiki/S-adenosyl_methionine (last access: February 25, 2014).
http://en.wikipedia.org/wiki/Osteoarthritis (last access: February 25, 2014).
http://glycoforum.gr.jp/science/word/proteoglycan/PGA00E.html (last access: February 25, 2014).
http://en.wikipedia.org/wiki/Glycosaminoglycan (last access: February 25, 2014).
http://en.wikipedia.org/wiki/Proteoglycans (last access: February 25, 2014).

3.4 A Man's Drug of Choice: Testosterone

D. A. Berry (2008): The science of doping. *Nature*, **454**, 692–693.
W. W. Franke, B. Berendonk (1997): Hormonal doping and androgenization of athletes: a secret program of the German Democratic Republic government. *Clin. Chem.*, **43**, 1262–1279.
http://en.wikipedia.org/wiki/Testosterone (last access: March 23, 2014).
http://en.wikipedia.org/wiki/Floyd_Landis_doping_case (last access: March 23, 2014).
http://jurisprudence.tas-cas.org/sites/CaseLaw/Shared%20Documents/1394.pdf (last access: March 23, 2014).
http://www.tas-cas.org/d2wfiles/document/4277/5048/0/Award20Devyatovskiy20&20Tsikhan20internet.pdf (last access: March 23, 2014).
http://www.wada-ama.org/Documents/World_Anti-Doping_Program/WADP-Prohibited-list/2014/WADA-prohibited-list-2014-EN.pdf (last access: March 23, 2014).

3.5 Doped Up and Ready to Win. Maybe

M. Enserink (2008): Does doping work? *Science*, **321**, 627.
W. W. Franke, B. Berendonk (1997): Hormonal doping and androgenization of athletes: a secret program of the German Democratic Republic government. *Clin. Chem.*, **43**, 1262–1279.
http://www.wada-ama.org/Documents/World_Anti-Doping_Program/WADP-Prohibited-list/2014/WADA-prohibited-list-2014-EN.pdf (last access: March 23, 2014).

3.6 Can Placebos Heal?

B. Brinkhaus, D. Pach, R. Luedtke, S. N. Willich (2008): Who controls the placebo? Introducing a Placebo Checklist for pharmacological trials. *Contemporary Clinical Trials*, **29**, 149–156.
A. R. Brunoni, M. Lopes, T. J. Kaptchuk, F. Fregni (2009): Placebo response of non-pharmacological and pharmacological trials in major depression: a systematic review and meta-analysis. *PLOS One*, **4**, e4824.
P. Enck, F. Benedetti, M. Schedlowski (2008): New insights into the placebo and nocebo responses. *Neuron*, **59**, 195–206.
D. G. Finniss, T. J. Kaptchuk, F. Miller, F. Benedetti (2010): Biological, clinical, and ethical advances of placebo effects. *Lancet*, **375**, 686–695.
T. J. Kaptchuk, J. M. Kelley, A. Deykin, P. M. Wayne, L. C. Lasagna, I. O. Epstein, I. Kirsch, M. E. Wechsler (2008): Do "placebo responders" exist? *Contemporary Clinical Trials*, **29**, 587–595.
G. S. Kienle, H. Kiene (1997): The powerful placebo effect: Fact or fiction? *J. Clin. Epidemiol.*, **50**, 1311–1318.
C. E. Margo (1999): The placebo effect. *Surv. Ophthalmol.*, **44**, 31–44.
K. M. Mattocks, R. I. Horwitz (2000): Placebos, active control groups, and the unpredictability paradox. *Biol. Psychiatry*, **47**, 693–698.
http://www.wired.com/medtech/drugs/magazine/17-09/ff_placebo_effect?currentPage=all (last access: February 25, 2014).
http://science.howstuffworks.com/life/placebo-effect.htm (last access: February 25, 2014)
http://en.wikipedia.org/wiki/Placebo_in_history (last access: February 25, 2014).

3.7 Synthetic Drugs vs. Natural Herbs

D. J. Newman, G. M. Cragg (2007): Natural products as sources of new drugs over the last 25 years. *J. Nat. Prod.*, **70**, 461–477.
O. Oniyangi, D. H. Cohall (2010): Phytomedicines (medicines derived from plants) for sickle cell disease. *Cochrane Database of Systematic Reviews*, CD004448.
R. C. Pandey, P. Tripathi, R. Misra, P. N. Gillette, T. Asakura (2005): Status of NICOSAN (TM)/HEMOXINI (TM), an extract of a mixture of plants, for the treatment of sickle cell disease (SCD). *In Vitro Cellular & Developmental Biology-Animal*, **41**, 24A–24A.
W. Sneader (2005): Drug discovery—A history. Wiley, Chichester.
http://www.combichemistry.com/principle.html (last access: February 25, 2014).

3.8 Bitter, Stronger: Herbal Tea Sweeteners

M. Viuda-Martos, Y. Ruiz-Navajas, J. Fernández-López, J. A. Pérez-Alvarez (2008): Functional properties of honey, propolis, and royal jelly. *J. Food Sci.*, **73**, R117–124.
http://en.wikipedia.org/wiki/Sucrose (last access: February 25, 2014).

3.9 Homeopathy: No Active Ingredient? No Side Effect?

D. Csupor, K. Boros, J. Hohmann (2013): Low Potency Homeopathic Remedies and Allopathic Herbal Medicines: Is There an Overlap? *PLOS One*, **8** (9), e74181, DOI: 10.1371/journal.pone.0074181
G. Köhler (1989): The Handbook of Homeopathy—Its Principles and Practice. Healing Arts Press, Rochester—Vermont.
S. Singh, E. Ernst (2009): Trick or Treatment: The Undeniable Facts about Alternative Medicine. W. W. Norton & Company, New York.
http://www.csicop.org/si/show/difference_between_hahnemann_and_darwin/ (last access: February 25, 2014).

3.10 Aromatherapy: Nice but Useless?

L. F. S. Brum, T. Emanuelli, D. O. Souza, E. Elisabetsky (2001): Effects of linalool on glutamate release and uptake in mouse cortical synaptosomes. *Neurochem. Res.*, **26**, 191–194.
B. Cooke, E. Ernst (2000): Aromatherapy: a systematic review. *British Journal of General Practice*, **50**, 493–496.
E. Elisabetsky, J. Marschner, D. O. Souza (1995): Effects of linalool on glutamatergic system in the rat cerebral cortex. *Neurochem. Res.*, **20**, 461–465.
R. S. Herz (2009): Aromatherapy facts and fictions: a scientific analysis of olfactory effects on mood, physiology and behavior. *Int. J. Neurosci.*, **119**, 263–290.
S. J. Hossain, K. Hamamoto, H. Aoshima, Y. Hara (2002): Effects of tea components on the response of GABA(A) receptors expressed in Xenopus oocytes. *J. Agric. Food Chem.*, **50**, 3954–3960.
M. Lis-Balchin (2006): Aromatherapy Science: A Guide for Healthcare Professionals. Pharmaceutical Press, London.
T. Ozek, N. Tabanca, F. Demirci, D. E. Wedge, K. H. C. Baser (2010): Enantiomeric distribution of some linalool containing essential oils and their biological activities. *Records of Natural Products*, **4**, 180–192.
S. Singh, E. Ernst (2009): Trick or Treatment: The Undeniable Facts about Alternative Medicine. W. W. Norton & Company, New York.
H. W. H. Tsang, T. Y. C. Ho (2010): A systematic review on the anxiolytic effects of aromatherapy on rodents under experimentally induced anxiety models. *Rev. Neurosci.*, **21**, 141–152.
http://onlinelibrary.wiley.com/o/cochrane/cochrane_search_fs.html?newSearch=true (last access: February 25, 2014. A query for keywords „aromatherapy" AND „anxiety", „aromatherapy" AND „lavender", respectively)

3.11 Innocent or Guilty? Chamomile

C. Andres, W. C. Chen, M. Ollert, M. Mempel, U. Darsow, J. Ring (2009): Anaphylactic reaction to chamomile tea. *Allergol. Int.*, **58**, 135–136.
C. L. Casterline (1980): Allergy to chamomile tea. *Jama-Journal of the American Medical Association*, **244**, 330–331.
P. A. G. M. De Smet, K. Keller, R. Hänsel, R. F. Chandler (1992): Adverse effects of herbal drugs 1. Springer-Verlag, Berlin.
C. Foti, E. Nettis, R. Panebianco, N. Cassano, A. Diaferio, D. P. Pia (2000): Contact urticaria from Matricaria chamomilla. *Contact Derm.*, **42**, 360–361.
R. Franke, H. Schilcher (2005): Chamomile—Industrial profiles. Taylor & Francis, Boca Raton.
R. Hänsel, O. Sticher (2010): Pharmakognosie—Phytopharmazie. Springer Verlag, Heidelberg. pp. 999–1007.
K. Lundh, M. Hindsen, B. Gruvberger, H. Moller, A. Svensson, M. Bruze (2006): Contact allergy to herbal teas derived from Asteraceae plants. *Contact Derm.*, **54**, 196–201.
K. Lundh, B. Gruvberger, H. Moller, L. Persson, M. Hindsen, E. Zimerson, A. Svensson, M. Bruze (2007): Patch testing with thin-layer chromatograms of chamomile tea in patients allergic to sesquiterpene lactones. *Contact Derm.*, **57**, 218–223.
F. Pereira, R. Santos, A. Pereira (1997): Contact dermatitis from chamomile tea. *Contact Derm.*, **36**, 307.
N. Reider, N. Sepp, P. Fritsch, G. Weinlich, E. Jensen-Jarolim (2000): Anaphylaxis to chamomile: clinical features and allergen cross-reactivity. *Clin. Exp. Allergy*, **30**, 1436–1443.
F. C. Thien (2001): Chamomile tea enema anaphylaxis. *Med. J. Aust.*, **175**, 54.
http://fermi.utmb.edu/cgi-bin/SDAP/sdap_02?dB_Type=0&allid=1 (last access: February 25, 2014).
http://www.phadia.com/en/Allergen-information/ImmunoCAP-Allergens/Weed-Pollens/Allergens/Camomile-/ (last access: February 25, 2014).

3.12 Depressed: Antidepressant Side Effects

T. Kuba, T. Yakushi, H. Fukuhara, Y. Nakamoto, S. T. Singeo Jr., O. Tanaka, T. Kondo (2011): Suicide-related events among child and adolescent patients during short-term antidepressant therapy. *Psychiatry Clin. Neurosci.*, **65**, 239–245.
A. C. Leon (2007): The revised warning for antidepressants and suicidality: Unveiling the black box of statistical analyses. *Am. J. Psychiatry.*, **164**, 1786–1789.
A. C. Leon, D. A. Solomon, C. Li, J. G. Fiedorowicz, W. H. Coryell, J. Endicott, M. B. Keller (2011): Antidepressants and risks of suicide and suicide attempts: a 27-year observational study. *J. Clin. Psychiatry.*, **72**, 580–586.
Z. Rihmer, X. Gonda (2011): Antidepressant-resistant depression and antidepressant-associated suicidal behaviour: the role of underlying bipolarity. *Depress. Res. Treat.*, doi:10.1155/2011/906462
U.S. Food and Drug Administration: Briefing document for Psychopharmacologic Drugs Advisory Committee, December 13, 2006, http://www.fda.gov/ohrms/dockets/ac/06/briefing/2006-4272b1-01-FDA.pdf (last access February 25, 2014).

3.13 Pre-Emptive Action: Vitamin C

C. Bevan: Myths and realities: does vitamin C work for the common cold? (http://www.clinicalcorrelations.org/?p=3670 last access: February 25, 2014).

G. F. Combs (2008): cited work

R. M. Douglas, H. Hemila (2005): Vitamin C for preventing and treating the common cold. *Plos Medicine*, **2**, 503–504.

R. M. Douglas, H. Hemila, E. Chalker, B. Treacy (2007): Vitamin C for preventing and treating the common cold. *Cochrane Database of Systematic Reviews*, CD000980.

F. R. Frankenburg (2009): Vitamin discoveries and disasters. ABC-CLIO, Santa Barbara.

M. Levine, Y. H. Wang, S. J. Padayatty, J. Morrow (2001): A new recommended dietary allowance of vitamin C for healthy young women. *Proc. Natl. Acad. Sci. U. S. A.*, **98**, 9842–9846.

3.14 Vitamin Megadosing: the Real Truth

A. C. Bronstein, D. A. Spyker, L. R. Cantilena, J. L. Green, B. H. Rumack, S. E. Heard (2007): Annual Report of the American Association of Poison Control Centers' National Poison Data System (NPDS): 25th Annual Report. *Clin. Toxicol. (Phila).*, **46**, 927–1057.

G. F. Combs (2008): cited work

Council Directive 90/496/EEC of 24 September 1990 on nutrition labelling for foodstuffs. http://eur-lex.europa.eu/LexUriServ/LexUriServ.do?uri=CELEX:31990L0496:EN:HTML (last access: February 25, 2014)

F. R. Frankenburg (2009): Vitamin discoveries and disasters. ABC-CLIO, Santa Barbara.

Vitamin and mineral intake recommendations for Europeans—which ones are in most need of review? http://www.eufic.org/article/en/page/FTARCHIVE/artid/Vitamin-mineral-intake-recommendations-Europeans/ (last access: February 25, 2014).

http://nobelprize.org/nobel_prizes/medicine/articles/carpenter/ (last access: February 25, 2014).

http://en.wikipedia.org/wiki/Megavitamin_therapy (last access: February 25, 2014).

3.15 Detox Hoax: Poison Removal from Your Body

http://scienceblogs.com/denialism/2008/03/28/detoxificationthe-pinnacle-of/ (last access: February 25, 2014).

http://www.skepdic.com/detox.html (last access: February 25, 2014).

http://www.chem1.com/CQ/FootBathBunk.html (last access: February 25, 2014).

http://sciencebasedpharmacy.wordpress.com/2013/01/04/the-detox-delusion/ (last access: February 25, 2014).

3.16 Cloudy Dreams in Green: Absinthe

Committee on Herbal Medicinal Products (HMPC) Public statement on the use of herbal medicinal products containing thujone EMA/HMPC/732886/2010. http://www.ema.europa.eu/docs/en_GB/document_library/Public_statement/2011/02/WC500102294.pdf (last access: February 25, 2014).

D. W. Lachenmeier (2010): Wormwood (*Artemisia absinthium L.*). A curious plant with both neurotoxic and neuroprotective properties? *J. Ethnopharmacol.*, **131**, 224–227.

K. M. Hold, N. S. Sirisoma, T. Ikeda, T. Narahashi, J. E. Casida (2000): alpha-Thujone (the active component of absinthe): gamma-aminobutyric acid type A receptor modulation and metabolic detoxification. *Proc. Natl. Acad. Sci. U. S. A.*, **97**, 3826–3831.

D. W. Lachenmeier, M. Uebelacker (2010): Risk assessment of thujone in foods and medicines containing sage and wormwood—Evidence for a need of regulatory changes? *Regul. Toxicol. Pharm.*, **58**, 437–443.

D. W. Lachenmeier, J. Emmert, T. Kuballa, G. Sartor (2006a): Thujone—Cause of absinthism? *Forensic Sci. Int.*, **158**, 1–8.

D. W. Lachenmeier, S. G. Walch, S. A. Padosch, L. U. Kroner (2006b): Absinthe—A review. *Crit. Rev. Food Sci. Nutr.*, **46**, 365–377.

J. Luaute (2007): Absinthism: the fault of doctor Magnan. *Evolution Psychiatrique*, **72**, 515–530.

S. A. Padosch, D. W. Lachenmeier, L. U. Kroener (2006): Absinthism: a fictitious 19th century syndrome with present impact. *Substance Abuse Treatment Prevention and Policy*, **1**, 14.

S. K. Ritter (2008): Absinthe myths finally laid to rest. *Chem. Eng. News*, **86**(18), 42–43.

B. Rychiak (2007): Total thujone and anethole concentrations in various alcoholic beverages, including pre-ban absinthe. *Dtsch. Lebensm.-Rundsch.*, **103**, 419–424.

I. E. Scholten, E. van der Linden, H. This (2008): The life of an anise-flavored alcoholic beverage: does its stability cloud or confirm theory? *Langmuir*, **24**, 1701–1706.

M. Stapleton (2008): Absinthe and thujone. *Chem. Eng. News*, **86**(24), 9–9.

http://en.wikipedia.org/wiki/Absinthe (last access: February 25, 2014).

3.17 Is Green Tea Caffeine-Free?

J. Bryan (2008): Psychological effects of dietary components of tea: caffeine and L-theanine. *Nutr. Rev.*, **66**, 82–90.

C. Cabrera, R. Artacho, R. Gimenez (2006): Beneficial effects of green tea—A review. *J. Am. Coll. Nutr.*, **25**, 79–99.

T. Kakuda (2002): Neuroprotective effects of the green tea components theanine and catechins. *Biol. Pharm. Bull.*, **25**, 1513–1518.

E. K. Keenan, M. D. A. Finnie, P. S. Jones, P. J. Rogers, C. M. Priestley (2011): How much theanine in a cup of tea? Effects of tea type and method of preparation. *Food Chem.*, **125**, 588–594.

N. Khan, H. Mukhtar (2007): Tea polyphenols for health promotion. *Life Sci.*, **81**, 519–533.

P. Mohanpuria, V. Kumar, S. K. Yadav (2010): Tea caffeine: Metabolism, functions, and reduction strategies. *Food Sci. Biotechnol.*, **19**, 275–287.

V. Preedy (2013): Tea in Health and Disease Prevention. Elsevier, London.

P. C. van der Pijl, L. Chen, T. P. J. Mulder (2010): Human disposition of L-theanine in tea or aqueous solution. *J. Funct. Foods*, **2**, 239–244.

http://en.wikipedia.org/wiki/Camellia_sinensis (last access: February 25, 2014).
http://www.thefragrantleaf.com/caffeine.html (last access: February 25, 2014).
http://www.holymtn.com/tea/caffeine_content.htm (last access: February 25, 2014).
http://en.wikipedia.org/wiki/Decaffeination#Decaffeinated_tea (last access: February 25, 2014).
http://en.wikipedia.org/wiki/Caffeine (last access: February 25, 2014).

3.18 Red Rice Remedy: Cholesterol Relief?

A. W. Alberts, J. Chen, G. Kuron, V. Hunt, J. Huff, C. Hoffman, J. Rothrock, M. Lopez, H. Joshua, E. Harris, A. Patchett, R. Monaghan, S. Currie, E. Stapley, G. Albersschonberg, O. Hensens, J. Hirshfield, K. Hoogsteen, J. Liesch, J. Springer (1980): Mevinolin—a highly potent competitive inhibitor of hydroxymethylglutaryl-coenzyme-a reductase and a cholesterol-lowering agent. *Proc. Natl. Acad. Sci. U.S.A. Biol. Sci.*, **77**, 3957–3961.

D. J. Becker, R. Y. Gordon, S. C. Halbert, B. French, P. B. Morris, D. J. Rader (2009): Red yeast rice for dyslipidemia in statin-intolerant patients a randomized trial. *Ann. Intern. Med.*, **150**, 830–839.

A. Endo (1979): Monacolin-K, a new hypocholesterolemic agent produced by a Monascus species. *J. Antibiot.*, **32**, 852–854.

R. Y. Gordon, D. J. Becker (2011): The role of red yeast rice for the physician. *Curr. Atheroscler. Rep.*, **13**, 73–80.

R. Y. Gordon, T. Cooperman, W. Obermeyer, D. J. Becker (2010): Marked variability of monacolin levels in commercial red yeast rice products: buyer beware! *Arch. Intern. Med.*, **170**, 1722–1727.

R. Havel (1999): Dietary supplement or drug? The case for Cholestin—Reply. *Am. J. Clin. Nutr.*, **70**, 106–108.

D. Heber (1999): Dietary supplement or drug? The case for Cholestin. *Am. J. Clin. Nutr.*, **70**, 106–106.

D. Heber, I. Yip, J. M. Ashley, D. A. Elashoff, R. M. Elashoff, V. L. W. Go (1999): Cholesterol-lowering effects of a proprietary Chinese red-yeast-rice dietary supplement. *Am. J. Clin. Nutr.*, **69**, 231–236.

M. Journoud, P. J. H. Jones (2004): Red yeast rice: a new hypolipidemic drug. *Life Sci.*, **74**, 2675–2683.

M. Klimek, S. Wang, A. Ogunkanmi (2009): Safety and efficacy of red yeast rice (Monascus purpureus) as an alternative therapy for hyperlipidemia. *Pharm. Therap.*, **34**, 313–327.

J. Liu, J. Zhang, Y. Shi, S. Grimsgaard, T. Alraek, V. Fonnebo (2006): Chinese red yeast rice (Monascus purpureus) for primary hyperlipidemia: a meta-analysis of randomized controlled trials. *Chin. Med.*, **1**, 4.

M. Peraica, A. Domijan, M. Miletic-Medved, R. Fuchs (2008): The involvement of mycotoxins in the development of endemic nephropathy. *Wien. Klin. Wochenschr.*, **120**, 402–407.

3.19 What Exactly Does Aloë Cure?

Aloe vera gel monograph. In: WHO Monographs on Selected Medicinal Plants—Volume 1. 1999 (http://apps.who.int/medicinedocs/en/d/Js2200e/6.html (last access: February 25, 2014).

M. D. Boudreau, F. A. Beland (2006): An evaluation of the biological and toxicological properties of Aloe barbadensis (Miller), Aloe vera. *J. Environ. Sci. Health., Part C Environ. Carcinog. Ecotoxicol. Rev.*, **24**, 103–154.

L. Cuttle, J. Pearn, J. R. McMillan, R. M. Kimble (2009): A review of first aid treatments for burn injuries. *Burns*, **35**, 768–775.

E. Dagne, D. Bisrat, A. Viljoen, B. E. Van Wyk (2000): Chemistry of Aloe species. *Curr. Org. Chem.*, **4**, 1055–1078.

K. Eshun, Q. He (2004): Aloe vera: A valuable ingredient for the food, pharmaceutical and cosmetic industries—A review. *Crit. Rev. Food Sci. Nutr.*, **44**, 91–96.

O. M. Grace, M. S. J. Simmonds, G. F. Smith, A. E. van Wyk (2009): Documented utility and biocultural value of Aloe L. (Asphodelaceae): A Review. *Econ. Bot.*, **63**, 167–178.

J. H. Hamman (2008): Composition and applications of Aloe vera leaf gel. *Molecules*, **13**, 1599–1616.

M. Y. K. Leung, C. Liu, J. C. M. Koon, K. P. Fung (2006): Polysaccharide biological response modifiers. *Immunol. Lett.*, **105**, 101–114.
R. Maenthaisong, N. Chaiyakunapruk, S. Niruntraporn, C. Konqkaew (2007): The efficacy of aloe vera used for burn wound healing: A systematic review. *Burns*, **33**, 713–718.
S. R. Mentreddy (2007): Review—Medicinal plant species with potential antidiabetic properties. *J. Sci. Food Agric.* **87**, 743–750.
Y. I. Park, S. K. Lee (2006): New perspectives on Aloe. Springer, New York.
T. Reynolds, A. C. Dweck (1999): Aloe vera leaf gel: a review update. *J. Ethnopharmacol.*, **68**, 3–37.
B. K. Vogler, E. Ernst (1999): Aloe vera: a systematic review of its clinical effectiveness. *Br. J. Gen. Practice*, **49**, 823–828.

3.20 Island Pleasure for Businessmen: Noni

R. Heinicke (1985): Xeronine, a new alkaloid, useful in medical, food and industrial fields. United States Patent 4,543,212, issued September 24, 1985.
W. McClatchey (2002): From Polynesian healers to health food stores: changing perspectives of Morinda citrifolia (Rubiaceae). *Integr. Cancer. Ther.*, **1**, 110–120; discussion 120.
A. D. Pawlus, D. A. Kinghorn (2007): Review of the ethnobotany, chemistry, biological activity and safety of the botanical dietary supplement *Morinda citrifolia* (noni). *J. Pharm. Pharmacol.*, **59**, 1587–1609.
O. Potterat, M. Hamburger (2007): *Morinda citrifolia* (Noni) fruit—phytochemistry, pharmacology, safety. *Planta Med.*, **73**, 191–199.
M. Y. Wang, B. J. West, C. J. Jensen, D. Nowicki, C. Su, A. K. Palu, G. Anderson (2002): *Morinda citrifolia* (Noni): a literature review and recent advances in Noni research. *Acta Pharmacol. Sin.*, **23**, 1127–1141.
http://en.wikipedia.org/wiki/Tahitian_Noni (last access: February 25, 2014).
http://www.bilder-hochladen.net/files/1gml-8.jpg (last access: February 25, 2014).

3.21 Castor Oil: A Painful Story

R. Hänsel, O. Sticher (2010): Pharmakognosie—Phytopharmazie. Heidelberg: Springer Medizin Verlag.
D. S. Ogunniyi (2006): Castor oil: A vital industrial raw material. *Bioresour. Technol.*, **97**, 1086–1091.
http://en.wikipedia.org/wiki/Castor_oil (last access: February 25, 2014).
http://en.wikipedia.org/wiki/Ricin (last access: February 25, 2014).

3.22 The Oldest Sugar-Free Candy: Licorice

C. Fiore, M. Eisenhut, E. Ragazzi, G. Zanchin, D. Armanini (2005): A history of the therapeutic use of liquorice in Europe. *J. Ethnopharmacol.*, **99**, 317–324.
R. Hänsel, O. Sticher (2010). Pharmakognosie—Phytopharmazie. Springer Medizin Verlag, Heidelberg, pp. 833–834.
P. A. Lachance (ed.) (1997): Nutraceuticals: Designer Foods III. Garlic, soy and liquorice. Food & Nutrition Press, Trumbull, Connecticut, pp. 243–263.

Magyary-Kossa Gy (1921): A hazai gyógynövények hatása és orvosi használata. Athenaeum Irodalmi és Nyomdai Rt., Budapest, pp. 133–134. (The effect and medical use of Hungarian medicinal plants—in Hungarian)
http://www.enotes.com/how-products-encyclopedia/licorice (last access: February 25, 2014).
http://en.wikipedia.org/wiki/Liquorice (last access: February 25, 2014).
http://ec.europa.eu/food/fs/sc/scf/out186_en.pdf (last access: February 25, 2014).
http://en.wikipedia.org/wiki/Liquorice_(confectionery) (last access: February 25, 2014).
http://www.haribo.com/planet/hu/info/core/history/first_prosperities.php (last access: February 25, 2014).

3.23 Cannon Balls for Your Health: Antibiotics

C. R. Craig, R. E. Stitzel (2003): Chemotherapy. Chap. VII. In: modern pharmacology with clinical applications, Sixth Edition. Lippincott Williams & Wilkins. pp. 509–676.
K. J. Ryan, C. G. Ray (Eds.) (2004): Sherris medical microbiology: an introduction to infectious diseases. McGraw-Hill Medical.
http://nobelprize.org/nobel_prizes/medicine/laureates/1945/index.html (last access: February 25, 2014).
http://en.wikipedia.org/wiki/Penicillin (last access: February 25, 2014).
http://en.wikipedia.org/wiki/Daptomycin (last access: February 25, 2014).

3.24 Hearty Soybeans: Phytoestrogens

R. Bolanos, A. Del Castillo, J. Francia (2010): Soy isoflavones versus placebo in the treatment of climacteric vasomotor symptoms: systematic review and meta-analysis. *Menopause*, **17**, 660–666.C. Castelo-Branco, M. J. Cancelo Hidalgo (2011): Isoflavones: effects on bone health. *Climacteric*, **14**, 204–211.
C. R. Cederroth, J. Auger, C. Zimmermann, F. Eustache, S. Nef (2010): Soy, phyto-oestrogens and male reproductive function: a review. *Int. J. Androl.*, **33**, 304–316.
A. M. Chen, W. J. Rogan (2004): Isoflavones in soy infant formula: A review of evidence for endocrine and other activity in infants. *Annu. Rev. Nutr.*, **24**, 33–54.
Y. N. Clement, I. Onakpoya, S. K. Hung, E. Ernst (2011): Effects of herbal and dietary supplements on cognition in menopause: A systematic review. *Maturitas*, **68**, 256–263.
A. M. Duncan, W. R. Phipps, M. S. Kurzer (2003): Phyto-oestrogens. *Best Pract. Res. Clin. Endocrinol. Metabol.*, **17**, 253–271.
F. Holguin, M. M. Tellez-Rojo, M. Lazo, D. Mannino, J. Schwartz, M. Hernandez, I. Romieu (2005): Cardiac autonomic changes associated with fish oil vs soy oil supplementation in the elderly. *Chest*, **127**, 1102–1107.
Y. W. Hwang, S. Y. Kim, S. H. Jee, Y. N. Kim, C. M. Nam (2009): Soy food consumption and risk of prostate cancer: a meta-analysis of observational studies. *Nutr. Cancer Int. J.*, **61**, 598–606.
K. A. Jackman, O. L. Woodman, C. G. Sobey (2007): Isoflavones, equol and cardiovascular disease: Pharmacological and therapeutic insights. *Curr. Med. Chem.*, **14**, 2824–2830.
S. W. Luckey, J. Mansoori, K. Fair, C. L. Antos, E. N. Olson, L. A. Leinwand (2007): Blocking cardiac growth in hypertrophic cardiomyopathy induces cardiac dysfunction and decreased survival only in males. *Am. J. Physiol. Heart Circul. Physiol.*, **292**, H838–H845.
M. A. Mendez, M. S. Anthony, L. Arab (2002): Soy-based formulae and infant growth and development: A review. *J. Nutr.*, **132**, 2127–2130.
M. J. Messina (2003): Emerging evidence on the role of soy in reducing prostate cancer risk. *Nutr. Rev.*, **61**, 117–131.

M. Messina (2010): Soybean isoflavone exposure does not have feminizing effects on men: a critical examination of the clinical evidence. *Fertil. Steril.*, **93**, 2095–2104.

I. C. Munro, M. Harwood, J. J. Hlywka, A. M. Stephen, J. Doull, W. G. Flamm, H. Adlercreutz (2003): Soy isoflavones: A safety review. *Nutr. Rev.*, **61**, 1–33.

E. Ricci, S. Cipriani, F. Chiaffarino, M. Malvezzi, F. Parazzini (2010): Soy isoflavones and bone mineral density in perimenopausal and postmenopausal western women: a systematic review and meta-analysis of randomized controlled trials. *J. Womens Health*, **19**, 1609–1617.

B. L. Stauffer, J. P. Konhilas, E. D. Luczak, L. A. Leinwand (2006): Soy diet worsens heart disease in mice. *J. Clin. Invest.*, **116**, 209–216.

R. S. Tomar, R. Shiao (2008): Early life and adult exposure to isoflavones and breast cancer risk. *J. Environ. Sci. Health., Part C Environ. Carcinog. Ecotoxicol. Rev.*, **26**, 113–173.

D. Turck (2007): Soy protein for infant feeding: what do we know? *Curr. Opin. Clin. Nutr. Metab. Care*, **10**, 360–365.

Y. Vandenplas, E. De Greef, T. Devreker, B. Hauser (2011): Soy infant formula: is it that bad? *Acta Paediatrica*, **100**, 162–166.

A. Warri, N. M. Saarinen, S. Makela, L. Hilakivi-Clarke (2008): The role of early life genistein exposures in modifying breast cancer risk. *Br. J. Cancer*, **98**, 1485–1493.

A. H. Wu, M. C. Yu, C. Tseng, M. C. Pike (2008): Epidemiology of soy exposures and breast cancer risk. *Br. J. Cancer*, **98**, 9–14.

http://psychcentral.com/news/archives/2006-01/uoca-sdw122805.html (last access: February 25, 2014).

3.25 Willow Fever: Aspirin

T. Akao, T. Yoshino, K. Kobashi, M. Hattori (2002): Evaluation of salicin as an antipyretic prodrug that does not cause gastric injury. *Planta Med.*, **68**, 714–718.

From experience to design—The science behind Aspirin. http://www.creatingtechnology.org/biomed/aspirin.htm (last access: February 25, 2014).

D. Jeffreys (2005): Aspirin: The Remarkable Story of a Wonder Drug. Bloomsbury, New York.

P. N. P. Rao, E. E. Knaus (2008): Evolution of nonsteroidal anti-inflammatory drugs (NSAIDs): Cyclooxygenase (COX) inhibition and beyond. *J. Pharm. Pharmaceut. Sci.*, **11**, 81–110.

http://en.wikipedia.org/wiki/Acetylsalicylic_acid (last access: February 25, 2014).

http://nobelprize.org/nobel_prizes/medicine/laureates/1982/vane.html (last access: February 25, 2014).

3.26 Pill Secrets: the Efficiency of Drugs

L. M. Brewster, G. A. van Montfrans, J. Kleijnen (2004): Systematic review: antihypertensive drug therapy in black patients. *J. Ann. Intern. Med.*, **19**, 614–27.

S. Cox Gad (2009): Clinical Trials Handbook. Wiley, New Jersey.

C. D. Furberg, B. Pitt (2001): Withdrawal of cerivastatin from the world market. Curr. Control. Trials Cardiovasc. Med., 2, 205–207.

P. Glue, R. P. Clement (1999): Cytochrome P450 enzymes and drug metabolism—basic concepts and methods of assessment. *Cell. Mol. Neurobiol.*, **19**, 309–323.

3.27 In Quest of Active Ingredients: Herbs

D. R. Hadden (2005): Goat's rue—French lilac—Italian fitch—Spanish sainfoin: Galega officinalis and metformin: the Edinburgh connection. *J. R. Coll. Physicians Edinb.*, **35**, 258–260.
http://www.pharmacology2000.com/learning2.htm (last access: February 25, 2014).

3.28 Getting Rid of Acids: Alkaline Cures

F. Manz F (2001): History of nutrition and acid-base physiology. *Eur. J Nutr.*, **40**, 189–199.
R. Rylander, T. Tallheden, J. Vormann (2009): Acid-base conditions regulate calcium and magnesium homeostasis. *Magnes. Res.*, **22**, 262–265.
http://en.wikipedia.org/wiki/Robert_Young_(author) (last access: February 25, 2014).
http://en.wikipedia.org/wiki/Edgar_Cayce (last access: February 25, 2014).
http://www.quackwatch.org/11Ind/young3.html (last access: February 25, 2014).
http://www.intelihealth.com/IH/ihtPrint/WSIHW000/24479/408/465747.html?d=dmtHMSContent&hide=t&k=basePrint (last access: February 25, 2014).
http://www.ams.usda.gov/AMSv1.0/getfile?dDocName=STELPRDC5072951 (last access: February 25, 2014).
http://www.phmiracleliving.com/ (last access: February 25, 2014).

3.29 "...upon the Face of the Waters": Mineral and Tap Water

S. Keresztes, E. Tatár, V. G. Mihucz, I. Virág, C. Majdik, G. Záray (2009): Leaching of antimony from polyethylene terephthalate (PET) bottles into mineral water. *Sci. Total Environ.*, **407**, 4731–4735.
http://www.huffingtonpost.com/dr-susanne-bennett/drinking-water_b_1680027.html (last access: February 25, 2014).
http://www.aquasanastore.com/water-you_b02.html (last access: February 25, 2014).
http://www.mayoclinic.org/healthy-living/nutrition-and-healthy-eating/expert-answers/tap-water/faq-20058017 (last access: February 25, 2014).
http://www.epa.gov/envirofw/html/icr/gloss_dbp.html (last access: February 25, 2014).
http://people.chem.duke.edu/~jds/cruise_chem/water/wattap.html (last access: February 25, 2014).

3.30 The Illusive Qualities of Plant Preparations

S. Balayssac, S. Trefi, V. Gilard, M. Malet-Martino, R. Martino, M. Delsuc (2009): 2D and 3D DOSY H-1 NMR, a useful tool for analysis of complex mixtures: Application to herbal drugs or dietary supplements for erectile dysfunction. *J. Pharm. Biomed. Anal.*, **50**, 602–612.
Y. Chen, L. Zhao, F. Lu, Y. Yu, Y. Chai, Y. Wu (2009): Determination of synthetic drugs used to adulterate botanical dietary supplements using QTRAP LC-MS/MS. *Food Addit. Contam. Part A. Chem. Anal. Control. Expo. Risk Assess.*, **26**, 595–603.

C. Corns, K. Metcalfe (2002): Risks associated with herbal slimming remedies. *Journal of the Royal Society for the Promotion of Health*, **122**, 213–219.

Csupor D., Szekeres A., Kecskeméti A., Vékes E., Veres K., Micsinay Á., Szendrei K., Hohmann J. (2010): [Dietary supplements on the domestic market adulterated with sildenafil and tadalafil—in Hungarian]. *Orvosi Hetilap*, **151**, 1783–1789.

Directive 2002/46/EC of the European Parliament and of the Council of 10 June 2002 on the approximation of the laws of the Member States relating to food supplements. http://eur-lex.europa.eu/LexUriServ/LexUriServ.do?uri=CELEX:32002L0046:EN:HTML (last access: February 25, 2014).

S. R. Gratz, C. L. Flurer, K. A. Wolnik (2004): Analysis of undeclared synthetic phosphodiesterase-5 inhibitors in dietary supplements and herbal matrices by LC-ESI-MS and LC-UV. *J. Pharm. Biomed. Anal.*, **36**, 525–533.

C. M. Gryniewicz, J. C. Reepmeyer, J. F. Kauffman, L. F. Buhse (2009): Detection of undeclared erectile dysfunction drugs and analogues in dietary supplements by ion mobility spectrometry. *J. Pharm. Biomed. Anal.*, **49**, 601–606.

W. F. Huang, K. C. Wen, M. L. Hsiao (1997): Adulteration by synthetic therapeutic substances of traditional Chinese medicines in Taiwan. *J. Clin. Pharmacol.*, **37**, 344–350.

B. Jasiurkowski, J. Raj, D. Wisinger, R. Carlson, L. Zou, A. Nadir (2006): Cholestatic jaundice and IgA nephropathy induced by OTC muscle building agent superdrol. *Am. J. Gastroenterol.*, **101**, 2659–2662.

S. H. Kim, J. Lee, T. Yoon, J. Choi, D. Choi, D. Kim, S. W. Kwon (2009): Simultaneous determination of anti-diabetes/anti-obesity drugs by LC/PDA, and targeted analysis of sibutramine analog in dietary supplements by LC/MS/MS. *Biomed. Chromatogr.*, **23**, 1259–1265.

R. J. Maughan (2005): Contamination of dietary supplements and positive drug tests in sport. *J. Sports Sci.*, **23**, 883–889.

C. S. Nergard, K. Myhr (2009): Hepatic adverse events associated with the use of herbal drugs in Norway ("the Fortodol case"). *Drug Safety*, **32**, 50.

R. N. K. Padinjakara, K. Ashawesh, S. Butt, R. Nair, V. Patel (2009): herbal remedy for diabetes: two case reports. *Exp. Clin. Endocrinol. Diabetes*, **117**, 3–5.

W. T. Poon, S. W. Ng, C. K. Lai, Y. W. Chan, W. L. Mak (2008): Factitious thyrotoxicosis and herbal dietary supplement for weight reduction. *Clin. Toxicol.*, **46**, 290–292.

J. J. W. Ros, M. G. Pelders, P. A. G. M. De Smet (1999): A case of positive doping associated with a botanical food supplement. *Pharmacy World & Science*, **21**, 44–46.

J. Vaysse, S. Balayssac, V. Gilard, D. Desoubdzanne, M. Malet-Martino, R. Martino (2010): Analysis of adulterated herbal medicines and dietary supplements marketed for weight loss by DOSY 1H-NMR. *Food Addit. Contam. Part A Chem. Anal. Control Expo. Risk Assess.*, **27**, 903–916.

E. Wooltorton (2002): Hua Fo tablets tainted with sildenafil-like compound. *Can. Med. Assoc. J.*, **166**, 1568–1568.

3.31 The Antioxidant Story

B. Buijsse, E. J. Feskens, D. Schlettwein-Gsell, M. Ferry, F. J. Kok, D. Kromhout, L. C. P. G. M. de Groot, SENECA Investigators (2005): Plasma carotene and alpha-tocopherol in relation to 10-y all-cause and cause-specific mortality in European elderly: the Survey in Europe on Nutrition and the Elderly, a Concerted Action (SENECA). *Am. J. Clin. Nutr.*, **82**, 879–886.

B. Goldacre (2009): cited work, pp. 99–111.

I. Jialal, S. Devaraj (2000): Vitamin E supplementation and cardiovascular events in high-risk patients. *N. Engl. J. Med.*, **342**, 1917–1917.

P. E. Milbury, A. C. Richer (2008): Understanding the antioxidant controversy—scrutinizing the „fountain of youth". Praeger Publishers, Westport.

S. K. Osganian, M. J. Stampfer, E. Rimm, D. Spiegelman, J. E. Manson, W. C. Willett (2003): Dietary carotenoids and risk of coronary artery disease in women. *Am. J. Clin. Nutr.*, **77**, 1390–1399.

M. Rautalahti, D. Albanes, J. Virtamo, P. R. Taylor, J. K. Huttunen, O. P. Heinonen (1997): Beta-carotene did not work: aftermath of the ATBC study. *Cancer Lett.*, **114**, 235–236.

C. A. Rice-Evand, L. Packer (1998): Flavonoids in health and disease. Marcel Dekker, New York.

P. M. Rowe (1996): CARET and ATBC refine conclusions about beta-carotene. *Lancet*, **348**, 1369–1369.

E. C. Stuart, M. J. Scandlyn, R. J. Rosengren (2006): Role of epigallocatechin gallate (EGCG) in the treatment of breast and prostate cancer. *Life Sci.*, **79**, 2329–2336.

H. Yao, W. Xu, X. Shi, Z. Zhang (2011): Dietary flavonoids as cancer prevention agents. *J. Environ. Sci. Health., Part C Environ. Carcinog. Ecotoxicol. Rev.*, **29**, 1–31.

3.32 Stomach Ache: Baking Soda

A review of patents on effervescent granules. http://www.pharmainfo.net/reviews/review-patents-effervescent-granules (last access: February 25, 2014).

R. R. Berardi (2006): Self-directed options for preventing or treating heartburn. US Pharmacy Review, http://www.touchbriefings.com/pdf/1895/Berardi_edit.qxp.pdf (last access: February 25, 2014).

W. Grant Thompson (2009): Antacids. http://www.aboutgerd.org/library/download/520 (last access: February 25, 2014).

3.33 Does Hot Pepper Cause Ulcer?

O. M. E. Abdel-Salam, J. Szolcsányi, G. Mozsik (1997): Capsaicin and the stomach. A review of experimental and clinical data. *J. Physiol. (Paris)*, **91**, 151–171.

F. Borrelli, A. A. Izzo (2000): The plant kingdom as a source of anti-ulcer remedies. *Phytother. Res.*, **14**, 581–591.

S. Evangelista, C. A. Maggi, A. Meli (1987): Involvement of capsaicin-sensitive mechanism(s) in the antiulcer defense of intestinal-mucosa in rats. *Proc. Soc. Exp. Biol. Med.*, **184**, 264–266.

H. d. S. Falcao, J. A. Leite, J. M. Barbosa-Filho, P. F. de Athayde-Filho, M. C. d. O. Chaves, M. D. Moura, A. L. Ferreira, A. B. A. de Almeida, A. R. Monteiro Souza-Brito, Formiga Melo Diniz, Margareth de Fatima, L. M. Batista (2008): Gastric and duodenal antiulcer activity of alkaloids: a review. *Molecules*, **13**, 3198–3223.

I. Inada, H. Satoh (1996): Capsaicin-sensitive sensory neurons are involved in bicarbonate secretion induced by lansoprazole, a proton pump inhibitor, in rats. *Dig. Dis. Sci.*, **41**, 785–790.

J. Y. Kang, K. G. Yeoh, H. P. Chia, H. P. Lee, Y. W. Chia, R. Guan, I. Yap (1995): Chili—protective factor against peptic-ulcer. *Dig. Dis. Sci.*, **40**, 576–579.

J. Y. Kang, K. G. Yeoh, K. Y. Ho, R. Guan, T. P. Lim, S. H. Quak, A. Wee, D. Teo, Y. W. Ong (1997): Racial differences in Helicobacter pylori seroprevalence in Singapore: Correlation with differences in peptic ulcer frequency. *J. Gastroenterol. Hepatol.*, **12**, 655–659.

M. Kinoshita, E. Kume, H. Tamaki (1995): Roles of Prostaglandins, nitric-oxide and the capsaicin-sensitive sensory nerves in gastroprotection produced by ecabet sodium. *J. Pharmacol. Exp. Ther.*, **275**, 494–501.

H. Mimaki, S. Kagawa, M. Aoi, S. Kato, T. Satoshi, K. Kohama, K. Takeuchi (2002): Effect of lafutidine, a histamine H-2-receptor antagonist, on gastric mucosal blood flow and duodenal HCO_3^- secretion in rats: Relation to capsaicin-sensitive afferent neurons. *Dig. Dis. Sci.*, **47**, 2696–2703.

S. Onodera, M. Tanaka, M. Aoyama, Y. Arai, N. Inaba, T. Suzuki, A. Nishizawa, M. Shibata, Y. Sekine (1999): Antiulcer effect of lafutidine on indomethacin-induced gastric antral ulcers in refed rats. *Jpn. J. Pharmacol.*, **80**, 229–235.

M. N. Sathyanarayana (2006): Capsaicin and gastric ulcers. *Crit. Rev. Food Sci. Nutr.*, **46**, 275–328.

H. Yamamoto, S. Horie, M. Uchida, S. Tsuchiya, T. Murayama, K. Watanabe (2001): Effects of vanilloid receptor agonists and antagonists on gastric antral ulcers in rats. *Eur. J. Pharmacol.*, **432**, 203–210.

K. G. Yeoh, J. Y. Kang, I. Yap, R. Guan, C. C. Tan, A. Wee, C. H. Teng (1995): Chili protects against aspirin-induced gastroduodenal mucosal injury in humans. *Dig. Dis. Sci.*, **40**, 580–583.

F. Y. Zeyrek, E. Oguz (2005): In vitro activity of capsaicin against *Helicobacter pylori*. *Ann. Microbiol.*, **55**, 125–127.

http://en.wikipedia.org/wiki/Scoville_units (last access February 25, 2014).

3.34 Cocaine—Drug, Narcotic, or What?

D. Nutt, L. A King, W. Saulsbury, C. Blakemore (2007): Development of a rational scale to assess the harm of drugs of potential misuse. *Lancet*, **369**, 1047–1053.

http://www.wada-ama.org/Documents/World_Anti-Doping_Program/WADP-Prohibited-list/2014/WADA-prohibited-list-2014-EN.pdf (last access: March 23, 2014).

http://www.drugabuse.gov./PDF/RRCocaine.pdf (last access: February 25, 2014).

4.1 Is the Use of Cyanogen Bromide Forbidden in Hospitals?

Kovács L. (2008): A brómcián (Cyanogen bromide—in Hungarian). *Magy. Kém. Lapja*, **63**, 97–98. http://www.mke.org.hu/content/view/130/98/ (last access: February 25, 2014).

L. Kovács (2009): Cyanogen bromide. An update. In *e-EROS Encyclopedia of Reagents for Organic Synthesis*, 2nd ed., L. A. Paquette, D. Crich, P. L. Fuchs, G. Molander Eds., John Wiley and Sons, Chichester. http://onlinelibrary.wiley.com/doi/10.1002/047084289X.rc269.pub2/abstract (last access: May 19, 2014).

MTV Híradó (Hungarian MTV news), February 15, 2008. 7:30 PM, http://www.mtv.hu/videotar/?id=17975 (last access: February 25, 2014).

Rejtélyes dobozt találtak az egri kórházban—vizsgálják az anyagot (Mysterious box found in Eger hospital—in Hungarian. *MTI/Magyar Nemzet*, February 15, 2008. http://hirszerzo.hu/belfold/56856_rejtelyes_dobozt_talaltak_az_egri_korhazban (last access: February 25, 2014).

Vaklárma volt a brómcián-ügy? [Cyanogen bromide affair—a false alarm?—in Hungarian] *HírTV*, February 23, 2008. http://www.hirtv.hu/belfold/?article_hid=200821 (last access: February 25, 2014).

http://en.wikipedia.org/wiki/Cyanogen_bromide (last access: February 25, 2014).

4.2 Poison in Groundwater: Arsenic

T. Agusa, J. Fujihara, H. Takeshita, H. Iwata (2011): Individual variations in inorganic arsenic metabolism associated with AS3MT genetic polymorphisms. *Int. J. Mol. Sci.*, **12**, 2351–2382.

Barotányi Z. (2009): Arzén az ivóvízben—A zöld színű szörny (Arsenic in drinking water—The green monster—in Hungarian, *Magyar Narancs*, 2009. március 5. http://www.mancs.hu/index.php?gcPage=/public/hirek/hir.php&id=18500 (last access: February 25, 2014).

Bod T. (2010): Eddig kudarcos történet a dél-alföldi ivóvízminőség-javító program. Halálos veszély: arzén a vízben [The water remediation in the Great Hungarian Plain so fair is a failure story. Mortal danger: arsenic in drinking water—in Hungarian]. *168 óra*, 2010. november 25. http://www.168ora.hu/itthon/tobb-a-kelletenel-avagy-arzen-a-vizben-65133.html (last access: February 25, 2014).

P. Bucsky (2013): Láthatatlan veszély (Invisible danger—In Hungarian). *Figyelő*, 45, 30–31.

J. Emsley (2003): cited work, pp. 40–46.

Pénzes A. (2009): Halálos méreg, vagy a hosszú élet titka?! (Deadly poison or the secret of long life?—in Hungarian). *Magy. Kém. Lapja*, **64**, 341–346.

J. Timbrell (2005): cited work, pp. 118–120, 221–227.

MR1 Kossuth Rádió, 180 perc, August 17, 2007 http://www.mdche.u-szeged.hu/~kovacs/2007-08-17-k08_Arzen.rm (in Hungarian, last access: February 25, 2014).

http://en.wikipedia.org/wiki/Arsenic (last access: February 25, 2014).

http://en.wikipedia.org/wiki/Arsenic_poisoning (last access: February 25, 2014).

http://en.wikipedia.org/wiki/Lewisite (last access: February 25, 2014).

http://en.wikipedia.org/wiki/Arsphenamine (last access: February 25, 2014).

http://en.wikipedia.org/wiki/Britannicus (last access: February 25, 2014).

http://www.szennyviztudas.bme.hu/tartalom/arz%C3%A9n-hazai-iv%C3%B3v%C3%ADzben (last access: February 25, 2014)—in Hungarian.

http://viz.kalauzolo.hu/arennal-szennyezett-15millio-ivoviz/ (last access: February 25, 2014).

http://www.nyugatijelen.com/jelenido/a_viz_is_osszekot.php (last access: February 25, 2014).

4.3 Don't Touch the Spilled Mercury!

F. A. Lopez, A. Lopez-Delgado, I. Padilla, H. Tayibi, F. J. Alguacil (2010): Formation of metacinnabar by milling of liquid mercury and elemental sulfur for long term mercury storage. *Sci. Total Environ.* **408**, 4341–4345.

A. Santoro (2006): Mercury spill decontamination incident at the Rockefeller University. *J. Chem. Health Safety*, **13**, 30–37.

http://www.actontech.com/hgx6.htm (last access: February 25, 2014).

http://userpage.chemie.fu-berlin.de/~tlehmann/sonderab/quecksilber.html (last access: February 25, 2014).

http://en.wikipedia.org/wiki/Surface_tension (last access: February 25, 2014).

http://en.wikipedia.org/wiki/Lotus_effect (last access: February 25, 2014).

http://en.wikipedia.org/wiki/Mercury_%28element%29 (last access: February 25, 2014).

4.4 Was DDT of More Harm Than Use?

B. Brandt (February 28, 1944): Typhus in Naples. *Life*, 36–37.
R. Carson (1962): Silent spring. Houghton Mifflin, Boston.
J. M. Conlon (2006): The historical impact of epidemic typhus. http://entomology.montana.edu/historybug/typhus-conlon.pdf (last access: February 25, 2014).
J. G. Edwards (2004): DDT: A case study in scientific fraud. *J. Am. Physicians Surgeons*, **9**, 83–88.
J. Duffus, H. Worth (2001): DDT—A case study. IUPAC Educators' Resource Material. http://old.iupac.org/publications/cd/essential_toxicology/IUPACDDTcase.pdf (last access: February 25, 2014).
M. M. Hecht (2002): Bring back DDT, and science with it! http://www.21stcenturysciencetech.com/articles/summ02/DDT.html (last access: February 25, 2014).
H. I. Miller (2010): DDT can stymie malaria-carrying mosquitoes in Haiti. http://www.haitiantruth.org/ddt-can-stymie-malaria-carrying-mosquitoes-in-haiti/ (last access: February 25, 2014).
D. Roberts, R. Tren, R. Bate, J. Zambone (2010): The excellent powder: DDT's political and scientific history. Dog Ear Publishing, Indianapolis. http://www.theexcellentpowder.org/index.php (last access: February 25, 2014).
J. Timbrell (2005): cited work, pp. 90–98, 99, 109, 133–134.
C. M. Wheeler (1946): Control of typhus in Italy 1943–1944 by use of DDT. *Am. J. Pub. Health*, **36**, 119–129. http://ajph.aphapublications.org/cgi/reprint/36/2/119.pdf (last access: February 25, 2014).
http://en.wikipedia.org/wiki/DDT (last access: February 25, 2014).
http://books.google.hu/books?id=U1QEAAAAMBAJ&pg=PA36&lpg=PA36&dq=8&ved=0CFYQ6AEwBw#v=onepage&q&f=false (last access: typhu+naples+1944&source=bl&ots=aQD0NZgUYz&sig=UBKDWshEbzrtlcl78pVQFBpWSSk&hl=hu&ei=26umTdnEGcyVswbK5JSqDA&sa=X&oi=book_result&ct=result&resnum=February 25, 2014).

4.5 Could Dioxin be the Most Toxic Substance?

R. Eisler (2000): Handbook of chemical risk assessment, vol. 2. Lewis Publishers, Boca Raton. pp. 1021–1056.
J. Emsley (1994): cited work, pp. 173–203.
H. Kiviranta (2005): Exposure and human PCDD/F and PCB body burden in Finland. Ph. D. dissertation, Kuopio.
Simsa P., Járási Zs. É., Fülöp V. (2007): A környezetszennyező dioxinvegyületek mint az endometriosis és egyéb betegségek kórokai (Polluting dioxins as causative agents of of endometriosis and other diseases—in Hungarian). *Orvosi Hetilap*, **148**, 1745–1750.
J. T. Tuomisto, J. Tuomisto, M. Tainio, M. Niittynen, P. Verkasalo, T. Vartiainen, H. Kiviranta, J. Pekkanen (2004): Risk-benefit analysis of eating farmed salmon. *Science* **305** (5683), 476–477. http://www.sciencemag.org/content/305/5683/476 (last access: February 25, 2014).
J. Timbrell (2005): cited work, pp. 106–107, 120–126, 257–258, 311.
http://en.wikipedia.org/wiki/Polychlorinated_dibenzodioxin#Dioxin_exposure_incidents (last access: February 25, 2014).
http://en.wikipedia.org/wiki/Median_lethal_dose (last access: February 25, 2014).
http://en.wikipedia.org/wiki/Toxicity (last access: February 25, 2014).

4.6 Was Napoleon Murdered with Arsenic?

M. Clemenza, E. Fiorini, L. Guerra, C. Herborg, M. Labra, E. Orvini, A. Piazzoli, E. Previtali, F. Puggioni, A. Santagostin (2008): Misure con attivazione neutronica sulla presenza di arsenico nei capelli di Napoleone Bonaparte e di suoi famigliari (Neutron activation measurement of arsenic in hair samples of Bonaparte Napoleon and his family members—in Italian). *Il Nuovo Saggiatore*, **24**(1–2), 19–30.
J. Emsley (2008): cited work, pp. 214–215.
I. M. Kempson, D. A. Henry (2010): Determination of arsenic poisoning and metabolism in hair by synchrotron radiation: the case of Phar Lap. *Angew. Chem. Internatl. Ed.*, **49**, 4237–4240.
L. Tolstoy: War and peace (translator undeclared) http://www.fullbooks.com/War-and-Peace19.html (last access: March 23, 2014).

4.7 The Resin Wars: Formaldehyde

H.-D. Belitz, W. Grosch, P. Schieberle (2009): cited work, p. 788.
E. Block, R. Deorazio (1994): Chemistry in a salad bowl: Comparative organosulfur chemistry of garlic, onion and shiitake mushrooms. *Pure Appl. Chem.*, **66**, 2205–2206.
T. P. Coultate (2002): cited work, p. 311.
D. J. Hanson (2010): Formaldehyde in clothing. *Chem. Eng. News*, **88**(36), 51–54.
G. Hess (2011): Doubting EPA on formaldehyde. *Chem. Eng. News*, **89**(16), 10.
J. F. Liu, J. F. Peng, Y. G. Chi, G. B. Jiang (2005): Determination of formaldehyde in shiitake mushroom by ionic liquid-based liquid-phase microextraction coupled with liquid chromatography. *Talanta*, **65**, 705–709.
J. B. Morris (2006): Chapter 16. Nasal toxicology. In: H. Salem, S. A. Katz (2006): Inhalation toxicology, CRC Press, Boca Raton, 2nd edn. pp. 349–371.
J. Schwarcz (2009): cited work, pp. 217–219.
T. Shibuya, S. Shimada, H. Sakurai, H. Kumagai (2005): Mechanism of inhibition of platelet aggregation by lenthionine, a flavor component from shiitake mushroom. IFT Annual Meeting: Presentation 54G–9. http://ift.confex.com/ift/2005/techprogram/paper_29006.htm (last access: February 25, 2014).
Tyihák E. (2002): Formaldehid ciklus az élővilágban (Formaldehyde cycle in the living world—in Hungarian). *Biokémia*, **26**, 2–9.
http://en.wikipedia.org/wiki/Formaldehyde (last access: February 25, 2014).
http://en.wikipedia.org/wiki/Bakelite (last access: February 25, 2014).
http://en.wikipedia.org/wiki/Shiitake_mushroom (last access: February 25, 2014).
http://en.wikipedia.org/wiki/Lenthionine (last access: February 25, 2014).
http://www.food.gov.uk/science/research/foodcomponentsresearch/phytoestrogensresearch/t05-t06programme/t05t06projectlist/t05027project/ (last access: February 25, 2014).

4.8 The Great Hungarian Red Mud Deluge

Gy. Bánvölgyi, T. Minh Huan (2009): Dewatering, disposal and utilization of red mud: state of the art and emerging technologies. *Newsletter. A biannual publication of The International Committee for the Study of Bauxite, Alumina, Aluminium (ICSOBA)*, **2**, 14–29. https://www.yumpu.com/en/document/view/9042102/de-watering-disposal-and-utilization-of-red-mud-state-redmudorg (last access: February 25, 2014). www.chem.elte.hu/system/files/ELTE_vo-rosiszap_lexikonja.pdf (last access: February 25, 2014).

Bojtár B. E. (2010): A repedés (The fissure—in Hungarian). *Magyar Narancs* XXII. (41) (October 14). http://www.mancs.hu/index.php?gcPage=/public/hirek/hir.php&id=22362 (last access: February 25, 2014).

A. Gelencsér, N. Kováts, B. Turczi, Á. Rostási, A. Hoffer, K. Imre, I. Nyírő-Kósa, D. Csákberényi-Malasics, Á. Tóth, A. Czitrovszky, A. Nagy, Sz. Nagy, A. Ács, A. Kovács, Á. Ferincz, Zs. Hartyáni, M. Pósfai (2011): The red mud accident in Ajka (Hungary): characterization and potential health effects of fugitive dust. *Environ. Sci. Technol.*, **45**, 1608–1615.

Commission Regulation (EC) No 574/2004 of 23 February 2004 amending Annexes I and III to Regulation (EC) No 2150/2002 of the European Parliament and of the Council on waste statistics http://eur-lex.europa.eu/LexUriServ/LexUriServ.do?uri=OJ:L:2004:090:0015:0047:EN :PDF (last access: February 25, 2014).

ELTE Kémiai Intézet kislexikonja a vörösiszap-katasztrófával kapcsolatos fogalmak magyarázatára (Chemical cyclopaedia of the Chemistry Department of Eötvös University on concepts related to the red mud catasrophe- in Hungarian). October 18, 2010.

S. Everts (2010): Recycling red mud. *Chem. Eng. News*, **88**(43), 50–51.

D. Herard (2010): The Hungarian Toxic Red Sludge Spill and Determining Public Accountability. http://digitalcommons.fiu.edu/cgi/viewcontent.cgi?article=1006&context=drr_student (last access: February 25, 2014).

Index/MTI (2011): A vörösiszap veszélyes hulladék (Red mud is a dangerous waste—in Hungarian). http://index.hu/kulfold/2011/01/12/a_vorosiszap_veszelyes_hulladek/ (last access: February 25, 2014).

Joób S. (2010): A vörösiszap nem hömpölyög (Red mud does not roll on—in Hungarian). *Index*, 2010. October 5, 16:46. http://index.hu/belfold/2010/10/05/vorosiszap/ (last access: February 25, 2014).

Matkovich I. (2010): Devecser másfél hónap múltán—Végletek között (Devecser after one and half months. Between extremities—in Hungarian). *Magyar Narancs*, XXII. (47) (November.25). http://www.mancs.hu/index.php?gcPage=/public/hirek/hir.php&id=22644 (last access: February 25, 2014).

Matkovich I. (2011): A vörösiszap-katasztrófa helyszíne egy év múltán. Házak a magasban. (The red mud catastrophe after a year—in Hungarian) *Magyar Narancs*, XXIII. évf. 38. szám (2011.09.22). http://magyarnarancs.hu/belpol/a_vorosiszap-katasztrofa_helyszine_egy_ev_multan_-_hazak_a_magasban-76952 (last access: November 26, 2013).

Palla J., Rőczei I. (2008): Eljárás kohászati vashordozó és egyéb fém- vagy fémoxidtartalmú melléktermékek és vas vagy egyéb fém vagy fémoxidhordozó veszélyes hulladékok ipari alapanyaggá alakítására (A process for transformation of metallurgical by-products containing iron and other metals or metal oxides or hazardous wastes containing iron, other metals or metal oxides into industrial base material—in Hungarian). Hungarian patent P0600317.

Pallagi A. (2011): „Safety first!" Munkavédelem és vörösiszap-kezelés Ausztrália egyik legnagyobb timföldgyárában. (Safety first! Labour safety and red mud handling in one of the largest alumina factory of Australia—in Hungarian). *Magy. Kém. Lapja*, **66**, 86–87.

Rényi P. D., Vári Gy. (2010): A vörösiszap-katasztrófa miatti felelősség—Elterelő hadművelet (The responsibiliy of red mud catastrophe. Red herring—in Hungarian). *Magyar Narancs*, XXII. (42) (October 21). http://www.mancs.hu/index.php?gcPage=/public/hirek/hir.php&id=22411 (last access: February 25, 2014).

Ritz F. (ed.) (2010): Különszám a vörösiszapról (Special issue on red mud—in Hungarian). *Hírlevél. MKE Richter munkahelyi csoport.* 2010. október 28. 1–35.

S. Ruyters, J. Mertens, E. Vassilieva, B. Dehandschutter, A. Poffijn, E. Smolders (2011): The red mud accident in Ajka (Hungary): plant toxicity and trace metal bioavailability in red mud contaminated soil. *Environ. Sci. Technol.*, **45**, 1616–1622.

J. Somlai, V. Jobbágy, J. Kovács, S. Tarján, T. Kovács (2008): Radiological aspects of the usability of red mud as building material additive. *J. Haz. Mat.*, **150**, 541–545.

Somlai J., Kovács J., Sas Z., Bui P., Szeiler G., Jobbágy V., Kovács T. (2010): Vörösiszap-tározó sérülésével kapcsolatos sugárterhelés becslése (Estimation of radiation exposure as a consequence of red mud accident—in Hungarian). *Magy. Kém. Lapja*, **65**, 378–379.

J. Szépvölgyi (2011): A chemical engineer's view of the red mud disaster. *Nachrichten aus der Chemie*, **59**, May, Central Europe special issue, V–VII.

Szabó G. (2010): A vörösiszap-katasztrófa kérdőjelei (Questions of red mud catastrophe—in Hungarian). *HVG online*. October 14. http://hvg.hu/hvgfriss/2010.41/201041_a_vorosiszapkatasztrofa_kerdojelei_szakadas#utm_source=hirkereso&utm_medium=listing&utm_campaign=hirkereso_2010_10_14 (last access: February 25, 2014).

Szabó G. (2011): MAL-bírság az államosítás árnyékában. Jog, erő (The fine of MAL Zrt in the shadow of nationalization. Law and force—in Hungarian). *HVG*, **33** (38), 60–62. http://hvg.hu/hvgfriss/2011.38/201138_malbirsag_az_allamositas_arnyekaban_jog_ero (last access: November 26, 2013).

Szépvölgyi J., Kótai L. (2011): Az ajkai vörösiszap-ömlés I. A vörösiszap képződése, tulajdonságai és tárolása (The red mud disaster of Ajka 1. The formation, properties and storing of red mud—in Hungarian). *Magy. Kém. Lapja*, **66**, 2–8.

Szépvölgyi J., Kótai L. (2012): Az ajkai vörösiszap-ömlés II. A vörösipar hasznosítási és feldolgozási lehetőségei (The red mud disaster of Ajka 2. The possibilities of utilization and processing of red mud—in Hungarian). *Magy. Kém. Lapja*, **67**, 362–368.

Vári Gy. (2010): Repedések a rendszeren (Fissures in the system—in Hungarian). *Magyar Narancs*, XXII. (41) (October 14). http://www.mancs.hu/index.php?gcPage=/public/hirek/hir.php&id=22369 (last access: February 25, 2014).

Vörösiszap katasztrófa: következmények és tapasztalatok. Konferencia a Magyar Tudományos Akadémián (The red mud disaster: consequences and lessons. Conference at the Hungarian Academy of Sciences, Budapest, March 1, 2011—in Hungarian) http://www.katasztrofavedelem.hu/index2.php?pageid=szervezet_tudomany_konferencia_details&konf_id=5 (last access: February 25, 2014).

C. Zanbak (2010): Failure mechanism and kinematics of Ajka tailings pond accident, October 4, 2010. http://www.infomine.com/library/publications/docs/Zanbak2010.pdf (last access: February 25, 2014).

http://www.wise-uranium.org/mdafko.html (last access: February 25, 2014).
http://en.wikipedia.org/wiki/Ajka_alumina_plant_accident (last access: February 25, 2014).
http://mta.hu/mta_hirei/tajekoztato-a-kolontari-vorosiszap-tarozo-kornyezeteben-vegzett-vizsgalatokrol-125761/ (last access: February 25, 2014).
http://mta.hu/mta_hirei/osszefoglalo-a-vorosiszap-katasztrofa-elharitasarol-a-karmentesitesroles-a-hosszu-tavu-teendokrol-125859/ (last access: February 25, 2014).
http://www.uottawa.ca/services/ehss/ionizconversion.htm (last access: February 25, 2014).
http://www.redmud.org (last access: February 25, 2014).

4.9 Death in a Single Grain

A. Conan Doyle: A Case of Identity. http://www.eastoftheweb.com/cgi-bin/version_printable.pl?story_id=CaseIden.shtml (last access: March 23, 2014)

4.10 A Closer Look at Acids

N. N. Greenwood, A. Earnshaw (1984): cited work, pp. 633, 634, 961–962.
G. B. Kauffman (1988): The Brønsted-Lowry acid-base concept. *J. Chem. Educ.*, **65**, 28–31.
A. A. Shaffer (2006): Let us give Lewis acid-base theory the priority it deserves. *J. Chem. Educ.*, **83**, 1746–1749.

4.11 O, the Fresh Smell of Ozone

N. N. Greenwood, A. Earnshaw (1984): cited work, pp. 707–712.
T. Mann (1974): cited work, p. 265.
J. Verne: Dr. Ox's Experiment. http://ebooks.adelaide.edu.au/v/verne/jules/v52dr/ (last access: March 23, 2014)

4.12 The Diamond Thermometer

R. L. Wolke, M. Parrish (2010): cited work, pp. 227–229.
http://www.abazias.com/diamondblog/diamond-education/why-diamonds-are-called-ice (last access: March 23, 2014)

4.13 Everlasting Love: Diamonds

V. Blank, M. Popova, G. Pivovarov, N. Lvova, K. Gogolinsky, V. Reshetov (1998): Ultrahard and superhard phases of fullerite C60: comparison with diamond on hardness and wear. *Diamond Relat. Mater.*, **7**, 427–431.
H. Y. Chung, M. B. Weinberger, J. B. Levine, A. Kavner, J.-M. Yang, S. H. Tolbert, R. B. Kaner (2007): Synthesis of ultra-incompressible superhard rhenium diboride at ambient pressure. *Science*, **316**, 436–439.
A. Farkas (2002): Oxford dictionary of catchphrases. Oxford University Press. Oxford. p. 61.
http://www.adiamondisforever.com/ (last access: February 25, 2014).

4.14 Is pH 5.5 Neutral?

CRC Handbook of Chemistry and Physics. 87th Edition (2006–2007), pp. 8–80.
http://www.ciao.co.uk/Johnson_s_pH_5_5_Cleansing_Milk__6235107 (last access: February 25, 2014).
http://www.sebamedusa.com/ph-55.html (last access: February 25, 2014).

4.15 Sorrel Soup Scare: Oxalic Acid

J. Emsley (1998): cited work, pp 6–9.
M. Farré, J. Xirgu, A. Salgado, R. Peracaula, R. Reig, P. Sanz (1989): Fatal oxalic acid poisoning from sorrel soup, *Lancet*, **334**, 1524.
H. Tallquist, I. Väänanen (1960): Death of a child from oxalic acid poisoning due to eating of rhubarb leaves. *Ann. Paediatr. Fenn.*, **6**, 144–147.
H. This, M.-O. Monchicourt (2010): cited work, p. 91.
T. Mann (1974): cited work, p. 713–714.
http://www.tititudorancea.com/z/oxalic_acid.htm (last access: February 25, 2014).
http://oxalicacidinfo.com/ (last access: 2011. május 21).

4.16 The Mysteries of Magnetism

E. Danieli, J. Perlo, B. Blümich, F. Casanova (2010): Small magnets for portable NMR spectrometers, *Angew. Chem. Internatl. Ed.*, **49**, 4133–4135.
http://en.wikipedia.org/wiki/Magnetism (last access: March 23, 2014)

4.17 Water: The Extravagant Liquid

N. N. Greenwood, A. Earnshaw (1984): cited work, pp. 2–3, 726–742.
J. Verne: Off on a Comet or Hector Servadac (translated by E. E. Frewer). http://jv.gilead.org.il/roberts/index.html (last access: March 23, 2014).
http://www.lsbu.ac.uk/water/ (last access: February 25, 2014).

4.18 Fire in Space

G. Sparrow (2007): Spaceflight: The Complete Story from Sputnik to Shuttle—And Beyond. DK Publishing, New York.
http://www.klte.hu/~wwwinorg/essays/essay050.html (last access: February 25, 2014).
http://spaceflight.nasa.gov/history/shuttle-mir/history/h-f-linenger-fire.htm (last access: February 25, 2014).
http://www.braeunig.us/space/specs/salyut.htm (last access: March 23, 2014).

4.19 Is Natural Gas Toxic?

J. Emsley (2008): cited work, pp. 136–154.
http://en.wikipedia.org/wiki/Carbon_monoxide_poisoning (last access: March 23, 2014).
http://www.nhs.uk/conditions/carbon-monoxide-poisoning/Pages/Introduction.aspx (last access: March 23, 2014).

4.20 Respiratory Affairs: Carbon Dioxide

J. J. Grbic (2011): 'Cool it'. *Chem. Eng. News*, **89**(13), 40.
B. Lomborg (2007): Cool it: The Skeptical Environmentalist's Guide to Global Warming. Cyan and Marshall Cavendish
B. Lomborg (2011): A roadmap for the planet. *Newsweek*, Jun 12, pp. 24–27.
http://www.mofetta.eu/ (last access: February 25, 2014).
http://www.newstatesman.com/environment/2010/09/interview-gay-climate (last access: February 25, 2014).
http://news.yahoo.com/s/yblog_upshot/20100831/sc_yblog_upshot/noted-anti-global-warming-scientist-reverses-course (last access: February 25, 2014).

4.21 The True Death Toll of Chernobyl

J. Bohannon (2005): Panel puts eventual Chornobyl death toll in thousands. *Science*, **309**, 1663.
http://www-pub.iaea.org/MTCD/publications/PDF/Pub1239_web.pdf (last access: February 25, 2014).
http://www.iaea.org/Publications/Booklets/Chernobyl/chernobyl.pdf (last access: February 25, 2014).

4.22 Were There Nuclear Explosions in Chernobyl or Fukushima?

http://www.youtube.com/watch?v=zGhOW3DLlow (last access: March 23, 2014).
http://en.wikipedia.org/wiki/Chernobyl_disaster (last access: March 23, 2014).
http://en.wikipedia.org/wiki/Fukushima_Daiichi_nuclear_disaster (last access: March 23, 2014).

4.23 What Was the Gulf War Syndrome?

D. Hanson (2008): Panel validates gulf war ills health: Exposure to pesticides and a prophylactic drug caused vets' illness. *Chem. Eng. News*, **86**(47), 8.
Research Advisory Committee on Gulf War Veterans' Illnesses (2008): *Gulf War Illness and the health of gulf war veterans: Scientific findings and recommendations.* Washington, D.C., U.S. Government Printing Office.http://www.va.gov/RAC-GWVI/docs/Committee_Documents/GWIandHealthofGWVeterans_RAC-GWVIReport_2008.pdf (last access: February 25, 2014).
http://en.wikipedia.org/wiki/Gulf_war (last access: February 25, 2014).
http://en.wikipedia.org/wiki/Gulf_War_syndrome (last access: February 25, 2014).
http://en.wikipedia.org/wiki/Pyridostigmine (last access: February 25, 2014).
http://en.wikipedia.org/wiki/Pyridostigmine_bromide (last access: February 25, 2014).
http://mestinon.com/index.jspf (last access: February 25, 2014).

4.24 The Strange Case of Bisphenol A

R. Alleyne (2010): Can shopping receipts make you impotent? *The Telegraph*, 30th June 2010, http://www.telegraph.co.uk/health/healthnews/7861156/Sex-and-shopping-how-retail-therapy-really-is-bad-for-mens-health-and-fertility.html (last access: February 25, 2014).
EFSA (2006): Opinion of the Scientific Panel on Food Additives, Flavourings, Processing Aids and Materials in Contact with Food on a request from the Commission related to 2,2-BIS(4-HYDROXYPHENYL)PROPANE (Bisphenol A). Question number EFSA-Q-2005-100 Adopted on 29 November 2006. *The EFSA Journal* **428**, 1–75.
J. G. Hengstler, H. Foth, T. Gebel, P.-J. Kramer, W. Lilienblum, H. Schweinfurth, W. Völkel, K.-M. Wollin, U. Gundert-Remy (2011): Critical evaluation of key evidence on the human health hazards of exposure to bisphenol A. *Crit. Rev.Toxicol.*, **41**, 263–291.
T. Mendum, E. Stoler, H. VanBenschoten, J. C. Warner (2010): Concentration of bisphenol A in thermal paper. *Green Chem. Lett. Rev.*, 1–6.
S. K. Ritter (2011a): Debating BPA's toxicity. *Chem. Eng. News*, **89**(23), 14–19.

Sources and Further Reading 359

S. K. Ritter (2011b): Exposure routes confound BPA debate. *Chem. Eng. News*, **89**(23), 20–22.
S. K. Ritter (2011c): BPA is indispensible for making plastics. *Chem. Eng. News*, **89**(23), web exclusive, http://pubs.acs.org/cen/coverstory/89/8923cover4.html (last access: February 25, 2014).
B. Yin, M. Hakkarainen (2011): Oligomeric isosorbide esters as alternative renewable resource plasticizers for PVC. *J. Appl. Polym. Sci.*, **119**, 2400–2407.
http://en.wikipedia.org/wiki/Bisphenol_A (last access: February 25, 2014).
http://en.wikipedia.org/wiki/Precautionary_principle (last access: February 25, 2014).
http://reason.com/archives/1999/04/01/precautionary-tale (last access: February 25, 2014).
http://www.reuters.com/article/2010/11/25/us-eu-health-plastic-idUSTRE6AO3MS20101125 (last access: February 25, 2014).
http://www.senseaboutscience.org/for_the_record.php/21/can-shopping-receipts-make-you-impotent-the-telegraph-30th-june-2010 (last access: February 25, 2014).
http://www.senseaboutscience.org/for_the_record.php/42/bisphenol-a-in-food-packaging (last access: February 25, 2014).
http://en.wikipedia.org/wiki/Phytoestrogens (last access: February 25, 2014).

4.25 Death in the Danube: Cyanide Spills

Cyanide spill 'worst disaster since Chernobyl', Chemistry and Industry, 21st February, 2000. 124
 http://findarticles.com/p/articles/mi_hb5255/is_2000_Feb_21/ai_n28765356/?tag=mantle_skin;content (last access: February 25, 2014).
Death in the Danube, *Newsweek*, 28th February, 2000, p. 5.
R. Koenig (2000): Wildlife deaths are a grim wake-up call in Eastern Europe, *Science*, **287**, 1737–1738.
http://www.cameco.com/common/pdf/media/news_releases/1999/kumtor.pdf (last access: February 25, 2014).
http://www.claim-gv.org/docs/morancyanidepaper.pdf (last access: February 25, 2014).

4.26 Sparkling Cyanide: the Ultimate Poison?

I. Asimov (1958): The Death Dealers. Avon Publications, New York.
J. Emsley (2008): cited work, pp. 155–178.
http://www.asimovreviews.net/Books/Book028.html (last access: March 23, 2014).
http://www.topyaps.com/top-10-deadliest-poisons/ (last access: February 25, 2014).

4.27 Poly… What? Plastics vs. Natural Materials

J. Emsley (2010): cited work, pp. 102–125.
P. Le Couteur, J. Burreson (2004): cited work, pp. 71–86, 105–122.
Nemes S. (2007): Műanyagok megújuló nyersanyagokból (Plastics from renewable raw materials—in Hungarian). *Magy. Kém. Lapja*, **62**, 15–20.
http://en.wikipedia.org/wiki/Green_chemistry (last access: February 25, 2014).
http://en.wikipedia.org/wiki/Polylactic_acid (last access: February 25, 2014).
http://en.wikipedia.org/wiki/Plastic (last access: February 25, 2014).

http://en.wikipedia.org/wiki/Polyester (last access: February 25, 2014).
http://www.plasticseurope.org/documents/document/20101028135906-final_plasticsthefacts_26102010_lr.pdf (last access: February 25, 2014).
http://www.pentrace.net/penbase/Data_Returns/full_article.asp?id=264 (last access: February 25, 2014).

4.28 The Difference Between a Rose and a Cornflower

H.-D. Belitz, W. Grosch, P. Schieberle (2009): cited work, pp. 829–832.
M. Elhabiri, P. Figueiredo, K. Toki, N, Saito, R. Brouillard (1997): Anthocyanin–aluminium and—gallium complexes in aqueous solution. *J. Chem. Soc., Perkin Trans. 2*, 355–362.
M. Shiono, N. Matsugaki, K. Takeda (2005): Phytochemistry: Structure of the blue cornflower pigment. *Nature* **436**, 791–791.
K. Yoshida, M. Mori, T. Kondo (2009): Blue flower color development by anthocyanins: from chemical structure to cell physiology. *Nat. Prod. Rep.*, **26**, 884–915.
http://www.uni-regensburg.de/Fakultaeten/nat_Fak_IV/Organische_Chemie/Didaktik/Keusch/p26_anth-e.htm (last access: February 25, 2014).

4.29 The Erin Brockovich Mystery: Chromium Salts

R A Anderson (2005): Chromium. In: *Encyclopedia of human nutrition*. 2nd ed. Vol. 1. (Eds: B. Caballero, L. Allen, A. Prentice), Elsevier, Amsterdam, pp. 396–401.
D. A. Bender (2006): Benders' dictionary of nutrition and food technology. Woodhead Publishing Ltd., Cambridge, 8th edn., pp. 116, 216.
H.-D. Belitz, W. Grosch, P. Schieberle (2009): cited work, p. 426.
C. Bourotte, R. Bertolo, M. Almodovar, R. Hirata (2009): Natural occurrence of hexavalent chromium in a sedimentary aquifer in Urânia, State of São Paulo, Brazil. *An. Acad. Bras. Cienc.*, **81**, 227–242.
G. A. Fine: The Goliath effect: corporate dominance and mercantile legends. *J. Am. Folklore*, **98**, 63–84.
J. H. Duffus, H. G. J. Worth (Eds.) (2006): Fundamental toxicology. 2nd ed. RSC, Cambridge. p. 18.
A. R. Gonzalez, K. Ndung'u, A. R. Flegal (2005): Natural occurrence of hexavalent chromium in the Aromas red sands aquifer, California. *Environ. Sci. Technol.*, **39**, 5505–5511.
S. A. Katz, B. Ballantyne, H. Salem (2006): 23. The inhalation toxicology of chromium compounds. In: *Inhalation toxicology*. 2nd ed. (Eds: H. Salem, S. A. Katz) CRC Press, Boca Raton, 543–564.
K. Ndung'u, S. Friedrich, A. R. Gonzalez, A. R. Flegal (2010): Chromium oxidation by manganese (hydr)oxides in a California aquifer. *Appl. Geochem.*, **25**, 377–381.
http://en.wikipedia.org/wiki/Erin_Brockovich (last access: February 25, 2014).
http://www.imdb.com/title/tt0195685/ (last access: February 25, 2014).
http://www.brockovich.com/index.html (last access: February 25, 2014).
http://www.npr.org/2010/12/13/131967600/erin-brockovich-ii-activist-returns-to-aid-town (last access: February 25, 2014).
http://en.wikipedia.org/wiki/Hinkley,_California (last access: February 25, 2014).
http://en.wikipedia.org/wiki/Hinkley_groundwater_contamination (last access: February 25, 2014).

http://www.ewg.org/news/cancer-risk-of-chrome-6-in-drinking-water (last access: February 25, 2014).
http://www.allmoviephoto.com/photo/albert_finney_julia_roberts_erin_brockovich_001.html (last access: February 25, 2014).
http://www.boston.com/news/nation/articles/2010/12/13/survey_shows_unremarkable_cancer_rate_in_ca_town/ (last access: February 25, 2014).
http://www1.ucsc.edu/currents/02-03/06-09/gonzalez.html (last access: February 25, 2014).
http://danzarrella.com/the-goliath-effect.html# (last access: February 25, 2014).

Glossary of Terms

J. Schwarcz (2009): cited work, p. 149.
J. Timbrell (2002): cited work, pp. 9–15, 64, 174–175.
J. Timbrell (2005): cited work, pp. 36–39.
http://en.wikipedia.org/wiki/Parts-per_notation (last access: February 25, 2014).
http://en.wikipedia.org/wiki/Half-life (last access: February 25, 2014).
http://en.wikipedia.org/wiki/Biological_half-life (last access: February 25, 2014).
http://en.wikipedia.org/wiki/Median_lethal_dose (last access: February 25, 2014).

Index

предéльно допустúмая концентрáция, 314

A
A-A-A Diet™, 208
Abortion, 185, 250
Absinthe, 169, 170–172
Absinthin, 170
Absinthism, 171, 172
Absolute zero, 278
Absorption, 175, 201, 203, 210, 227
Abstinence, 67
Academy Award, 307
A Case of Identity, 264
Accident
   domestic, 5
   industrial, 6, 16, 102, 249
Accum, Friedrich Christian (Frederick), 116
Acemannan, 180
Acesulfame, 87
Acetaminophen, 128
Acetic acid, 44, 45, 293
Acetylcholine, 293
   esterase, 293, 294
Acetylsalicylic acid, O-, 92, 139, 198, 199, 200, 201, 232
Acid, 94, 143, 182, 229, 265, 266, 275, 302
   battery, 265
   Lewis, 266
   rain, 32
Acid Alkaline Diet, 208
Acid-base
   balance, 211
   concept, 265
   indicator, 39
   property, 274
   theory, 266
Acid-forming
   diet, 208
   food, 208, 209

Acne, 86, 163
Active comparator study, 203
Addison's disease, 238
Adenosine, 105
Adenosylmethionine, S-, 130
Ad hominem argument, 289
ADI *See* Allowable daily intake, 42
Adrenal gland, 157
Adrenaline, 105, 293, 316
Adrià, Ferran, 114
Adsorption, 243
Advertising Age, 270
Aflatoxin, 35, 60, 61, 62
   B1, 61
A fool and his money are soon parted, 179
Africa, 141, 173, 186
   South, 117
   West, 104
Agent Orange, 248
AGENT *See* Antibody-guided enzyme nitrile therapy, 127
AGFA *See* Aktiengesellschaft für Anilinfabrikation, 38
Aggrecan, 127, 128
Aging, 222
Aglycone, 188, 315
Agriculture, 5, 7, 60, 108, 245
Agrippina, Julia Augusta, 237, 252
AGW *See* Arbeitsplatzgrenzwert, 314
AIDS, 181, 182
Air, 21, 37, 49, 58, 76, 168, 235, 238, 255, 263, 267, 269, 274, 282, 283, 285–287, 299
   cell, 112, 113
   freshener, 255
   pollution, 107, 224, 314, 316
Ajka, 259
Aktiengesellschaft für Anilinfabrikation, 38
Alamogordo, 290

Alaska, 106
ALA *See* Aminolevulinic acid, 5-, 28
Albedo, 32
Albinos, 51
Albumen
    exterior, 112
    internal, 112
    middle, 112
Alchemist, 178, 242
Alchemy, 242
Alcohol
    consumption, 4, 64–67, 68, 224
    content, 65, 68, 169, 171, 172
Alcoholism, 69, 171
Aldehyde, 143, 256
Alexander VI (pope), 252
Alfred Wegener Institut für Polar- und Meeresforschung, 30
Algae, 109, 222
    green, 30
    toxic, 32
Aliso Viejo, 1
Alkali metals, 23
Alkaline, 26, 210, 258
    diet, 208, 209, 211
    food, 208, 211
Alkalizing food, 208
Alkaloid, 107, 139, 173, 182, 185, 198, 205, 314
    sesquiterpene, 170
    xanthine, 105
Allergen, 136, 151–153, 311
Allergy, 39, 86, 94, 99, 102, 151–153, 255
Alleyne, Richard, 295
Alligator, 247
Allopathy, 311
Allotrope, 21, 267, 318
Allowable daily intake, 42
All that glitters isn't gold, 215
Almond, 122, 301
Aloë
    barbadensis, 178
    vera, 178–181
Alpha-Tocopherol Beta-Carotene study, 226
Altitude sickness, 287
Alum, 306, 319
Alumina, 258, 260–262
Aluminate base, 261
Aluminum, 23, 24, 259, 261, 262, 269, 277, 306, 319
    hydroxide, 261
    oxide, 261
Alzheimer's disease, 67, 93, 137, 224
Amalgam, 243

AMA *See* American Medical Association, 126
Ambrosia artemisiifolia, 151
America
    Central, 54, 103, 104
    North, 14, 88, 141, 195, 196
    South, 104, 141, 173, 232
American Biologics, 126
American Cancer Society, 126
American Christian College, 123
American Medical Association, 126
America Online, 122
Ames, Bruce N., 7, 247
Amine, 38
Amino acid, 51, 58, 83, 88, 150, 174, 190, 196, 257
Aminoglycosides, 191
Amino group, 39
Aminolevulinic acid, 5-, 28
Ammonia, 51, 211, 236, 280
Ammonium
    chloride, 189
    sulfate, 51
Amphetamine, 233
Amsterdam, 104, 106
Amygdalin, 122, 124
Amylase, 177
Amylopectin, 100
Amylose, 100
Anandamide, 230
Anaphylactic shock, 151, 311
Anaphylaxis, 311
Ancient Rome, 28, 237, 252
Androgen, 247
Anemia, 28
    sickle-cell, 141, 319
Angiogenesis, 226
Ang Khak, 176
Anglesite, 28
Anhydrobase, 306
Aniline, 38, 39
    dye, 38
Animal test, 60, 62, 93, 149, 150, 157, 170, 176, 178, 181, 230, 255, 294, 312
Anise, 73, 141, 169, 189
Anisotropic, 271
Annelids, 30
Anode, 168
Antacid, 136, 203, 228, 229, 232
Antarctic, 22, 31
Anthecotulide, 151
Anthemis cotula, 152
Anthocyanidin, 305
Anthocyanin, 305, 306
    heavenly blue, 306

Index

Anthrax vaccine, 293
Antibacterial, 38, 39, 85, 144, 151, 181, 190, 193, 211
Antibiotics, 17, 137, 190–193
   synthetic, 192
Antibody, 51, 257, 311
   guided enzyme nitrile therapy, 127
   monoclonal, 17
Anticaking agent, 299
Anticancer, 50, 67, 73, 92, 124, 125, 137, 139, 159, 181, 208, 211, 316
Antidepressant, 136, 154, 155
Antidote, 293, 301
Anti-estrogenic, 133
Antifungal, 83, 198
Antigen, 151, 258
   Amb 1, 151
Antihypertension medication, 133
Anti-inflammatory, 73, 91–93, 127, 130, 139, 144, 151, 152, 180, 198, 200, 201
Antimicrobial, 84, 85, 86
Antimony, 171
   poisoning, 214
   trichloride, 171
   trioxide, 214
Antioxidant, 36, 46, 65–67, 73, 84, 91, 158, 168, 210, 222–227
Antiscorbutic, 157
Antispasmodic, 151, 198, 201
AOL *See* America Online, 122
Apácatorna, 258
Aphrodisiac, 169, 221
Appetite
   decreased, 165
   lack of, 141
   loss of, 165
Apple, 35, 89, 159, 273
Apricot, 122, 124
Aquifer, 241
Arabia, 101
Arabica, 105, 116
Arbeitsplatzgrenzwert, 314
Archaea, 30
Arctostaphylos uva-ursi, 142
Area under the curve, 120
Argentina, 152
Arginine, L-, 257
Argon, 59
Aristotle, 279
Armageddon, 281
Armi tunnel, 285
Aromachology, 149
Aromatherapy, 148, 149
Arrhenius, Svante, 265

Arsenate, 238
Arsenic, 190, 214, 215, 237, 238–241, 252–254, 262, 275, 301
   contamination, 241
   (III) oxide, 238, 252
   poisoning, 237, 240, 252
   white, 238, 252
Arsenite, 238
Arsenobetaine, 239, 240
Arsenolite, 238
Arsine, 238
Arsphenamine, 137, 239
Artemisia
   absinthium, 169
   annua, 170
   dracunculus, 170
   pontica, 170
   vulgaris, 170
Arthritis, 4, 164, 210
Artichoke, 42, 168, 176
Artificial substance, 1, 7, 132
Ascorbic acid, L-, 35, 36, 122, 130, 156, 157, 158, 159, 160, 224, 230, 265
Ashes to Ashes, 271
Asia, 14, 72–74, 104, 111, 172, 173, 187, 195, 213
   East, 195, 197
   Southeast, 176, 182
Asimov, Isaac, 302
Asparagus, 42
Aspartame, 83, 87, 88, 89, 189
Aspergillus, 61, 62
   flavus, 8
   terreus, 177
Aspirin, 4, 7, 92, 94, 139, 198–201, 232
Assassination at St. Helena, 252
Association
   false negative, 10
   false positive, 10
Assyrians, 187
Astaxanthin, 222
Asteraceae, 151, 152
Asthma, 35, 106, 136, 151, 255
Astragalus gummifer, 101
ATBC *See* Alpha-Tocopherol Beta-Carotene study, 226
At Eternity's Gate, 153
Athens, 148, 287
Atherosclerosis, 66, 182, 316
Atomic
   bomb, 290
   explosion, 290
   mass, 317
   nucleus, 291

number, 23, 26
weight, 24
A Treatise on Adulterations of Food and Culinary Poisons, 116
AUC *See* Area under the curve, 120
Aum Shinri Kyo, 301
Austria, 26, 27, 212, 216, 288, 298
Austro-Hungarian Empire, 238
Autism, 10
Aviation, 106
Avocado, 149
Avogadro constant, 317
A Whiff of Death, 302
AWI *See* Alfred Wegener Institut für Polar- und Meeresforschung, 30
Ayers, 270
Ayurveda, 27
Azoreductase, 39
Aztec Empire, 54

B
Baby, 50, 62, 194, 197, 202, 285, 296, 297
Bacterioplankton, 30
Bacterium, 12, 17, 30, 35, 41, 43, 59, 60, 80, 82, 84, 85, 88, 112, 138, 146, 148, 190–192, 194, 231, 241, 256, 268
Badische Anilin- und Soda-Fabrik, 38
Baekeland, Leo Hendrik, 255, 302
Bakelite, 255, 302
Baker's yeast, 82
Baking powder, 53, 94
Bakonyi, Zoltán, 260
Bald eagle, 246
Baldness, 249
Ball mill, 243
Balvano, 284
Bangladesh, 241, 252
Bardeen, John, 158
Bärendreck, 187
Barham, Peter, 113
Barite, 264
Barium, 263, 264
  bisulfate, 264
  sulfate, 263, 264
Bark beetles, 245
Barley, 263
Barskaun river, 300
Barzelay, David, 114
Base, 229, 259, 265, 266
  Lewis, 266
BASF *See* Badische Anilin- und Soda-Fabrik, 38
Batista government, 124
Batrachotoxin, 8

Bauxite, 261, 262, 263
Bayer, Carl Josef, 261
Bayer technology, 261
Bay leaves, 73
Bay of Bengal, 241
Bean, 42, 54, 100, 185, 196
Bearberry, 36, 142
Beard, Howard H., 125
Beard, John, 124, 125
Beauchamp, Gary K., 91
Beck, Ulrich, 6
Bee
  Melipona, 54
  sting, 4
Beecher, Henry, 135
Beer, 62, 65, 104, 115, 296
Beetroot, 42, 276
Beijing Olympics, 130
Belarus, 288
Belgium, 81, 100, 249
Beni-Koji, 176
Benzaldehyde, 122
Benzalkonium chloride, 85
Benzene, 36, 37
Benzethonium chloride, 84, 85, 86
Benzoic acid, 35, 36, 94
Benzyl alcohol, 1
Benzylpenicillin, 191
Bergamottin, 84
Berkeley Daily Gazette, 125
Berlin, 79, 116
Berlin Olympics, 131
Bern, 129
Bernard, Claude, 206
Bertazzi, Pier Alberto, 250
Beryllium, 24
  oxide, 8
Berzelius, Jöns Jacob, 320
Beta-2 agonists, 133
Beta-glucosidase, 122
Better safe than sorry, 197, 297
Bezukhov, Pierre, 253
Bible, 15, 101, 279
Bifidobacterium, 60
Big Pharma, 153
Bile, 144, 166–168
Bilobalide, 218
Biobutanol, 111
Biodegradability, 303
Biodiesel, 107, 109, 110, 111, 186
Bioequivalence, 120
Bioethanol, 107, 108, 111
Bioethylene, 304
Biofuel, 106, 107, 111

Index

Biogas, 111
Biogasoline, 111
Biohydrogen, 110
Biomass, 12, 79, 106, 107, 110, 111, 304, 320
Biomethane, 111
Biomethanol, 110
Biorefinery, 108
Biotechnology, 16, 17, 18, 110
Biotin, 162
Birch
   sap, 82
   wood, 82
Bird
   fish-eating, 246
   of prey, 246
Bismuth, 25
Bisphenol A, 9, 295
Bitterness, 81, 144, 198
Blackcurrant seed, 76
Blastoderm, 112
Black pepper, 73
Bleach, 181, 236, 262
Blindness, 88, 110
Blofeld, Ernst Stavro, 271
Blood
   clot, 66, 77, 92, 200, 201
   coagulation, 66
   group, 12, 80
   letting, 145
   plasma, 52, 85, 120, 121
   platelet aggregation, 66, 67, 200, 256
   red ~ cell, 287, 319
   vessel, 105, 151, 176, 293
Blowgun dart, 206
Blue acid *See* Hydrogen cyanide, 265
Blue-baby syndrome, 41
Blueberry, 35
Blurred vision, 165
Body mass index, 64
Böker, Hella, 133
Bolivia, 232
Bond, James, 270, 271, 278
Bone, 28, 127, 130, 165, 209
   deformation, 128
   formation, 216
   softening, 161
   subchondral, 127
Book of explanation, 242
Borage, 76
Bordered White, 245
Borgia, Cesare, 252
Borgia, Lucrezia, 252
Boron, 23
Bos *bubalus*, 46
Boston, 296

Boston Tea Party, 173
Botox, 7, 301
Botulinum toxin, xiii, 7, 251, 301
Botulism, 5
Bouchard's nodes, 127
BPA *See* Bisphenol A, 295
Brabant, 46
Bradford Foundation, 126
Bradford, Robert, 126
Brain, 10, 67, 77, 105, 136, 150, 174, 286, 294
   stimulant, 169
   tumor, 249
Bran, 60, 62
Brazil, 104, 108, 304
Bread, 115, 116, 194
   raisin, 116
Bremerhaven, 30
Britannicus, 237, 252
British East India Company, 73
British Medical Journal, 70
Broccoli, 42, 71, 226
Brochotrix thermosphacta, 59
Brockovich, Erin, 307, 309, 310
Bromide ion, 236
Bromine, 235
Bronchitis, 149
Brønsted, Johannes Nicolaus, 266
Brussels sprout, 226
Bubalus bubalis, 46
Buchner, Johann Andreas, 199
Budapest, 27, 103
Buddha, 173
Buffer, 211
Bulgarian secret service, 301
Bundesinstitut für Risikobewertung, 36
Bunge, Gustav, von, 69
Bupalus piniaria, 245
Burk, Dean, 126
Butanol, 1
Butchering, 13, 57
Butter, 45–47, 50, 320
Butter yellow, 39
Butyl mercaptan, tert-, 284

C
Cabbage, 42, 157, 276
   Chinese, 42
   red, 38, 71
   Savoy, 42, 226
Cadmium, 262
   oxide, 8
   sulfide, 8
Café, 103
Caffeine, 1, 7, 103–106, 116, 172, 173, 174, 175, 220, 221, 228, 232

Calcite, 111
Calcium, 42, 45, 113, 130, 165, 211, 216, 246, 277, 299
  carbonate, 111, 229
  hydrogen phosphate, 94
  hypochlorite, 12
  ion, 215, 230, 306
  oxalate, 276
  phosphate, 94
California, 1, 30, 124, 125, 307, 309, 310
California Cancer Advisory Council, 125
California Cancer Registry, 309
California condor, 246
Camellia sinensis, 142, 173
Cameroon, 193, 287
Canada, 124, 251
Canary Islands, 178
Cancer, 9, 61, 67, 90, 92, 103, 105, 122–127, 137, 141, 182, 184, 191, 197, 203, 205, 208, 209, 210, 213, 224, 225, 239, 247, 250, 255, 256, 288, 307, 309
  bone, 249
  breast, 67, 139
  colon, 49
  gastrointestinal, 167, 249
  larynx, 255
  lung, 226, 309
  pancreas, 255
  prevention, 226
  prostate, 226
  risk of, 7
  skin, 249
  stomach, 5, 252
  urinary bladder, 52, 249
Candida, 201, 209
Canis familiaris, 251
Cannabinoids, 133, 234
Caoutchouc, 303
Capsaicin, 230, 231, 232
Capsaicinoid, 230
Capsicum
  annuum, 230
  chinense, 230
  frutescens, 230
Caramel, 189
Carat, 101
  metric, 101
Carbenoxolone, 188
Carbohydrate, 60, 79, 80, 82, 86, 98, 112, 304, 305, 315
  complex, 79
  simple, 79
Carbol (phenol) fuchsin, 40

Carbon, 19, 271, 300
  atom, xii
  activated, 13
  loading, 241
  tetrachloride, 235
Carbon-12 (isotope), 317
Carbon-14 (isotope), 315
Carbon dioxide, 30, 32, 33, 36, 59, 80, 82, 94, 110, 113, 211, 215, 216, 228, 229, 236, 272, 274, 282, 284, 285, 286, 287, 303
Carbonic acid, 211, 212, 216, 274
Carbon monoxide, 59, 110, 284, 285
  detector, 285
  poisoning, 284, 285
Carboxylic
  acid, 74
  anhydrides, 237
Carboxymyoglobin, 58, 59
Carcinogenic, 7, 36, 39, 60–62, 87, 238, 247, 256, 294, 314
Carcinogenicity, 61, 247
CARC *See* Chemical agent resistant coating, 293
Cardamom, 73
Cardiac
  arrest, 193
  failure, 5
  glycoside, 313
Cardiovascular
  benefit, 68
  disease, 40, 49, 50, 64, 77, 90, 91, 127, 176, 195, 224–226, 308, 316
  risk, 84
  system, 103
CARET *See* Carotene and Retinol Efficacy Trial, 226
Cargill, 108
Caribbean, 79, 84, 104
Cariogenicity, 83
Carnitine, 158
Carob
  gum, 93, 100
  seed, 101
  tree, 101
Carotene, 93, 226, 227
  α, 226
  β, 226
Carotene and Retinol Efficacy Trial, 226
Carotenoid, 84, 91, 222, 224, 226, 227, 305
Carrot, 71, 226
Carson, Rachel, 246
Cartilage, 127, 128, 129, 130
  articular, 127

hyaline, 127
Cartilage-specific proteoglycan core protein, 127
Casein, 302
Cassava, 123
Cassette tape, 277
Castellani's paint, 40
Casticin, 147
Castor
 beans, 185, 186
 oil, 57, 184, 185, 186
Castorp, Hans, 267
Castro, Fidel, 124
Cat, 52, 106
Catalase, 223, 224
Catharanthus roseus, 139
Cathode, 168
Cauliflower, 42, 276
Cayce, Edgar, 208
Cayenne, 230
Celery, 7, 42
Cell phone, 295
Celluloid, 302
Cellulose, 79, 107, 108, 111
Centaurea cyanus, 306
Central European University, 260
Central nervous system, 67, 88, 95, 105, 106, 150, 154
Centro Panamericano de Ingeniería Sanitaria y Ciencias del Ambiente, 12
CEPIS *See* Centro Panamericano de Ingeniería Sanitaria y Ciencias del Ambiente, 12
Cereal, 62
 whole grain, 60
Cerium, 25
Cerivastatin, 202
Cerussite, 28
Cesium, 25
CEU *See* Central European University, 260
Ceylon, 104
CFCCT *See* Committee for Freedom of Choice in Cancer Therapy, 126
Chaff, 240
Chaffee, Roger B., 282
Chain, Ernst Boris, 191
Chalaza, 112
Chain reaction, 291
Chamomile, 151–153, 218
 Argentine, 152
 German, 151
 stinking, 152
Charcoal, 166, 173, 318
Chard, 42

Chasteberry, 147
Cheese, 115
 cream, 101
 Parmesan, 96
Chemical agent resistant coating, 293
Chemical Weapons Convention, 237
Chemistry & Industry, 298
Chemophobia, 1, 2, 103, 247
Chemotherapeutic agent, 190, 191
Chemotherapy, 190, 191
Chernobyl, 288, 290, 292, 298
Cherry, 37, 189
Chervil, 42
Chevreul, Michel Eugène, 45
Chewing gum, 82
Chicago, 126, 290
Chicken, 246
 embryo, 245
Chicory, 42, 116
Children, 29, 39, 41, 42, 51, 53, 69, 70, 77, 82, 100, 101, 164, 185, 194, 216, 277, 289, 297
Chili pepper, 62, 73, 230
China, 14, 51, 78, 103, 104, 110, 148, 173, 174, 176, 196, 238
Chinese Restaurant Syndrome, 95, 96
Chitterlings, 62
Chloracne, 249, 250
Chloride ion, 97, 210, 251
Chlorination, 12, 13, 214, 215
Chlorine, 12, 24, 110, 213, 214, 250, 251, 277, 304
 dioxide, 210
Chlorite, 210
Chloroacetic acid, 214
Chloroform, 214
Chlorogenic acid, 116
Chloroperoxidase, 12
Chlorophenols, 12
Chlorophyll, 32, 91, 169, 209, 305, 306
Chlorous acid, 210
Chocolate, 48, 51, 54, 103, 106, 115, 209, 223, 296
Cholecalciferol, 161
Cholera, 11, 12, 241
Cholesterol, 43, 49, 64, 66, 77, 91, 164, 176, 177, 196, 202, 316
Choline, 293
Chondrodendron tomentosum, 206
Chondroitin, 128, 129, 130
 sulfate, 127, 128
Christian Brothers Contracting Corporation, 122
Chromate ion, 308, 309

Chromatogram, 218, 311, 316
Chromatography, 311, 316
  gas, 131, 220
  high-performance liquid, 220, 311
  thin-layer, 220
Chromium, 262, 307–310, 319
  dioxide, 277
  hexavalent, 308
  trivalent, 308
Chrysoidine, 38
Chymotrypsin, 123
Cigarette, 37
Cinchona, 139, 198
  tree, 145
Cinnabar, 242
Cinnamon, 73, 189
Cisplatin, 316
Citric acid, 35, 44, 45, 53, 189, 210, 211, 229, 242, 265, 266
Citrinin, 178
Citronellal, (+)-, 57
Citrus
  maxima, 84
  paradisi, 84
  sinensis, 84
Citrus red, 39
Civil organization, 11, 13, 214, 261
CLA *See* Conjugated linoleic acids, 50
Claudius, 237, 252
Claviceps purpurea, 314
Clear Lake, 245
Clinical trial, 126, 129, 134, 146, 149, 155, 162
Clover
  subterranean, 194
  white, 194
Cloves, 73
Coal, 104, 110, 255, 284, 287, 291
  gas, 284
Coalite Chemicals, 249
Cobalt, 277, 278
  (II) chloride, 8
Coca, 232
Coca-Cola, 232, 274
Cocaine, 232, 233, 234
Cochrane Collaboration, The, 128, 159
Cochrane Library, 146, 147, 149
Cocoa, 54, 62, 103–105, 223
  butter, 47, 48, 50
  content, 48
Coconut, 106, 109
  oil, 46
  plantation, 106
Codeine, 139

Codex Alimentarius, 11
Cofactor, 223
Coffee, 62, 103–106, 115, 116, 209, 214, 296
  adulteration, 116
  instant, 116
Collagen, 130, 158
Colloid, 311
Colloidal sol, 100, 311
Colluricincla megarhyncha, 8
Colombia, 12, 232
Colon, 49, 88
Columbo, Lieutenant, 271
Columbus, Christopher, 79, 104, 230
Committee for Freedom of Choice in Cancer Therapy, 126
Common cold, 151, 156, 158–160, 163, 209
Common hazel, 141
Common kestrel, 246
Common ragweed, 151, 152
Computer
  monitor, 29
  simulation, 15
Conalbumin, 113
Concrete, 269, 290
Confectionery, 51, 82, 187
Conjugated linoleic acids, 50
Connor, John, 278
Conspiracy theory, 153, 197
Constipation, 100, 136, 165, 167, 185
Consumentenbond, 117
Consumer Product Safety Commission, 255
Contact angle, 243, 244
Contact dermatitis, 151, 312
Contact lense, 295
Contergan, 202
Continental Blockade, 199
Continental drift, 30
Contraceptive pill, 194, 197
Contrast agent, 264
Contreras, Ernesto, 124
Cook, James, 182
Cool It. The Skeptical Environmentalist's Guide to Global Warming, 288
Copper, 36, 42, 99, 113, 158, 171, 269, 298, 299, 300
  acetate, 171
  ion, 158
  sulfate, 171
Coral tree, 101
Coriander, 72, 73
Coriandrum sativum, 72
Corn, 62, 76, 82, 108, 130
  oil, 91
Cornell University, 108

Cornflower, 306
Coronium, 24
Corsica, 253, 254
Cortés, Hernán, 54, 104
Corylus avellana, 141
Cosmetics, 2, 4, 7, 39, 54, 100, 148, 180, 186, 274, 301
Cotton, 303
Cottonseed, 130
Cough, 82, 139
Coumestan, 194
Court of Arbitration for Sport, 131, 132
Cousteau, Jacques-Yves, 30
Cow
    dung, 53
    tallow, 46
    udder, 46
COX-1 *See* Cyclooxygenase-1, 91
COX-2 *See* cyclooxygenase-2, 91
CPSC *See* Consumer Product Safety Commission, 255
Crab, 128
Cranberry, 36
Cress, 42
Crimean War, 156
Croissant, 116
Crop rotation, 108, 115
Crustaceans, 30
Crutzen, Paul J., 22
Cryolite, 261
Cryptoxanthin, β-, 226
Crystal violet, 40
Csáky, István, 193
CSI Las Vegas, 287
Ctenochaetus striatus, 8
Cuba, 124
Cucumber, 42
Cuevas de la Araña, 78
Cui prodest?, 213
Cuisine
    Asian, 95
    Indian, 73
    Japanese, 114
Culbert, Michael, 125, 126
Culinary luddism, 117
Cumin, 72
Cuminum cyminum, 72
Curare, 206
Curcuma longa, 72
Curcuminoid, 73
Curie, Marie, 158
Curie temperature, 277
Curing, 35, 41, 54, 60, 115, 155, 198, 203
Curry, 71, 72, 73

Currywurst, 73
Cuticle, 112
Cyamopsis tetragonolobus, 100
Cyan, 235, 300
Cyanamides, 237
Cyanate ion, 236
Cyanates, 237
Cyanide
    ion, 236, 298–300
    poisoning, 122, 123, 301
    pollution, 298
    spill, 298, 299, 300
Cyanidin, 305, 306
Cyanine, 306
Cyanobacteria, 30
Cyanocobalamine, 300
Cyanogen bromide, 235–237
Cyanogen chloride, 237
Cyanuric acid, 51–53
Cyclamate, 87
Cyclohexanedimethanol, 1,4-, 296
Cyclooxygenase-1, 91
Cyclooxygenase-2, 91
Cyprus, 287
Cysteine, 223
Cytochrome oxidase, 300

D

Daewoo Shipbuilding, 107
Daidzein, 194, 218
Dam, Henrik, 162
Dandelion, 243
Danger, 3, 21, 60, 94, 106, 134, 172, 208, 211, 227–229, 233, 264, 289, 302
Danube river, 298
Daptomycin, 192
Dartmouth College, 12
Darwin, Charles, 28
DDD, 245
DDE, 245, 247
DDT, 9, 245–248, 250
Dean, Jacqui, 1
De Beers, 270
Debrecen, xii
Decoction, 143
Defoliant, 248
Deforestation, 32
Degas, Edgar, 171
Dehydroascorbic acid, 156
Delirium, 170
Delphinidin, 306
Dementia, 90, 93, 164
Demethylase, 258
Demethylation, 257, 258

Dendrobatidae, 8
Deoxynivalenol, 4-, 60, 62
Dependence
  physiological, 233
  psychological, 233
Depression, 83, 149, 153–155, 182
Dermatan sulfate, 127
Der Zauberberg, 256, 267, 275
Desert bluebell, 306
Desquamation, 315
Detergent, 262
Detoxification, 166–169, 214, 315
Devecser, 258
Devil's claw, 141
Devyatovskiy, Vadim, 130
DHA *See* Docosahexaenoic acid, 76
DHMO *See* Dihydrogen monoxide, 1
Diabetes, 5, 40, 86, 100, 144, 181, 220, 224, 276
Dialdehyde, 91, 92
Diamagnetic, 277, 278
Diamond, 19, 101, 268–272
Diamonds are Forever, 270
Diarrhea, 28, 51, 82, 100, 101, 117, 136, 167, 186, 191, 196, 216, 314
  osmotic, 83
Diatoms, 30, 32
Diazepam, 313
Dibenzodioxins, 100, 250
Dibenzofurans, 251
Dicarboxylic acid, 276
Dichlorophenoxyacetic acid, 2,4-, 248
Dichromate ion, 308
Diclofenac, 127
Dicyanamides, 237
Diesel, 107, 109, 110
  renewable, 110
Dietary supplement, 50, 69, 84, 85, 127, 132, 161, 176–178, 180, 196, 197, 209, 219–222
Diethyl ether, 235
Digitalis, 139
Diglyceride, 43, 44
Digoxin, 313
Dihydrogen monoxide, 1
Dihydrogen trioxide, 257
Dilatancy, 320
Dill, 42
Dimethylfuran, 2,5-, 111
Dimethyl sulfide, 32
Dimethyl sulfone, 129
Dimethyl terephthalate, 296
Dioxin, 100, 102, 103, 117, 248–251
Disaccharide, 79, 143
Disinfectant, 52, 142, 181, 211, 237, 255, 268

Disinfection, 12, 13, 237
Distillation, 89, 148
  steam, 54
Diuretic, 133, 167
Dizziness, 165, 286
DNA, 67, 95, 191, 255
Dobson, Gordon, 21
Dobson unit, 21
Docosahexaenoic acid, 74, 76, 77
Dog, 52, 83, 106
Doisy, Edward A., 162
Domagk, Gerhard, 38
DON *See* Deoxynivalenol, 4-, 62
Dopamine, 105, 154
Doping, 106, 130–134, 220, 234
  agent, 132, 133
  test, 133
  violation, 130–132
Dose-response
  curve, 313
  relationship, 7, 312
Dosha, 27
Double blind study, 68, 96, 136, 147, 314
Douglas, Mary, 6
Down syndrome, 250
Doyle, Sir Arthur Conan, 264
Drepanocytosis, 319
Drink
  alcoholic, 68, 88, 104, 110, 228
  soft, 35, 37, 88, 89, 212, 228, 285
Dr. Ox's Experiment, 268
Drug (narcotic), 133, 134, 233, 314
  testing, 131
  trafficker, 232
Drug (pharmaceutical), 4, 38, 68, 91, 93, 120, 135–137, 139, 153–155, 177, 188, 198, 203, 229, 312, 314, 318
  anti-fever, 200
  artificial, 190
  blockbuster, 119, 139
  counterfeit, 122
  development, 135, 137, 138, 141
  generic, 119, 120, 121, 122, 200
  illegal, 197, 220
  original, 120
  orphan, 141
  plant-derived, 139, 141
  reference, 120
  safety, 128
  synthetic, 84, 91, 141, 193, 218, 220
Drug Price Competition and Patent Term Restoration Act, 119
Drummond, Sir Jack Cecil, 157
Dr. Watson, 264

Drying oil, 76
Dumas, Jean-Baptiste André, 51
Duphar, 249
DuPont Co., 22, 303
Dust, 31, 262, 263
DU *See* Dobson unit, 21
DVD, 295
Dvicesium, 24
Dvitellurium, 24
Dye, 39, 176, 235, 262, 306
   artificial, 57
   azo, 37, 38, 39
   fluorescent, 51
   natural, 37
   synthetic, 37, 38
   triphenylmethane, 37, 39
Dynamo, 212
Dysentery, 241

E
E85, 108
E110, 38
E121, 38
E160, 93
E210-E213, 35
E260, 93
E262, 93
E300, 93
E336, 94
E341, 94
E410, 93, 100
E412, 100
E413, 100
E415, 100
E418, 100
E420, 81
E421, 81
E470a, 45
E470-E480, 44
E471, 45
E472, 44
E472a, 45
E472b, 45
E472c, 45
E472d, 45
E472e, 45
E473, 45
E474, 44, 45
E475, 45
E477, 45
E500, 94
E536, 299, 301
E626-E635, 95
E967, 81, 82
E968, 81, 82
Ear
   ache, 149
   infection, 82
Ear blight, 63
Earth, 20, 21, 22, 24, 30, 32, 79, 104, 115,
   117, 238, 251, 268, 279, 280, 282,
   283, 287, 290, 291, 320
Earthquake, 12, 287, 290
EAR *See* Estimated average requirement, 161
East
   Far, 197, 231
   Middle, 101, 294
Ebers Papyrus, 198
Ecuador, 12
Eczema, 86, 149
ED_{50} *See* median effective dose, 312
EDC *See* Endocrine disrupting chemical, 247
Edema, 315
Edwards, J. Gordon, 246, 247
E factor *See* Environmental factor, 16
EFSA *See* European Food Safety Authority, 11
Eger, 235
Egg, 18, 30, 100, 111–115, 209
   chicken, 112
   consumption, 111
   duck, 112
   hard-boiled, 111, 113
   pores, 112
   shell, 111, 112, 246
   soft-boiled, 111, 113
   white, 111–114
   yolk, 111, 113, 114
Eggplant, 42
Egypt, 148
Ehrlich, Paul, 137, 190, 239
Eicosapentaenoic acid, 74, 76
Eijkman, Christian, 162
Einstand law, 260
Ekaaluminum, 23
Ekaboron, 23
Ekacadmium, 25
Ekacerium, 25
Ekacesium, 25
Ekaiodine, 25
Ekamanganese, 24
Ekamolybdenum, 25
Ekaniobium, 25
Ekasilicon, 23
Ekatantalum, 24
Elaidic acid, 49
Elba
   island of, 253, 254
Electricity, 107, 110, 111, 261, 270

Electrolysis, 168, 214, 261
Electromagnet, 278
Elephant grass, 108
ELISA, 51
El Niño, 12
Emulsifier, 44–46
Emulsion
    oil-in-water, 113
    water-in-oil, 46
Enantiomer, 150
Endive, 42
Endo, Akira, 176
Endocrine
    disrupting chemical, 247, 296
    gland, 294
Endogenous, 88, 131, 194, 258
Endosperm, 60
Enema, 27
Energy, 32, 35, 79, 80, 106–110, 113, 132, 242, 243, 287, 291, 292, 300, 303, 307
    light, 80
    balance, 108
    crop, 108
    density, 110
    drink, 132
    independence, 109
    production, 107
    solar, 96, 110
    supply, 107
Engel, Ernst, 116
England, 5, 72, 79, 104, 198, 199, 251
Enlightenment, 137
Enquiry into Plants, 187
E-number, 11, 93
Environmental
    catastrophe, 258
    factor, 16
    footprint, 212, 213
    pollution, 2, 4, 6, 16, 102, 213, 258, 303
    toxicology, 309
Environmental Assessment Agency, 111
Environmental Protection Agency, 9, 30, 251
Environmental Working Group, 310
EPA *See* Environmental Protection Agency; *See* Eicosapentaenoic acid, 9
Ephedrine, 220
Epicatechin-3-O-gallate, 1
Epidemiological study, 64, 65, 69, 92, 293, 314, 317
Epidemiology, 247
Epigallocatechin-3-O-gallate, 1, 225
Epilepsy, 121, 165, 170, 171
Epinephrine, 316

Epitestosterone, 131
Equol, 194
Erectile dysfunction, 220
Erection, 220
Ergocalciferol, 161
Ergotism, 60, 314
Ergotoxicosis, 314
Ergot poisoning, 314
Erin Brockovich, 307, 309
Erinoid, 302
Erithacus rubecula, 245
Erythrina, 101
Erythritol, 81, 83
Erythropoetin, 133
Erythropoietin, 133
Erythroxylon coca, 232
Escherichia coli, 17
Esophagus, 274
Essential oil, 54, 148, 149, 169, 170, 171, 192
Ester, 43–47, 74, 88, 148, 315
    fruit, 88, 315
Estimated average requirement, 161
Estradiol, 67, 68, 193, 194, 297
Estrogen, 62, 67, 193, 194, 196, 197, 247, 295, 296, 297
Ethane, 284
Ethanol, 63, 65, 66, 108, 110, 304
Ether (supposed element), 24, 27
Ethics
    new, 15
    old, 16
Ethiopia, 81, 103
Ethoxymethylfurfural, 5-, 111
Ethyl alcohol *See* Ethanol, 89
Ethylene, 213, 303, 304
Ethylenediaminetetraacetic acid, 242
Ethyl mercaptan, 284
Euphorbia resinifera, 232
Europe, 11, 13, 14, 46, 54, 62, 72, 73, 79, 93, 94, 101, 103, 104, 108, 119, 141, 162, 171, 172, 173, 180, 182, 187, 189, 195–198, 212, 235, 239, 242, 249, 253, 258, 261, 301–304
    Central, 11
    Eastern, 230, 289
    Mediterranean, 90
    Northern, 79, 104
    Western, 104, 239
European Commission, 42, 62, 100
European Food Safety Authority, 11, 53, 162, 296
European Union, 13, 14, 15, 46, 59, 62, 63, 89, 103, 119, 141, 171, 177, 179, 184, 189, 212, 215, 220, 239, 242, 259, 264, 314

EU *See* European Union, 10
Evening primrose, 76
Evidence-based medicine, xi
Exogenous, 131, 132, 224, 227, 311
Expectorant, 55, 187
Explosion
  chemical, 290, 292
  thermal, 290, 292
Exposure, 3, 7, 9, 13, 29, 30, 37, 59, 60, 76, 88, 92, 142, 161, 195, 197, 215, 239, 242, 247, 250, 255, 263, 288, 289, 295–297, 309, 314
  time, 313

F
Faisal I (king), 238
FAO *See* Food and Agriculture Organization, 11
Fat, 43, 45–50, 74, 77, 80, 100, 103, 107, 109, 130, 162, 164, 166, 186, 211, 223, 226, 227, 245, 317
  animal, 46
  chicken, 109
  content, 46, 50
  modified, 44
  saturated, 47, 49
  semisynthetic, 44
  tissue, 50
  trans, 49, 50
  unsaturated, 47, 49
  veal, 49
Fatty acid, 43–45, 47, 49, 50, 74–78, 82, 88, 90–93, 109, 130, 148, 177, 196, 211, 315
  cis, 48, 50
  methyl ester, 109
  monounsaturated, 74, 90
  omega, 47
  omega-3, 74–78
  omega-6, 74–78, 130
  omega-9, 74, 75, 78
  polyunsaturated, 50, 74, 75, 90
  saturated, 46–49
  trans, 48–50
  unsaturated, 49
FDA *See* Food and Drug Administration, 7
Feed (fodder), 60, 108, 156
Fehér, Miklós, 193
Fenfluramine, 220
Fennel, 42, 73, 141, 169
Fenugreek, 73
Fermentation, 17, 66, 88, 89, 101, 108, 174, 176, 185, 285, 286
Fermi, Enrico, 290

Ferromagnetic, 277, 278
Fertility, 28, 106, 295
Fertilization, 31, 32, 62, 115
Fertilizer, 41, 42
  artificial, 40, 41, 42
  natural, 42
  synthetic, 41
Fire, 281, 282, 283
Fischer, Hermann Emil, 175
Fish, 7, 30, 72, 76, 77, 90, 95, 130, 209, 239, 245, 298, 299, 317
  ciguateric, 8
  freshwater, 78
  larvae, 16
  oil, 46, 77, 103
  raw, 12
  salted, 156
  sauce, 95
  sea, 78, 239, 240
Fishing
  deep-sea, 5
Flatulence, 82, 136, 141, 196
Flavonoid, 54, 66, 84, 190, 222, 224–226, 305, 306, 315
Flax, 194
Flegal, Russell, 309
Fleming, Alexander, 139, 190
Fleming, Ian, 271
Flood, 5, 12, 119, 258, 259
Florey, Howard Walter, 190
Flour, 95, 116
  whole grain, 60
Fluoride, 216
Foé, Marc-Vivien, 193
Folic acid, 162, 164, 165
Food
  additive, 4, 10, 11, 35, 37, 38, 44, 45, 63, 86, 93, 95, 101, 115, 189, 301, 314
  adulteration, 51, 115–117
  dye, 35, 37
  price index, 107
  safety, 1, 6, 11, 115, 117
  staple, 39, 111, 123, 196
  supply chain, 117
Food and Agriculture Organization, 11
Food and Drug Administration, 7, 11, 36, 50, 53, 122, 130, 146, 177, 215, 296
Food and its adulterations, 116
Food Standards Agency, 37, 63, 163
Formaldehyde, 51, 255–258, 302
Formalin, 255
Formula (chemical), xii, 79, 91, 98, 126, 141, 180, 220, 238, 239, 264, 272
Forshufvud, Sten, 252

Foxglove, 139
France, 65, 69, 79, 103, 106, 170, 212, 252, 314
Francesco I de' Medici, 238
Francium, 24
Franco-Prussian War, 46
Franklin, Benjamin, 28
Freedom from cancer, 126
Free radical, 168, 222–226, 260
Freeze-drying, 115
French fries, 116
French rose, 306
Friedrich Bayer & Co., 199
Frølich, Theodor, 156
Fructose, 79, 81, 86, 98, 143
FSA *See* Food Standards Agency, 163
Fuchsine, 39
Fuel, 57, 106, 108–111, 283, 287
 fossil, 238, 303
 jet, 106, 109
 liquid, 111
 rocket, 110
Fukushima Daiichi, 290
Fumigant, 237
Fumonisins, 60, 62
Fungicide, 63
Fungus, 7, 62, 82, 84, 94, 108, 138, 139, 146, 176, 177, 190–192, 201, 209, 314
Fur, 38
Furan, 111, 304
Furocoumarin, 84, 86
Fusariosis, 62
Fusarium
 culmorum, 62
 graminearum, 62

G
GABA *See* Gamma-aminobutyric acid, 154
Gadolinium, 277
Gagnaire, Pierre, 114
GAG *See* Glycosaminoglycan, 127
GAIT *See* Glucosamine/Chondroitin Arthritis Intervention Trial, 128
Galactose, 81, 100
Galalith, 302
Galápagos Islands, 30
Galega officinalis, 207
Galegine, 207
Galena, 28
Galgamácsa, 27
Gall
 bladder, 55
 stone, 55
Galliano, Chad, 114
Gallian Sea, 280, 281
Gallium, 23, 306
 arsenide, 238
Gambierdiscus toxicus, 8
Gamma-aminobutyric acid, 150, 154, 170, 174
Gamma-valerolactone, 110
Ganges river, 241
Gansu
 province of, 51
GAO *See* Government Accountability Office, 255
Garland, Judy, 313
Garlic, 42, 71, 73, 176, 192
Gasohol, 108
Gasoline, 29, 108, 110
Gastric
 acid, 144, 216, 228–231, 300
 juice, 88, 144
Gastrointestinal
 cramp, 229
 infection, 86
 motility, 293
 problem, 165
 tract, 88, 123, 166, 264
Gastronomy
 European, 81
 molecular, 114
Gattefossé, René-Maurice, 149
Gauguin, Paul, 169
Geber, 242
Gelatin, 320
Gellan gum, 100, 102
Gem, 269
Generally regarded as safe, 83
Genistein, 194, 218
Genotoxicity, 61
Gentiobiose, 79
Gentlemen Prefer Blondes, 270
George III, 238
Geraniol, 1
Gerasch, Sylvia, 106
Germ, 60
German Federal Institute for Risk Assessment, 36
Germanium, 23, 262
German Society of Toxicology, 297
Germany, 27, 31, 38, 46, 73, 100, 104, 106, 111, 117, 212, 249, 251, 298, 314
 East, 131, 134
Germinal disk, 112
Ghee, 46
GH *See* Growth hormone, 132
Giddens, Anthony, 6
Gin, 89

Ginger, 73
Ginkgo, 218
Ginkgo biloba, 218
Ginkgolid, 218
Glasow, Niko von, 202
Glasse, Hannah, 73
Glass (material), 5, 28, 29, 96, 243, 269
GLA *See* Linolenic acid, γ-, 76
Globalization, 102, 117, 152
Global warming, 32, 287
Globin, 57, 58
Globulin, 113
Gloop, 320
Glucosamine, 128, 129
Glucosamine/Chondroitin Arthritis Intervention Trial, 128
Glucose, 79–83, 86, 98, 100, 101, 129, 143, 177, 180, 220, 300, 306, 308, 315
Glucose Tolerance Factor, 308
Glucuronic acid, 187
Glurch, 320
Glutamic acid, 88, 95, 96, 149
Glutamine, L-, 174
Glutathione, 223, 224, 257
Gluten, 96
Glycemic index, 60, 82, 315
Glycerol, 43–45, 47, 74, 109, 110, 148, 185, 186, 315
Glycitein, 194
Glycocorticosteroids, 133
Glycogen, 80
Glycopeptides, 191
Glycosaminoglycan, 127
Glycoside, 87, 200, 201, 315
  cyanogenic, 122
Glycyrrhetinic acid, 188
Glycyrrhiza glabra, 187
Glycyrrhizin, 188
Glycyrrhizinic acid, 187–189, 190
Goat's rue, 207
Gödel, Kurt, 288
Godnose, 157
Gold, 146, 173, 269, 277, 300, 301
  mine, 298
Gold, Lois, 247
Goliath effect, 309
Gonzalez, Ana, 309
Good agricultural practice, 62, 63
Goop, 320
Gout, 28
Government Accountability Office, 255
Gram, 320
Grape, 66, 285, 305
  juice, 28, 62, 66, 67, 69
  seed oil, 91
Grapefruit, 66, 84, 85, 86, 192, 226
Graphite, 19, 270
GRAS *See* Generally regarded as safe, 82
Great yellow gentian, 144
Grebe, 245
Greece, 91
Greenhouse effect, 32, 287
Greenland, 30
Greenpeace, 11–13, 31, 247, 259, 262, 289
Green pepper, 42, 226
Green S, 40
Greentech-Vekerd Inc., 26
Gregson, Jessica, 238
Gribble, Gordon W., 12
Grissom, Virgil I., 282
Groundwater, 240, 241, 309
Growth hormone, 132
Guaiacol, 54
Guangxu (emperor), 238
Guanidines, 237
Guar
  bean, 100
  gum, 100, 102, 103
Guinea fowl, 112
Guinea pig, 156, 157, 251
Gulf War
  First, 292, 293
  Syndrome, 292–294
Gull, 245
Gustaf V, 139
Gut, 60, 88, 194
GVL *See* Gamma-valerolactone, 110

H

Haarman and Reimer, 56
Habañero, 230
Hahnemann, Samuel, 145, 311
Hair, 51, 165, 186, 239, 252–254
Haiti, 12, 79
Half-life, 52, 251, 297, 315, 316
Hall, Charles Francis, 238
Hallucination, 171, 232, 233
Hallucinogen, 169
Halogens, 23, 235
Ham, 115
Hamblin, Terry J., 70
Hammer throw, 130
Hamster, 251
Hardening (of oils), 46, 49
Hard tack, 156
Harman, Denham, 222
Harpagophytum procumbens, 141
Hartung, Thomas, 15

Harvesting, 13, 218
Hashish, 134
Hassall, Arthur Hill, 116
Hata, Sahachiro, 137, 239
Hawaii, 182
Haworth, Sir Walter Norman, 158, 162
Hazard, 1, 3, 115, 273
Hazardous waste, 17, 27, 243, 259, 260, 262
HDL *See* High density lipoprotein, 49
Headache, 28, 55, 94, 96, 105, 165, 292, 315
Heart
 attack, 65, 92, 200
 disease, 64, 65, 77, 123, 193, 197, 208, 216
 failure, 139, 193
 ischemic ~ disease, 64, 67, 316
 muscle, 66, 190, 316
 sudden ~ failure, 193, 194, 197
Heat conduction, 268
Heavy metals, 27, 32, 218, 262
Heberden's nodes, 127
Hector Servadac, 280
Heinicke, Ralph, 182
Helichrysum italicum, 74
Helicobacter pylori, 231
Helios Airways, 287
Helium, 24, 280
Heme, 57, 58
Hemicellulose, 107, 108
Hemin, 58
Hemoglobin, 28, 41, 57, 58, 284
Hende, Csaba, 260
Heparin, 127
Hepatic necrosis, 83
Hepatitis, 210, 241
Herb, 137, 151, 166, 170, 176, 205, 207, 218
Herbal product, 217–219, 221
Herbert, Victor, xi
Herbicide, 52, 115
Heroin, 134, 232
Herring, 78
 freshwater, 78
Hesperidin, 84
Hexacyanoferrate(II) ion, 299
Hexanol, 1
Hexenal, *trans*-2-, 91
Hexuronic acid, 157
Hibiscus, 169
High density lipoprotein, 49, 316
High fructose corn syrup, 81
High Production Volume chemical, 9
Hindenburg dirigible, 292
Hindi, 31, 72
Hingis, Martina, 234
Hinkley, 307, 308, 309, 310

Hippocrates of Kos, 148
Hippocratic Oath, 169
Hiroshima, 290
Hitler, Adolf, 131
HIV, 210
HLB *See* Hydrophilic-lipophilic balance, 186
HMF *See* Hydroxymethylfurfural, 5-, 304
Hoagland, Cornelius, 94
Hoagland, Joseph C., 94
Hodgkin, Dorothy Crowfoot, 162
Hodgkin's disease, 249
Hoffer, Abram, 164
Hoffmann, Felix, 199
Hollywood, 282, 286
Holmes, Sherlock, 264
Holst, Axel, 156
Holy Fire, 314
Holy thistle, 144
Homeopathic Pharmacopeia, 146
Homeopathy, 144–147, 311
Honey, 78, 81, 86, 113, 114, 142–144, 320
Hong Kong, 110, 117
Hong Qu, 176
Hooper's Medical Dictionary, 135
Hopkins, Frederick G., 162
Hormone, 67, 105, 130, 131, 133, 193, 194, 308, 316
Horn, 303
Horse, 106, 239
Hortensia, 306
Hoskins, John, 11
HPLC *See* Chromatography, high-performance liquid, 218
HPV *See* High Production Volume chemical, 9
Humic acids, 12
Humming bird, 54
Hungarian Aluminum Production and Trade Company Ltd., 258
Hungary, 10, 26, 27, 86, 103, 212, 216, 220, 230, 235, 238–240, 246, 258, 259, 261, 263, 273, 298
HVG (magazine), 260
Hyaluronan, 127
Hydrangea macrophylla, 306
Hydrocarbon, 109, 110, 148, 284
Hydrochloric acid, 228, 265, 266, 304
Hydrochlorothiazide, 133
Hydrogen, 24, 81, 82, 109, 110, 111
 atom, xii, 48, 280
 bond, 52, 280
 bromide, 236, 237
 chloride, 266
 cyanide, 122, 124, 127, 235–237, 265, 299–302

Index 379

economy, 110
fluoride, 280
ion, 265, 266, 272, 273, 318
-oxygen explosion, 292
peroxide, 223, 257
sulfide, 114, 280, 301, 302
Hydrogenation, 46, 49, 50, 186
Hydrolysis, 43, 143, 188, 201
Hydrophilic-lipophilic balance, 186
Hydroquinone, 142
Hydroxide ion, 168, 272, 273
Hydroxyanthracene, 180
Hydroxyl
  group, 43, 44, 65, 186, 225, 305, 315, 319
  radical, 223, 224
Hydroxylamine, 237
Hydroxymethylfurfural, 5-, 304
Hyperactivity, 39, 106
Hyperkeratosis, 238
Hyperplasia, 52
Hypersensitivity, 151–153, 312
Hypertension, 139, 149
Hypertrophic cardiomyopathy, 193
Hyperventilation, 286
Hypervitaminosis, 162, 165
Hypobromite ion, 236
Hypochlorite ion, 236
Hypoglycemia, 83
Hypoiodite ion, 236
Hypovitaminosis, 164

I

Ibsen, Henrik, xi
Ibuprofen, 91–93, 127
Ice, 18, 19, 20, 172, 268, 269, 280, 281
  synthetic, 20
Ice cream, 101, 194
Idaho Falls, 1
Ifrita kowaldi, 8
IgNobel prize, 53, 54
Ignose, 157
Illés, Zoltán, 260
Immune system, 9, 62, 99, 151, 159, 160, 208, 247, 294, 311
Immunoglobulin, 151
Immunostimulant, 83
Immunotoxic, 294
Impotence, 249, 295
India, 14, 27, 31, 72, 73, 78, 173, 241, 304
India Glycols Ltd., 100
Indicative occupational exposure limit value, 314
Indium, 25
Industrial revolution, 32, 77

Industry
  aerospace, 110
  aviation, 110
  chemical, 1, 2, 5, 13, 14, 16, 18, 38, 52, 103, 296, 318
  construction, 5
  dye, 38
  petroleum and gas, 16
  pharmaceutical, 16, 17, 38, 63, 136, 137, 138, 153, 156, 186, 192
Infant, 39, 41, 42, 50, 62
Infertility, 28, 194
Inflammation, 50, 52, 151, 190
Infusion, 142, 186
Ingestion, 52, 83, 96, 123, 189, 235, 301, 312
Inhalation, 149, 150, 235, 255, 263, 309, 312
inland taipan, 8
Inorganic
  element, 146
  nitrate, 41
  salt, 1, 17
  substance, 276
  waste, 17
Insect, 7, 63, 101, 146, 245, 246, 294
Insecticide, 245
In Situ Iron Studies, 33
Insomnia, 105, 136, 165
Insulin, 83, 100, 308
Intelligence, 30
Interhalogens, 235
International Association of Cancer Victims/ Victors and Friends, 124
International Atomic Energy Agency, 288
International Biozymes Ltd., 124
International Medical Center of Japan, 53
International Olympic Committee, 131
International Space Station, 283
International System of Units, 317
Interventional study, 224, 226, 316, 318
Intestine
  large, 82
  small, 82
Intraperitoneal, 8
Intravenous, 186, 206
Inuit, 77
Iodine, 113
Iodine-123 (isotope), 316
IOELV *See* indicative occupational exposure limit value, 314
Ipoly river, 298
Ipomoea tricolor, 306
Iraq, 238
Iridium, 25
Iron, 113, 216

anode, 168
content, 69, 70, 71
fertilization, 33
(II), 41, 58, 168, 298, 299
(II) hydroxide, 168
(III), 41, 58, 168, 241, 306
(III) hydroxide, 168
(II) sulfate, 298, 299, 300
ions, 32, 57, 114, 158
oxide, 262
(oxyhydr)oxides, 241
production, 262
salt, 30, 31, 32
IRONEX I, 30
Irritability, 105
Isoamyl alcohol, 1
Isoascorbic acid, D-, 156
Isoflavone, 194, 196, 197, 218
Isomer, 245
   cis, 48
   optical, 56
   trans, 48
Isomerization
   cis-trans, 49
Isotope, 24, 292
Israel, 124
Italy, 26, 27, 69, 91, 186, 187, 189, 212, 249, 314
Itching, 165, 312, 314
Itten, Johannes, 57
Ivory Coast, 104

J
Jābir ibn Hayyān, Abu Mūsā, 242
James, Robbie, 193
Japan, 54, 67, 95, 173, 174, 176, 196, 251
Jatropha, 109
Jaundice, 165
Jedlik, Ányos, 212
Jellyfish, 32
Jinping, Geng, 51
John Beard Memorial Foundation, 124, 125
John Birch Society, 126
John Snow, 12
Joint, 28, 127, 128, 129, 130
   inflammation, 141
   pain, 127, 128, 150, 292
Joint Research Centre, 15
Joób, Sándor, 259
Josephine (empress), 253, 254
JRC *See* Joint Research Centre, 15
Junjie, Gao, 51
Jurgens, Anton Paul, 46

K
Kapha, 27
Karil, 71
Karrer, Paul, 162
Kelp, 96
Kelsey, Donald R., 296
Kennedy, Edward, 126
Keratan sulfate, 127, 128
Kerátion, 101
Ketchup, 37, 102, 115
Khamisiyah, 294
Kidney, 28, 51, 52, 105, 165, 166, 191, 216, 229, 276, 294, 308
   damage, 52, 62, 178, 191, 229, 294
   failure, 50
   stone, 51, 52, 53, 165, 216, 276
King, Harold, 206
Kirsch, W., 302
Kisberzseny, 258
Kittler, Glenn, 124
Kjeldahl, Johan, 51
Kohlrabi, 42
Kolontár, 258, 259, 260
Kontinentalverschiebung, 30
Korea, 73, 196
Krebs, Ernst Theodore, Jr., 123–126
Krebs, Ernst Theodore, Sr., 123–125
Krebs, Hans Adolf, 123
Krypton, 25
Kuhn, Richard, 162
Kurara, 101
Kurti, Nicholas, 114
Kuwait, 292
KwaZulu-Natal, 247
Kyanite, 235
Kyrgyzstan, 300

L
L'Absinthe, 171
Lactation, 100, 141, 142
Lactic acid, 44, 45
Lactobacterium, 60
Lactose, 51, 81, 125, 145
   intolerance, 51
Laetrile, 122–126
Lafutidine, 232
Lake Apopka, 247
Lake Michigan, 245
Lake Monoun, 287
Lakritz, 187
Landis, Floyd, 131, 132
Lanthanide, 25
Lanthanum, 25

Larnaca, 287
Laser, 271
Laudan, Rachel, 117
Lausanne, 131
Lavender, 149, 150
　　oil, 149, 150
Laxative, 27, 40, 82, 101, 167, 180, 184, 185, 186, 216, 220
LC_{50} See median lethal concentration, 313
LCt_{50}, 313
LD_{50}, 7, 52, 123, 206, 251, 276, 301, 312, 313, 316
LDL See Low density lipoprotein, 49
Lead, 27, 28, 29
　　battery, 26, 27
　　carbonate, 28
　　exposure, 28
　　(II) acetate, 28
　　intake, 28
　　poisoning, 28
　　re-processing, 26, 27
　　sugar of, 28
　　sulfate, 28
　　sulfide, 28
　　tetraethyl, 29
Leather, 38, 319
Lecithin, 46, 196
Lecoq De Boisbaudran, Paul-Émile, 23
LED See Light emitting diode, 238
Leek, 42
Legume, 62, 101
Lemon, 156, 157, 159, 209
Lenthionine, 255, 256
Lentil, 71
Lentinus edodes, 256
Lerner, Irving J., 122
Lettuce, 42
Leukemia, 139, 249, 255
Leukotriene, 76, 200
Lewis, Gilbert, 266
Lewisite, 239
Lichen, 82
Licorice, 187, 188, 189
Ligand, 316
Light emitting diode, 238
Lignan, 194
Lignin, 54, 107
L'Île mystérieuse, 43
Límite de exposición profesional, 314
Limonoid, 84
Linalool, 150
Linalyl acetate, 150
Linamarin, 123, 127
Lind, James, 156

Linoleic acid, 50, 74–76, 90
Linolenic acid
　　α, 74–77, 196
　　γ, 76
Linseed oil, 76
Lipid, 47, 113, 315
Lipophilicity, 227
Lipovitellin, 113
Lipovitellinin, 113
Liquefied petroleum gas, 284
Liquorice, 187
Lisozyme, 113
Lithium, 25
　　perchlorate, 283
Litvinenko, Alexander, 301
Live and Let die, 278
Liver, 61, 62, 80, 83, 84, 103, 105, 165, 167, 176, 203, 218, 294, 308
　　chicken, 71
　　damage, 165, 218
Livetin, 113
　　γ, 113
　　β, 113
Locust, 101
Locusta of Gaul, 237
Locust bean, 101
　　gum, 100, 101
LOHAFEX experiment, 32, 33
Lomborg, Bjørn, 287
London, 12, 73, 103, 106, 116, 301
Loos, Anita, 270
Loren, Karl, 130
Lotus effect, 243
Louse, 245
Lovastatin, 177
Love Canal, 249
low density lipoprotein, 316
Low density lipoprotein, 49, 113
Lowe-Porter, H. T., 256, 267, 275
Lowry, Thomas Martin, 266
Loyola University, 126
LP disc, 255
LPG See Liquefied petroleum gas, 284
LSD. See Lysergic acid diethylamide 232
Lubricant, 100, 186
　　natural, 127
Lucas, Edgar, xi
Ludwigshafen, 249
Luminal, 51
Lunar landing, 286
Lung, 58, 103, 106, 211, 286, 308, 309
Lutein, 226
Lycopene, 226
Lysergic acid diethylamide, 8

M

Macaca mulatta, 251
Mackerel, 78
Macrolide, 191
MAC-waarde *See* Maximaal Aanvaarde Concentratie, 314
Madagascar, 54, 139
Madagascar periwinkle, 139
Maersk, 107
Magnan, Valentin, 171
Magnesium, 42, 45, 91, 111, 216, 230
   carbonate, 111
   ion, 215, 230, 306
Magnet, 277, 278
Magnetic resonance imaging, 278
Magnetite, 277
Magyar Narancs (magazine), 260
Maitotoxin, 8
Maize, 63
MAK *See* Maximale Arbeitsplatz-Konzentration, 314
Malaria, 139, 145, 198, 199, 210, 245, 247
Maltose, 81
MAL *See* Hungarian Aluminum Production and Trade Company Ltd., 258
Mammal, 245
Mammillary layer, 112
Man of Bicorp, 78
Manganese, 99, 113, 215
   dioxide, 309
Manihot esculenta, 123
Manioc, 123
Manner, Harold W., 126
Mannitol, 36, 81, 82, 83
Mannose, D-, 100
Mann, Thomas, 255, 256, 267, 275
Manufacturing
   brick, 262
   building material, 5
   clothes and shoe, 5
   furniture, 5
   margarine, 49
   paper, 16
   vehicle, 5
Manure, 41, 54
Margaric acid, 45
Margarine, 45, 46, 47, 49, 50, 77, 115
Marggraf, Andreas, 79
Marijuana, 170, 232
Marker (substance), 90, 224, 237
Marketing, 6, 46, 60, 63, 90, 109, 117, 120, 122, 124, 130, 178, 179, 182, 201, 208, 217, 219, 223
Markhot Ferenc Hospital, 235
Markov, Georgi, 185, 301
Marmite, 96
Martin, John, 30
Marzipan, 122
Masala, 73
   chaat, 73
   chicken tikka, 73
   garam, 73
   tandoori, 73
Masking agent, 133
Masry, Ed, 307
Mass spectrometry, 131
Matricaria recutita, 151
Matricin, 153
Matthew
   Gospel of, 15, 101
Maximaal Aanvaarde Concentratie, 314
Maximale Arbeitsplatz-Konzentration, 314
Mayonnaise, 102, 114
McDonald's, 117
McLachlan, Ed, 2
McNaughton, Andrew George Latta, 124
McNaughton, Andrew R.L., 124
McQueen, Steve, 126
Meadowsweet, 199
Meat
   canned, 44
   consumption, 77
   cured, 57, 60
   dry, 57
   kangaroo, 50
   minced, 59
   poultry, 57
   raw, 57, 58
Media, 1, 2, 4, 5, 10, 11, 30, 31, 41, 43, 88, 130, 132–134, 184, 193, 223, 232, 238, 263, 268, 272, 273, 287, 289, 296, 298
Median effective dose, 312, 313
Median lethal concentration, 313
Median lethal dose, 312, 313
Mediterranean, 78, 90, 93
   diet, 64, 66, 69, 90, 92, 317
   Sea, 90, 317
Mège-Mouriès, Hippolyte, 45
Melamine, 50, 51–53, 116, 237, 255
Melanine, 51
Melon, 42, 226
Meloxicam, 127
Melyrid beetle, 8
Membrane, 97, 319
   filtration, 13
   inner, 112
   mucous, 150, 238

Index 383

outer, 112
porous, 112
semi-permeable, 319
Mendeleev, Dmitri Ivanovich, 23–26
Mendelevium, 26
Mental dysfunction, 106
Mentha × piperita, 54, 55
Menthol, 54, 56, 57
  (-)-, 56, 57
Mercadé-Prieto, Ruben, 114
Mercuric sulfide, 242, 243
Mercurisorb™, 242
Mercury, 8, 242, 243, 262
  (II) chloride, 8
Messalina, Valeria, 237
Meta-analysis, 42, 128, 317
Metabolism, 38, 82, 84, 86, 89, 167, 203, 220, 308
Metacinnabarite, 242, 243
Meta-cresol, 56
Metal, 27, 146, 242, 261–263, 269, 270, 277, 295, 298, 303, 306
  noble, 23
Metamphetamine, 233
Metastable, 270, 271
Metformin, 207
Methane, 82, 280, 284
Methanol, 88, 89, 110, 171
  economy, 110
Methemoglobin, 41, 58
Methemoglobinemia, 58
Methionine, L-, 257
Methyl alcohol *See* Methanol, 88
Methylascorbinogen, N-, 257
Methylation, 256, 257, 258
Methyl chloride, 12
Methylsulfonylmethane, 129
Metmyoglobin, 58, 60
Metronidazole, 191
Mevastatin, 177
Mexico, 104, 122
Mg/kg of body weight, 52, 83, 197, 206, 251, 276, 296
Microgram, 320
Microinfarction, 67
Microorganism, 17, 35, 63, 268
Microrisk, 4
Middle Ages, 41, 78, 134, 314
Midgley, Thomas, 29
Migraine, 149
Milan, 249
Milk
  breast, 166
  condensed, 114

powder, 50
sugar, 51, 81
tainting scandal, 51
Milk-alkali syndrome, 229
Milk thistle, 167, 218
Milligram, 320
Mineral, 27, 28, 60, 99, 100, 111, 143, 146, 196, 209, 212, 214–216, 219, 238, 241, 260
Ming dynasty, 176
Mining, 13
  coal, 5
  offshore oil and gas, 5
  open-pit, 5
Minot, George R., 162
Mircene, 57
Miscanthus, 108
Miscarriage, 28, 194
Mitochondrium, 35, 36, 300
Mode of action, 88, 155
Mofette, 287
Mojave desert, 307
Mol, 317
Molar concentration, 317
Molasses, 71, 81, 98–100
Mold, 35, 60, 61, 94, 190
Mol/dm^3, 273, 317
Molina, Mario J., 22
Molybdenum, 25
Monacolin K, 176, 177
Monascus ruber, 176
Monell Chemical Senses Center, 91
Monoglycerides, 43
Monosaccharide, 79, 98, 122
Monosodium glutamate, 95
Monroe, Marylin, 270
Monsanto, 249
Montreal Protocol, 22
Morinda, 182, 183
Morinda citrifolia, 182
Morning glory, 306
Morocco, 117
Morphine, 134–136, 139, 199, 205, 232
Morren, Charles François Antoine, 54
Morrone, John A., 124
Mortality rate, 5
Mosquito, 198, 245
Moss Landing Marine Laboratories, 30
Mother tincture, 146, 147
Mouse, 251
Mouthwash, 82
MRI *See* Magnetic resonance imaging, 278
MSG *See* Monosodium glutamate, 95
MSM *See* Methylsulfonylmethane, 129

Mt. Everest, 287
Mt. Pinatubo, 30
Mucopolysaccharide, 111
MUFA *See* Fatty acid, monounsaturated, 74
Müller, Paul Hermann, 245
Murphy, William P., 162
Murraya koenigii, 73
Muscle
    atrophy, 127
    growth, 220
    pain, 150, 165, 177, 292
    relaxant, 206
    smooth, 106, 151
    tissue, 57
Mushroom, 42, 256
    poisoning, 167
Muslims, 78
Mussel, 239
Mussolini, Benito, 186
Mustard
    agent, 294
    seed, 73
Must (grape juice), 66, 285
Myasthenia gravis, 293
Mycotoxin, 60, 62, 63, 115, 166
Myers, John Peterson, 296
Myoglobin, 57, 58, 59, 60

N

Nanogram, 320
Naples, 245
Napoleon, 79, 199, 237, 252–254
Napoleonic Wars, 79
Napoleon II, 253, 254
Napoleon III, 45
Napoleon Murdered?, 252
Narcotic, 133, 166, 205, 232–234, 314
NASA, 282
National Cancer Institute, 122, 126
National Park of Hortobágy, 27
Natural
    background, 289
    catastrophe, 290
    compound, 139, 191, 192, 205, 207
    decomposition, 45
    gas, 110, 284, 287, 302, 309
    laws, 268
    material, 6
    process, 12, 45, 241, 309, 310
    product, 6, 10, 85, 177
    source, 13, 35, 56, 138, 227
    substance, 7, 9, 27, 37, 44, 45, 60, 296
    supply, 28

Naturally-Occurring Radioactive Materials, 263
Nausea, 165, 202, 315
NCI *See* National Cancer Institute, 122
N'Dungu, Kuria, 309
Nelson, Lord, 156
Neohesperidin DC, 87
Neosalvarsan, 137
Neotame, 87
Nero, 237, 252
Nerve agent, 293, 294
Nerve gas, 293, 301
    poisoning, 294
Nervousness, 105
Neste Oil, 109
Netherlands, 46, 73, 104, 111, 117, 189, 249
Neuron, 230, 231, 245
Neurotoxic, 170, 171, 294
Neurotransmitter, 105, 139, 141, 154, 155, 174, 205, 245, 293
Neutralization, 223, 228
Neutron, 253, 291
    activation analysis, 253
Nevada, 125
Newborn, 297
New Mexico, 290
Newsweek, 298
New World, 79, 305
New York (city), 122
New York (state), 29
New Zealand, 1, 51
Nickel, 46, 49, 277
Nicosan, 141
Nicotine, 7, 251
Nicotinic
    acid, 164, 165
    amide, 164, 165
Nigeria, 86
Nilson, Lars Fredrik, 23
Niobium, 25
Nitrate, 41, 42
Nitric acid, 265, 276
Nitric oxide, 60
Nitriles, 237
Nitrite, 41, 60
Nitrocellulose, 249
Nitro (city), 249
Nitrogen, 38, 41, 42, 51, 59, 108, 127, 205, 206, 282, 300
Nitrosamine, 60
NMR *See* Nuclear magnetic resonance, 278
No adverse effect level, 313
NOAEL *See* Non-observable adverse effect level, 52

Nobel Prize, 22, 53, 139, 156, 158, 159, 162, 176, 191, 200, 245
Noble gases, 23
Nocebo, 121, 136, 203
No effect level, 313
NOEL *See* non-observable adverse effect level, 313
Nomen est omen, 232, 235
Nomilin, 84
Non-Hodgkin's lymphoma, 249
Noni, 182, 183, 184
Non-observable adverse effect level, 313
Non-observable effect level, 313
Non omne, quod nitet, aurum est, 215
Non-steroidal anti-inflammatory drugs, 92, 127, 198
Norepinephrine, 154, 158
Norman, Chris, 268
Norman (city), 309
Normann, Wilhelm, 46
NORM *See* Naturally-Occurring Radioactive Materials, 263
North Carolina, 296
North Pole, 22
Norway, 14, 59, 304
Nosodes, 146
NSAID *See* Non-steroidal anti-inflammatory drugs, 127
Nuclear
　accident, 290
　bomb, 2
　energy, 110, 288
　explosion, 290, 291
　fission, 291, 292
　fusion, 291, 292
　industry, 29, 289
　magnetic resonance, 278
　power plant, 5, 289, 290
　reactor, 253, 290, 292
　weapon, 158, 292
Nucleic acids, 107, 223
Nucleosides, 95
Nutella, 114
Nutmeg, 73
Nylon, 237, 302, 304

O

Obesity, 40, 81, 82, 86, 117, 209
Observational study, 67, 224, 226, 318
Occupational Safety and Health Administration, 255
Ocean fertilizing experiment, 30, 32
Ochratoxin, 60, 62
　A, 62

Oetker, August, 94
Ogiri-igbo, 185
Oklahoma, 123, 309
Olah, George, 110
Old World, 79, 230
Olea europaea, 90
Oleic acid, 74, 90
Oleocanthal, 91, 92, 93
Oligosaccharide, 79
Olive
　oil, 66, 90, 91, 115, 148, 149
　　extra virgin, 91
　　refined, 90, 91
　　virgin, 90, 91, 93
　tree, 90
Onion, xi, 42, 71
Onsen tamago, 114
Oobleck, 320
Ooze, 320
Opium Wars, 173
Opren, 4
Oradea, 26, 27
Orange, 37, 66, 159
Orchidaceae, 54
Organic
　acid, 44, 189, 229, 305
　compound, xii, 12, 13, 29, 213, 239, 315, 318, 320
　farming, 27
　gardening, 6, 7
　material, 30, 54, 267
　matter, 79, 241
　phosphate, 247
　solvent, 17, 18, 90, 175
　substance, 51, 214, 239, 256, 276
　substrate, 237
　vegetables, 42
　waste, 17
Organization for the Prohibition of Chemical Weapons, 237
Orpiment, 238
O sancta simplicitas!, 217
OSHA *See* Occupational Safety and Health Administration, 255
Osmium, 25
Osmosis, 97, 215
　reverse, 319
Osteoarthritis, 127, 128, 130
Osteoporosis, 105, 195, 209
OTA *See* Ochratoxin A, 62
Oudry, M., 175
Ovalbumin, 113
Ovoglobulin, 113
Ovomucin, 113

Ovomucoid, 113
Ovotransferrin, 113
Oxalic acid, 71, 274–277, 320
    poisoning, 276
Oxidative
    damage, 222–224, 227
    stress, 222, 224, 226
Oxonium ion, 272
Oxygen, 21, 30, 32, 38, 58, 80, 133, 168, 208, 210, 225, 267, 268, 280, 282–287, 292, 299, 300, 318
    singlet, 257
Oxymyoglobin, 58
Oxyuranus microlepidotus, 8
Oyster sauce, 95
Ozone, 13, 21, 22, 23, 32, 257, 258, 267, 268, 318
    hole, 21–23, 267
    layer, 21, 267
    shield, 21
Ozonolysis, 13, 318

P

Pacific Gas & Electric Company, 307
Pacific Ocean, 37
Pacific Tropical Botanical Garden Bulletin, 182
Pacific yew, 139
Paclitaxel, 139, 141, 208
PAHO *See* Pan American Health Organization, 12
Painkiller, 4, 205, 232
Paint, 28, 38, 96, 100, 253, 293, 309, 311, 320
Pakistan, 148
Palisade layer, 112
Palla, János, 262
Palladium, 25
Palm, 109
Palmitic acid, 45
Panacea, 179, 180, 198
Pan American Health Organization, 12
Pan American Mines, 124
Panchakarma, 27
Pangamic acid, 123, 125
Panicum virgatum, 108
Pantothenic acid, 162, 164
Papaverine, 139
Papaver somniferum, 205
Papaya, 226
Paper, 12, 17, 18, 38, 54, 91, 100, 262, 296, 303
    heat-responsive, 295
    industry, 16
    mill, 16

Paprika, 115, 116, 158, 159, 230
Paracelsus, 275
Paracetamol, 8
Paraffin wax, 110
Paraformaldehyde, 255
Paramagnetic, 277, 278
Parasite, 84, 146, 198
Paresthesias, 314
Paris, 103
Pariza, Michael, 50
Parkes, Alexander, 302
Parmesan syndrome, 96
Paroxetine, 121
Parsley, 42, 123
Parsnip, 42
Parts
    per billion, 312
    per million, 312
    per notations, 312, 318, 319
    per trillion, 249, 312
Pastry, 116
Pathogen, 99, 112, 146, 149, 190–192, 198, 209, 214, 215, 241
Pauling, Linus, 158, 159, 164
PB *See* Pyridostigmine bromide, 293
PCDF *See* pentachlorodibenzo[b,d]furan, 2,3,4,7,8-, 251
Pea, 42, 276
Peanut, 62, 76, 91, 108, 320
    butter, 209
    oil, 46, 91, 149
Pectin, 88, 89
Peer Gynt, xi
Pelargonidin, 305
Penicillin, 17, 139, 190, 191, 239
Penicillium notatum, 139, 190
Pentachlorodibenzo[b,d]furan, 2,3,4,7,8-, 251
Pentachlorophenol, 100, 102
Pentobarbital, 167
Peonidin, 305, 306
Peppermint, 54
    oil, 55
Pepperoncini, 230
Pepsin, 144
Perch, 78
Peregrine falcon, 246
Perfume, 55, 148
Periodic table, 23, 25, 26, 298
Peristalsis, 144
Peroxidase, 223, 224, 258
Persia, 78
Peru, 11, 12, 232
Pesticide, 7, 11, 32, 102, 237, 246, 247, 249, 293, 294

natural, 7
organophosphate, 294
pyrethroid, 247
residue, 218, 247
Pet, 106, 250
Pet food scandal, 52
Petroleum, 109, 302, 303, 304
crude, 106, 109
refining, 110
PET *See* Poly(ethylene terephthalate), 213
PG&E *See* Pacific Gas & Electric Company, 307
pH, 36, 38, 53, 58, 113, 209, 210, 211, 228, 229, 230, 261–263, 272–274, 286, 305, 306, 318
Phacelia campanularia, 306
Phacelianin, 306
Phar Lap, 238, 252
Pharmaceutical
company, 119, 120, 199, 202
research, 119
Pharmacodynamics, 203, 318
Pharmacokinetics, 203, 318
Pharmacological study, 64, 318
Pharmacology, 205, 311
Phase diagram, 19, 20, 270, 271
Pheasant, 246
Phenol, 40, 255
Phenolic, 319
Phenolic, 91
Phenolphthalein, 40
Phenylalanine, 83, 88
Pheophytin, 91
Philadelphia, 91
Philippines, 30
pH Miracle, 208
Phosphodiesterase, 220
Phosphoric acid, 94, 265, 274
Phosphorus, 42, 113
white, 8
Phosphovitin, 113
Photosynthesis, 32, 80
Phylloxera plague, 170
Phytoestrogen, 193–197, 296, 297
Phytoplankton, 30
Phytosterol, 91
Phytotherapy, 84, 318
Picogram, 320
Picophytoplankton, 32
Pilchard, 78
Pimentel, David, 108
Pimpinella anisum, 169
Piper, Peter W., 36
Piria, Raffaele, 199

Pirrhotite, 277
Pistachio, 62
Pitta, 27
Pitt, William, 28
Placebo, 121, 128, 134–137, 144, 145, 147, 150, 155, 159, 160, 202, 203, 314
Plague, 148
Plankton, 30, 32
Plant
medicinal, 84, 156, 219
oil, 50, 76, 77, 109
Plastic, 2, 17, 20, 27, 51, 237, 255, 282, 283, 295, 296, 302, 303, 304
Platinum, 25
Pneumonia, 190
Poirot, 81
Poison, xi, 7, 27, 60, 61, 166, 167, 206, 237, 276, 294, 300, 301, 312
Poisoning, 162, 167, 185, 220, 251, 252, 254, 259, 284
Poland, 95
Polarstern, 31, 32
Pollen, 143, 151, 152, 153
Polonium, 24
Polonium-210 (isotope), 301
Polyacrylonitrile, 303
Polycarbonate, 295, 296
Polyethylene, 302, 303, 304
Poly(ethylene terephthalate), 213, 303
Polylactic acid, 303
Polymer, 9, 302, 303, 304
natural, 303, 304
Polynesia, 78
Polyphenol, 65–67, 86, 224
Polypropylene, 303
Polysaccharide, 79, 88, 100, 101
Polystyrene, 302, 303
Poly(vinyl chloride), 303
Pomace, 90
Pomegranate, 222
Pomelo, 84
Popeye, 41, 69, 70, 71, 276
Poplar, 108
Poppy, 139, 205, 320
Pork, 57, 62
Porphyria cutanea tarda, 249
Post-traumatic stress disorder, 293
Potassium, 35, 42–45, 94, 188, 299, 300, 301, 306
cyanide, 276, 300, 301
hexacyanoferrate(II), 301
hydrogen tartrate, 94
nitrate, 41
sorbate, 35
tartrate, 94

Potato, 42, 108, 116, 159
Potentization, 145, 146, 147
Ppb *See* Parts per billion, 7
Ppm *See* Parts per million, 28
Ppt *See* Parts per trillion, 36
Precautionary principle, 14, 197, 297
Precious stone, 27, 268, 270–272
Pregnancy, 5, 77, 106, 161, 202
Premenstrual syndrome, 106, 147, 163
Preservative, 35, 41, 46, 85, 94, 95, 115, 166, 176, 253
Priestley, Joseph, 212
Product stewardship, 14
Prohibition, 170
Prontosil rubrum, 38
Prooxidant, 226
Propane, 110, 284
Propolis, 192
Prostaglandin, 76, 200
Protactinium, 24
Protein, xii, 17, 46, 51, 57, 58, 62, 83, 88, 95, 102, 107, 111, 113, 114, 116, 127, 128, 130, 132, 139, 144, 151, 152, 158, 185, 190, 193, 196, 203, 211, 223, 302, 311, 319
Proteoglycan, 127
Protocyanin complex, 306
Prout's hypothesis, 24
Provitamin, 226
Prussian blue, 235
Pseudobase, 305, 306
Pseudomonas, 59
Psoriasis, 163
Psychosis, 155, 315
PTFE, 303
PTSD *See* Post-traumatic stress disorder, 293
Puberty, 197
Pudding, 44
Puerta, Antonio, 193
PUFA *See* Fatty acid, polyunsaturated, 74
Pungency, 91, 230
Purine, 28
PVC, 302, 303, 304
Pyridostigmine bromide, 293
Pyridoxine, 164

Q
Qīrāṭ, 101
Quail, 246
Quality factor, 18
Quaternary ammonium salts, 85, 86
Quercetin, 1, 66
Quicksand, 312
Quina *See* Cinchona, 198

Quinic acid, 306
Quinine, 199
Quinolones, 191

R
Rabbit, 250
RAC-GWVI *See* Research Advisory Committee on Gulf War Veterans' Illnesses, 293
Radiation, 21, 29, 215, 225, 288, 289, 301
  infrared, 32, 287
  ionizing, 8
  sickness, 288
Radioactivity, 215, 262, 263
Radish, 42
Radium, 215, 263
Radon, 215, 263
Rain forest, 108
Raisin, 62
Rapeseed, 109
  oil, 109
Rapid Alert System for Food and Feed, 63
Rappe, Christopher, 103
Rare earth metal, 277
RASFF *See* Rapid Alert System for Food and Feed, 63
Rat, 52, 67, 87, 88, 123, 149, 171, 251, 255, 296, 309, 312
Raw materials, 51, 54, 57, 106–109, 139, 146, 181, 196, 218, 261, 262
Rāzī
  Muhammad ibn Zakariyā, 103
RDA *See* Recommended daily allowance, 161
REACH *See* Registration, Evaluation, Authorisation, and Restriction of Chemicals, 13
Reaction
  equilibrium, 272
  mechanochemical, 243
Reagan, Ronald, 119
Realgar, 238
Receptor, 67, 136, 147, 149, 150, 170, 194, 201, 203–206, 229, 230, 232, 245, 316
Recommended daily allowance, 161
Red Cross, 210
Red mud, 13, 258–263, 273
Red sludge, 258–263
Registration, Evaluation, Authorisation, and Restriction of Chemicals, 13
Reiff, Fred M., 12
Renaud, Serge, 64
Renewables, 110, 304

Index 389

Research Advisory Committee on Gulf War Veterans' Illnesses, 293
Research octane number, 110
Research Triangle Institute, 296
Resin, 17, 51, 255
　epoxy, 295
Resiniferatoxin, 232
Respiratory
　infection, 82
　tract, 262
Responsible care, 14
Resveratrol, trans-, 66, 67, 68, 69, 194, 222
Rhenium, 24
　diboride, 271
Rhesus monkey, 251
Rheumatism, 137
Rhine Valley, 60
Rhodium, 25
Rhubarb, 276, 277
Riboflavin, 164
Rice, 18, 62, 115, 177, 178, 241
　red, 176, 177, 178
　red yeast, 176, 177
　wine, 176, 177
Richardson, John, 125
Ricin, 185, 301
Ricinin, 185
Ricinoleic acid, 185, 186
Ricinus communis, 184
Rio Grande do Sul, 304
Risk, 2, 3, 7, 9, 10, 12, 14, 28, 29, 35, 37, 40, 50, 52, 59, 62, 63, 67, 77, 82, 92, 96, 97, 106, 112, 123, 154, 165, 167, 176, 193, 196, 209, 213, 219, 221, 224, 227, 228, 232, 239, 255, 256, 284, 288, 289, 297, 301, 308
　assessment, 4, 15, 102, 103, 115, 162, 251, 288, 289
　factor, 49, 64, 65, 127, 293
　perception, 5, 6
　relative, 65, 319
Risk-cost-benefit analysis, 4, 205
RJ-4, 111
RNA, 95
Roberts, Julia, 307
Robin, 245
Robusta, 105, 116
Rock climbing, 3
Rockefeller Foundation, 245
Röczei, István, 262
Rodent, 88
Roman Empire, 28
Romania, 27, 298
RON *See* Research octane number, 110

Rosa gallica, 306
Rosehip, 142, 156, 159
Rothamsted, 251
Rovida, Costanza, 15
Rowland, F. Sherwood, 22
Royal Navy, 156
Royal Nedalco, 108
Royal Society, 198
RSSL, 51
Rubber
　natural, 303
　synthetic, 302
　vulcanized, 302, 303
Rubidium, 25
Ruby, 271
Russia, 104, 288, 314
Ruthenium, 25

S

Saal, Frederick S. vom, 296
Sabatier, Paul, 46
Saccharin, 83, 87, 88
Saccharose, 79, 80, 87
Saccharum saturni, 28
Safety factor, 62, 171, 314
Safflower, 130
Saffron, 115
Sainsbury's, 36
Saint Anthony's Fire, 314
Saint Helena
　island of, 252, 253
Salad dressing, 102
Salad rocket, 42
Salicin, 139, 199, 200
Salicylic acid, 7, 94, 199, 200
Saligenin, 200
Salinization, 263
Saliva, 144, 166, 230
Salix alba, 199
Salmon, 78, 103
Salt
　mine, 98
　mined, 96, 97
　pan, 98
　sea, 96, 97
　table, 18, 52, 97, 269, 299
Saltpeter, 41
　Chilean, 41
Salvarsan, 137, 190, 239
Samarium, 278
SAM-e *See* Adenosylmethionine, S-, 130
SAM *See* Adenosylmethionine, S-, 130
Sand, 269, 293, 320
San Francisco, 123

Sanger, Frederick, 158
Sanlu brand, 52
Sansa, 90
Sapa, 28
Saponification, 43
Saponin, 187
Sardines, 115
Sarin, 293, 294, 301
Scandium, 23
Scheele, Carl Wilhelm, 51, 265, 275
Schönbein, Christian Friedrich, 267
Science (journal), 298
Scoville scale, 230
Scurvy, 156, 158, 160
Scythian root *See* Licorice, 187
Seaweed, 96
Seconal, 313
Sedative, 150, 167, 170, 198, 201, 313
Segar, Elzie Crysler, 69
Sekisui Chemical Company, 54
Selenium, 223
Self-dissociation, 272, 273
Semicarbazide-sensitive amino oxidase, 258
Semiconductor, 238
Serbia, 298
Serrano, 230
Sertürner, Friedrich, 199
Sesame oil, 46
Sesquiterpene lactone, 151, 152
Seveso, 102, 249
Shakespeare, William, 275
Sharpe, Richard M., 295
Shaving foams, 255
Sheep, 131, 194, 250
Shennong (Emperor), 173
Shermer, Michael, 10
Shibata, Keita, 306
Shibata, Yuji, 306
Shiitake, 255, 256
Ship
    building, 4
    cargo, 107
    container, 107
    research, 31
Shopping receipt, 295, 296
Shrimp, 239
Sibutramine, 220
Sickling, 141, 319
Side effect, 4, 5, 82, 84, 87, 94, 95, 102, 121, 122, 127, 131, 133, 136, 144, 146, 147, 151, 162, 165, 166, 177, 178, 189–193, 196, 198, 202, 203, 205, 207, 226, 229, 245
Sigmoid curve, 312

Sildenafil, 220, 221
Silent killer, 284
Silent Spring, 246
Silica gel, 242
Silicate, 32, 33, 235, 242
Silicon, 23
Silicone, 303
Silk, 38, 303
Silver, 143, 146, 242, 269, 277, 298
Silybum marianum, 167, 218
Silymarin, 218
Singapore, 110
Sirtuin, 67
Skate, 18, 19, 20
Skin
    abnormality, 165, 182, 238
    absorption, 150
    discoloration, 165
    infection, 40
    inflammation, 151
    irritation, 151
    lesion, 94
    massage, 149
    softening, 186
Skynet, 278
Sleeping sickness, 190, 239
Slovakia, 95
Slovenia, 26
Slow Food, 116, 117
Small and medium-sized enterprises, 14
SME *See* Small and medium-sized enterprises, 14
Smoking, 4, 35, 64, 115, 224, 256
Snake, 146
    bite, 7
    poison, 7
    small-scaled, 8
Snow, John, 12
Soap, 45, 262
Soda, 212
    baking, 93, 94, 228, 229
    water, 212
Sodium, 45, 181, 216, 261, 277
    acetate, 93, 302
    aluminum hydrosilicates, 262
    benzoate, 36
    carbonate, 212
    chloride, 96, 97, 168, 228, 229
    chlorite, 210, 211
    cyanide, 8
    di~ hydrogen phosphate, 94
    hydrogen carbonate, 53, 94, 228, 229
    hydroxide, 236, 261–263
    hypochlorite, 181

Index 391

ion, 97
methylate, 50
molybdate, 8
nitrate, 41
nitrite, 7
silicate, 112
succinate, 188
thiosulfate, 242
Soft tissue sarcoma, 249
Soil, 26, 41, 42, 54, 62, 63, 79, 108, 163, 217, 245, 251, 263, 306, 309
Sola dosis facit venenum, 7
Soluble glass, 112
Soman, 293
Somatotropin, 132
Some Hearts are Diamonds, 268
Somlóvásárhely, 258
Sommer, Frank, 295
Sorafenib, 137
Sorbic acid, 35, 95
Sorbitol, 81–83, 296
Sorrel, 276
  wood-, 275, 276
Sorrowing Old Man, 153
Southampton report, 39
Southampton University, 39
South Pole, 22
Soviet Union, 185, 290
Soy, 62, 193, 194, 196, 197
  sauce, 95
Soybean, 193–197, 218
  oil, 46
Spacecraft, 282, 286
Spaceship, 271, 282
  Apollo 1, 282
  Apollo 13, 286
Space station
  Mir, 283
  Salyut 6, 283
Spain, 78, 91, 104, 106, 187, 314
Spanish flu, 123
Spasm, 55, 139, 167, 182, 314
Spasmolytic, 229
Speciation, 238
Sperm motility, 106
Sperm whale, 58
Sphingomonas elodea, 102
Spice, 71–73, 170, 176, 219
Spinach, 41, 42, 57, 69, 70, 71, 226, 276
Spiraea ulmaria, 199
Spirit
  cider, 89
  fruit, 89
  fruit marc, 89

grape marc, 89
perry, 89
wine, 89
Spitteler, Adolph, 302
Sports, 106, 130–234, 290, 295
Squalene, 91
Squamous cell carcinoma, 255
Squash, 42, 226
Stabilizer, 45
Stanford University, 126
Starch, 81, 100, 107, 177
Stasi, 134
Stearic acid, 45, 74
Steelmaking, 110
Stereoisomer, 156, 170
Steroid
  anabolic, 131, 133, 134, 220
Stevia, 87
Stilbenes, 194
Stimulant, 103, 133, 182, 205, 220, 232,–234
St. John the Baptist, 100
Stomach
  ache, 28, 103, 141, 228, 229
  irritation, 94, 199
  mucus, 230, 231
  ulcer, 188, 229, 231, 232, 252
Stone Age, 2, 28
Stone, Edward, 198
Stratosphere, 21
Straw, 41, 82, 205, 240
Strawberry, 189
Streptococcus mutans, 82
Stroke, 5, 200, 208
Strontium, 25
Strychnine, 205, 301
Styria, 239
Subcutaneously, 8
Sucralose, 83, 87
Sucrose, 79, 80–82, 86, 87, 98, 99, 143, 145, 188, 189, 190
Suffocation, 285, 286, 287, 300
Sugar
  alcohol, 81–83, 86
  beet, 79, 98
  birch, 80, 82
  brown, 81, 98–100
  cane, 45, 79, 98, 143
  caramelized, 116
  consumption, 81
  fruit, 79, 81
  grape, 79, 81, 98, 285
  invert, 81, 143
  poisoning, 86
  raw, 81

table, 79, 81
white, 98–100
Sugar beet, 79, 98, 108
Sugar cane, 71, 78, 79, 98, 99, 108, 304
Suicide, 154, 155, 283
Sulfonamide, 38
Sulfur, 130, 146, 215, 223, 242, 243
    dioxide, 35
Sulfuric acid, 27, 266
Sulfur-mercury theory, 242
Sunburn, 4
Sunflower, 76, 109
    oil, 90, 91, 93, 148
    seed, 62, 130
Sunset yellow, 38
Superoxide
    dismutase, 223, 224
    radical, 223
Surface tension, 186, 242, 243
Surgery, 4, 53, 127, 206, 278
    plastic, 7
Süssholzwurzel, 187
Sustainability, 304
Sutton, Mike, 70
Sweat, 166
Sweden, 12, 23, 139
Sweetener
    artificial, 83, 86, 87, 296
    bulk, 81
    intense, 81
Sweetness, 78, 81, 83, 87, 190
Sweet potato, 42
Sweet root *See* Licorice, 187
Swenberg, James A., 255
Swindled. The dark history of food fraud, from poisoned candy to counterfeit coffee, 115
Switchgrass, 108
Switzerland, 14, 100, 129, 131, 132, 304
Synthetic
    analog, 163
    compound, 205
    material, 208, 303
    medicine, 137, 141, 205, 208
    pollutant, 247
    procedure, 141, 199
    process, 140
    product, 191
    substance, 220
Syphilis, 137, 190, 239
Szamos river, 298
Szeged, xii
Szent-Györgyi, Albert, 156–159, 162, 164, 230
Szilard, Leo, 290

T
Tabasco, 230
Tadeka, Kosaku, 306
Tahitian Noni International, 182
Takasago, 56
Tamil, 71
Tang dynasty, 176
Tannic acids, 66, 319
Tannin, 1, 173, 174, 319
Tantalum, 25
Tapioca, 123
Tarragon, 170
Tartaric acid, 44, 45, 94, 229, 265, 266
Tartrazine, 39
Taste bud, 83
Taxus brevifolia, 139
TCDD *See* Tetrachlorodibenzo[b,e][1,4]dioxin, 2,3,7,8-, 250
TDI
    tolerable daily intake, 62, 171, 296, 314, 319, 320
Tea, 103
    black, 173–175
    green, 172–175, 222, 225
    herbal, 141–144, 173
    nursing, 141
    oolong, 174, 175
    plantation, 173
    pu-erh, 174
    shrub, 104
    white, 174, 175
Technetium, 24
Teflon, 303
Television, 6, 29, 96, 288, 298
Tellurium, 25
Tennison, Lord, 28
Teratogenic, 62, 294
Terminator series, 278
Terpene, 57, 83, 111
Testosterone, xii, xiii, 130, 131, 247, 295
Tetétlen, 26
Tetracyclines, 191
Tetrahydrocannabinol, 170
Tetramethyl-1,3-cyclobutanediol, 2,2,4,4-, 296
Tetrodotoxin, 251
Textile, 255
Thalidomide, 202
Thallium, 262
Thaumatin, 87
Theaflavin, 173
The Angel Makers, 238
Theanine, L-, 174
The Art of Color, 57
Thearubigin, 173

Index

The Community Environmental Legal Defense Fund, 9
The Daily Telegraph (newspaper), 295
The Death Dealer, 302
The Forme of Cury, 72
Thein, 173, 175
The Independent (newspaper), 35
The Magic Mountain, 256, 267, 275
The Magnificent Seven, 126
The Mysterious Island, 43
Theobromine, 105, 106, 173
Theophrastus, 187
Theophylline, 105, 106, 173
Therapeutic index, 313
The Rise of the Machines, 278
Thermal conductivity, 114, 269
Thermodynamics, 268, 272
Thermometer, 114, 268
    mercury, 242
The Skeptical Environmentalist, 288
The Spy who Loved Me, 278
The Towering Inferno, 126
Thiocyanates, 237
This, Hervé, 114
Thixotropy, 100, 312, 320
Threshold limit value, 314
Throat
    irritation, 91
    lozenge, 82
    sore, 86, 144
Thrombosis, 256
Thromboxane, 76, 200
Thujone, 170
    (+)-3-, 170
    (-)-3-, 170
Thyroid gland, 216
Tijuana, 124, 126
Tikhon, Ivan, 130
Till, Rochelle W., 296
Times Beach, 249
Tin, 25
Tisane, 142, 173, 207
Tisza river, 298
Tiszazug murders, 238
Titanium, 262, 319
TI *See* therapeutic index, 313
TLV *See* threshold limit value, 314
Tocopherol, 227
Todd, Alexander R., 162
Tokyo, 301
Tolerable daily intake, 62, 171, 251, 296, 297, 314
Tolerable upper intake level, 162
Tolstoy, Leo, 252
Tomato, 42, 71, 96, 226, 227, 276

Tongue, 82, 83
Too much of a good thing, 40, 97, 160, 275
Toothpaste, 55, 82, 101, 114
Tornado, 5
Toronja, 12
Toulouse-Lautrec, Henri de, 169
Tour de France, 131
Toxicity, 7, 9, 15, 28, 30, 39, 82, 88, 126, 127, 164, 165, 171, 181, 238, 251, 274, 277, 284
Toxicology, 103, 246, 247, 274, 275, 302
Toxic Substances Control Act, 9
Toxin, 7, 62, 166, 168, 169, 315
    F-2, 62
    fungal, 178
    Fusarium, 60, 62, 63
    natural, 7, 115
Trafalgar
    Battle of, 156
Tragacanth gum, 100, 101
Tramadol, 205
Transesterification, 50
Transportation, 4, 29, 37, 59, 73, 98, 106, 107, 109, 111, 117, 213, 284
Travel
    by air, 4
    by bicycle, 4
    by bus, 4
    by car, 4
    by motorcycle, 4
    by train, 4
Tree, deciduous, 257
Trichloroisocyanuric acid, 237
Trichloromethane, 214
    derivative, 12
Trichlorophenoxyacetic acid, 2,4,5-, 248
Trichothecenes, 62
Trifolium
    repens, 194
    subterraneum, 194
Triglyceride, 43, 47, 49, 90, 93, 315
Trimanganese, 24
Triple E, 107
Triterpenoid, 187
Troposphere, 21, 22
Trout, 78, 209
TSCA *See* Toxic Substances Control Act, 9
Tsetse fly, 190
Tsunami, 290
Tuberculosis, 190, 254
Tubocurare, 206
Tubocurarine, D-, 206
Tuna, 78
Tungsten, 25
Tuomisto, Jouni T., 103

Turmeric, 72, 73, 168
Turnip, 42
Tüskevár, 258
Tylenol, 128
Tyndall effect, 311
Typhus, 241, 245
Tyrol, 239
Tyrosine, 51

U
Ubiquinone, 224
UK, 63, 65, 116, 187, 249, 251
Ukraine, 288, 290
Ultracentrifuge, 113
Ultraviolet
  irradiation, 13, 36
  light, 62
  radiation, 21, 22, 36, 268
UL *See* Tolerable upper intake level, 162
Umami, 95
Umbrella Murder, 301
United Kingdom, 5, 163, 302
United Nations, 11, 288
University of California at Berkeley, 7
University of California, Santa Cruz, 309
University of Cincinnati, 249
University of Illinois, 123
University of Milano, 250
University of Minnesota, 122
University of Nottingham, 70
University of Oxford, 114
University of Szeged, 158
University of Wisconsin, 50
Urânia, 309
Uranium, 291
  depleted, 293, 294
Uranium-238 (isotope), 316
Urea, 255, 320
Ureas, 237
  seleno-, 237
  thio-, 237
Uric acid, 28, 36, 52, 53, 276
Urinary
  bladder, 52
  tract, 167
    infection, 86, 142
Urination, 35
Urine, 36, 41, 52, 82, 125, 131, 160, 166, 167, 211, 238, 297
US, 5, 7, 12, 13, 29, 36, 37, 49, 50, 52, 53, 65, 101, 106, 108, 109, 122, 124, 125, 127, 139, 146, 166, 182, 183, 196, 197, 212, 215, 216, 246, 251, 255, 264, 270, 294, 307, 309

Army, 248, 293, 294
Usare l'olio di ricino, 186
UV-meter, 274, 277
UV radiation *See* Ultraviolet radiation, 21

V
Vaccine, 10, 293
Vaccinium macrocarpon, 36
Vacuole, 305, 306
Vale, Jason, 122
Valencia, 78
Valeur limite d'exposition professionnelle, 314
Valore limite di soglia, 314
Vanadium, 262
Van der Bergh, Simon, 46
Vane, John Robert, 200
Van Gogh, Vincent, 153, 169
Vanilla, 54
Vanilla planifolia, 54
Vanillin, 53, 54, 56, 57
  ethyl, 54
Vanilloid receptor, 230
Vanillyl amine, 230
Varietas dēlectat, 215
Vata, 27
Vega, César, 114
Vegemite, 96
Vegetable, 10, 12, 41, 42, 57, 69, 70, 71, 72, 74, 82, 90, 91, 93, 116, 130, 163, 168, 176, 186, 209, 211, 225–227, 276, 296, 317
Vekerd, 26, 27
Veracruz, 104
Verguin, François-Emmanuel, 39
Verlaine, Paul, 169
Vermouth, 170, 172
Verne, Jules, 43, 268, 280
Very large and very small quantities, 317, 318, 320
Viagra®, 220
Vibrio cholerae, 12
Vietnam War, 248
Vinblastine, 139
Vincristine, 139, 141
Vindesine, 139
Vinegar, 67, 93, 94, 156
Vinorelbine, 139
Vinyl chloride, 304
Virgin Atlantic Airways, 106
Virus, 146
Viscose, 303
Viscosity, 186, 311, 320
Vis vitalis theory, 7, 276, 320
Vitamer, 156, 161, 165

Index 395

Vitamin
 A, 70, 162, 164, 165, 226
 B_1, 161–164
 B_2, 162–164
 B_3, 162–165
 B_5, 163, 164
 B_6, 99, 162–165
 B_7, 163, 164
 B_9, 164
 B_{12}, 162–165, 300
 B_{15}, 123
 B_{17}, 122–127
 C, 17, 35, 36, 42, 60, 93, 122, 130, 142, 156–164, 224, 227, 257, 265
 D, 130, 161, 162, 164, 165, 211
 E, 162, 164, 165, 224, 225, 227
 F, 75
 fat-soluble, 164
 K, 162, 164
 K_1, 163
 overdose, 164, 165, 166
 water-soluble, 84, 164
Vitelline, 112
Vitex agnus-castus, 147
Vízi, Béla, 302
VLEP *See* valeur limite d'exposition professionnelle, 314
Vodka, 89
Volcano, 287
Vomiting, 165, 315
Vomitoxin, 62
Vorsorgeprinzip, 14, 297
VX (nerve gas), 8

W

WADA *See* World Anti-Doping Agency, 132
Wales, 5
War and Peace (novel), 252
Waste water, 9, 13, 16, 17, 54
Water, 269
 alkaline, 217
 carbonated, 212, 216, 217
 hexagonal, 210
 mineral, 116, 209, 212–217
 pi, 217
 piped, 241
 super-cooled, 281
 tap, 32, 116, 212–217
 virtual, 18
Water buffalo, 46
Watermelon, 42
Waxing temperature, 109
Wegener, Alfred, 30

Well
 deep, 241
 tube, 241
WEL *See* workplace exposure limit, 314
Wenhua, Tian, 51
Wesley, John, 28
West Indies, 79
West Virginia, 249
Wettability, 243
Wetting, 243
Wheat, 18, 62, 96, 108, 116
 dinkel, 62
Whipple, George H., 162
Whisky, 28, 89
White II, Edward H., 282
WHO *See* World Health Organization, 11
Wiedemann–Franz law, 269
Wilde, Oscar, 169
Williams, Charles D. H., 113
Willis, Bruce, 281
Willow
 bark, 94, 139, 198–201
 white, 199
Willstätter, Richard, 306
Wilson, Bee, 115
Wimbledon, 234
Windaus, Adolf Otto Reinhold, 162
Wine, 104, 115
 cellar, 284, 285
 consumption, 64, 65, 67, 222
 Port, 28
 red, 63, 64, 65, 66, 67, 68, 69, 194, 222
 white, 64, 65, 66, 216
Winkler, Clemens, 23
Winnie the Pooh, 86
Wisconsin, 46
Wishful thinking, 179, 205
Wöhler, Friedrich, 276, 320
Wolff, Emil Theodor von, 69
Wood, 76, 80, 251, 255, 269, 282, 303
Woodward, Robert Burns, 162
Wool, 38, 303
Workplace exposure limit, 235, 314
World Anti-Doping Agency, 106, 132, 233
World Bank, 288
World Health Organization, 11, 28, 36, 42, 53, 64, 97, 180, 215, 228, 239, 245, 247, 251, 263
World War
 I, 123, 212, 235, 237, 249, 277
 II, 46, 73, 117, 124, 135, 191, 245, 249, 290, 293
World Wide Fund for Nature, 247

Wormwood
annual or sweet, 170
common, 144, 170
grand, 169, 170
Roman, 170
WWF *See* World Wide Fund for Nature, 247

X

Xanthan gum, 100, 101
Xanthomonas campestris, 101
Xenobiotic, 166, 167, 168
Xenon, 25
Xeronine, 182, 184
X-ray, 29, 127, 224, 264
Xylan, 82
Xylitol, 80–83
Xylose, 82, 108

Y

Yamamoto, Mayu, 53
Yeast, 35, 82, 96, 108, 177, 222, 285
cells, 36
Yellow fever, 245

Yin and *yang,* 242
Yoghurt, 50, 89, 100, 116
Yohimbine, 221
Young, Robert O., 208
Ytterbium, 25
Yttrium, 25
Yujun, Zhang, 51
Yushchenko, Viktor, 249

Z

Zay, Andrea, 260
Zearalenone, 60, 62
Zeaxanthin, 226
Zeidler, Othmar, 245
Zeitgeist, 6, 150
Zero-g, 283
Zinc, 42, 242
Zirconium, 292, 319
Zohner, Nathan, 1
Zola, Émile, 169, 285
Zooplankton, 12, 30, 32
Zsáka, 26, 27

CPSIA information can be obtained at www.ICGtesting.com
Printed in the USA
LVOW02*0829301114
416256LV00001B/8/P